Lecture Notes in Computer Science 3649

Commenced Publication in 1973
Founding and Former Series Editors:
Gerhard Goos, Juris Hartmanis, and Jan van Leeuwen

Wil M.P. van der Aalst   Boualem Benatallah
Fabio Casati   Francisco Curbera (Eds.)

Business Process Management

3rd International Conference, BPM 2005
Nancy, France, September 5-8, 2005
Proceedings

 Springer

Volume Editors

Wil M.P. van der Aalst
Eindhoven University of Technology
Faculty of Technology and Management (PAV D2)
Department of Information Systems
P.O. Box 513, NL-5600 MB, Eindhoven, The Netherlands
E-mail: w.m.p.v.d.aalst@tm.tue.nl

Boualem Benatallah
The University of New South Wales
School of Computer Science and Engineering
Sydney 2052, Australia
E-mail: boualem@cse.unsw.edu.au

Fabio Casati
Hewlett-Packard
1501 Page Mill Rd, MS 1142, Palo Alto, CA, 94304, USA
E-mail: fabio.casati@hp.com

Francisco Curbera
IBM T.J. Watson Research Center
19 Skyline Drive, Hawthorne, New York 10532, USA
E-mail: curbera@us.ibm.com

Library of Congress Control Number: 2005931138

CR Subject Classification (1998): H.3.5, H.4.1, H.5.3, K.4.3, K.4.4, K.6, J.1

ISSN   0302-9743
ISBN-10   3-540-28238-6 Springer Berlin Heidelberg New York
ISBN-13   978-3-540-28238-9 Springer Berlin Heidelberg New York

Springer is a part of Springer Science+Business Media

springeronline.com

© Springer-Verlag Berlin Heidelberg 2005
Printed in Germany

Typesetting: Camera-ready by author, data conversion by Scientific Publishing Services, Chennai, India
Printed on acid-free paper   SPIN: 11538394   06/3142   5 4 3 2 1 0

Preface

This volume contains the proceedings of the 3rd International Conference on Business Process Management (BPM 2005), organized by LORIA in Nancy, France, September 5–8, 2005.

This year, BPM included several innovations with respect to previous editions, most notably the addition of an industrial program and of co-located workshops. This was the logical result of the significant (and still growing) industrial interest in the area and of the broadening of the research communities working on BPM topics.

The interest in business process management (and in the BPM conference) was demonstrated by the quantity and quality of the paper submissions. We received over 176 contributions from 31 countries, accepting 25 of them as full papers (20 research papers and 5 industrial papers) while 17 contributions were accepted as short papers. In addition to the regular, industry, and short presentations invited lectures were given by Frank Leymann and Gustavo Alonso. This combination of research papers, industrial papers, keynotes, and workshops, all of very high quality, has shown that BPM has become a mature conference and the main venue for researchers and practitioners in this area.

We would like to thank the members of the Program Committee and the reviewers for their efforts in selecting the papers. They helped us compile an excellent scientific program. For the difficult task of selecting the 25 best papers (14% acceptance rate) and 17 short papers each paper was reviewed by at least three reviewers (except some out-of-scope papers).

We would like to acknowledge the splendid support of the local organization (in particular Claude Godart, Olivier Perrin, and Daniela Grigori). We also thank Chris Bussler as workshop chair. Special thanks (as always!) go to Eric Verbeek: he did a great job in compiling the final versions of the papers into this LNCS volume. Finally, we would like to mention the excellent co-operation with Springer during the preparation of this volume. We hope you will find the papers in this volume interesting and stimulating.

June 2005

Wil van der Aalst, Boualem Benatallah, Fabio Casati, and Francisco Curbera
Editors
BPM 2005

Organization

Organizing Committee

Sami Bhiri
Chris Bussler (Workshops Chair)
Anne-Lise Charbonnier
Antoinette Courrier
Francisco Curbera (Industrial Chair)
Armelle Demange
Walid Gaaloul

Claude Godart (General Chair)
Daniela Grigori (Publicity Chair)
Adnene Guabtni
Olivier Perrin (Local Organizing Chair)

Program Committee

Wil van der Aalst, The Netherlands (Co-chair)
Vijay Atluri, USA
Karim Baina, Morocco
Nouredine Belkhatir, France
Ladjel Bellatreche, France
Boualem Benatallah, Australia (Co-chair)
Athman Bouguettaya, USA
Thierry Bouron, France
Mokrane Bouzeghoub, France
Chris Bussler, Ireland
Fabio Casati, USA (Co-chair)
Malu Castellanos, USA
François Charoy, France
Dickson K.W. Chiu, Hong Kong, China
Jen-Yao Chung, USA
Leonid Churilov, Australia
Francisco Curbera, USA
Peter Dadam, Germany
Jörg Desel, Germany
Jan Dietz, The Netherlands
Eric Dubois, Luxembourg
Marlon Dumas, Australia
Schahram Dustdar, Austria
Johan Eder, Austria
Marie Christine Fauvet, France

Dimitrios Georgakopoulos, USA
Claude Godart, France
Paul Grefen, The Netherlands
Mohand Said Hacid, France
Kees van Hee, The Netherlands
Arthur ter Hofstede, Australia
Geert-Jan Houben, The Netherlands
Stefan Jablonski, Germany
Gerti Kappel, Austria
Hassan Khorshid, USA
Kwang-Hoon Kim, Korea
Akhil Kumar, USA
Lea Kutvonen, Finland
Dan Marinescu, USA
Olivera Marjanovic Australia
Michael Maximilien, USA
Anne Ngu, USA
Andreas Oberweis, Germany
Maria Orlowska, Australia
Helen Paik, Australia
Mike Papazoglou, The Netherlands
Cesare Pautasso, Switzerland
Barbara Pernici, Italy
Fethi Rabhi, Australia
Krithi Ramamritham, India
Manfred Reichert, Germany
Hajo Reijers, The Netherlands

Michael Rosemann, Australia
Yucel Saygin, Turkey
Karsten Schulz, Germany
Robert James Steele, Australia
Aixin Sun, Australia
Farouk Toumani, France

François Vernadat, France
Mathias Weske, Germany
Jian Yang, Australia
Liangzhao Zeng, USA
Yanchun Zhang, Australia
Leon Zhao, USA

Referees

Michael Adams
M. Salman Akram
Lachlan Aldred
Samuil Angelov
Danilo Ardagna
Emilia Cimpian
Feras Dabous
Peter Dadam
Helga Duarte
Nguyen Dung
Rik Eshuis
Matthias Faerber
Douglas Foxvog
Walid Gaaloul
Manolo Garcia
Juan Miguel Gomez
Daniela Grigori
Adnene Guabtni
Dan Hong
Jens Hündling
Stéphane Jean
Gabriel Juhas
Kirsten Keferstein
Agnes Koschmider
Dimitre Kostadinov
Ulrich Kreher
Dominik Kuropka
Emily (Rong) Liu
Xumin Liu
Robert Lorenz
Zaki Malik
Udo Mayer
Nikolay Mehandjiev
Christian Meiler
Marco Mevius
Harald Meyer

Adrian Mocan
Stefano Modafferi
Sascha Mueller
Tariq Al Naeem
Christian Neumair
Alex Norta
Justin O'Sullivan
Phillipa Oaks
Bart Orriens
Chun Ouyang
Hagen Overdick
Veronika Peralta
Olivier Perrin
Horst Pichler
Pierluigi Plebani
Rodion Podorozhny
Jan Recker
Abdelmounaam Rezgui
Stefanie Rinderle
Florian Rosenberg
Nick Russell
Sonia Sanlaville
Pano Santos
Brahmananda Sapkota
Arnd Schnieders
Hilmar Schuschel
John Shepherd
Philippe Thiran
German Vega
Jochem Vonk
Jerry Wang
Aklouf Youcef
Qi Yu
Surendra Sarnikar
Sherry Sun
Harry Wang

Table of Contents

Research Papers

Industrial Papers

Short Papers

Modeling and Analysis of Mobile Service Processes by Example of the Housing Industry

Volker Gruhn[1], André Köhler[1], and Robert Klawes[2]

[1] University of Leipzig, Chair of Applied Telematics / e-Business,
Klostergasse 3, 04109 Leipzig, Germany
{gruhn, koehler}@ebus.informatik.uni-leipzig.de
[2] Deutsche Bank AG, Alfred-Herrhausen-Allee 16-24, 65760 Eschborn, Germany
robert.klawes@db.com

Abstract. This article describes the method of Mobile Process Landscaping by example of a project in which the service processes of a company from the housing industry were analyzed regarding their mobile potential. This analysis was conducted with the aim to organize these processes more efficiently in order to realize cost savings. Therefore, the method of Mobile Process Landscaping, which is introduced in this article, was used. The method refers to the stage of requirements engineering in the software process. It is shown how the initial situation of the company was analyzed, which alternative process models on the basis of mobility supporting technology were developed and how these alternatives were economically evaluated. Furthermore, it is shown how first restrictions for the software and system design were made on the basis of one process model. Finally, it is shown how the Mobile Process Landscaping method can be used to verify whether the adoption of mobility supporting technology is suitable to obtain a defined goal and which requirements such a solution needs to fulfill.

Keywords: business process modeling and analysis, mobile business processes.

1 Introduction

Since the availability of broadband radio networks and the receded costs for appropriate devices the use of mobility-supporting technology has become an interesting opportunity for companies to optimize selected business processes and to increase their efficiency. Mobile business processes are characterized by a high degree of mobility of the involved persons and by a lack of knowledge about the next location of the person. Often a connection to IT-systems of the company would be desirable. In such processes, media-breaks, long processing times, inefficient routes and lacks of information can be observed. The use of mobility supporting technology offers the opportunity to solve these problems. But therefore a systematic analysis of the professional requirements on the basis of business processes is necessary.

This article deals with the method of Mobile Process Landscaping (MPL) by whose use the described tasks can be handled. Referring to the software process the activities and their results can be assigned to the requirements engineering. The use of

W.M.P. van der Aalst et al. (Eds.): BPM 2005, LNCS 3649, pp. 1–16, 2005.

this method is shown by example of the technical service processes taken from a company of the housing industry.

In chapter 2 the MPL method is explained. First, mobile business processes are defined (2.1.). Afterwards, the structure of the method (2.2.) as well as its aim (2.3.) are described. Subsequently, an overview about related work is given (2.4). Chapter 3 shows the usage of the MPL method by example of a company of residential trade and industry. That chapter corresponds to the structure of the MPL method as explained in section 2.3. Chapter 4 draws a conclusion.

2   Mobile Process Landscaping Method

2.1 Mobile Business Processes

The term „business process" was defined by numerous authors ([1], [2], [3], [4], [5], [6]). Below, we follow the commonly used definition of Davenport [2] according to which a business process can be understood as „a specific ordering of work activities across time and place, with a beginning, an end, and clearly identified inputs and outputs: a structure for action."

A business process can be decomposed in different levels into sub-processes. If a sub-process is not decomposable it is called "activity." Thus, a business process can be understood as an abstract description of workflows in a company. The actual occurrence of such a business process in reality is called a business process instance.

In the following, only business processes with a specific distribution structure and thus a certain mobility of the process-executing person are considered. We suppose that mobility is given when for at least one activity an "uncertainty of location" exists. This assumption is based on the concept of "location uncertainty" by Valiente and van der Heijden [7], according to which the place of the execution of an activity can be different in different instances of the business process or the place can change during the execution of an activity. Thus, we deal with a mobile activity within a business process. Because multiple mobile activities are conceivable, and a mobile activity often affects the whole business process, the complete business process is called "mobile business process".

Furthermore, it can be noticed that the "uncertainty of location" is externally determined. This assumption implies that the location uncertainty is caused by external factors and that the process-executing person has therefore no freedom of choice regarding the place of the process execution. Beyond, often a cooperation with external resources (from the process-point of view) is needed during the execution of the process. This fact restricts the term "mobile business process" to the necessity of cooperation with external resources within the considered activity, for instance caused by the need for communication or coordination with other persons or interaction with other objects.

Considering this, we propose the following definition: A mobile business process is a business process, in which

a. at least one person is involved, which executes its tasks in different locations,
b. the actual location of the task-execution is known just vague and/or just short before the beginning of the task,

c. this uncertainty (b) is determined externally and can not be fully controlled by the process-executing person.

On the basis of this definition, two conclusions can be drawed. First, the definition implies that the assigned task causes the mobility of the involved person. The mobile worker need to appear physically on the specified location because there exists a resource (damaged device, customer) necessary for the solution of his task. Second, it is not relevant for this definition whether mobile information technology is used or not. In fact, mobile information technology will be the key for the realization of an efficient work flow in the majority of the cases.

2.2 Structure of the Method

Subsequently, the structure of the method is explained. Figure 1 shows the essential steps of the method. First, the company needs to define the objective which is to be achieved by use of this method. Usually, the goal is to optimize the process parameter (personnel) costs, the duration of the process or the quality of the produced goods and services.

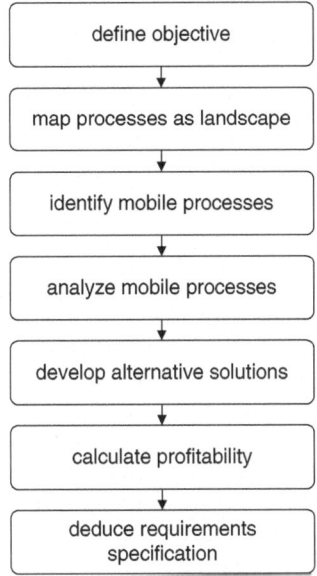

Fig. 1. Steps of the MPL-Method

As soon as the objective is defined one can start with acquiring the processes and depicting them as process landscape. The process landscape shows relations between the main business processes and allows its user to recognize dependencies between processes very early. During the next step, those sub-processes characterized by a high degree of mobility of the process-executing person need to be identified. Therefore, the method provides an assessment scheme which is explained in chapter 3.3. If mobile processes are identified, an analysis is necessary.

During this step, shortcomings in the process flow resulting from the mobility of the process-executing person can be discovered. On the basis of these insights new process versions can be developed in order to avoid the recognized shortcomings. This can be conducted by constraining or supporting the mobility within the process. Subsequently, the different alternatives need to be evaluated by an economically point of view. If a positive decision for the realization of one alternative has been reached, first requirements specifications and restrictions for the software architecture can be deduced. Further information about the Mobile Process Landscaping method can be found in ([8], [9], [10], [11]).

2.3 Aim of the MPL Method

By the use of the MPL method within the business process model of a company the following can be achieved:

- the discovery of mobile processes in the business process model,
- the analysis of potential for optimization by the support or elimination of the mobile processes,
- the development of alternative solutions based on the use of mobility-supporting technology,
- the economic evaluation of those alternatives and
- the deduction of general conditions and requirements for the software and system design for the selected alternative.

On the basis of these results alternative solutions can be evaluated according to the companies strategic goals. The architecture of the resulting system can be developed on the foundation of clear professional guidelines.

2.4 Related Work

A number of recent publications show that efficiency and effectiveness of certain activities can be improved through the use of mobile technologies ([7], [12], [13]). The mentioned examples are case studies describing successfully released solutions in different companies. However, how these companies choose the described business processes and activities for the use of mobile technologies remain open questions.

Frequently, a technology-driven approach can be observed for realising potential benefits, which adjusts processes corresponding to the available features of certain mobile devices. But often, a large number of complex processes with many involved people prevails, e.g. in large companies and corporate groups. Such an approach may then lead to wrong decisions, especially in the long term. In our opinion, the process of decision-making about use and design of a mobile information system needs to be systematic and comprehensible.

Beyond, the question for the quality of mobility is an important one in this context. As stated in [14] user mobility is often distinguished into personal and terminal mobility. Particularly, the movement of the terminal and therewith the terminal mobility has come to the fore in the recent years. In contrast, the personal mobility has received less attention. In our opinion, it is necessary to focus research on this topic because of the constraints for software development resulting from certain requirements of the specific kind of mobility.

Kakihara and Sorensen [19] discuss the notion of mobility in distinguishing spatial, temporal and contextual mobility. This is an important contribution to the definition of mobility in this article.

Dustdar and Gall [16] as well as Sairamesch et al. [17], [18] describe frameworks for distributed and mobile collaboration which can be used to develop software architectures for mobile systems. These frameworks are of particular interest for the MPL method because it is the methods aim to provide a systematic deduction of constraints for the software architecture on the basis of the process model.

Gupta and Moitra [20] introduce an highly interesting technology integration approach for pervasive IT Infrastructure. The aim of this approach is partly the same as of the MPL method (focusing on the protection of investments and maximizing the returns) but has a more generic character.

3 The Use of MPL in an Industry Case

In the following, the application of the MPL method is shown by example of a municipal company called LWB, located in Leipzig, Germany. Its main task is to assure a socially acceptable apartment supply for a great number of citizens. For this purpose, LWB builds and maintains apartments particular in the low price segment. These apartments are mainly located in multi-storey houses which are affected by vacancy more than averagely.

The economic situation in the concerned real estate market is characterized by a considerable oversupply of apartments. Because of the large share of vacant apartments (approx. 17%) as well as the continuous migration of prosperous inhabitants into the suburbs, landlords compete for the lowest rents on the market. In this situation the company LWB is confronted with high losses of revenue and high costs for maintenance. In order to overcome this situation the company induced a variety of steps, basically to lower the costs. In this context, an examination and analysis of the internal workflow was planned in order to discover potential for optimization.

The company started quite fast to focus on its maintenance processes because of the very large number of apartments (approx. 12,000) and therefore the very large number of process recurrences. Beyond, these processes are characterized by a high degree of mobility of the process-executing person. These facts promised a high potential for optimization. Therefore, the company LWB asked the University of Leipzig to conduct an analysis by the use of the MPL method. The precise proceeding during this project and the achieved results are introduced in the following. The explanations are aligned on the structure of the method as described in Figure 1.

3.1 Defining Objective

The company aimed at preferably high cost-savings by an optimization of the business processes with a high degree of mobility of the process-executing person. The use of mobile technology therefore was favored. The analysis should be conducted for the business process "Technical Service." As a result, a couple of alternative technical

and process solutions as well as an economical evaluation and first requirement specifications of them was expected.

3.2 Mapping Processes as Landscape

3.2.1 Depiction of Processes

The business process "Technical Service" consists of the sub-processes "Maintenance," "Administration," "Allocation of Costs," "Billing" and "Recording of Consumption Values." These business processes are depicted as a process landscape (PL)
to recognize their essential relations to each other (Figure 2). The sub-processes are
evaluated by the level of value added and the assumed degree of mobility.

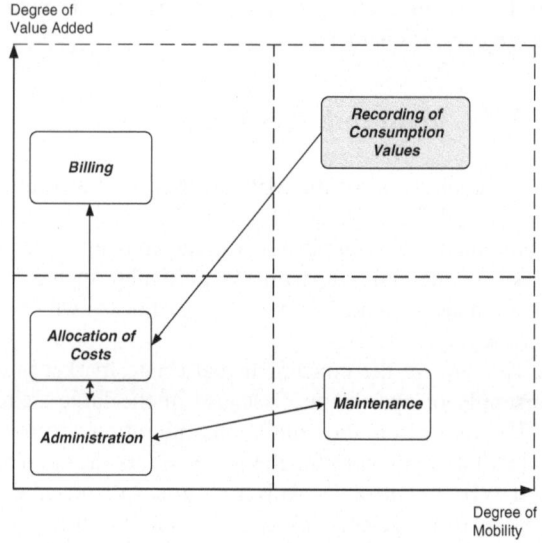

Fig. 2. Figure 1. PL "Technical Service"

The sub-processes "Allocation of Costs" and "Administration" were classified as
slightly value adding and not mobile. The sub-process "Billing" was classified as
value adding but not mobile. The sub-process "Maintenance" is extremely mobile but
just slightly value adding due to the fact that necessary repairs need to be achieved but
no revenue can be associated to them. The sub-process "Recording of Consumption
Values" is extremely mobile and value adding. Because of this classification the sub-
process "Recording of Consumption Values" was examined. Figure 3 shows the detailed sub-process. The notation in this figure as well as in the following ones is used
according to the Business Process Modeling Notation [15]. The sub-process „Recording of Consumption Values" contains the recording, the transportation and the
processing of the consumption values for water and heating. They are measured by
appropriate meters for each apartment.

The recording of the consumption values is executed by a subsidiary company of
the LWB called WSL. The process starts when the LWB assigns the WSL to record

the consumptions in all properties of the company. This assignment is done once a year. The execution of the whole process takes three months for approximately 12.000 apartments in 240 properties. A property is a real estate which contains 50 apartments on average. The LWB sends lists with addresses and tenant data to the WSL to prepare the recording of the consumption values. Subsequently, the WSL processes the acquired lists electronically for the upcoming tasks (process data records). On basis of this information dates for the recording are arranged and efficient tours for the inspection of the apartments are planned (plan tour). Furthermore, for each daily tour paper lists with addresses and tenant data are created. These preparing tasks take about one week.

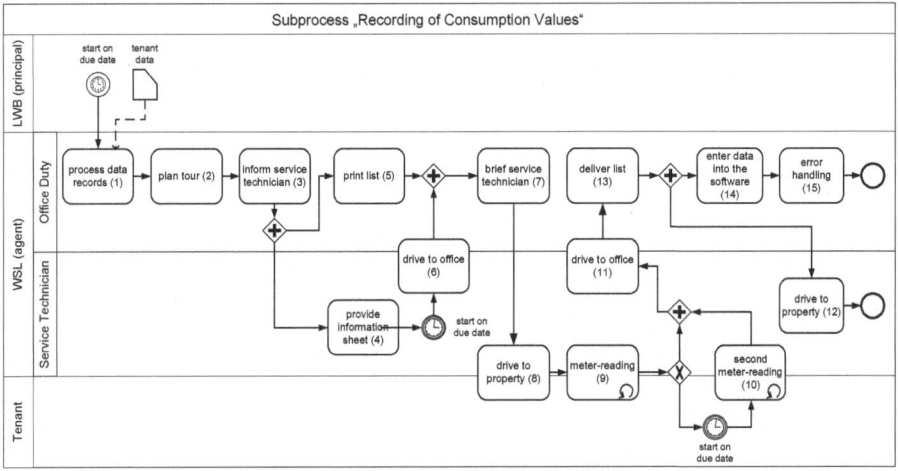

Fig. 3. Recording of consumption values

After completing the tour planning the service technicians get informed about their individual tours (inform service technician). Afterwards, they drive to their assigned properties and attach an information sheet with the recording date to each entrance (provide information sheet). Shortly before the date of recording the service technicians drive to the office duty to get recording lists as well as information about the recording procedure (brief service technician). Then they drive to the properties according to their tour plans and start the recording (meter-reading).

For each apartment an average of four values need to be recorded. The service technician walks from apartment to apartment and records by hand each value as well as the tenant name and the identification number of each device on his list. This activity takes approximately ten minutes per customer. After the recording the service technician has to drive back to office duty and deliver the recording lists. There the staff transcript the data into the appropriate software.

3.2.2 Shortcomings of the Process
Within the process different shortcomings could be found. In approximately ten percent of the cases the tenant is absent, which leads to a repetition of the process starting with a new appointment. If the tenant is absent again the process is repeated a third

time at the tenant's own expense. Beyond, there is the danger of transcription errors due to media breaks. They can be caused by inaccurate recordings or by accidentally mixing up the numbers. The company WSL estimates that approximately five percent of the recording lists are wrong without anybody noticing it. Additionally, the staff estimates that around ten percent of the lists are unreadable. In such cases a request by telephone and an adjustment are necessary. Apart from its vulnerability for errors the double-recording causes high efforts. Because of the large number of apartments and the necessary personnel effort the recording of the consumption values is conducted just once a year. In that way, defective or manipulated meters can be recognized just once a year.

Table 1. Costs of "Technical Service"

Activity	duration (minutes)	number of executions per property	duration per property (minutes)	cost per property (EUR)	total cost (EUR)
process data records	900	0,00	3,75	1,12	268,44
plan tour	1500	0,00	6,25	1,86	447,41
inform service technician	10	0,00	0,04	0,07	17,98
provide information sheet	5	5,00	25,00	7,46	1789,63
print list	30	0,00	0,13	2,12	508,95
drive to office	30	0,33	10,00	2,98	715,85
brief service technicians	5	0,33	1,67	0,50	119,31
drive to property	30	0,33	10,00	2,98	715,85
meter-reading	10	50,00	500,00	149,14	35792,64
second recording	12	5,00	60,00	17,90	4295,12
drive to office	30	0,33	10,00	2,98	715,85
drive to property	30	0,33	10,00	2,98	715,85
deliver list	5	0,00	0,02	0,07	16,49
enter data into software	2	50,00	100,00	29,83	7158,53
error handling	6	5,00	30,00	8,95	2147,56
			766,85	*230,94*	*55425,47*

3.2.3 Costs of the Process

Table 1 shows the distribution of the process costs over the single activities. The figured costs of the activities refer to the execution of the activity for one property.

For this calculation only personnel costs are considered. Therefore, an hourly rate of 17.90 EUR was assumed. The whole personnel costs for the process "Recording of Consumption Values" account for approximately 55,500 EUR.

3.3 Identification of Mobile Processes

Not every business process is suitable for an optimization by the use of mobile technology. Because of that potentially mobile sub-processes need to be identified by means of different criteria. Therefore, the MPL method provides an assessment scheme in order to evaluate each activity by different criteria. For the evaluation a scale from 0 (not true) to 2 (true) is used. The small scale limits the subjective discretion of the conducting person.

Table 2. Evaluation of the activities

		General Potential					Mobile Potential					
		creation of value	number of executions	customer satisfaction	occurrence of media breaks	weighted sum of General Potential	involved persons meet in specified location	involved persons are spatially separated	activity in motion	estimated data amount	weighted sum of mobile potential	sum
weight		1,0	1,5	0,3	1,8		1,5	1,0	0,3	0,8		
No.	Activity											
1	process data records	1,0	0,0	0,0	1,0	**2,8**	0,0	0,0	0,0	2,0	**1,6**	**4,4**
2	plan tour	1,0	0,0	0,0	1,0	**2,8**	0,0	0,0	0,0	2,0	**1,6**	**4,4**
3	inform service technician	0,0	0,0	0,0	0,0	**0,0**	2,0	0,0	0,0	0,0	**3,0**	**3,0**
4	provide information sheet	0,0	0,0	1,0	1,0	**2,1**	0,0	0,0	0,0	1,0	**0,8**	**2,9**
5	print list	1,0	0,0	0,0	1,0	**2,8**	0,0	0,0	0,0	1,0	**0,8**	**3,6**
6	drive to office	0,0	1,0	0,0	0,0	**1,5**	0,0	0,0	2,0	0,0	**0,6**	**2,1**
7	brief service technician	0,0	0,0	0,0	0,0	**0,0**	2,0	0,0	0,0	0,0	**3,0**	**3,0**
8	drive to property	0,0	1,0	0,0	0,0	**1,5**	0,0	0,0	2,0	0,0	**0,6**	**2,1**
9	meter reading	2,0	2,0	2,0	2,0	**9,2**	2,0	0,0	0,0	2,0	**4,6**	**13,8**
10	second recording	2,0	2,0	2,0	2,0	**9,2**	2,0	0,0	0,0	2,0	**4,6**	**13,8**
11	drive to office	0,0	1,0	0,0	0,0	**1,5**	0,0	0,0	2,0	0,0	**0,6**	**2,1**
12	drive to property	0,0	1,0	0,0	0,0	**1,5**	0,0	0,0	2,0	0,0	**0,6**	**2,1**
13	deliver list	1,0	0,0	0,0	0,0	**1,0**	2,0	0,0	0,0	0,0	**3,0**	**4,0**
14	enter data into software	2,0	2,0	0,0	2,0	**8,6**	0,0	0,0	0,0	2,0	**1,6**	**10,2**
15	error handling	2,0	2,0	2,0	2,0	**9,2**	0,0	2,0	0,0	2,0	**3,6**	**12,8**

The criteria can be divided into two different groups. The first group contains universal criteria (General Potential) showing general potential for optimization. They are:

- creation of value,
- number of executions,
- importance for customer satisfaction and
- occurrence of media breaks.

The second group of criteria (Mobile Potential) allows to assess whether an activity is particularly influenced by the mobility of the process-executing person. They are:

- involved persons meet in specified location,
- involved persons are spatially separated,
- activity in motion and
- estimated amount of data.

Table 2 shows the result of the evaluation with these criteria for the process "Recording of Consumption Values." The single criteria were weighted by the company LWB. Figure 4 shows the result of this analysis.

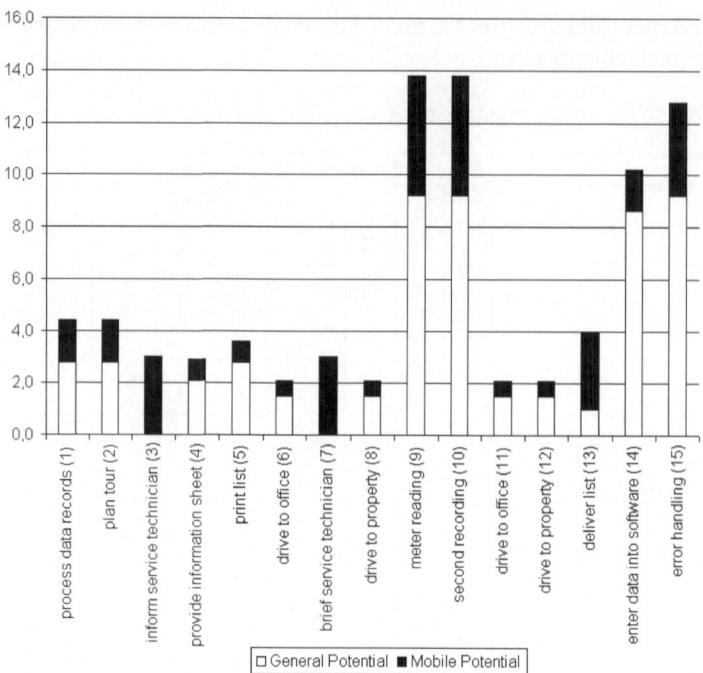

Fig. 4. Potential for optimization per activity

The activities "meter-reading," "second meter-reading," "enter data into software" and "error handling" are characterized by a particularly high potential for optimization. Interestingly, the general potential for optimization for these four activities is evenly distributed whereas the mobile potential for optimization causes differences in the evaluation. Furthermore, the activities "inform service technician," "brief service technician" und "deliver list" are (nearly) exclusively characterized by potential for optimization on the basis of the mobile criteria. On the basis of this analysis the development of a solution aiming at the activities with the biggest potential for optimization (9, 10, 14 and 15) was started. To do so, it is very important to interpret the term "Mobile Potential" in the right way. Activities with a high degree of mobile potential are characterized by the effects of physical mobility. This fact does not imply a need for implementing an activity support by means of mobility-supporting technology. It is rather a question of regarding the activities in the process-context and justifying IT-solutions on the whole process. Consequently, this can lead to a completely new structure of the activities. In particular, the IT-solution should have positive effects on the four processes named. The following section shows three alternative solutions which were developed on the basis of this analysis.

3.4 Developing Alternative Solutions and Calculation of Profitabilty

The starting point for the development of a mobile solution are activities which involve the meter-reading (9, 10). Therefore, an electronical recording seems to be an opportunity in order to avoid media breaks. It can be done either automatically or manually with the help of the process-executing person. The two propositions "Online Device Support" and "Mobile Device Support" are directed towards the support of the mobility of the service technicians. In contrast, the proposition "Remote Meter-Reading" focuses on the removal of the mobile activities.

3.4.1 Online Device Support

One approach for the solution of the outlined problems is the use of a mobile electronic device. It displays the form electronically and the service technician can enter the consumption values. By the use of a mobile radio network adapter the connection to a central server can be realized. The application can be designed browser-based due to its low complexity. In case that the radio network is not available the consumption values can be noted on paper and transcripted later.

Figure 5 shows the change in the sub-process which results from the implementation of the change. The grey activities are the ones which are dropped compared to the original solution. With this solution the sub-process would be much shorter and simpler in its structure than before. By this solution especially the media breaks between meter-reading and transcription into the software can be avoided.

Furthermore, activities dealing with creation, transport and analysis of the recording lists are avoided. This affects activity 5, 6, 7, 11, 12, 13, 14 and 15. As with this solution the service technician is always online the use of the mobile device for other purposes is possible. For example the support of the sub-process "Maintenance" is imaginable. In the current situation the service technician receives calls on his mobile phone from the office duty and gets informed about trouble messages from tenants. The distribution and the management of these tasks by the use of mobile devices is an interesting opportunity which could be hooked on the solution "Online Device Support".

The estimated costs for the new process according to the scheme showed in Table 1 amount to 45,000 EUR a year. Thereby, only personnel and mobile radio costs were taken into account. That way savings of approximately 10,000 EUR a year can be realized. In order to realize this solution an investment of around 40,000 EUR is needed.

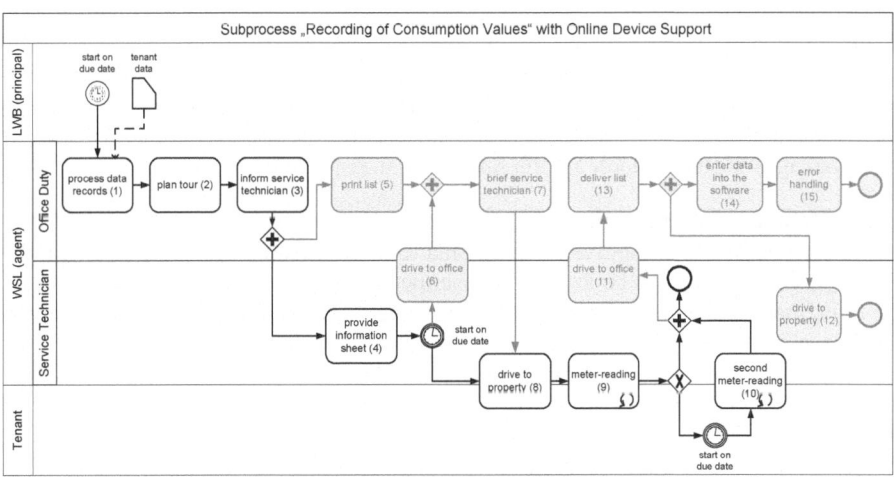

Fig. 5. "Online Device Support"

Thus, the project would be worthwhile about four years after its introduction. The essential advantages of this solution are:

- improvement of those activities with the highest potential for optimization,
- avoidance of media breaks between service technician and office duty,
- reduction of drives, the avoidance of handling of recording lists and
- creation of an important framework for the development of further applications which can support the service technician via the mobile device.

Disadvantages of this solution are:

- in some cases the application might not be available if the radio network is out of reach,
- potentially new sources for errors due to mistakes in handling the device or the software and
- potentially higher risk of process disruption due to failure of the devices or the software.

3.4.2 Mobile Device Support

In contrast, an alternative to the solution outlined above could be to retain from the use of a radio network. Then, software needs to be installed at the mobile device in order to support the meter-reading. The recorded consumption values are entered into the software by the service technician and synchronized with the central server.

In this case, the flow of the business process "Recording of Consumption Values" occurs as shown in Figure 5. Inside of the activity "meter-reading" the synchronization with the central server has to be conducted additionally. The synchronization has to be done using a docking station in a determined location (service technician's home, a company's branch).

Because in this solution no radio network costs occur, the costs per year for the whole process amount to approximately 40.000 EUR (personnel costs). This results in costs savings of about 15.000 EUR a year. To realize this solution an investment of approximately 40.000 EUR is needed. The project would turn worthwhile after three years. The financial savings are assessed as minor ones, but they are even higher than in the solution described above.

The advantages of this solution are the same as in the solution "Online Device Support". An additional advantage is the independence of radio networks, especially in basements and garages. Beyond, there are no costs for the use of radio networks. The disadvantages of this solution are the same as in the one described above.

3.4.3 Remote Meter-Reading

A further alternative is the use of an automatic meter-reading system. For this, a device is needed which can be attached to the meter and which has the capability to record the meter-value and to send it via a radio network. For each apartment a central module needs to be installed which sends the acquired data to a central server.

Within this alternative the business process "Recording of Consumption Value" becomes redundant because no human intervention is necessary except in the case of a malfunction. Beyond, the consumption values can be recorded as often as desired.

The estimation of costs of the process was not possible in this project. There is a wide variety of vendors offering suitable devices for the creation of an infrastructure for remote meter-reading. The replacement of all meters would need an investment of millions of EUR. Thus, the realization of this solution only makes sense in combination with the renovation or the new building of houses. Furthermore, prices for the needed devices are anticipated to be on the decrease for the next two years. Hence, an investment at this juncture seems not to be recommendable.

Thus, from the economical point of view this alternative is not realizable measured on the primary goal of cost-savings. Nevertheless, the advantages of such a solution should be named. These are:

- the complete avoidance of the business process "Recording of Consumption Values,"
- a considerable saving of personnel costs,
- the complete avoidance of former sources of errors and media breaks,
- the meter-reading in any desired period (detailed billing and forecasts are possible) and
- an early recognition of malfunctions at the meter (energy losses caused by malfunctions and thievery can be avoided).

Disadvantages of the solution are:

- very high investment needed and
- an additional effort for the maintenance of the devices is needed.

3.5 Deducing Requirements Specifications

On the basis of these results the company decided to realize the solution "Online Device Support." The determining factors were the costs-savings per year as well as the opportunities for further applications on the basis of the infrastructure of mobile devices and radio networks. On the basis of the professional requirements a requirements specification could be deduced. Thereby, it consciously remains an open question whether a software development is needed or if single components or complete products from specialized vendors can be used.

The draft of the system architecture is shown in Figure 6. The service technician is equipped with a mobile device. Therewith, he can access the mobile application provided by a service provider. The company WSL is connected via WAN with the service provider. The office duty can access all data via the intranet and its ERP-client.

The mobile application needs to provide the following functionality:

- creation of recording lists and tour plans (office duty),
- inquiry of the current state of recording (office duty),
- display of recording lists and tour plans (service technician) and
- recording of the consumption values for each apartment.

The mobile device needs to fulfil the special requirements of the service technicians. Therefore, the following requirements are defined:

- weight at maximum 500 gram,
- size at maximum (w/h/d) 90/200/50 mm,
- precipitation protection at minimum 100 cm,
- display size at minimum 320/240 pt (coloured),

- slightly water resistant,
- battery runtime at minimum 8 hours,
- docking station with car recharge adapter and
- large keypad with separate number field.

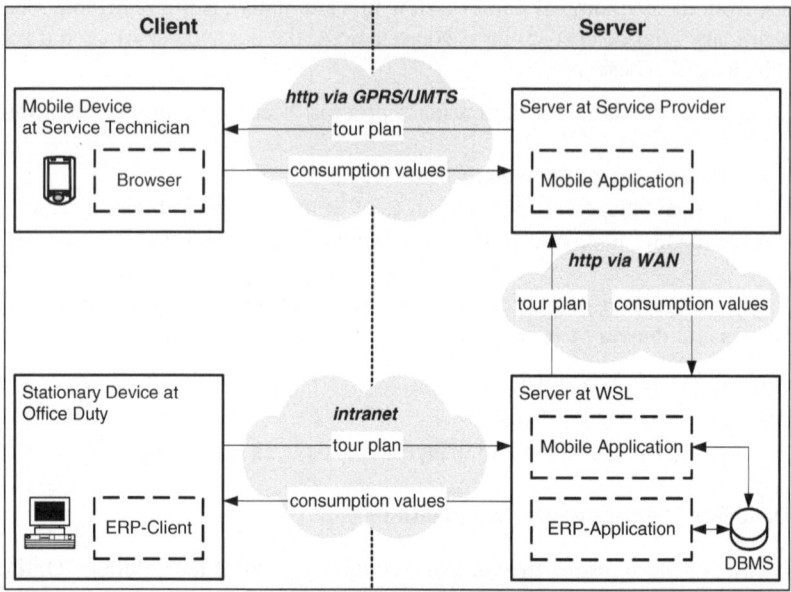

Fig. 6. System architecture

4 Summary

The development of an IT-solution for mobile business processes requires a detailed analysis of the professional requirements as well as an adaptation of the solution onto these requirements. With the described example it was shown that by the help of the MPL method the outlined tasks can be accomplished. The main feature of the method is the analysis of the business processes as well as their dependencies. Furthermore, it was shown how mobile solutions can be evaluated economically in order to justify the needed investment. If the decision for the realization of one alternative solution is made, comprehensive defaults for the system development can be defined on the basis of the detailed professional analysis.

Based on the shown MPL method, further research is planned. First, existing languages for the modeling and analysis of mobile business processes seem not to be suitable for the explicit modeling of mobility. In order to use the process model for the prediction of efficiency and performance of the solution the simulation of the transferred data volume as well as the necessary response time is needed. Second, the systematic deduction of general conditions for the system architecture of a mobile system from the process model would be desirable. Therefore, patterns for mobile business processes as well as corresponding classes of mobile system architectures could be helpful.

Acknowledgements

The Chair of Applied Telematics / e-Business is endowed by Deutsche Telekom AG.

References

1. Hammer, M., Champy, J.: Reengineering the corporation: a manifesto for business revolution. Brealey, London (1993)
2. Davenport, T. H.: Process innovation: reengineering work through information technology. Harvard Business School Press, Boston Mass. (1993)
3. Scheer A.: Business process engineering: reference models for industrial enterprises. Springer, Tokyo (1998)
4. Aalst, W. v. d., Hee, K. v.: Workflow Management: Models, Methods, and Systems. MIT Press, Cambridge, (2002)
5. Deiters, W.: Information Gathering and Process Modeling in Petri Net Based Approach. In: van der Aalst, W.M. et al. (eds.): Business Process Management – Models, Techniques, and Empirical Studies, Lecture Notes in Computer Science 1806, Springer, Berlin Heidelberg New York (2000), pp. 274–288
6. Noor, N. M. M., Papamichail, K. N., Warboys, B.: Process Modeling for Online Communications in Tendering Processes. In: Proceedings of the 29th EUROMICRO Conference 'New Waves in System Architecture', IEEE Computer Society (2003), pp. 17-24
7. Heijden, H. van der, Valiente, P.: Mobile Business Processes: Cases from Sweden and the Netherlands. SSE/EFI Working Paper Series in Business Administration, Stockholm School of Economics (2002)
8. Gruhn, V., Wellen, U.: Process Landscaping: Modeling Distributed Processes and Proving Properties of Distributed Process Models. In: Ehrig, H., Juhás, G., Padberg, J., Rozenberg, G. (Eds.): Unifying Petri Nets - Advances in Petri Nets, Lecture Notes in Computer Science 2128, Springer, Berlin Heidelberg New York (2001), pp. 103-125
9. Köhler, A., Gruhn, V.: Effects of Mobile Business Processes on the Software Process. In: Proceedings of 5th International Workshop on Software Process Simulation and Modeling, 26th International Conference on software engineering, IEE, Stevenage UK (2004), pp. 228-231
10. Köhler, A., Gruhn, V.: Analysis of Mobile Business Processes for the Design of Information Systems. In Bauknecht, K., Bichler, M., Pröll, B. (eds.): Proceedings of 5th International Conference on Electronic Commerce and Web Technologies, Lecture Notes in Computer Science 3182, Springer, Berlin Heidelberg New York (2004), pp. 238-247
11. Gruhn, V., Wellen, U.: Structuring Complex Software Processes by 'Process Landscaping'. In Conradi, R. (ed.): Proceedings of 7th EWSPT European Workshop on Software Process Technology, Lecture Notes in Computer Science 1780, Springer, London (2000), pp. 138-149
12. Jorstad, I., Thanh, D. v., Dustdar, S.: An Analysis of Service Continuity in Mobile Services. In Proceedings of the 13th IEEE International Workshops on Enabling Technologies: Infrastructure for Collaborative Enterprises, IEEE Computer Society, Washington, DC, USA (2004), pp. 121-126
13. Nielsen, C., Sondergaard, A.: Designing for mobility – an integration approach supporting multiple technologies. In: Proceedings of NordiCHI, Royal Institute of Technology, Stockholm, Sweden (2000), pp. 23-25
14. Thai, B., Wan, R., Seneviratne, A., Rakotoarivelo, T.: Integrated personal mobility architecture: a complete personal mobility solution. In Chlamtac, I. et al. (eds.): Journal of Mobile Networks and Applications, vol. 8, Kluwer Academic Publishers, Hingham (2003), pp. 27-36

15. White, S. A.: Business Process Modeling Notation, BPMI.org, 2003.
16. Dustdar, S., Gall, H.: Architectural concerns in distributed and mobile collaborative systems. In Hellwanger, H. et al. (eds.): Journal of Systems Architecture, Elsevier Science, Amsterdam (2003), pp. 457-473
17. Sairamesh, J., Goh, S., Stanoi, I., Padmanabhan, S., Li, C. S.: Disconnected processes, mechanisms and architecture for mobile e-business. In Chlamtac, I. et al. (eds.): Journal of Mobile Networks and Applications, vol. 9, Kluwer Academic Publishers, Hingham (2004), pp. 651-662
18. Sairamesh, J., Goh, S., Stanoi, I., Li, C. S., Padmanabhan, S.: Commerce and Business : Self-managing, disconnected processes an mechanisms for mobile e-business. In: Proceedings of the 2nd International Workshop in Mobile Commerce, International Conference on Mobile Computing and Networking, ACM Press, New York NY USA (2002), pp. 82-89
19. Kakihara, M., Sorensen, C.: Expanding the mobility concept. ACM SIGGROUP Bulletin, vol. 22, no. 3, ACM Press (2001), pp. 33-37
20. Gupta, P., Moitra, D.: Evolving a pervasive IT infrastructure: a technology integration approach. In: Journal of Personal and Ubiquitous Computing, vol. 8, no. 1, Springer, London (2004), pp. 31-41

An Organisational Perspective on Collaborative Business Processes

Xiaohui Zhao, Chengfei Liu, and Yun Yang

CICEC - Centre for Internet Computing and E-Commerce,
Faculty of Information and Communication Technologies,
Swinburne University of Technology,
Melbourne, VIC 3122, Australia
{xzhao, cliu, yyang}@it.swin.edu.au

Abstract. Business collaboration is about coordinating the flow of information among organisations and linking their business processes. It brings great challenge to keep participating organisations as autonomous entities in integrating business processes of these organisations seamlessly. To address this issue, we develop a new perspective on business collaborations with a novel concept called relative workflow, which defines what a participating organisation can perceive in collaboration. By incorporating a visibility control mechanism, relative workflows allow each organisation to define its own collaboration structure and behaviours. In this paper, we present a formal definition of relative workflows and related algorithms for generating relative workflows, along with a discussion on how to perform tracking over relative workflows.

1 Introduction

Recent years have seen the trend of global business collaboration urgently requiring organisations to dynamically form virtual organisation alliances. The business processes of different organisations need to be integrated seamlessly to adapt the continuously changing business conditions and to stay competitive in the global market [1]. To enable such business collaboration, research efforts have been put on improving current workflow technologies for supporting collaborative business processes [2-6]. Web service technology has also emerged partly for this purpose and has been deployed for implementing inter-organisational workflows [7,8].

Current inter-organisational workflow approaches mainly focus on modelling workflows from a *public view*, where a third-party designer or the leading organisation of a virtual organisation alliance defines the business collaboration structure and behaviours by choosing participating organisations and linking workflow processes of these organisations into an inter-organisational workflow process. These approaches work well with the assumption that there exists a third-party designer or a leading organisation that can see certain level of details of all participating organisations. However, this assumption is too restrictive. Though several organisations may be involved in the same collaborative business process for a defined business goal, the relationship between each pair of participating organisations could be different. As such, the visibility between participating

W.M.P. van der Aalst et al. (Eds.): BPM 2005, LNCS 3649, pp. 17–31, 2005.

organisations may be *relative*, rather than *absolute* as adopted in the public view approaches. Besides, the predominant view of a third-party designer or a leading organisation may put participating organisations in a passive position. This violates that each participating organisation acts as an autonomous entity in business collaboration. Moreover, the pre-fixed business collaboration in the public view approaches may not be applicable to those applications where the partner relationship is not fixed.

Aiming at solving these problems, we develop a new perspective on business collaboration based on a novel concept called *relative workflow*, to support participating organisations as autonomous entities. A collaborative business process is represented as a series of relative workflow processes, each of which is defined from the perspective of an individual participating organisation. This allows each organisation, as an autonomous entity, to design its own collaboration structure and behaviours. The third party or leading-organisation-oriented inter-organisational workflow design can be distributed into multiple one-party oriented relative workflow process design. Different visibility constraints can then be defined for different organisations to reflect the fine granularity of visibility control between participating organisations.

The remainder of this paper is organised as follows. In Section 2, we analyse, with a motivating example, the business collaboration requirements that are not well supported by current approaches. Section 3 presents the formal definitions relevant to relative workflow processes. Section 4 introduces a procedure and algorithms for generating relative workflow processes. Section 5 addresses how to perform workflow tracking among organisations. Section 6 discusses the advantages of relative workflows and their applied domains. Related work and concluding remarks are given in Section 7 and Section 8, respectively.

2   Requirement Analysis with Motivating Example

Traditional inter-organisational workflow design approaches streamline business processes contributing to a common business goal, yet belonging to different organisations, into a public workflow process. As discussed earlier, this procedure has the following problems.

The first problem is that the collaboration choreography of all participating organisations is determined by a third party designer or a leading organisation. Following this approach, each organisation behaves in the collaboration passively as a worker does in a pipeline workshop. We find that in many cases, a participating organisation expects to choose its own partner organisations and define inter-organisational workflow processes by itself to adapt its own collaboration objectives and benefits rather than delegate to a third-party designer or a leading organisation. Actually, it is not always possible to find an appropriate third party designer or a leading organisation.

The second problem is the coarse granularity of visibility control. As the public inter-organisational workflow process is open to each involved organisation, either excessive information has to be disclosed or required collaboration information is not provided sufficiently. In the former, some private business information may be disclosed unwillingly to an involved organisation with a distant partner relationship. In the latter, business processes belonging to involved organisations cannot be integrated seamlessly.

The third problem may be caused by pre-determined collaboration choreography of participating organisations. This may not be applicable to some business collaboration scenarios, where the partner relationship between participating organisations may be changed in an ad hoc manner.

We believe that business collaboration should be decided from the view of each individual organisation, i.e., an organisation defines its collaboration structure and behaviours by following corresponding contracts with proper partner organisations, and may change them later by updating existing contracts or signing new contracts. In this way, each organisation acts as a highly autonomous collaboration participant.

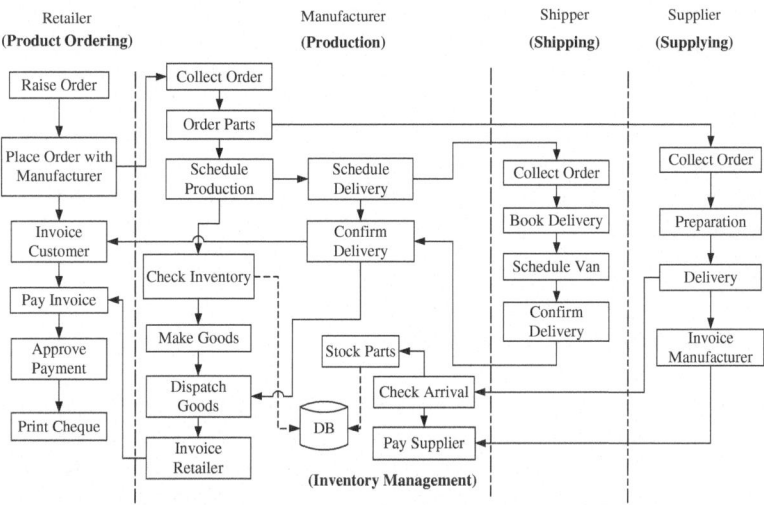

Fig. 1. Inter-organisational workflow process example (modified from [9])

Figure 1 shows business collaboration among a retailer, a manufacturer, a shipper and a supplier, from a public view. Five intra-organisational workflow processes and their interaction are shown in the figure. When a 'Product Ordering' process of a retailer sends a product order to a manufacturer, the 'Production' process of the manufacturer may hold this order until it has collected enough orders from more than one 'Product Ordering' process for the purpose of batch production. Before it starts producing products, the manufacturer needs to order necessary parts from suppliers, which will interact with the 'Inventory Management' process of the manufacturer later for arrival checking and invoice/payment processing. Also, the manufacturer needs to contact shippers to book the delivery of products, and simultaneously checks inventory with the 'Inventory Management' process through the corporate database within the same organisation. Finally, the retailer receives the products from the shipper, and pays the invoice to the manufacturer.

In this example, all the participating organisations have the global knowledge of the whole collaboration process, which is somehow pre-determined and may be defined by a third party designer or a leading organisation such as the manufacturer. Once the collaborative process has been defined, each participating organisation acts passively and loses more or less its autonomy. It will be difficult for an organisation to change its collaboration structure and behaviours, for instance, to start a new partner relationship or to terminate an existing partner relationship. Besides, the global knowledge of the whole collaboration process gives no chance to define a close or distant partner relationship between participating organisations. For example, from Figure 1, we can clearly see that the views from a retailer and a manufacturer on the collaborative process are different. While a manufacturer has a close partner relationship with all other participating organisations, a retailer, however, only has a close partner relationship with a manufacturer via a proper purchase/supply contract. A retailer may not need to know, and actually should not know the manufacturer's partner relationships, say, with a supplier. At the same time, a retailer may need to have some knowledge about a shipper of the manufacturer so that tracking on the delivery of products may be made possible. We may also need to allow that a manufacturer changes partner relationships with suppliers and shippers for better services. All these are not well supported in the public view approaches.

In this paper, we propose a new approach to enable the participating organisations as autonomous entities. The different views from individual organisations are well supported by the concept of relative workflows. This approach also provides visibility control with the finer granularity and allows the easy change of partner relationships in business collaboration.

3   Relative Workflow Processes

In this section, we define a relative workflow, which is based on a visibility control mechanism, to support the requirements discussed in Section 2. In our context, a collaborative workflow process consists of several intra-organisational workflow processes of participating organisations and their interaction. We call these intra-organisational workflow processes as local workflow processes.

Definition 1 (Local Workflow Process). A local workflow process lp is defined as a directed acyclic graph (T, R), where T is the set of nodes representing the set of tasks, and $R \subseteq T \times T$ is the set of arcs representing the execution sequence.

Definition 2 (Organisation). An organisation g is defined as a set of local workflow processes $\{lp^1, lp^2, \dots , lp^n\}$. An individual local workflow process lp^i of g is denoted as $g.lp^i$.

As the owner of local workflow processes, a participating organisation may wish to protect the critical or private business information of some workflow tasks from disclosing to other participating organisations, during business collaboration. According to the two most important behaviours in the context of collaborative workflows, i.e. workflow tracking and business interaction, we define the following three values for the visibility of workflow tasks as listed in Table 1.

Table 1. Visibility values

Visibility value	Explanation
Invisible	A task is said invisible to an external organisation, if it is hidden from that organisation.
Trackable	A task is said trackable to an external organisation, if that organisation is allowed to trace the execution status of the task.
Contactable	A task is said contactable to an external organisation, if the task is trackable to that organisation and the task is also allowed to send/receive messages to/from that organisation for the purpose of business interaction.

Definition 3 (Visibility Constraint). A visibility constraint vc is defined as a tuple (t, v), where t denotes a workflow task and $v \in \{$ Invisible, Trackable, Contactable $\}$. A set of visibility constraints $\mathcal{VC}$ defined on a workflow process lp is represented as a set $\{vc:(t, v) \mid \forall t\ (t \in lp.\mathcal{T})\}$.

Example 1. Based on Figure 1, two sets of visibility constraints are given as follows:
$\mathcal{VC}_1 = \{$ ('Raise Order', Invisible), ('Place Order with Manufacturer', Contactable), ('Invoice Customer', Contactable), ('Pay Invoice', Contactable), ('Approve Payment', Invisible}), ('Print Cheque', Invisible)}.
$\mathcal{VC}_2 = \{$ ('Collect Order', Contactable), ('Order Parts', Invisible), ('Schedule Production', Trackable), ('Schedule Delivery', Trackable), ('Confirm Delivery', Contactable), ('Check Inventory', Invisible), ('Make Goods', Trackable), ('Dispatch Goods', Trackable), ('Invoice Retailer', Contactable)}.

These two sets are defined on the 'Product Ordering' and 'Production' processes, respectively.

Definition 4 (Perception). A perception $p_{g_1}^{g_0.lp}$ of an organisation g_0's local workflow process lp from another organisation g_1 is defined as ($\mathcal{VC}$, $\mathcal{MD}$, f), where

- $\mathcal{VC}$ is a set of visibility constraints defined on $g_0.lp$.
- $\mathcal{MD} \subseteq \mathcal{M} \times \{$ in, out $\}$, is a set of the message descriptions that contains the messages and the passing directions. $\mathcal{M}$ is the set of messages used to represent inter-organisational business activities.
- $f\colon \mathcal{MD} \to g_0.lp_{g_1}.\mathcal{T}$ is the mapping from $\mathcal{MD}$ to $g_0.lp_{g_1}.\mathcal{T}$, and $g_0.lp_{g_1}$ is the *perceivable workflow process* of $g_0.lp$ from g_1.

The generation of $g_0.lp_{g_1}$ from $g_0.lp$ will be discussed in the next section.

Example 2. Again, based on Figure 1, the perception of the retailer's 'Product Ordering' process from the manufacturer, and the perception of the manufacturer's 'Production' process from the retailer are given, respectively, as follows:
$p_{Manufacturer}^{retailer.productOrdering} = ($ $\mathcal{VC}_1$,
{('Order of Products', out), ('Confirmation of Delivery Date', in), ('Invoice', in) },
{('Order of Products', out) $\to$ 'Place Order with Manufacturer', ('Confirmation of Delivery Date', in) $\to$ 'Invoice Customer', ('Invoice', in)$\to$ 'Pay Invoice'});
$p_{retailer}^{Manufacturer.production} = ($ $\mathcal{VC}_2$,
{('Order of Products', in), ('Confirmation of Delivery Date', out), ('Invoice', out) },

{('Order of Products', in) → 'Collect Order', ('Confirmation of Delivery Date', out) → 'Confirm Delivery', ('Invoice', out) → 'Invoice Retailer'}).

where $\mathcal{VC}_1$ and $\mathcal{VC}_2$ are defined in Example 1.

Definition 5 (Relative Workflow Process). A relative workflow process $g_1.rp$ perceivable from an organisation g_1 is defined as a directed acyclic graph ($\mathcal{T}$, $\mathcal{R}$), where

$\mathcal{T}$ is the set of the tasks perceivable from g_1, which is a union of the following two parts:

- $\underset{k}{\cup} g_1.lp^k.\mathcal{T}$, the union of the task sets of all $g_1.lp^k$.

- $\underset{i}{\cup}\underset{j}{\cup} g_i.lp^j_{g_1}.\mathcal{T}$, the union of the task sets of all perceivable workflow processes of $g_i.lp^j$ from g_1.

$\mathcal{R}$ is the set of arcs perceivable from g_1, which is a union of the following three parts:

- $\underset{k}{\cup} g_1.lp^k.\mathcal{R}$, the union of the arc sets of all $g_1.lp^k$.

- $\underset{i}{\cup}\underset{j}{\cup} g_i.lp^j_{g_1}.\mathcal{R}$, the union of the arc sets of all perceivable workflow processes of $g_i.lp^j$ from g_1.

- L, the set of messaging links between local and perceivable workflow processes, consisting of two parts:

  - L_{intra}, the set of intra-organisational messaging links that connect tasks belonging to different local workflow processes, and is defined on
  $$\underset{i}{\cup}\underset{j}{\cup} g_1.lp^i.\mathcal{T} \times g_1.lp^j.\mathcal{T} \text{ where } i \neq j .$$

  - L_{inter}, the set of inter-organisational messaging links that connect tasks between a local workflow process and a perceivable workflow process, and is defined on
  $$\underset{i}{\cup}\underset{j}{\cup}\underset{k}{\cup}\left(g_1.lp^k.\mathcal{T} \times g_i.lp^j_{g_1}.\mathcal{T} \cup g_i.lp^j_{g_1}.\mathcal{T} \times g_1.lp^k.\mathcal{T}\right).$$

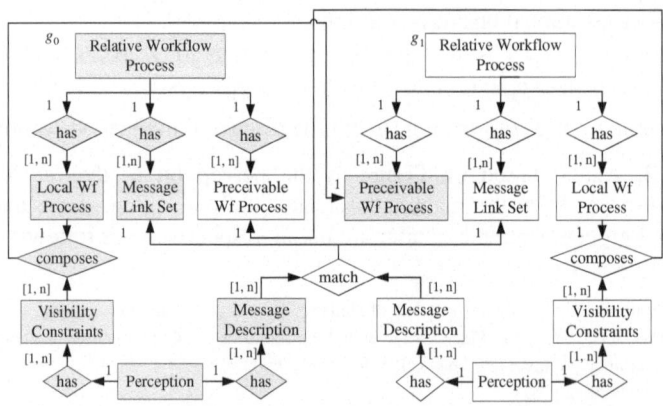

Fig. 2. Relative workflow model

Figure 2 illustrates how the components of the relative workflow model are related across organisations. Given the discussion and definition of the relative workflow process above, a necessary procedure for an organisation, g_0 or g_1, to generate relative workflow processes is to define the perceptions on local workflow processes. This step includes defining visibility constraints, message links and matching functions. Once the perceptions on local workflow processes of its partner organisations have been defined, a relative workflow process can be generated by two more steps: composing tasks and assembling relative workflow processes.

The purpose of composing tasks is to hide some private tasks of local workflow processes. We choose to merge invisible tasks with the contactable or trackable tasks into composed tasks, if not violating the structural validity; otherwise, those invisible tasks are combined into a dummy task. According to the perception defined from g_1, a local workflow process of g_0 after this step becomes a perceivable workflow process for g_1.

In the step of assembling relative workflow processes, an organisation, say g_1, assembles its local workflow processes with perceivable workflow processes from its partner organisations, say g_0, into a relative workflow process, as shown in Figure 2.

The details are to be discussed in the following section.

4   Generating Relative Workflow Processes

4.1 Defining Perceptions

A perception can be derived by analysing and decomposing a commercial contract between organisations in connection to certain business collaboration. Unlike a contract, a perception is defined from the perspective of one organisation on the local workflow processes of other participating organisations. To represent a business

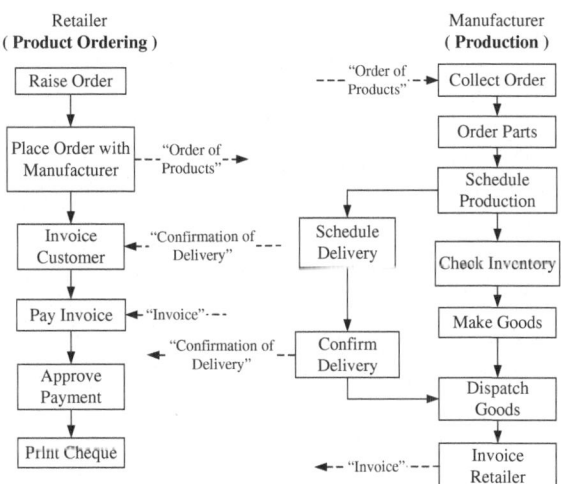

Fig. 3. Local workflow processes

interaction between an organisation g_0 and other participating organisations $g_1, \ldots,$ g_m, two sets of such perceptions are required: $\mathcal{PS}_1$, the set of the perceptions defined on $g_0.lp^1, \ldots, g_0.lp^{n_0}$ from $g_1, \ldots, g_m$, i.e. $\{\ p_{g_1}^{g_0.lp^1}, \ldots, p_{g_1}^{g_0.lp^{n_0}}, \ldots, p_{g_m}^{g_0.lp^1}, \ldots,$ $p_{g_m}^{g_0.lp^{n_0}}\ \}$; and $\mathcal{PS}_2$, the set of the perceptions defined on all local workflow processes of $g_1, \ldots, g_m$ from g_0, i.e. $\{\ p_{g_0}^{g_1.lp^1}, \ldots, p_{g_0}^{g_1.lp^{n_1}}, \ldots, p_{g_0}^{g_m.lp^1}, \ldots, p_{g_0}^{g_m.lp^{n_m}}\ \}$.

Figure 3 shows the 'Product Ordering' process and the 'Production' process in Figure 1, where the dashed arrows denote the message descriptions. To represent the business interaction between these two processes, we can define the perception $p_{Manufacturer}^{Retailer.productOrdering}$ of the retailer's 'Product Ordering' process from the manufacturer and the perception $p_{Retailer}^{Manufacturer.production}$ of the manufacturer's 'Production' process from the retailer, which are already given in Example 1.

4.2 Composing Tasks

In this step, a local workflow process needs to hide its invisible tasks by composing them with proper contactable or trackable tasks for creating the corresponding perceivable workflow process. The algorithm is given below.

For the simplicity of discussion, we only consider composing one local workflow process lp of the organisation g_0 from another organisation g_1. Furthermore, we conduct a pre-processing on all split/join structures of lp such that for all those branches consisting of only invisible tasks, a dummy task is created to delegate these branches.

Algorithm 1. Task Composition

Input: $lp = g_0.lp$, the organisation g_0's local workflow process lp before composition
$p = p_{g_1}^{g_0.lp}$, the perception of g_0's lp from g_1

Output: $lp' = g_0.lp_{g_1}$, the perceivable workflow process composed from lp for g_1, according to $p_{g_1}^{g_0.lp}$.

$lp' = lp$;
$\mathcal{VT} = \{$ all the visible tasks of lp, defined in $p\}$;
// connect invisible tasks
while $(\exists t, t' \in (lp'.\mathcal{T}\text{-}\mathcal{VT})) ((t, t') \in lp'.\mathcal{R}) \wedge seq(t) \wedge seq(t'))$ // $seq(t)=(indegree(t)=1 \wedge outdegree(t)=1)$
$\{$ $t^\circ = t + t'$;
    $lp'.\mathcal{T} = lp'.\mathcal{T} \cup \{t^\circ\} - \{\ t, t'\}$;
    $lp'.\mathcal{R} = lp'.\mathcal{R} - \{(\ t, t')\}$;
    replace t, t' in $lp'.\mathcal{R}$ with t°;
$\}$
// downward composition with incoming interaction tasks
while $((\exists t \in \mathcal{VT}(p'.f^{-1}(t) = (m, \text{in}) \wedge outdegree(t) = 1) \wedge (\exists t' \in (lp'.\mathcal{T}\text{-}\mathcal{VT}))((t, t') \in lp'.\mathcal{R} \wedge indegree(t')=1))$
$\{$ $t^\circ = t + t'$;
    $\mathcal{VT} = \mathcal{VT} \cup \{t^\circ\} - \{t\}$;
    $lp'.\mathcal{T} = lp'.\mathcal{T} \cup \{t^\circ\} - \{t', t\}$;
    $lp'.\mathcal{R} = lp'.\mathcal{R} - \{(t, t')\}$;
    replace t, t' in $lp'.\mathcal{R}$ with t°;
$\}$

```
// upward composition with outgoing interaction tasks
while ((∃t∈ 𝒱𝒯(p′.f⁻¹(t)=(m, out)∧indegree(t) =1)∧(∃t′∈ (lp′.𝒯-𝒱𝒯))((t′,t)∈ lp′.ℛ∧outdegree(t′)=1))
{      t°=t+t′;
       𝒱𝒯= 𝒱𝒯∪{t°}-{t};
       lp′.𝒯= lp′.𝒯∪{t°}-{t′, t};
       lp′.ℛ= lp′.ℛ-{(t′, t)};
       replace t, t′ in lp′.ℛ with t° ;
}
```

Algorithm 1 first keeps composing each pair of neighbouring sequential invisible tasks into one invisible task, then downward composes invisible tasks with incoming interaction tasks and upward composes invisible task with outgoing interaction tasks. Figure 4 shows the results of task composition: (a) is the perceivable 'Product Ordering' process of the retailer from the manufacturer; and (b) is the perceivable 'Production' process of the manufacturer from the retailer, where the dashed rectangles denote invisible tasks.

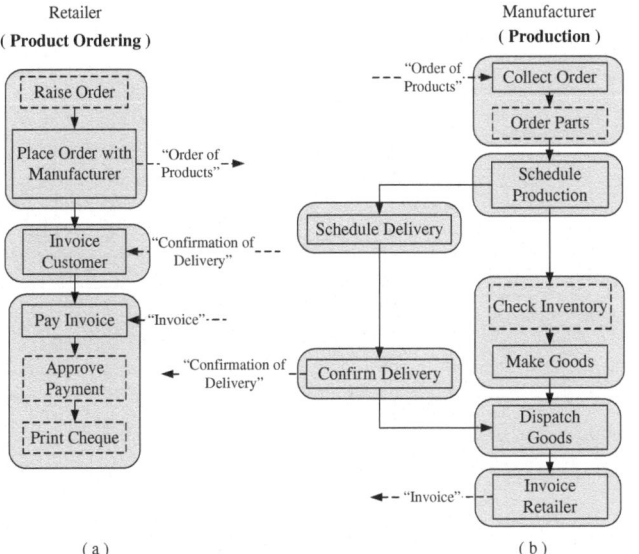

Fig. 4. Perceivable workflow processes

4.3 Assembling Relative Workflow Processes

In this step, proper local workflow processes and perceivable workflow processes are connected together by linking the corresponding interaction operations. This can be achieved by matching the message descriptions defined in the perceptions, using the algorithm below. Similarly, for the simplicity of discussion, we only consider matching one local workflow process lp of the organisation g_0 from another organisation g_1.

Algorithm 2. Local Workflow Process Matching

Input: $lp' = g_0.lp_{g_1}$, the perceivable workflow process composed from g_0's local workflow process lp.

$p = p_{g_1}^{g_0.lp}$, the perception of g_0's lp from g_1

$ps = \{ p_{g_0}^{g_1.lp^1} , \ldots , p_{g_0}^{g_1.lp^{1_n}} \}$, the set of perceptions defined on g_1's perceivable workflow processes from g_0

Output: $\mathcal{L}$, the set of generated messaging links

$\mathcal{L} = \varnothing$;

for each $t \in lp'.\mathcal{T}$

  if $\exists md(p.f(md) = t)$ **then** {

    $md_1 = p.f^{-1}(t)$;

    for each $p° \in ps$

      for each $md_2 \in p°.\mathcal{MD}$

        if md_1 **matches** md_2 **then** $\mathcal{L} = \mathcal{L} \cup \{(t, p°.f(md_2), md_1)\}$;

}

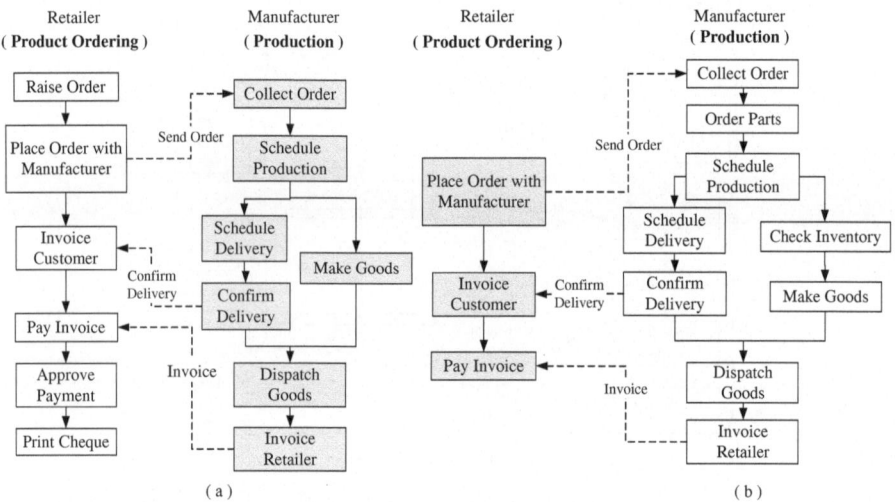

Fig. 5. Relative workflow processes

By one message description md_1 matches another message description md_2 in Algorithm 2, we mean that they have the same message, and one has passing direction 'in' while the other has 'out'. With the set $\mathcal{L}$ of generated messaging links, we can now finally assemble relative workflow processes. As shown in Figure 5: (a) is the relative workflow process perceivable from the retailer; and (b) is the relative workflow process perceivable from the manufacturer, where the dashed connecting arrows denote the generated message links. Different participating organisations may have different views to the same inter-organisational workflow process. This reflects our *relativity* characteristics.

5   Tracking on Business Collaborations

Once a relative workflow process is generated at the process level, we may perform workflow tracking of its instances, through complex interactions between business process instances of participating organisations. Tracking on business collaboration means the ability to track and report on the events happened to a specific business collaborative process during its execution. In our context, we are referring to the status enquiries on a relative workflow instance and other related relative workflow instances.

As we can see from the relative workflow assembling algorithm, i.e. Algorithm 2, a relative workflow process is created for a specific organisation. However, as a relative workflow process is a part of a collaborative workflow process, this organisation may track the status, for instance, of an outsourced job, and the tasks related to this job. Sometimes, these related tasks may belong to non-neighbouring organisations.

Referring to the motivating example in Figure 1, the 'Confirm Delivery' task of the manufacturer's 'Production' process cannot send the confirmation of delivery to the retailer's 'Product Ordering' process until it receives the confirmation of delivery from the shipper's 'Shipping' process. But from the retailer's relative workflow process about the product purchase/supply collaboration shown in Figure 5 (a), the retailer cannot see this dependency between the manufacturer and the shipper directly. Then, what if the shipper permits the retailer's tracking on its 'Shipping' process? Can the relative workflow model support tracking beyond neighbouring organisations? We address this issue in this section.

As mentioned in Section 3, the 'trackable' visibility value is dedicated to represent the trackability of a task to a specific organisation. Based on the motivating example, the shipper may permit the retailer's tracking by signing a contract, from which the corresponding perception can be defined with some tasks set to 'trackable' for the retailer. For example, here we suppose the shipper defines the following visibility constraints for the retailer in the perception $p_{Retailer}^{Shipper.Shipping}$,

$\mathcal{VC}_3$ = {('Collect Orders', Trackable), ('Book Delivery', Invisible), ('Schedule Van', Invisible), ('Confirm Delivery', Trackable) };

With such visibility constraints, the retailer can obtain the perceivable 'Shipping' workflow process, which is shown in Figure 6 (a).

Given the visibility constraints defined by the shipper for the manufacturer in the perception $p_{Manufacturer}^{Shipper.Shipping}$,

$\mathcal{VC}_4$ = {('Collect Orders', Contactable), ('Book Delivery', Trackable), ('Schedule Van', Invisible), ('Confirm Delivery', Contactable)},
the manufacturer can obtain a relative workflow process about the product delivery collaboration, as shown in Figure 6 (b).

Now, the manufacturer's 'Production' process may become a bridge between the retailer and the shipper because it exists in both the retailer's relative workflow process from the manufacturer in Figure 5 (a) and the manufacturer's relative workflow process from the shipper in Figure 6 (b). As the 'Shipping' process is also perceivable from the retailer, it is possible for the retailer to track the related tasks in the shipper's 'Shipping' process via the manufacturer.

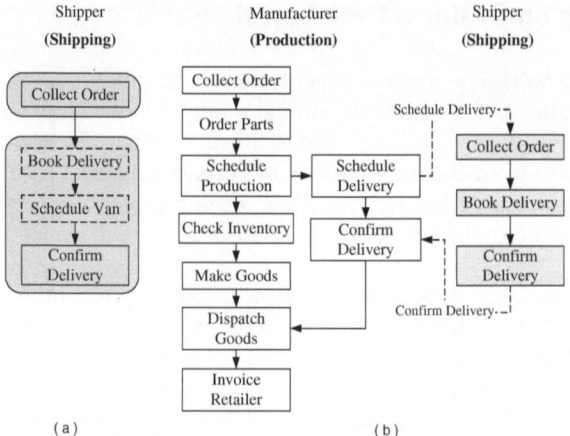

Fig. 6. 'Shipping' process perceivable from different organisations

However, such a bridging procedure requires a composite view from the retailer to the shipper by composing the view of the retailer from the manufacturer and the view of the manufacturer from the shipper. This step has to be handled by the manufacturer, since only the manufacturer has the knowledge of all necessary visibility constraints and relative workflow processes. First, we start from the manufacturer's relative workflow process in Figure 6 (b). We can see that the 'Collect Order' task of the 'Shipping' process has the 'Schedule Delivery' message link with the 'Schedule Delivery' task of the manufacturer's 'Production' process. And the 'Schedule Delivery' task is also set trackable to the retailer (refer to $\mathcal{VC}_2$ defined in Example 1). Therefore, the 'Schedule Delivery' message link can be kept as original, because the tasks connected by it are perceivable to both the manufacturer and the retailer. Similarly, the same result can be achieved when composing the 'Confirm

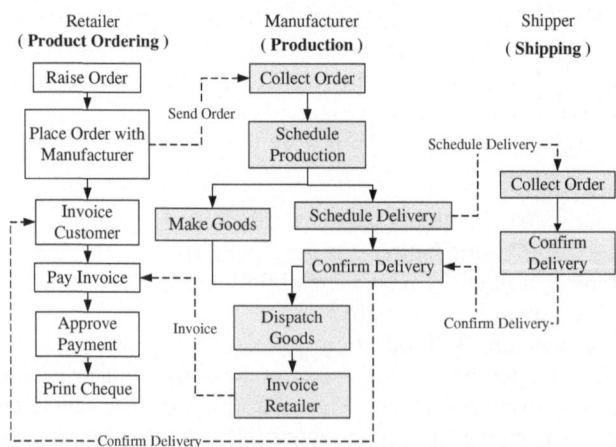

Fig. 7. Tracking structure

Delivery' message link. Finally, at the site of the retailer, the tracking structure can be extended by connecting the perceivable 'Shipping' workflow process to its original relative workflow process in Figure 5 (a), with the composed message links. The extended tracking structure is given in Figure 7.

Once such a tracking structure is derived, the execution information of all involved workflow processes can be collected by propagating along this structure.

6 Discussions

With the proposed relative workflows, organisation centred business collaboration can be easily achieved. In this collaboration scheme, an individual organisation can actively choose partner organisations, and assemble proper 'off-the-shelf' perceivable workflow processes from partner organisations with its own workflow processes into a relative workflow process. This relative workflow process forms part of a collaborative workflow process for specific business collaboration. This collaboration scheme has the following appealing features:

1. *Support of high autonomy in collaborations.*

    As an autonomous entity, each organisation is in charge of defining the collaboration structure and behaviours with its partner organisations to fulfil its own business planning and management, without being forced to adapt the restrictions and irrationalities caused by the design of a third party designer or a leading organisation anymore. Therefore, each organisation owns the full control of its business collaboration.

2. *Support of flexible collaborations.*

    The proposed collaboration scheme can support business collaborations among loosely-coupled organisations in a dynamic or temporary manner. With the help of this scheme, a participating organisation is now able to easily redefine its collaboration structure and behaviours on the fly, e.g., to change partner organisations, to alter requirements for business collaboration with partner organisations, etc.

3. *Support of information protection.*

    The visibility control mechanism prevents the private information disclosure at the task level or at the process level. Participating organisations are now able to control the level of information revealing to different participating organisations accordingly.

As we can see, a collaborative workflow process modelled in the public view approaches can be modelled in our relative workflow approach by a series of relative workflow processes with more advantages described above. Moreover, it may also support some applications that the public view approaches cannot cope with. One such an application is to support transient supply chains. In current e-marketplaces or other information portals, buyer, supplier, seller and distributor organisations can exchange their trading information and find trading partners. These sorts of collaborations are most likely to be dynamic and temporary, because a partner relationship is usually decided by means of price matching, bidding or auctions, and it will terminate as soon as the trading finishes. As discussed earlier, our relative workflow approach can support it very well. Another application is a virtual organisation alliance consisting of small-to-medium sized enterprises (SMEs), where SMEs join a virtual community to share business services from each other. Each organisation in such an open alliance is aware of the services utilisable, and also needs to publish its business services to other organisations. Such a dual-awareness requirement can be well supported using visibility control based perceptions. In addition, the "bottom-up" building mechanism of relative workflows suits this kind of alliances perfectly.

7 Related Work

Chiu et al. [10] borrowed the notion of 'view' from federated database systems, and employed a virtual workflow view for the inter-organisational collaboration instead of the real instance, to hide internal information. Our relative workflows approach extracts the explicit visibility constraints from the commercial contracts to restrict the information disclosure. Different from the workflow view model, the relative workflow approach distributes the macro business collaboration into interactions between neighbouring organisations, and these interactions are performed by the relative workflows designed from the perspective of individual organisations.

van der Aalst and Weske [4] proposed a "top-down" approach for inter-organisational workflow processes and adopted a public-to-private method to formalise the partition process. In this paper, we take a "bottom-up" approach to build up relative workflow processes from each individual organisation first, then to represent a collaborative business process as a series of relative workflow processes.

Schulz and Orlowska [3] developed a cross-organisational workflow architecture, on the basis of communication between the entities of a view-based workflow model. In comparison, our relative workflow approach defines perceptions from the view of each participating organisation. The relative workflow processes can be dynamically generated by linking the local workflow processes using perceptions.

The CrossFlow project [2] aimed to support cross-organisational workflow management in a dynamic virtual enterprise, with the cooperation based on dynamic service outsourcing specified in electronic contracts. However, the contracts in this project did not include explicit visibility parameters. Compared with this work, our relative workflow approach provides a more systematic support in visibility control.

A business contract specification language (XLBC) was introduced in [11] to formally link the Component Definition Language (CDL) specification of business object based workflow systems. A brief discussion on object visibility specified by contracts was also given in that research. Nevertheless, no more detailed work in that regard could be found. Our relative workflows use perceptions to define a specific visibility of each workflow process to different organisations. Based on this visibility control mechanism, support of some advanced features, such as flexible collaborations, autonomy in collaborations, are now available.

8 Conclusions

This paper has presented a new approach on inter-organisational business collaboration, by proposing a novel concept called relative workflow. In this approach, each organisation acts as an autonomous entity with the full control of choosing its partner organisations and defining its collaboration structure and behaviours. Instead of defining a collaborative business process as a whole, each participating organisation may define its relative workflow processes from its own perspective for business collaboration. Associated with a relative workflow process, a set of visibility constraints are defined for interaction and tracking. In this paper, both the formal definitions of relative workflows and the algorithms for generating relative workflows have been presented. The tracking on relative workflow processes has also been discussed.

In the future, we plan to prototype this work in the Web service environment and to refine the relative workflow architecture to better support collaborative business processes.

Acknowledgements

The work reported in this paper is partly supported by the Australian Research Council discovery project DP0557572.

References

1. Osterle, H., Fleisch, E.Alt, R.: Business Networking - Shaping Collaboration between Enterprises. Springer Verlag (2001)
2. Grefen, P., Aberer, K., Ludwig, H.Hoffner, Y.: CrossFlow: Cross-Organizational Workflow Management for Service Outsourcing in Dynamic Virtual Enterprises. Data Engineering, 24(1) (2001) 52-57
3. Schulz, K. and Orlowska, M.: Facilitating Cross-organisational Workflows with a Workflow View Approach. Data & Knowledge Engineering, 51(1) (2004) 109-147
4. van der Aalst, W. and Mathias, W.: The P2P Approach to Inter-organizational Workflows. Advanced Information Systems Engineering, Proceedings (2001) 140-156
5. Wetzel, I. and Klischewski, R.: Serviceflow beyond Workflow? IT Support for Managing Inter-organizational Service Processes. Information Systems, 29(2) (2004) 127-145
6. Groiss, H. and Eder, J.: Workflow Systems for Inter-organizational Business Processes. SIGGroup Bulletin, 18(3) (1997) 23-26
7. Business Process Execution Language for Web Services (BPEL4WS) Ver1.1. http://www.ibm.com/developerworks/library/ws-bpel/ (2003)
8. Zhao, X., Liu, C.Yang, Y.: Web Service based Architecture for Workflow Management Systems. Database and Expert Systems Applications, Proceedings, LNCS 3180 (2004) 34-43
9. Anderson, M. and Allen, R.: Workflow Interoperability - Enabling E-Commerce. www.aiim.org/wfmc/mainframe.htm (1999)
10. Chiu, D., Cheung, S., Karlapalem, K., et al.: Workflow View Driven Cross-organizational Interoperability in a Web-service Environment. Web Services, E-Business, and the Semantic Web, LNCS 2512 (2002) 41-56
11. van den Heuvel, W. and Weigand, H.: Cross-Organizational Workflow Integration using Contracts. ACM Conference on Object-Oriented Programming, Systems, Languages, and Applications (2000)

Mining Hierarchies of Models: From Abstract Views to Concrete Specifications

Gianluigi Greco[1], Antonella Guzzo[2], and Luigi Pontieri[2]

[1] Dept. of Mathematics, UNICAL,
Via P. Bucci 30B, 87036, Rende, Italy
[2] ICAR-CNR, Via P. Bucci 41C, 87036 Rende, Italy
ggreco@mat.unical.it, {guzzo, pontieri}@icar.cnr.it

Abstract. Process mining techniques have been receiving great attention in the literature for their ability to automatically support process (re)design. The output of these techniques is a concrete workflow schema that models all the possible execution scenarios registered in the logs, and that can be profitably used to support further-coming enactments. In this paper, we face process mining in a slightly different perspective. Indeed, we propose an approach to process mining that combines novel discovery strategies with abstraction methods, with the aim of producing hierarchical views of the process that satisfactorily capture its behavior at different level of details. Therefore, at the highest level of detail, the mined model can support the design of concrete workflows; at lower levels of detail, the views can be used in advanced business process platforms to support monitoring and analysis. Our approach consists of several algorithms which have been integrated into a systems architecture whose description is accounted for in the paper as well.

1   Introduction

The difficulties encountered in the design of complex workflows have recently stimulated the development of process mining techniques [1,2,3,4,5,6,7,8], whose aim is to automatically derive a model for the process at hand, based on log data collected during its past enactments. Notably, when a large number of activities and complex behavioral patterns are involved in the analysis, process mining may be a rather trickish task, and the discovered model might fail in representing the process in a clear and concise manner. Indeed, process mining algorithms are generally designed to maximize the accuracy of the mined model, i.e., they equip the model with as many variants as they are required to support all the registered logs; therefore, the resultant schema is well-suited for supporting the enactment, but is less useful for a business user who wants to monitor and analyze the business operation at some appropriate level of abstraction.

To overcome this limitation, we propose an approach to process mining that produces a hierarchical process model which satisfactorily captures the behavior of the process at hand, by providing different views at different level of details. Roughly speaking, the model is essentially a tree such that the root encodes the

W.M.P. van der Aalst et al. (Eds.): BPM 2005, LNCS 3649, pp. 32–47, 2005.

most abstract view, which has no pretension of being an executable workflow, whereas any level of internal nodes encodes a refinement of such an abstract model, in which some specific details are introduced.

The capability of discovering a modular and expressive description for a process can be a valid help in designing, monitoring, and analyzing process models, and can pave the way for effectively reusing, customizing and semantically consolidating process knowledge. And, in fact, the need and the usefulness of process hierarchies/taxonomies has already emerged in several applicative contexts, and process abstraction is currently supported in some advanced platforms for business management (e.g, iBOM [9], ARIS [10]), in which the designer can manually define the relationships among the abstract and the actual process.

In the literature, the definition of process hierarchies was first considered in [11], envisaging a repository of process descriptions for supporting both design and sharing of process models. The notion of process specialization/generalization (w.r.t some suitable behavioral semantics) has been investigated for different modelling formalisms, such as Object Behavior Diagrams [12], UML diagrams [13], process-algebra specifications and Petrinets [14,3], DataFlow diagrams [15]. Recently, some abstraction techniques aiming at summarizing complex processes have been proposed in [9,16].

The main distinguishing feature of our approach with respect to the proposals cited above is the combination of mining and abstraction methods for automatically producing a hierarchical process model. This entails that no substantial human intervention is required while abstracting process schemas, so that software modules implementing the algorithms described in the paper represent a valuable add on for advanced process management platforms. In more details, the contribution of the paper is as follows:

- In Section 3, we introduce a top-down clustering algorithm that generates a hierarchy of workflow schemas, by inducing each of them from a homogeneous cluster of traces. Since, at each step, the algorithm greedily splits the cluster equipped with the least sound schema, schemas at the leaves of the hierarchy effectively model different usage-scenarios for the process.
- The whole hierarchy build by means of clustering is of great value in structuring different execution classes into an effective taxonomical view. In Section 4, we propose an algorithm for obtaining a taxonomy of schemas, by producing, for each non-leaf node of the hierarchy, an abstract schema generalizing all those associated with the children.
- In Section 5, we present an abstraction algorithm and some associated metrics, which are meant to support the above generalization algorithm by properly replacing groups of "specific" activities with "higher-level" activities.
- Finally, in Section 6, we sketch the architecture of a system implementing the whole approach, and discuss some concluding remarks and future works.

2   Formal Framework

In this section, we introduce the basic notions and notation for formally representing workflow models, which will be exploited in the rest of the paper. The *control flow graph* of a process P is a tuple $\langle A, E, a^0, F \rangle$, where: A is a finite set of *activities*, $E \subseteq (A - F) \times (A - \{a^0\})$ is a relation of precedences among activities, $a^0 \in A$ is the starting activity, $F \subseteq A$ is the set of final activities. A control flow graph defines the potential orderings according to which the activities of P can be executed; it is often enriched with some kind of constraints imposing further restrictions on the executions.[1] For any a activity of the workflow schema, the split constraint for a is: *(S.i) AND-split* if a activates all of its successor activities, once completed; *(S.ii) OR-split*, if a may activate any number (non-deterministically chosen) of its successor activities, once completed; *(S.iii) XOR-split* if a activates exactly one out of all its successor activities, once completed. The join constraint for a is: *(J.i) AND-join* if a can be executed only after all of its predecessors have notified a to start; *(J.ii) OR-join*, if a can be executed as soon as one of its predecessors notifies a to start.

Let P be a process. A *workflow schema* for P, denoted by $\mathcal{W}(P)$, is a tuple $\langle A, E, a^0, F, \mathcal{C} \rangle$, where $\langle A, E, a^0, F \rangle$ is a control flow graph for P, and $\mathcal{C}$ is a set of constraints for the activities in A. Fig. 1 shows a possible workflow schema for the *OrderManagement* process of handling customers' orders in a business company. Constraints are drawn by means of labels beside the tasks – e.g., accept order is an *and-join* activity as it must be notified by its predecessors that both the client is reliable and the order can be supplied correctly.

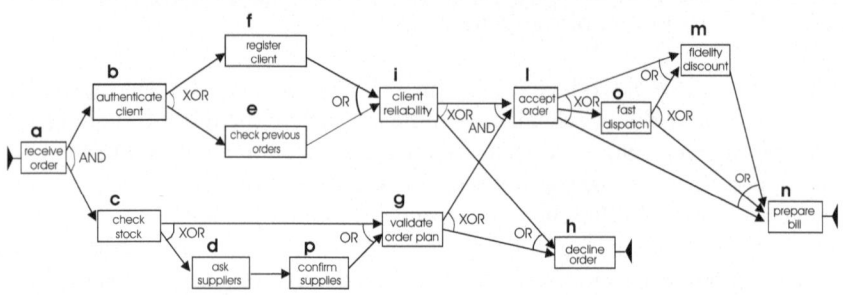

Fig. 1. Workflow schema for the sample *OrderManagement* process

Each time a workflow is enacted in a workflow management system, it produces an *instance*, i.e., a suitable subgraph of the schema, containing both initial and final activity, that satisfies all the constraints. Actually, many process-oriented commercial systems store partial information about the various instances of a process, by tracing some events related to the execution of its activities. In particular, the *logs* kept by most of such systems simply consist of

---

[1] We do not refer to any specific syntax proposed for expressing constraints; rather, we deal with some basic features occurring in the most typical workflow systems.

sequences of event occurrences, which, in general, cannot allow to reconstruct the structure of all workflow instances. Let A_P be the set of task identifiers for the process P; then, a *workflow trace s over A_P* is a string in A_P^*, representing a task sequence. For instance, in our running example, a trace can be encoded by the string `acbgih`. A *workflow log for P*, denoted by $\mathcal{L}_P$, is a bag of traces over A_P, i.e., $\mathcal{L}_P = [\, s \mid s \in A_P^* \,]$.

We next formalize the relationship between traces and instances. Let I be an instance of a workflow schema $\mathcal{W}$, and s be a trace in $\mathcal{L}_P$. Then, s is *compliant with $\mathcal{W}$ through I*, denoted by $s \models^I \mathcal{W}$, if the last activity of s is a final activity w.r.t. to $\mathcal{W}$ and there exists a topological sort s' of I such that s is a prefix of s'. Furthermore, s is simply said to be *compliant with $\mathcal{W}$*, denoted by $s \models \mathcal{W}$, if there exists an instance I such that $s \models^I \mathcal{W}$.

Finally, the following functions allow to evaluate the degree of conformance of $\mathcal{W}$ w.r.t. a given log $\mathcal{L}_P$: *(i) soundness$(\mathcal{W}, \mathcal{L}_P)$*, expressing the percentage of instances of $\mathcal{W}$ which have some corresponding traces in $\mathcal{L}_P$, and *(ii) completeness$(\mathcal{W}, \mathcal{L}_P)$*, which measures the percentage of traces in $\mathcal{L}_P$ that are compliant with $\mathcal{W}$. It is worth noticing that both soundness and completeness should be considered during the process mining task, in order to discover a schema that satisfactorily model the input traces.

3 Mining Hierarchies of Workflow Schemas

Our approach to discover expressive process models at different level of details is articulated in two phases. First, we mine a hierarchy of workflow schemas, by means of a hierarchical top-down clustering algorithm, called `Hierarchy Discovery`. Then, we visit the mined model in a bottom-up way, i.e., from the leaves to the root, and we restructure it at several levels of abstraction, by means of the algorithm `BuildTaxonomy`. Details on the former phase are reported in this section, whereas details on the latter are reported in Section 4.

3.1 Algorithm `HierarchyDiscovery`

A process mining technique that is specifically tailored for complex process, involving lots of activities and exhibiting different variants has been presented in [8]. It relies on the idea of explicitly representing all the possible usage scenarios by means of a collection of different, specific, workflow schemas, in order to obtain a modular representation of the process itself, which is yet sounder than a single workflow schema mixing all of them. We here propose a new algorithm that extends the one presented in [8] by allowing the computation of hierarchical process models rather than simple collections of workflow schemas. The mined model is now meant to be a hierarchy of workflow schemas that collectively represent the process at different levels of granularity and abstraction: the set of schemas corresponding to children of any node v represents the same set of execution as v, but in a more detailed and sounder way, as different subclass of executions are separately described. We next formalize the notion of hierarchical model.

Input: A set of log traces $\mathcal{L}_P$, two natural numbers $maxSize$ and k, a threshold γ.
Output: A schema hierarchy for P.
Method: Perform the following steps:
```
 1  W_0 := mineWFschema(L_P);
 2  WS := {W_0};
 3  Traces[W_0] := L_P;    // Traces[W_i] refers to the log traces modelled by W_i, ∀W_i ∈ WS
 4  T := ⟨{ v_0}, ∅, v_0 ⟩;
 5  λ(v_0) := W_0;
 6  while |WS| ≤ maxSize and soundness(⟨WS, T, λ⟩, L^P) < γ do
 7      let W_q be the least sound "leaf" schema ᵃ and v_q=λ^{-1}(W_q) be its associated node in T;
 8      let n=|WS| be the number of schemas currently stored in WS;
 9      ⟨L_{n+1}, ..., L_{n+k}⟩ := partition-FB(Traces[W_q]);
10      if k > 1 then
11          for h = 1..k do
12              W_{n+h} := mineWFschema(L_{n+h});
13              WS := WS ∪ {W_{n+h}};
14              Traces[W_{n+h}] := L_{n+h};
15              T.V := T.V ∪ {v_{n+h}};    T.E := T.E ∪ {(v_q, v_{n+h})};
16              λ(v_{n+h}) := W_{n+h};
17          end for
18      end if
19  end while
20  return ⟨WS, T, λ⟩;
```

---

[a] i.e., $W_q = argmin_{W \in \mathcal{WS}}\{soundness(W, traces(W)) \mid \lambda^{-1}(W)$ is a leaf of $T\}$

Fig. 2. Algorithm `HierarchyDiscovery`

Definition 1. Let $\mathcal{L}_P$ be a set of log traces for a process P. Then, a *schema hierarchy* for P is a tuple $\mathcal{H} = \langle \mathcal{WS}, T, \lambda \rangle$, such that:

- $\mathcal{WS}$ is a set of workflow schemas for P;
- $T = \langle V, E, v_0 \rangle$ is a tree, where V (resp. E) denotes the set of vertices (resp. edges), and $v_0 \in V$ is the root;
- $\lambda : V \mapsto \mathcal{WS}$ is a bijective function associating each vertex $v \in V$ with a workflow schema $\lambda(v)$ in $\mathcal{WS}$;

Soundness and completeness of $\mathcal{H}$ are defined as follows: *(i) soundness($\mathcal{H}$, $\mathcal{L}_P$)* is the percentage of the instances modelled by the schemas associated with the leaves of T that have some corresponding trace in $\mathcal{L}_P$, *(ii) completeness($\mathcal{H}$, $\mathcal{L}_P$)* is the percentage of traces in $\mathcal{L}_P$ that are compliant with at least one schema associated with a leaf of T. $\square$

Notice that for each vertex v in V, the set S_v of the schemas associated with the children of v, i.e., $S_v = \{\lambda(v_i^c) \mid (v, v_i^c) \in E\}$, is essentially meant to model the same set of instances modelled by $\lambda(v)$, but in a sounder way. Therefore, the union of all the schemas associated with the leaves constitute, as a whole, the soundest model for the process.

Given a log $\mathcal{L}_P$, we can discover a schema hierarchy for P by recursively partitioning the traces in $\mathcal{L}_P$ into clusters, according to the different behavioral patterns they exhibit, and building a schema for each of these clusters. This is accomplished by the algorithm `HierarchyDiscovery` (see Fig. 2), where the function `mineWFschema` is exploited for discovering each single workflow schema in the hierarchy. Some possible implementations of `mineWFschema` are discussed in [1,2,3,4,5,6,7,8], and essentially consist in discovering precedence relationships and constraints that involve the activities.

The meaning of the other input parameters is as follows: γ is a (lower) threshold for the soundness of the mined hierarchy H, while $maxSize$ and k bound the total number of nodes in H and their out-degrees, respectively. Notice that we here assume that the discovered model must have maximal completeness. Obviously, we can straightforwardly extend the approach to discovering not fully complete models, e.g., by introducing a threshold for completeness and using some implementation of mineWFschema taking account for such a threshold.

The algorithm starts by building a workflow schema W_0 (Line 1) which is a first attempt to represent the behavior captured in the log traces, and which will be the only component of $\mathcal{WS}$ (Line 2). The schema $\mathcal{WS}_0$ is associated with the whole log via the auxiliary structure $Traces$ (Line 3), which enables for recording the set of traces each discovered schema was derived from. Moreover, the tree T is initialized with a single node (its root) v_0, which is associated with W_0 by properly setting the function λ (Lines 4-5).

In order to produce a more accurate model, we greedily chose to refine the least sound schema W_q in $\mathcal{WS}$ and to derive a set of more refined schemas (Lines 7-18) as children of node corresponding to W_q. To this purpose, the set of traces modelled by the selected schema W_q is partitioned through the procedure partition-FB (Line 9) into a set of clusters which, in a sense, are more homogeneous from a behavioral viewpoint. Roughly speaking, the procedure mainly relies on the discovery of frequent rules representing behavioral patterns that were unexpected with respect to W_q. Such rules are then used to map the traces into a feature space, where classical clustering methods can be applied (see [8] for more details).

For each new cluster L_{i+h} a specific workflow schema W_{i+h} is extracted, by using again function mineWFschema, and added to $\mathcal{WS}$ (Lines 10-11). Moreover, W_{i+h} is associated with the cluster L_{i+h} it was induced from, and with a new node in the tree, which is a child of the node corresponding to the refined schema W_q (Lines 14-16). The whole process of refining a schema can then be iterated in a recursive way, by selecting again the least sound leaf schema in the current hierarchy, until the desired value γ of soundness has been achieved or too many schemas (i.e., $maxSize$ or more) are already in $\mathcal{WS}$ (Line 6).

Example 1. In order to provide some insight on how the algorithm works, we report a few notes on its behavior when used to mine a synthesized log. To this purpose 100,000 traces for the workflow schema shown in Fig. 1 were randomly generated by means of the generator described in [17]. Notably, in the generation of the log, we also required that task m could not occur in any execution trace containing f, and that task o could not appear in any trace containing d and p, thereby modelling the intuitive restriction that a fidelity discount in never applied to a new customer, and that a fast dispatching procedure cannot be performed whenever some external supplies were asked for. These additional constraints allow us to simulate the presence of different usage scenarios that cannot be captured by a simple workflow schema.

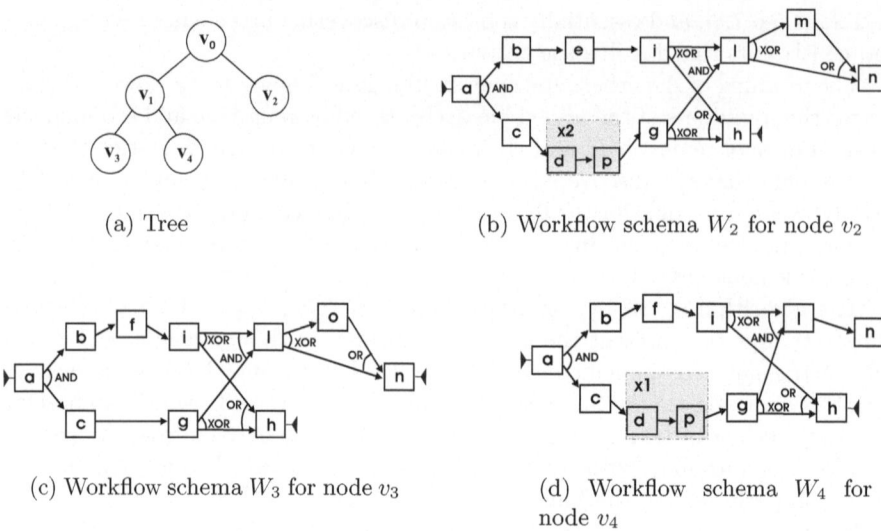

(a) Tree

(b) Workflow schema W_2 for node v_2

(c) Workflow schema W_3 for node v_3

(d) Workflow schema W_4 for node v_4

Fig. 3. Hierarchy generated by `HierarchyDiscovery` (details for leaf schemas only)

The output of `HierarchyDiscovery`, for $maxSize = 5$ and $\gamma = 0.85$, is the schema hierarchy reported in Fig. 3.(a), where each node logically corresponds to both a cluster of traces and a workflow schema induced from that cluster by means of traditional algorithms for process mining. Thus, node v_0 corresponds to the whole set of traces and to an associated (mined) workflow. Actually, the algorithm `HierarchyDiscovery` finds that the schema of v_0 is not as sound as required by the user, and therefore partitions the traces by means of a clustering algorithm (k-means in the implementation). In the example, we fix $k = 2$ and the algorithm generates two children v_1 and v_2; then, v_2 is not further refined (due to its high soundness), while traces associated with v_1 are split again into v_3 and v_4. At the end, the schemas associated with the leaves of the tree are those shown in the Figure. As a matter of fact, schemas W_0 and W_1 (associated with v_0 and v_1, respectively) are only preliminary attempts to model executions that are, indeed, modelled in a sounder way by the leaf schemas. Nevertheless, the whole hierarchy is an important result as well, for it somehow structures the discovered execution classes, and is a basis for deriving a schema taxonomy representing the process at different abstraction levels, as it will be discussed in Section 4.   □

4   Restructuring Schema Hierarchies

In the second phase of our approach, we exploit the schema hierarchy produced by `HierarchyDiscovery`, in order to restructure it for producing a description of the process at different levels of details. Intuitively, leaf nodes stand for concrete usage scenarios, whereas non-leaf nodes are meant to represent suitable generalizations of the different process models corresponding to their children. Relations among activities are next formalized by means of abstraction dictionaries.

4.1 Abstraction Relationships

Let A be a set *activities*. An abstraction dictionary for A is a tuple $\mathcal{D} = \langle \mathcal{I}sa, \mathcal{P}artOf \rangle$, such that $\mathcal{D}.\mathcal{I}sa \subseteq A \times A$, $\mathcal{D}.\mathcal{P}artOf \subseteq A \times A$ and, for each $a \in A$, $(a, a) \notin \mathcal{D}.\mathcal{P}artOf$ and $(a, a) \notin \mathcal{D}.\mathcal{I}sa$. Roughly speaking, for two activities a and b, $(b, a) \in \mathcal{D}.\mathcal{I}sa$ indicates that b is a refinement of a; conversely, $(b, a) \in \mathcal{D}.\mathcal{P}artOf$ indicates that b is a component of a.

Given two activities a and a', we say that a *generalizes* a' w.r.t. a given abstraction dictionary $\mathcal{D}$, denoted by $a \uparrow^{\mathcal{D}} a'$, if there is a sequence of activities $a_0, a_1, .., a_n$ such that $a_0{=}a'$, $a_n{=}a$ and $(a_i, a_{i-1}) \in \mathcal{D}.\mathcal{I}sa$ for each $i = 1..n$; we call such a sequence a *genpath* from a' to a with length n. Moreover, the *generalization distance* between a and a' w.r.t. $\mathcal{D}$, denoted by $dist_G^{\mathcal{D}}$, is the minimal length of the *genpaths* connecting a' to a, or vice-versa. As a special case, we assume that $dist_G^{\mathcal{D}}(a, a) = 0$ for any activity a. Finally, the *most specific generalization* of two activities x and y w.r.t. $\mathcal{D}$, denoted by $msg^{\mathcal{D}}(x, y)$, is the closest activity, if there exists one, that generalizes them both, i.e., $msg^{\mathcal{D}}(x, y) = argmin_z\{dist_G^{\mathcal{D}}(x, z) + dist_G^{\mathcal{D}}(y, z) \mid z \uparrow^{\mathcal{D}} x$ and $z \uparrow^{\mathcal{D}} y\}$.

Given two activities a and a' and an abstraction dictionary $\mathcal{D}$, we say that a *implies* a' w.r.t. $\mathcal{D}$, denoted by $a \longrightarrow^{\mathcal{D}} a'$, if $(a', a) \in \mathcal{D}.\mathcal{I}sa$ or $(a', a) \in \mathcal{D}.\mathcal{P}artOf$ or, recursively, there exists an activity x such that $a \longrightarrow^{\mathcal{D}} x$ and $x \longrightarrow^{\mathcal{D}} a'$. The set of activities implied by a w.r.t. $\mathcal{D}$ is referred to as $impl^{\mathcal{D}}(a)$, i.e., $impl^{\mathcal{D}}(a) = \{a' \mid a \longrightarrow^{\mathcal{D}} a'\}$. An activity a is then said to be *complex* if there exists at least one activity x such that $a \longrightarrow^{\mathcal{D}} x$; otherwise, a is a *basic* activity. In other words, complex activities represent higher level concepts defined by aggregating or generalizing basics activities actually occurring in real process executions.

The above relationship between activities is a basic block for building taxonomies that can significantly reduce the efforts for comprehending and reusing process models, for they structuring process knowledge into different abstraction levels. Let $\mathcal{W}_1$ and $\mathcal{W}_2$ be two workflow schemas over the sets of activities A_1 and A_2, respectively. Then, we say that $\mathcal{W}_2$ *specializes* $\mathcal{W}_1$ ($\mathcal{W}_1$ *generalizes* $\mathcal{W}_2$) w.r.t. a given abstraction dictionary $\mathcal{D}$, denoted by $\mathcal{W}_2 \prec^{\mathcal{D}} \mathcal{W}_1$, if (i) for each activity a_2 in A_2 there exists at least one activity a_1 in A_1 such that $a_1 \longrightarrow^{\mathcal{D}} a_2$, and (ii) there is no activity b_1 in A_1 such that $a_2 \longrightarrow^{\mathcal{D}} b_1$.

The output of the restructuring of a schema hierarchy is an abstraction dictionary and a schema taxonomy as for formalized below.

Definition 2. Let $\mathcal{D}$ be an abstraction dictionary for the activities of a given process P, and $\mathcal{H} = \langle \mathcal{WS}, T, \lambda \rangle$ be a schema hierarchy for P. Then, $\mathcal{H}$ is a *schema taxonomy* for P w.r.t. $\mathcal{D}$ if for any pair of nodes v and v_c in V such that $(v, v_c) \in T.E$ (i.e., v_c is a child of v) $\lambda(v) \prec^{\mathcal{D}} \lambda(v_c)$. $\square$

4.2 Algorithm BuildTaxonomy

In Fig. 4 we illustrate an algorithm, called BuildTaxonomy, for restructuring a schema hierarchy into a schema taxonomy, representing the process at hand at several abstraction levels. The algorithms takes in input a schema hierarchy $\mathcal{H}$ and produces a taxonomy $\mathcal{G}$ and an abstraction dictionary $\mathcal{D}$, which $\mathcal{G}$ has been build according to. Roughly speaking, the basic task allowing for such a generalization consists in replacing groups of "specific" activities, appearing in the schemas to be generalized, with new "virtual" activities which represent them at a higher level of abstraction. In this way, a more compact description of the process is obtained, where portion of the actual workflow are represented at a lower level of granularity. Indeed, during such a restructuring process, the abstraction dictionary $\mathcal{D}$ is required to maintain the relationships between the activities that were abstracted and the new higher-level concepts replacing them.

Input: A schema hierarchy $\mathcal{H} = \langle \mathcal{WS}, T, \lambda \rangle$;
Output: A schema taxonomy $\mathcal{G}$, an abstraction dictionary $\mathcal{D}$;
Method: Perform the following steps:
  1   **let** $T = \langle V, E, v_0 \rangle$, and **let** $\mathcal{D} := \emptyset$;
  2   $Done := \{ v \in V \mid \not\exists v' \in V \text{ s.t. } (v, v') \in E \}$;   // Done initially contains the leaves of T;
  3   **while** $\exists v \in V$ such that $v \notin Done$, and $\{c \mid (c, v) \in E\} \subseteq Done$ **do**
  4   **let** $ChildSchs = \{ \lambda(c) \mid v \in V \text{ and } (v, c) \in E \}$, i.e., the schemas of all v's children;
  5     $\lambda'(v) := \texttt{generalizeSchemas}(ChildSchs, \mathcal{D})$;
  6     $Done := Done \cup \{v\}$;
  7   **end while**
  8   $\mathcal{G} := \langle \mathcal{WS}, T, \lambda' \rangle$;
  9   $\texttt{normalizeDictionary}(\mathcal{G}, \mathcal{D})$;
  10 **return** $(\mathcal{G}, \mathcal{D})$;

Procedure $\texttt{generalizeSchemas}($ $\mathcal{WS} = \{W_1, ..., W_n\}$: set of workflow schemas,
                            var $\mathcal{D}$: abstraction dictionary): workflow schema;
  g1   **let** $W_h = \langle A_h, E_h, a_h^0, F_h, \mathcal{C}_h \rangle$ for $h = 1..n$;
  g2   **let** $I = \bigcap_{i=1}^{n} A_i$;
  g4 $\overline{W} := \langle \bigcup_{i=1}^{n} A_i, \bigcup_{i=1}^{n} E_i, \bigcup_{i=1}^{n} a_i^0, \bigcup_{i=1}^{n} F_i, \emptyset \rangle$;
  g5 $\texttt{mergeConstraints}(\overline{W}, \{\mathcal{C}_h \mid h = 1..n\})$;
  g6   **for each** $i = 1..n$ **do**
  g7 $\texttt{abstractActivities}(A_i\text{-}I, \overline{W}, \mathcal{D})$;
  g8   **end for**
  g9 $\texttt{abstractActivities}(\overline{W}.A\text{-}I, \overline{W}, \mathcal{D})$;
  g10 **return** $\overline{W}$;

Fig. 4. Algorithm BuildTaxonomy

The algorithm works in a bottom-up fashion (Line 2-7): starting from the leaves of the input hierarchy, it produces, for each non-leaf node v, a novel workflow schema that generalizes all the schemas associated with the children of v. Notably, such a schema is meant to accurately represents only the features that are shared by all the subsets of executions corresponding to the children of v, while abstracting from specific activities, which are actually merged into new high-level (i.e., *complex*) activities. Such a generalization task is carried out by providing the procedure generalizeSchemas with the schemas associated with the children of v, along with the abstraction dictionary $\mathcal{D}$, initially empty (Line 5). As a result, a new generalized schema is computed and assigned to v through the function λ'; moreover, $\mathcal{D}$ is updated to suitably relate the activities that were abstracted with the complex ones replacing them in the generalized schema.

As a final step, after the schema taxonomy $\mathcal{G}$ has been computed, the algorithm also restructures the abstraction dictionary $\mathcal{D}$ by using the procedure

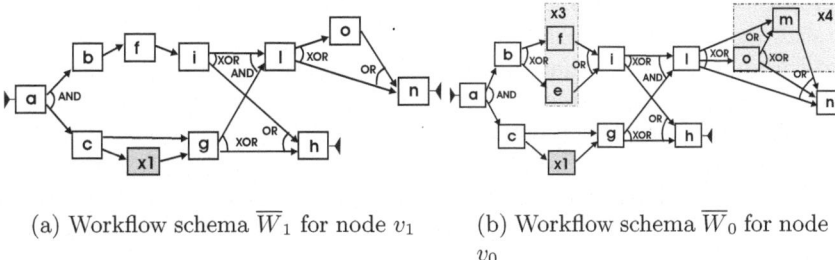

(a) Workflow schema $\overline{W}_1$ for node v_1 (b) Workflow schema $\overline{W}_0$ for node v_0

Fig. 5. Generalized workflow schemas in the resulting taxonomy

normalizeDictionary (Line 9), which actually removes all "superfluous" activities that were created during the generalization. In particular, this step will eliminate any complex activity a not appearing in any schema of $\mathcal{G}$, which can be abstracted into another, higher-level, complex activity b, provided that this latter can suitably abstract all the activities implied by a.

Clearly enough, the effectiveness of the technique depends on the way the generalization of the activities and the updating of the dictionary are carried out. Procedure generalizeSchemas (reported in Fig. 4 as well) first merges all the input workflow schemas into a preliminary workflow schema $\overline{W}$ (Line g4), which represents all the possible flow links in the input workflows by roughly performing the union of their corresponding control flow graphs. Subsequently, the set of constraints of $\overline{W}$ (initially empty) is populated by suitably combining the constraints specified in the input schemas, by means of procedure mergeConstraints (Line g5); as a matter of fact, this latter procedure derives a split (resp., join) condition for each activity a of $\overline{W}$, based on the split (resp., join) conditions a is associated with in each input schema, yet taking into account all the control flow relationships a takes part to, in the involved schemas.

The main task in the generalization process is performed by repeatedly applying the procedure abstractActivities, which transforms $\overline{W}$ by merging activities in the reference set it receives as the first parameter, and by updating the associated constraints and the abstraction dictionary $\mathcal{D}$ as well. In particular, abstractActivities is first applied for merging only activities that derived from the same input schema – at step i only activities coming from the i-th schema can be merged (Line g7). A further application of abstractActivities is then performed to possibly abstract any non-shared activity in the current schema, independently of its origin. Due to its relevance to the generalization algorithm, abstractActivities is illustrated in details in Section 5; however, we conclude this description by providing an intuition on its behavior.

Example 2. Consider again the schema hierarchy shown in Fig. 3. Then, algorithm BuildTaxonomy starts generalizing from the leafs, thus first processing the schemas W_3 and W_4 associated with v_3 and v_4, respectively. The result of this generalization is the schema $\overline{W}_1$ shown in Fig. 5.(a), which is obtained by first merging all the activities and flow links contained in either W_3 or W_4, and by then performing a series of abstractions steps over all non-shared activities, namely o, d and p. As we shall formalize in Section 5, in general, we iteratively abstract a pair of activities into a complex one, trying to minimize the number of spurious flow links that their merging introduces between the remaining activities, and yet considering their mutual similarity w.r.t. the contents of the abstraction dictionary. When deriving the schema $\overline{W}_1$, only the activities d and p are abstracted, by aggregating them both into the new complex activity x_1; consequently, d and p are replaced with x_1, while the pairs (d, x_1) and (p, x_1) are inserted in the $\mathcal{P}artOf$ relationship. The schema $\overline{W}_1$ is then merged with the schema W_2 associated with v_2, and a new generalized schema, shown in Fig. 5.(b), is derived for the root v_0. In fact, when abstracting activities coming from W_2, d and p are aggregated again together, into a new complex activity x_2; however, in a subsequent step x_2 is incorporated into x_1, as these two complex activities have the same set of sub-activities and the same control flow links. Furthermore, the activities e and f are aggregated into the complex activity x_3, while m and o are aggregated into x_4. As a consequence, the pairs $(e, x_3), (f, x_3), (m, x_4)$ and (o, x_4) are added to the $\mathcal{P}artOf$ relationship.   □

5   Abstracting Workflow Activities

In this section, we discuss the implementation of the `abstractActivities` procedure. To this aim, we preliminary introduce some metrics that we exploit for singling out those activities that can be safely abstracted into higher-level ones.

5.1   Matching Activities for Abstraction Purposes

We next describe a series of functions which are meant to provide different ways for evaluating how much two activities are suitable to being abstracted by a single higher-level activity. Roughly speaking, $sim_P^{\mathcal{D}}$ and $sim_G^{\mathcal{D}}$ aims at capturing semantical affinities based on the contents of a given abstraction dictionary $\mathcal{D}$; on the contrary, sim^E just compares two activities from a topological viewpoint according to a set E of control flow edges.

While merging tasks in a workflow schema, a major concern is to limit the creation of spurious control flow paths among the remaining activities in the workflow schema, yet admitting to lose some precedence relationships involving the abstracted ones. In this respect, we focus on two cases that can lead to a meaningful merging without upsetting the topology of the control flow graph, as formalized in the following definition.

Definition 3. Given a set of edges E, we say that an (unordered) pair of activities (x, y) is *merge-safe* if one of the following conditions holds:

a) x and y are directly linked by some edges in E and after removing these edges no other path exists connecting x and y, i.e., $\{(x, y), (y, x)\} \cap E \neq \emptyset$ and $\{(x, y), (y, x)\} \cap (E - \{(x, y), (y, x)\})^* = \emptyset$

b) there is no path in E connecting x and y, i.e., $\{(x, y), (y, x)\} \cap E^* = \emptyset$

where E^* denotes the transitive closure of E.   □

Notably, only in the case (b) of Definition 3 the merging of x and y may lead to spurious dependencies among other activities in the schema. Indeed, this happens when there are two other activities z and w such that $(z, w) \notin E^*$, and either $\{(z, x), (y, w)\} \subseteq E$ or $\{(z, y), (x, w)\} \subseteq E$.

By the way, a straight way for preventing this problem, consists in requiring that at least one of the following conditions holds: (i) $\mathcal{P}_x = \mathcal{P}_y$, (ii) $\mathcal{S}_x = \mathcal{S}_y$, (iii) $\mathcal{P}_x \subseteq \mathcal{P}_y$ and $\mathcal{S}_x \subseteq \mathcal{S}_y$, (iv) $\mathcal{P}_y \subseteq \mathcal{P}_x$ and $\mathcal{S}_y \subseteq \mathcal{S}_x$, where $\mathcal{P}_a$ (resp. $\mathcal{S}_a$) denotes the set of predecessors (resp. successors) of activity a, according to the arcs in E. Actually, in order to also deal with the presence of complex activities in the set of predecessors (resp., successors), we extend the above expressions by replacing $\mathcal{P}_a$ (resp., $\mathcal{S}$) with $\mathcal{P}_a^+$ (resp., $\mathcal{S}_a^+$), defined as follows:

$$\mathcal{P}_a^+ = \bigcup_{b \in \mathcal{P}_a} impl(b) \qquad \mathcal{S}_a^+ = \bigcup_{b \in \mathcal{S}_a} impl(b)$$

However, the above requirements on the flow relationships of two activities could not allow for an appreciable level of abstraction. Therefore, we somehow incorporate them, in a smoothed way, into the function $sim^E(x, y)$, reported below, which is meant to evaluate a pair of activities according to the number of spurious flows that would be generated when merging them, in an inverse manner (i.e, the more spurious flows are introduced, the lower is the score):

$$sim^E(x, y) = \frac{\alpha(\mathcal{P}_x^+, \mathcal{P}_y^+) \times \alpha(\mathcal{S}_x^+, \mathcal{S}_y^+) + \beta(\mathcal{P}_x^+, \mathcal{P}_y^+) \times \beta(\mathcal{S}_x^+, \mathcal{S}_y^+)}{2}$$

where, for any two sets B and C, $\alpha(B, C) = \frac{|B \cap C|}{min(|B|, |C|)}$ and $\beta(B, C) = \frac{|B \cap C|}{|B \cup C|}$.

As a matter of facts, sim^E produces a maximal value whenever one of the "strong" conditions discussed before holds, and, in general, tends to attribute high similarity to activities matching in most of their predecessors (successors).

On the contrary, function $sim_P^{\mathcal{D}}$ provides a way for measuring "semantical" similarities between two activities x and y, based on the implied activities they actually share. It is defined as:

$$sim_P^{\mathcal{D}}(x, y) = \beta(impl^{\mathcal{D}}(x) \cup \{x\}, impl^{\mathcal{D}}(y) \cup \{y\})$$

Moreover, function $sim_G^{\mathcal{D}}$, which is instead devoted to compare two activities based on the generalization relationships recorded in $\mathcal{D}.\mathcal{I}sa$, is defined as follows:

$$sim_G^{\mathcal{D}}(x, y) = 1 - \frac{dist_G^{\mathcal{D}}(x, msg^{\mathcal{D}}(x, y)) + dist_G^{\mathcal{D}}(y, msg^{\mathcal{D}}(x, y))}{max\{dist_G^{\mathcal{D}}(a, b) \mid a, b \in A \text{ and } b \uparrow^{\mathcal{D}} a\}}$$

Finally, an overall score can be assigned to each pair of activities in order to rank them for abstraction purposes, as follows:

$$score^{\mathcal{D}, E}(x, y) = \begin{cases} 0, \text{ if } (x, y) \text{ is not a merge-safe pair of activities} \\ max\{sim^E(x, y), sim_P^{\mathcal{D}}(x, y), sim_G^{\mathcal{D}}(x, y)\}, \text{ otherwise} \end{cases}$$

5.2   Abstracting Activities

Fig. 6 provides a detailed description of procedure `abstractActivities`, that is meant to merge activities in S for a given schema $\bar{W}$ and to abstract them via higher-level, complex, activities. To this aim, besides $\bar{W}$ and S, the procedure takes in input an abstraction dictionary $\mathcal{D}$. As a result, it transforms $\bar{W}$ by replacing the abstracted activities with the associated complex ones, and modifies $\mathcal{D}$ in order to suitably record the performed abstraction transformations.

Procedure `abstractActivities`(S: set of activities; var $\bar{W} = \langle A, E, a^0, F, \mathcal{C}\rangle$: a workflow schema; var $\mathcal{D} = \langle \mathcal{P}artOf, \mathcal{I}sa\rangle$: abstraction dictionary;)

1 **let** $E' = \{(x,y) \in E \text{ s.t. } x \in S \text{ and } y \in S\}$;
2 $\langle m_1, m_2, p, mode\rangle :=$`getBestAbstraction`$(S, E', \mathcal{D})$;
3 **while** $p \neq \varepsilon$ **do**
4   **let** $ActuallyAbstracted = \{m_1, m_2\} - \{p\}$;
5   **if** $mode =$ ISA **then**
6      $\mathcal{I}sa := \mathcal{I}sa \cup \{(x, p) \text{ s.t. } x \in ActuallyAbstracted\}$;
7   **else**
8      $\mathcal{P}artOf := \mathcal{P}artOf \cup \{(x, p) \text{ s.t. } x \in ActuallyAbstracted\}$;
9   **end if**
10   `deriveConstraints`$(\mathcal{C}, m_1, m_2, p, E)$;
11   `arrangeEdges`$(E, ActuallyAbstracted, p)$;
12   $A := A - ActuallyAbstracted \cup \{p\}$;
13   $S := S - ActuallyAbstracted \cup \{p\}$;
14   $\langle p, m_1, m_2\rangle :=$`getBestAbstraction`$(S, E', \mathcal{D})$;
15 **end while**

Procedure `getBestAbstraction`(S: set of activities; E: set of activity pairs; $\mathcal{D}$: abstraction dictionary): a tuple in $S \times S \times \{\mathcal{I}sa, \mathcal{P}artOf, \varepsilon\} \times \mathcal{A}$; [a]

b1 **if** $|S| < 2$ **then**
b2   **return** $\langle \varepsilon, \varepsilon, \varepsilon, \varepsilon\rangle$;
b3 **else**
b4   **let** a and b be two activities s.t. $score(a,b) = max\{score^{\mathcal{D},E}(x,y) \mid x,y \in S\}$;
b5   **if** $score^{\mathcal{D},E}(a,b) < \rho$ **then return** $\langle \varepsilon, \varepsilon, \varepsilon, \varepsilon\rangle$;
b6   **else if** $sim_G^{\mathcal{D}}(a,b) \geq \rho^s$ **then return** $\langle a, b, \mathcal{I}sa, msg^{\mathcal{D}}(a,b)\rangle$;
b7   **else if** $impl^{\mathcal{D}}(b) \subseteq impl^{\mathcal{D}}(a)$ **then return** $\langle a, b, \mathcal{P}artOf, a\rangle$;
b8   **else if** $impl^{\mathcal{D}}(a) \subseteq impl^{\mathcal{D}}(b)$ **then return** $\langle a, b, \mathcal{P}artOf, b\rangle$;
b9   **else if** $sim_P^{\mathcal{D}}(a,b) \geq \rho^s$ **then return** $\langle a, b, \mathcal{I}sa, \text{a new activity}\rangle$;
b10   **else return** $\langle a, b, \mathcal{P}artOf, \text{a new activity}\rangle$;
b11   **end if**
b12 **end if**

---

[a] in any tuple $\langle m_1, m_2, M, p\rangle$ the procedure returns, m_1 and m_2 are the abstracted activity, p is the abstracting one, and M indicates the abstraction mode – $\mathcal{A}$ denotes the universe of all activities.

Fig. 6.   Procedure `abstractActivities`

The procedure `abstractActivities` works in a pairwise fashion by repeatedly abstracting two activities m_1 and m_2, both taken from S, by means of a complex activity p. All such activities are identified with the help of the function `getBestAbstraction` that returns a tuple indicating, besides p, m_1 and m_2, the kind of abstraction relationship to be used, i.e., $\mathcal{P}artOf$ or $\mathcal{I}sa$. As a special case, procedure `getBestAbstraction` will return the tuple $\langle \varepsilon, \varepsilon, \varepsilon, \varepsilon\rangle$ if there is no pair of activities in S that can be suitably abstracted. In such a case the condition $p = \varepsilon$ will hold, thus causing the termination of the abstraction procedure.

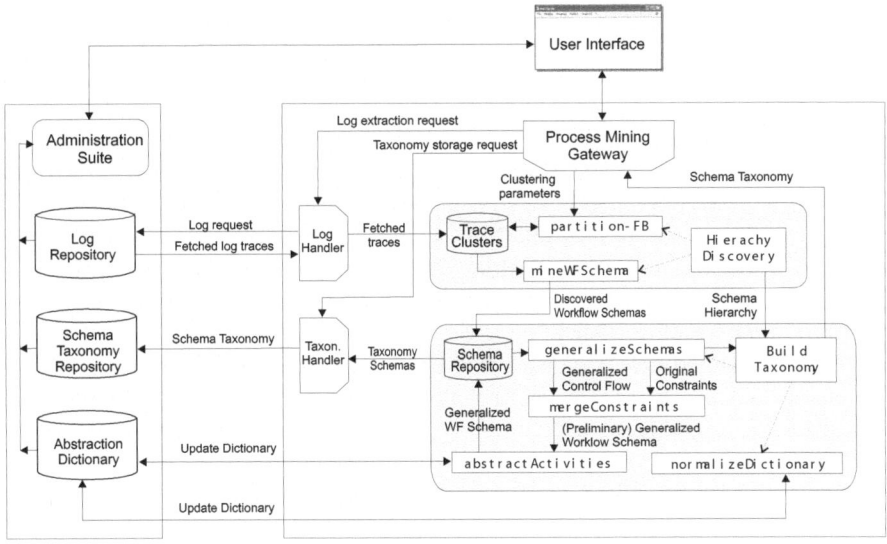

Fig. 7. System Architecture.

Otherwise, in the resulting tuple $\langle m_1, m_2, mode, p \rangle$, m_1 and m_2 denote the two activities to abstract, and p is the complex activity which will replace them both, while *mode* denotes which kind of abstraction must be stored in $\mathcal{D}$: aggregation, via the $\mathcal{P}artOf$ relationship, or specialization, via the $\mathcal{I}sa$ relationship.

Procedure getBestAbstraction, still shown in Figure 6, essentially relies on the matching measures defined in Section 5.1. In more detail, the procedure takes as input a set S of activities and an associated set E of control flow edges, along with an abstraction dictionary $\mathcal{D}$. If there is no *merge-safe* pair in S that receives a sufficient score (w.r.t. a threshold ρ), then getBestAbstraction returns the tuple $\langle \varepsilon, \varepsilon, \varepsilon, \varepsilon \rangle$ (Lines b2 and b5), simply meaning that no abstraction can be performed over the activities in S. Otherwise, the procedure computes a tuple whose elements, respectively, specify the two activities to be abstracted, the kind of abstraction relationship to be used (i.e., $\mathcal{P}artOf$ or $\mathcal{I}sa$), and the complex activity which will abstract both of them. As a matter of facts, the choice of the abstracting activity and of the abstraction mode is based again on the similarity values computed via $sim_P^{\mathcal{D}}$ and $sim_G^{\mathcal{D}}$. In principle, if either of these measures is above the threshold ρ^S, the two activities are deemed similar enough to be looked at as two variants of some activity that generalizes them both. In particular, if $sim_G > \rho^S$ such an activity already exists: that is $msg^{\mathcal{D}}(m_1, m_2)$, which is indeed returned in the resulting tuple (Line b6). Before considering the creation of a new activity for generalizing m_1 and m_2 (Line b9), we check whether one of them implies the other: in such a case the implied activity can be abstracted by the other via an aggregation relationship (Lines b7-b8); we can, indeed, exclude that the implied activity is a specialization of the other, since such a condition was tested previously (Line b6). If none of the above cases applies, the two activities are eventually abstracted by a new activity via aggregation (Line b10).

As concerning the remainder of procedure abstractActivities, since either m_1 or m_2 might coincide with p, the set *ActuallyAbstracted* is used to keep trace of which of them should be really abstracted, for it actually being distinct from p (Line 4). Procedure deriveConstraints (see Line 20) is then applied to suitably derive the split and join conditions for p, based on those of the activities m_1 and m_2 that are being merged into it. Notice that, in principle, a looser join (resp., split) condition might be computed for p than those associated with m_1 and m_2, whenever these latter activities do not exactly match in their predecessor (resp., successor) nodes and in their join (resp., split) conditions. For space reasons, we skip here a detailed description of this procedure. In order to properly replace the abstracted activities, the control flow graph is properly settled by using procedure arrangeEdges, which simply transfers the edges of the abstracted activities to p (Line 11). Finally, m_1 and m_2 are removed from both A and the reference set S (Lines 12-13), and a novel activity pair is searched for, in order to reiterate the whole abstraction procedure.

6   Discussion and Conclusions

We proposed a process mining approach that is meant to discover a hierarchical model representing the analyzed process through different views, at different abstraction levels. The approach consists of several mining and abstraction techniques, which are exploited in an integrated way. In particular, a preliminary schema hierarchy, accurately modelling the process at hand, is first discovered, by using a divisive clustering algorithm; the hierarchy is then restructured into a taxonomy, by equipping each non leaf node with an abstract schema that generalizes all the different schemas in the corresponding subtree.

The algorithms proposed in the paper have been implemented in JAVA and integrated into a stand-alone system architecture that is sketched in Fig. 7. For the sake of clarity and conciseness, major modules in the architecture are labelled with the names of the algorithms and procedures previously presented in the paper. Notably, different repositories are exploited to specifically manage the main kinds of information involved in the process mining task: log data, schema taxonomies, and abstraction relationships. Actually, a separate administration suite allows for effectively browsing and exploiting all such data. By the way, two further, "internal", repositories are used to maintain and share data on the trace clusters produced by the clustering algorithm and, respectively, the schemas generated during both the mining phase and the restructuring one. Currently, in order to offer the functionalities presented to a larger community of users, we are working at integrating the architecture into the *ProM* [18] process mining framework. At the time of writing, the hierarchical clustering module is already available as an additional, plug-in, component for *ProM*.

References

1. van der Aalst, W., Weijters, A., Maruster, L.: Workflow mining: Discovering process models from event logs. IEEE Transactions on Knowledge and Data Engineering (TKDE) **16** (2004) 1128–1142
2. van der Aalst, W., van Dongen, B., Herbst, J., Maruster, L., G.Schimm, Weijters, A.: Workflow mining: A survey of issues and approaches. Data and Knowledge Engineering **47** (2003) 237–267
3. van der Aalst, W., Hirnschall, A., Verbeek, H.: An alternative way to analyze workflow graphs. In: Proc. 14th Int. Conf. on Advanced Information Systems Engineering. (2002) 534–552
4. van der Aalst, W., van Dongen, B.: Discovering workflow performance models from timed logs. In: Proc. Int. Conf. on Engineering and Deployment of Cooperative Information Systems (EDCIS 2002). (2002) 45–63
5. Agrawal, R., Gunopulos, D., Leymann, F.: Mining process models from workflow logs. In: Proc. 6th Int. Conf. on Extending Database Technology (EDBT'98). (1998) 469–483
6. Cook, J., Wolf, A.: Automating process discovery through event-data analysis. In: Proc. 17th Int. Conf. on Software Engineering (ICSE'95). (1995) 73–82
7. Muth, P., Weifenfels, J., M.Gillmann, Weikum, G.: Integrating light-weight workflow management systems within existing business environments. In: Proc. 15th IEEE Int. Conf. on Data Engineering (ICDE'99). (1999) 286–293
8. Greco, G., Guzzo, A., Pontieri, L., Saccà, D.: Mining expressive process models by clustering workflow traces. In: Proc. 8th Pacific-Asia Conference (PAKDD'04). (2004) 52–62
9. Castellanos, M., Casati, F., Dayal, U., Shan, M.C.: ibom: A platform for business operation management. In: Proc. Intl. Conf. on Data Engineering (ICDE05). (2005)
10. IDS Prof. Scheer, G.: (Aris-tool set. version 2.0 manual.) Saarbrücken 1994.
11. Malone, T.W., et al.: Tools for inventing organizations: Toward a handbook of organizational processes. Management Science **45** (1999) 425–443
12. Stumptner, M., Schrefl, M.: Behavior consistent refinement of object life cycles. ACM Transactions on Software Engineering and Methodology **11** (2002) 92–148
13. Stumptner, M., Schrefl, M.: Behavior consistent inheritance in uml. In: Proc. 19th Int. Conf. on Conceptual Modeling (ER 2000). (2000) 527–542
14. Basten, T., van der Aalst, W.: Inheritance of behavior. Journal of Logic and Algebraic Programming **47** (2001) 47–145
15. Lee, J., Wyner, G.M.: Defining specialization for dataflow diagrams. Information Systems **28** (2003) 651–671
16. Liu, D.R., Shen, M.: Workflow modeling for virtual processes: an order-preserving process-view approach. Information Systems **28** (2003) 505–532
17. Greco, G., Guzzo, A., Manco, G., Saccà, D.: Mining frequent instances on workflows. In: Proc. 7th Pacific-Asia Conference (PAKDD'03). (2003) 209–221
18. ProM: http://www.daimi.au.dk/PetriNets/tools/db/promframework.html.

Flexible Business Process Management Using Forward Stepping and Alternative Paths

Mati Golani and Avigdor Gal*

Technion - Israel Institute of Technology
{iemati, avigal}@ie.technion.ac.il

Abstract. The abilty to continuously revise business practices is essential to organizations aiming at reducing their costs and increasing
their revenues. Rapid and continuous changes to business processes result in less control over the executed activities. As a result, the ability
of process designers to produce solid, well-validated workflow models is
limited. Workflow management systems (WfMSs), serving as the main
vehicle of business process execution, should recognize these risks and
become more dynamic to allow the required business flexibility. In this
paper, we propose a dynamic mechanism that allows backtracking and
forward stepping at an instance level. This mechanism analyzes the feasibility of applying certain modifications to running instances and provides
an efficient algorithm that avoids redundant operation activation. We believe that this mechanism can bolster the ability of a business process
management system to deal with unexpected situations and to resolve,
in runtime, scenarios in which such resolution both is called for and does
not violate any business process constraints. Throughout this paper, we
use the paradigm of Web services to demonstrate the capabilities of the
proposed mechanism.

1 Introduction

Rapidly changing business environments require organizations to continuously
revise their business practices, seeking better business opportunities and continuously aiming at reducing their costs and increasing their revenues. In the
last decade, businesses have turned to technological solutions to assist them in
this task. The use of electronic means to commerce, data mining, and customer
profiling are all recent technological developments that penetrate business activities. One of the most recent technological developments is the use of Web
services, components with a well-defined interface that are embedded in cross-
organizational business processes. Using Web services, the functional aspects of
business applications are encapsulated [19], with interfaces defined using standards such as BPEL4WS [3], and invocation controled using approaches such as
Service Oriented Architecture (SOA). Web services promise to deliver greater
choice and flexibility to business processes.

* The work of Gal was partially supported by two European Commission 6th Framework IST projects, QUALEG and TerreGov, and the Fund for the Promotion of
Research at the Technion.

W.M.P. van der Aalst et al. (Eds.): BPM 2005, LNCS 3649, pp. 48–63, 2005.

Rapid and continuous changes to business processes carry with it risks, due to shorter (or even nonexistent) design time and less control over the executed activities. As a result, the ability of process designers to produce solid, well-validated workflow models is limited. Workflow management systems (WfMSs), serving as the main vehicle of business process execution, should recognize these risks and become more dynamic to allow the required business flexibility. To illustrate this point, we next present two examples, involving Web services. First, we observe that the development of Web services is an ongoing task, and new and improved services are continuously replacing existing ones. Currently, WfMSs provide little support to the reexecution of successfully processed tasks for running instances, even if the gains from such reexecution outweigh the costs. As another example, observe that Web services merely provide syntactic information regarding their input, output and processing logic, through standards such as WSDL [20]. In most cases, such descriptions fail to convey all necessary constraints and restrictions. Modeling using Web services, therefore, is likely to make the validation of workflow models more difficult [8], and more exceptions at run-time are to be expected. Efficient exception handling is a fundamental component of WfMSs and is critical to their successful implementation in real-world scenarios [1].

The motivation for this work, as illustrated above, is the need for flexible and dynamic WfMSs to support the growing number of exceptions that cannot be designed a priori, due to poor design or the lack of sufficient information regarding the internal logic of Web services. In this paper, we propose a dynamic workflow mechanism that allows backtracking and forward stepping at an instance level. This mechanism works by analyzing the feasibility of applying certain modifications to running instances and implementing an efficient algorithm that avoids redundant operation activation as a result of instance modification. In particular, the introduction of a new Web service may trigger backtracking of an instance to an activity in which the new Web service is performed, and then forward stepping while utilizing, to the extent possible, previously executed activities. In the case of an exception, the proposed algorithm identifies a feasible alternative that avoids the failing activity or communication channel. In this scenario, again, we backtrack to an activity from which it is considered safe to step forward.

Our approach is based on and extends the approach described in [7]. Therefore, we provide a rollback mechanism followed by a forward execution. We extend the existing proposal by providing a precise location in the workflow graph for the rollback target and a semi-automated forward execution (involving relevant agents when needed). We believe that this mechanism can bolster the ability of a business process management system to deal with unexpected situations and to resolve, in runtime, scenarios in which such resolution is both called for and does not violate any business process constraints.

We base our workflow model on ADEPT WSM nets [13]. It combines a graphical representation with a solid formal foundation, taken from graph theory, allowing both reasoning and an execution engine [14].

Our specific contributions are as follows:

- We provide an analysis of a workflow model, based on WSM nets, that generates a conceptual framework in which backtracking and forward stepping can be evaluated and implemented.
- We provide efficient algorithms for alternative route identification (at design time or run time) and forward stepping (at run time), to allow dynamic modifications to workflows.
- We introduce the concept of a meta-process, an efficient and a fully automatic mechanism (at the WfMS level) for activating the proposed algorithms.

1.1   Related Work

Exception and modification handling – *i.e.*, the way a workflow system responds once an exception or a modification notification occurs – has been discussed in the literature for some time. The system may react in such cases either by terminating a process or handling an exception [10,4]. Sadiq et al. in [15] classified the latter option as one of the available modification policies (which was called *Adapt*) to a given change in a running process. This change is due to an unexpected exception, so the process should be handled differently than originally designed. Yet the authors do not define how to infer this modification.

Generally speaking, exception handling involves compensation flows [5]. Compensation flows provide rollback, a set of undo actions. These flows are predefined. If compensation does not exist, the workflow operator may be willing to accept inconsistencies in which a completed activity is not voided. For example, assume an activity provides a customer with bonus points, on the assumption that a purchase will be made. Then another activity is chosen, which also awards bonus points. For a successful termination of the workflow (*e.g.*, a sale), an operator may be willing to grant double bonus points in this case.

Eder et al. describe several types of compensation in [6], and provide a three-step mechanism to handle exceptions [6, workflow recovery] [7]. The first step entails rollback based on compensation type of activities in the workflow graph. In the next step, an agent determines whether to continue backward or to take an alternative path. The final step is a forward execution (which could lead to the same point of failure). In the event of rollback, existing work [7] does not specify the stop point, implying that this point represent the decision on whether to continue. However, in many cases the parameter which drives this decision has been set before this point. Furthermore, these mechanisms are static (*e.g.*, during build time) [5,12,6]. Our approach detects the actual/optimal stop point (via analysis), and can provide in run time an alternative execution that overtakes the failed activity.

A dynamic approach was presented by Hwang et al. [11]. Here, a failure recovery language supports multiple exceptions per activity, and applies rollback using ECP (end compensation point). This language uses the process parameters in order to determine its flow. The drawback of this approach is that it fails to make use of the user's insight (and output). It is impossible to accurately forecast all user intentions, and under different circumstances, individual users may make different choices based on the same input. Thus, the user's output is essential.

The rest of the paper is organized as follows: Section 2 introduces a motivating example. In Section 3 we present the workflow graph-based model. In Sections 4-6 we present forward stepping, alternative path, and parametric analysis mechanisms, respectively. Finally, in Section 7, we introduce the architecture for implementation, and show the use of meta-processes.

2   Illustrative Example

As an example, consider a process that handles registrations for package tours (see Figure 1). This process uses some local applications (member deals), as well as Web services (hotel registration: activity 0, flight reservation: activity 1) with some special offers available to gold members only (activities 5-8). We use two examples to illustrate the needs of a flexible business process. The first involves introduction of a new Web service that offers better hotels for cheaper prices, at a time when the process instance is already handling membership registration (activity 4), and hotel registration and flight ticket activities are already completed. The second event involves a failure of the special offer system for a gold customer (activity 6).

In the first example, nothing has gone wrong, but the "world" has changed (a new service has become available). Taking advantage of this new service may affect the customer's total cost, given penalties for canceling an existing order and the cost of creating a new one. Other activities or services may also need to be compensated or reexecuted as a result of this update. For instance, activity 1 may require compensation if the original flight dates were modified based on the new hotel reservations. These costs must be quantified before the customer decides whether to continue with the original plan or to use the newly available service.

In the second event, the system will benefit by using a different path that allows successful completion of the business process while bypassing the special offers. Consider the example depicted in Figure 1. Activities 0, 1, 5, and 7 were performed in this instantiation, yet a failure at activity 6 blocks the process and

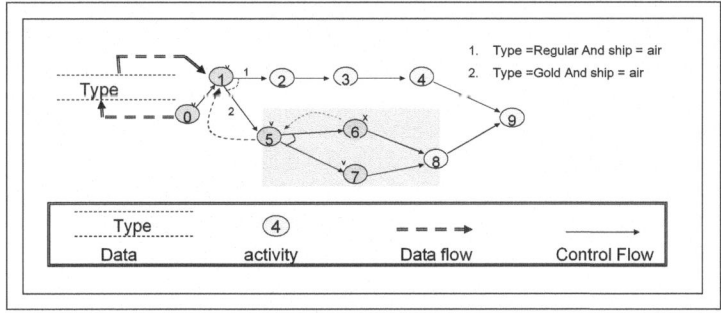

Fig. 1. An example of a business process

prevents its completion. Alternative paths to the current path, to be formally defined in Section 3, are those paths in the graph that possibly lead to a successful termination of the business process, yet do not contain activity 6. Since activity 1 leads to activities 2 and 5 using a Xor condition, a rollback procedure to activity 1 would enable use of an alternative path to 9 through activities 2, 3, and 4.

In this paper we address only processes in which the alternative path has a real semantic alternative meaning. That means that an executed instance along one path can be logically executed (albeit, at a possibly higher cost) along the alternative path as well (as in the Regular/Gold customer example).

3   Workflow Model

In this section we define basic constructs in workflow graphs, to be used later in the paper. The classification of workflow constructs is not new and has been discussed in various works (*e.g.*, [16]).

A workflow model can be described as a graph (ADEPT WSM net) $G(V, E)$ $(V = (V_a \cup V_d); E = (E_c \cup E_d))$, where V_a is a set of activities, V_d is a set of data parameters, E_c is a set of control edges, and E_d is a set of data edges. For simplicity, whenever possible, we will refer to the reduced graph $G' = G(V_a, E_c)$. Data flows (as appear in Figure 1) are discussed in Section 6.

An activity a in V_a has $in\text{-}degree(G', a)$ incoming edges and $out\text{-}degree(G', a)$ outgoing edges. Whenever it becomes clear from the context, we eliminate the graph reference and refer to $in\text{-}degree(a)$ and $out\text{-}degree(a)$. A *path* in G is a set of activities such that any two consecutive activities on the path are connected by an edge in E_c. We denote a path from a_i to a_j by $(a_i, \ldots, a_j)$. The *length* of a $path(a_i, \ldots, a_j)$ (denoted $length(a_i, \ldots, a_j)$) is the number of edges in $(a_i, \ldots, a_j)$. Finally, $Minlength(a_i, a_j)$ is the length of the shortest path in G that starts at a_i and ends at a_j.

We next define two graph constructs, namely splits and joins, based on the Workflow Management Coalition standard [17]. A Xor *split* a is a node (activity) with multiple outgoing edges ($out\text{-}degree(a)>1$), only one of which can be followed in the execution flow. The decision as to which edge to follow is based on the satisfaction of mutually exclusive conditions that are typically associated with the outgoing edges. Let $c_{a,a'}$ be a DNF (Disjunctive Normal Form) Boolean statement with a set of variables $Var(c_{a,a'},)$ that must be satisfied in order to pass from activity a to activity a'. Activity 1 in Figure 1 is an example of a Xor split. Each Xor split a is associated with a Xor join (e.g., activity 9 in Figure 1), an activity common to all paths that start from a. During runtime, when reaching a, the workflow engine evaluates the conditions on each of a's outgoing edges, and continues the execution along the edge whose associated condition is satisfied. The Xor join activity acts as a synchronization point in the execution.

An *And split* is a node with multiple outgoing edges whose execution flow follows all outgoing edges by parallel threading. Activity 5 in Figure 1 is an example of an And split. Threads of an And split a need to be synchronized at an *And join*, which is also a node in the graph that is common to all paths that start from a. Activity 8 is an example of an And join for activity 5.

Definition 1. *Xor split point* *Let $G' = (V, E)$ be a workflow graph, and a be an activity in V. A Xor split point of a is a Xor split a_i with a Xor join a_j such that a_i is a predecessor of a and a_j is a successor of a.*

Definition 2. *NXSP* *Nearest Xor split point of a $(NXSP(a))$, is a Xor split point of a, a_i, which satisfies that any other Xor split point of a (a_j) is also a Xor split point of a_i.*

And split point and *NASP* are similarly defined. Using the basic definitions given above, we now define blocks in a graph. Let $G' = (V, E)$ be a workflow graph and let a_i be a Xor split and a_j be the Xor join associated with a_i. A Xor *block* of a_i is a subgraph of G' induced by the nodes of all paths $(a_i, \ldots a_j)$ in G'. Similarly, given an And split a_i and the associated And join of a_i, a_j, an *And block* of a_i is a subgraph of G' induced by the nodes of all paths $(a_i, \ldots a_j)$ in G'. For example, the induced subgraph of activities $\{5, 6, 7, 8\}$ in Figure 1 (marked with grey rectangle) is an *And Block*. It models two threads that start after the execution of activity 5 and synchronize before the execution of activity 8.

Clearly, any activity a is within a Xor block defined by its Xor split point (can be null) and its associated Xor join. In particular, a is within a Xor block defined by $NXSP(a)$ and its associated Xor join.

Definition 3. *Alternative paths* *Let $G' = (V, E)$ be a workflow graph with a sink f, and let $P_1 = (a_i, \ldots, a_j)$ and $P_2 = (a_i, \ldots, a_k)$ be paths in G'. P_1 is an alternative to P_2 (and vice versa) if the following four conditions hold:*
1. *a_i is of type Xor Split.*
2. *There is no activity a in $V \backslash \{a_i\}$ such that a is in P_1 and a is also in P_2.*
3. *Any path $(a_j, \ldots f)$ in G' does not include an activity in P_2.*
4. *Any path $(a_k, \ldots f)$ in G' does not include an activity in P_1.*

It is worth noting that P_1 and P_2 share a common initial activity a_i. As an example, consider the alternative paths $(1, 2, 3, 4)$ and $(1, 5, 6, 7, 8)$ in Figure 1. Note that the paths $(5, 6)$ and $(5, 7)$ are **not** alternative paths, since activity 5 is not of type Xor Split (both activity 6 and activity 7 are part of the same And block).

The importance of *Xor blocks* in our analysis is related to the ability to provide an alternative paths analysis. In Figure 1, the *Xor Block* includes the entire graph save activity 0, and thus an alternative path for any activity (exluding activity 9) will start from activity 1. Therefore, once activity 6 fails, the *Xor Block* to which activity 6 belongs allows an alternative execution, using the paths that contains activities $\{1, 2, 3, 4\}$. We will present an algorithm for identifying alternative paths in Section 5.

We next discuss the normalization of Xor and And blocks. A normalized (Xor or And) block is a block in which neither the outgoing edges of the split activity, nor the incoming edges of the join activity, are connected to any activities outside the block. This property matches the WFMC definition (in interface 1) of *full-blocked workflows*. Formally,

Definition 4. *Normalized Block* *Let $G'(V, E)$ be a workflow graph with a source s and a sink f, and B a Block (either Xor or And) with split activity a_i and join activity a_j. B is* normalized *if a_j is on all paths $(a_i, \ldots, f)$ in G and a_i is on all paths $(s, \ldots, a_j)$ in G'.*

It is easy to show that if B is normalized, then $out\text{-}degree(G', a_i) = out\text{-}degree(B, a_i) = in\text{-}degree(G', a_j) = in\text{-}degree(B, a_j)$. For brevity, we refrain from presenting the algorithm for block normalization in this paper.

Given a workflow graph $G'(V, E)$, $Inst(G')$ represents an instance of G'. $Inst(G')$ encapsulates instance-related data, such as activity state and input/output parameter values. $Inst(G')$ is a DAG and loop constructs in $G'(V, E)$ are removed by duplicating loop blocks and re-labeling of activities.

An activity in $Inst(G')$ can be classified into one of the following states: uninitiated (yet, but on an execution path), void (on path that was not invoked), completed (finished on current path), compensated, or failed.

4   Forward Stepping

In this section we present an efficient algorithm for forward stepping. Consider a path that begins from activity a_i, and assume that activity a_i needs to be reexcuted due to exception or modification. Analyzing the state and dependencies of the activities (or Web services) that participate in a given process instance can help determine which activities have not yet been executed, and which need to be reexecuted. When we deal with exceptions, a stop (target) activity (a_x) is provided, so the forward stepping is executed until reaching this activity. In other scenarios, it is possible that no end point is given.

We assume *validity*, as follows. Activity a_i is *valid* if for a given input it has provided an output in the original instance, and this output is required for the forward stepping with the same input parameters values. Therefore, activities/services that have the same input as in the original process are *valid* and should not be reexecuted, but rather semantically executed at the workflow level without invoking the underling application/service (*e.g.*, given that flight tickets have been ordered and been approved in the original instance, then if the forward stepping invokes this activity with the same destination and dates as input, it can use the confirmed reservation from the previous execution). In this case the WF system is notified by the client that the activity was executed, while no application/service was invoked. The required execution mode of activity a_i is evaluated (during run time) below, using the following notation.

- $Inst(G')$ is the original instance
- $Inst'(G')$ is the new/modified instance
- $input(a, p, Inst(G'))$ returns the value of p, which is an input parameter to activity a in $Inst(G')$.

$$Exec(a_i) = \begin{cases} \textit{Reexecute} & \exists p\ input(a_i, p, Inst'(G')) \neq \\ & input(a_i, p, Inst(G')) \\ \\ \textit{Semantic} & \textit{Otherwise} \end{cases} \tag{1}$$

Algorithm 1 Forward stepping

Input: $G(V_a \cup V_d, E_c \cup E_d)$, a_i -first activity path, a_x - the stop activity (optional)
Output: *potentialList* - a list of potential activities to be reexecuted, *semanticList* - a list of activities to be semantically executed

add $a_i.successors$ into Q // Q is a Queue
put a_i into *visitedList*.
put $a_i.outputParameters$ into D.
while Q not empty **do**
   put $a_k = dequeue(Q)$ into *visitedList*
   if $\exists a_k.inputParameter \in D$ **then**
      add a_k to *potentialList*
      add $a_k.outputParameter$ to D
   else
      add a_k to *semanticList*
   end if
   if $a_k \neq a_x$ **then**
      add a_k's executed successor activities (as appear in $Inst(G)$) to Q. Add only activities that satisfy $predecessor(a) \subset visited$
   end if
end while
return *potentialList, semanticList*

The forward stepping algorithm is given in Algorithm 1. Looking at the process structure, this algorithm - given it doesn't use instance data - can be invoked asynchronically with runtime instances (*e.g.*, in advance). *PotentialList* holds potential activities (derived from the process structure) for reexecution, of which only those satisfying the *Reexecute* condition in Eq 1 should be reexecuted. During runtime, some of these activites may receive the same input values as in the original execution. Therefore, despite their dependecy on other reexecuted activities, we expect their previous output to be valid (due to the validity property). In such cases, semantic execution is sufficient, and there is no overhead cost for reexecution.

5   Alternative Paths Detection

In the case of an exception, undefined in advance, the workflow engine should rollback to an activity in the graph from which it can provide an alternative path to complete execution of the business process. We will refer to this activity as a *rollback point*. In an extended version of this work, we will elaborate on the heuristics of finding the best rollback point, and design time considerations in determining the suitability of alternative paths. This section details the necessary steps for rollback.

Definition 5. *Rollback point: Let a_i be an activity in a normalized workflow graph $G^` = (V, E)$ with a sink $f$. A rollback point of $a_i$ in a given instance $Inst(G^`)$ is an activity a_j that satisfies the following conditions:*

1. a_j was activated during $Inst(G')$ (i.e., $a'_j s$ state in $Inst(G')$ is "completed").
2. There is a path $P_1 = (a_j, \ldots, a_i)$ in $Inst(G')$ of which all activities in $P_1 \backslash a_i$ are in state "completed".
3. There is a path $P_2 = (a_j, \ldots f)$ in G', such that P_2 is an alternative path to P_1.

A *nearest rollback point* of a_i in a given instance $(Inst(G')$ of $G')$ is a rollback activity a_k such that $minlength(a_k, a_i) \leq minlength(a_l, a_i)$ for any rollback point a_l of a_i.

Theorem 1. *Let a_i be an activity in a workflow graph $G' = (V, E)$. The nearest rollback point of a_i is $NXSP(a_i)$.*

We refrain from presenting the proof of Theorem 1 in this paper due to space considerations.

Rollback can be classified into three types, namely *single threaded, parallel threaded*, and *hybrid*. We will define each of these types and specify the rollback activities needed for each type, using Theorem 1 as a guideline.

Single-threaded rollback is a rollback in which the failing activity falls within a single thread. This means that upon failure, the rollback procedure should be applied only to this thread. The following rollback activity should be taken in a single-threaded type:

$$a_j = NXSP(a_i). \text{ Rollback until reaching } a_j$$

Parallel-threaded rollback refers to a rollback in which the failing activity falls in one of multiple running threads. That means that there is an And split $(NASP(a_i))$ in the path$(NXSP(a_i),a_i)$. In this case, the rollback is performed for all parallel threads within the same And block, until the And split activity of the block containing the failing activity (marked as B_a) is reached. At this point it continues as single-threaded until reaching the nearest Xor split point. In the example given in Figure 1, there are two parallel threads running when activity 6 fails. The other thread, which executes activity 7, is forced to rollback until reaching activity 5, at which point the process continues as single-threaded. The following rollback activity should be taken in a parallel-threaded type:

$$\text{Rollback all current executing and completed activities}$$
$$\text{within } B_a \text{ and proceed rollback as single-threaded.}$$

Hybrid rollback is a rollback in which the failing activity a in $Inst(G')$ is part of a single thread, but some activities in $Inst(G')$ are part of an And block prior to the execution of a. For example, in Figure 1 assume that activity 8, which runs as single-threaded, fails. Since the process contains an And-block (activities 6 and 7 running in parallel), the rollback mechanism should apply to the entire And block and continue with the rollback until reaching $NXSP(8) = 1$. The following rollback activity should be taken in a hybrid-threaded type:

$a_j = NXSP(a_i)$. Rollback until reaching a_j. For each

And-block, rollback all activities in the block and continue.

Algorithm 2 summarizes the mechanism for rollback discussed above. The correctness of Algorithm 2 stems immediately from Theorem 1. It is worth noting that single-threaded rollback is a special case of hybrid-threaded rollback, and therefore the algorithm refers only to the latter.

Algorithm 2 Rollback

1: **Input:** G, $Inst(G)$ -Instantiation, a_i -activity from which the rollback starts.
2: **Output:** $Inst'(G)$ - revised instantiation. Rollback of activities is performed to a_i's nearest rollback point.
3: Process:
4: On the failure of activity a_i, $a_A = NASP(a_i)$ and $a_X = NXSP(a_i)$, if exist.
5: **if** $a_A = null$ **then**
6:     Rollback as Hybrid threaded.
7: **else if** $lenght(a_X, a_i) < lenght(a_A, a_i)$ **then**
8:     Rollback as Hybrid threaded.
9: **else**
10:     Rollback as parallel threaded.
11: **end if**

In case of a failure in one of the threads of an And block (e.g., activity 6), one needs to rollback other threads as well (e.g., activity 7 in Figure 1). However, there can be scenarios in which there is no need for rollback of the concurrent threads. In particular, if the failing activity occurs in a Xor block within an And block, an alternative path that does not require the rollback of all of the And block activities can be provided. This case is handled in Line 8 of the algorhithm.

$NASP$ and $NXSP$ can be pre-assigned by analyzing the graph at design time. At each rollback step the compensation activity is assumed to execute in $O(1)$ (a more refined approach which addresses more complicated executions is deferred to an extended version of this work). There are $minlength(NXSP(a_i), a_i)$ steps to be taken, which is bounded by the cardinality of E. Therefore, the algorithm complexity is $O(E)$.

6   Parametric Modification Analysis

This section discusses scenarios, such as exception handling, that are handled with alternative paths. Once an alternative path has been discovered (see Section 5), it is necessary to evaluate the pre-conditions for performing this new path, and to request a change of values to satisfy these pre-conditions. We therefore turn our attention to the data flow of a business process. As an example, consider once more Figure 1, which introduces a data flow of a single data item, *Type*. This data item is updated during the execution of activity 0, and is retrieved by activity 1. Using common notation [14], the data flow is marked using dashed double-line arrows. In what follows, we denote by $Update(var, a)$ the nearest predecessor of a in which the variable var has been updated. This information can be generated offline and kept with each node, so that accessing it can be done in $O(1)$.

The parametric modification analysis is performed in two steps. The first entails identifying a set of variables whose modification would allow the use of an alternative path. The next is to identify the agents that have assigned the original values to these variables, and to request a change that would allow the use of the alternative path. We here detail each of these steps.

6.1   Satisfying Changes

Going back to Figure 1, recall that the original path to be taken was the path (0, 1, 5, 6, 7, 8, 9). Once activity 1 has been performed, the decision on whether to continue to activity 2 or to activity 5 is based on a mutually exclusive condition (regular or gold customer). Therefore, it becomes evident that the condition that enables us to proceed to activity 2 cannot be satisfied unless some of the variables are assigned different values.

For a given instance $Inst(G')$, each variable var in $Var(c_{a,a'})$ is assigned a value. Let $D(c_{a,a'}, Inst(G'))$ be **a set of sets** of assignments of the type $var = val$ from $Inst(G')$, for which $c_{a,a'}$ **cannot be satisfied**. In the example given in Figure 1, $Var(c_{1,2}) = \{Type, Shipping\}$, and $D(c_{1,2}, Inst(G')) = \{Type = \text{``Regular''}\}$, since under this instance, $Type = \text{``Gold''}$.

Given an instance $Inst(G')$ with an assignment $var = val$, where var is in $Var(c_{a,a'})$, one may consider a modified instance $Inst'(G') = Inst(G')\setminus\{var = val\}\cup\{var = val'\}$ in which $var = val$ is replaced with $var = val'$. For example, a modified instance may include $Type = \text{``Regular''}$ instead of $Type = \text{``Gold''}$.

Definition 6. *minimal satisfying change: Let $c_{u,v}$ and $Inst(G')$ be defined as before and let*

$$D(c_{a,a'}, Inst(G')) = \{set1\{var_{11} = val_{11}, \ldots, var_{1n} = val_{1n}\}, set2\{var_{21} = val_{21}, \ldots, var_{2m} = val_{2m}\}, \ldots\}.$$ *Note that each set may be in different length, and some sets may share the same variables. A satisfying change to $Inst(G')$ is a set of assignments $\{var_{i1} = val'_{i1}, \ldots, var_{in} = val'_{in}\}$ such that $c_{a,a'}$ can be satisfied under*

$$Inst'(G') = Inst(G')\setminus\{var_{i1} = val_{i1}, \ldots, var_{in} = val_{in}\}\cup\{var_{i1} = val'_{i1}, \ldots, var_{in} = val'_{in}\}$$

A *minimal satisfying change* is a satisfying change such that L (Eq 2) is the minimal of all possible satisfying changes. The max function is required since a DNF expression contains sets of predicates. Each set contain simple predicates with an And relation between them. All predicates in this set have to be satisfied in order to satisfy the set.

$$L = \max_{i=1}^{n}\left(\text{minlength}\left(Update\left(\underset{i}{var}, a\right), a\right)\right) \tag{2}$$

Algorithm 3 Parametric modification algorithm

1: **Input:** Graph G', $Inst(G)$ -Instantiation, a_i -failed activity.
2: **Output:** $parameters$ - The parameters to be modified after Role approval.
3: Process:
4: **repeat**
5:    //execute over the nested Xor blockes
6:    $A1 = NXSP(a_i)$.
7:    **for** each $e_{j,k}$ (an outgoing edge from A1) that does not lead to a_i **do**
8:       get $C_{j,k}$
9:       get $Var(C_{j,k})$
10:    **end for**
11:    $L = getD(c, Inst(G))$ //list of satisfying changes.
12:    Sort L by increasing length, using equation 2.
13:    **for** $i = 0$ to L size **do**
14:       $parameters[var, val'] = L[i]$
15:       **for all** set of assignments **do**
16:          $var = val'$
17:       **end for**
18:       **if** all assignments accepted **then**
19:          return $(parameters)$
20:       **else**
         $continue$ // to the next satisfying change.
21:       **end if**
22:    **end for**
23:    $A1 = NXSP(A1)$ // next Xor split point
24: **until** $A1 = null$
25: return $null$

Definition 6 defines a minimal change to be the set of assignments in $Inst(G')$ that can satisfy $c_{a,a'}$. Consider, for example, $c_{1,2}$ that includes the following statement:

$$(Type = \text{“Regular”} \wedge Shipping = \text{“air”}) \vee (Destination = 972 \wedge City = \text{“TLV”})$$

and assume that all variables but $Type$ are updated before activity 0. In this case, $Var(C_{1,2}) = \{Type, Shipping, Destination, City\}$. Assume an instance in which $Type = \text{“Gold”}$, $Shipping = \text{“air”}$, $Destination - 33$, and $City = \text{“Nancy”}$. Therefore, $D(c_{1,2}, Inst(G')) = \{\{Type = \text{“Regular”}\}, \{Destination = 972, City = \text{“TLV”}\}\}$. The *minimal satisfying change* would be $\{Type = \text{“Regular”}\}$.

It is worth noting that Definition 6 minimizes the maximal number of activities for which rollback is needed. Such a definition seems reasonable when the rollback of any activity has the same cost, from the user's point of view. A more general approach would require the definition of a cost model to evaluate the impact of an activity rollback as well as a variable change, and to minimize this impact. We defer the introduction of this approach to the extended version of this work.

6.2   Variable Modification

Once the minimal satisfying change is computed, we can identify the variables that need to be modified. From the workflow model, using either a-priori information or mining procedures [9,2,18], one can identify the activity where those variables are modified. For example, in Figure 1 an exception occurs while executing activity 6. $NXSP$ is detected as activity 1 and the minimal satisfying change is $\{Type = \text{"Regular"}\}$. This change can be set in activity 0. Let a_i be an activity in which a variable change is required. There are two possible sources for updated values, as follows:

A user-defined value: This is a value inserted by the agent that executed a_i. In this case, the user will be presented with a request for reexecution of the activity, with the specific condition needed to allow execution of the alternative path.

A derived value: This is an expression whose input includes both data flow and user input. If the user input affects the data value in such a way that the desired data value is feasible, the user's approval is requested for reexecution of a_i with a specific range of valid input to allow the alternative path execution.

Upon approval, the business process rollbacks to a_X (see Section 5), and forward stepping is performed from a_X to $NXSP(a_i)$. Once $NXSP(a_i)$ is completed, the expression is evaluated again, but this time the evaluation result redirects the execution to the alternative path.

In the case the modifying change request is rejected, the next minimal satisfying change is checked, and so on, until all changes have been exhausted. Then, the second nearest rollback point is computed, and the same procedure is applied to it. The *second nearest rollback point* (based on Definition 5), can be computed recursively as the nearest rollback point of $NXSP(a_i)$. Thus, the algorithm will recursively compute the same actions over the next nested Xor block (*i.e.*, $NXSP(NXSP(a_i))$). This process is summarized in Algorithm 3.

In the worst case the algorithm will iterate over all Xor split points, scanning all graph edges ($O(E)$). For each edge we generate (line 11) a list L of cardinality $|L|$, sorted (line 11) in $O(|L|log|L|)$. Therefore, the total complexity of Algorithm 3 is $O(E|L|log|L|)$. It is worth noting that L may be exponential in $Var(C)$. However, under a reasonable assumption of rather simple conditions with a small $Var(c)$ with a constant upper limit, the algorithm complexity is $O(E)$.

6.3   Forward Stepping - Revisited

Given that a change to activity a_i was approved by the relevant agent, at least one of the output parameters (P_i) of an activity a_i must have been modified. The näive approach assumes that the process can move forward in a semantic manner (see Section 4) until reaching the relevant Xor split point (a_x). However, along the path (a_i, a_x) there are activities which may be affected by the modified value of one of the output parameters of a_i. An activity which uses a modified parameter value as input should not be executed semantically, since the output parameters values may have been revised based on the modified input data. The mechanism for such an approach has been discussed in Algorithm 1 (combined with Eq 1). In this case, the algorithm is executed with a known stop activity a_x.

7   Architectural Considerations

We used the core functionality of a WfMS system to orchestrate our solution. A meta process is a process that manages other processes. Figure 2 presents a description of a *meta process* for exception handling. Slight modifications are needed to generalize it to the more general case. In our case, once a problem/opportunity is monitored, the meta process is invoked and its activities are executed to provide the best solution. Each solution should be confirmed by the relevant agents prior to the semi-automated execution of the underlying process. Upon approval, those activities that are not affected by user input are semantically executed (assuming validity), while other activities are referred to their original responsible agent for execution. The result is a single meta process that can interact with all running processes using the system infrastructure and constructs, and that provides a transparent mechanism to handle such ad-hoc changes via backtracking and forward stepping.

A meta process invokes a monitor that acts as a special workflow client (see Figure 3(a)). The monitor receives modification notifications and exception-oriented messages (*e.g.*, work items), and in response creates an instance of a process that requests a parameter change from the relevant agent. If the reply (again as work item) is negative, then the monitor seeks the next available solution and makes another request to an appropriate agent. This continues iteratively until there are no more solutions to suggest (as discussed in Algorithm 3). Once a positive answer arrives at the monitor, it rolls back to the required activity a_y (or creates a new instance that imitates and semantically executes the original instance activities until reaching a_y), and then starts the reexection and semantic execution of the proceeding activities, until reaching an activity that was not on the original path.

A prototype was built over the ADEPT workflow system (see Figure 3(b)). The BP monitor reads messages from the work items list. An exception is stored in the *exception Store*, while the analyzer analyzes the process graph and creates a list of solutions (for this exception) sorted from best to worst according to an estimation function (one of many within a repository). This list is stored in the *solution store*. At each iteration the *exception store* requests a new solution from the *solution store*.

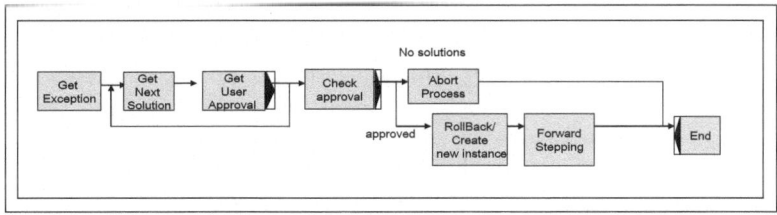

Fig. 2. The Meta-process description for exception handling

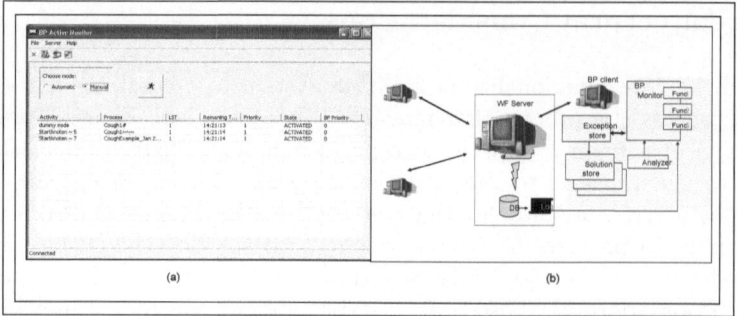

Fig. 3. (a) Business process monitor prototype   (b) Architecture

8   Conclusion

In this paper, we propose a mechanism for efficient management of felxible business processes – in particular, for forward stepping and backtracking. For illustration purposes, we demonstrate our techniques with two somewhat different scenarios. First, we show how changes in Web services can be dynamically embedded into a workflow model, minimizing the costs related to reexecution of previously performed activities. Second, we propose an improved exception handler using forward stepping, backtracking, and alternative paths. Alternative paths, while not the only possible exception handling tool, can serve in a broad variety of cases, and can be easily produced automatically, either in design time (serving as a recommendation) or at run-time (serving as a crisis management tool, in the absence of immediate valid solutions). Our goal is to develop an approach that allows (semi-) automatic dynamic management for arbitrarily complex business processes, balancing the difficulties faced by current workflow models and the control of a designer over the business process.

Future work involves a thorough analysis of utility in the context of reexecution and compensation of activities. Another intriguing direction is data integration for required services, since a replacement service may require data not needed by the original service.

References

1. A. Agostini and G. De Michelis. Improving flexibility of workflow management systems. In W. van der Aalst and J. Oberweis, editors, *BPM: Models, Techniques, and Empirical Studies*, pages 218–234. Springer Verlag, 2000.
2. R. Agrawal, D. Gunopulos, and F. Leymann. Mining process models from workflow logs. In O. Etzion and P. Scheuermann, editors, *Advances in Database Technology - EDBT'98, 6th international Conference on Extending Database Technology, Valencia, Spain, March 23-27, 1998,Proceedings*, Lecture Notes in Computer Science 1337, pages 469–483. Springer, 1998.

3. Specification: Business process execution language for web services version 1.1. http://www-128.ibm.com/developerworks/library/ws-bpel/.
4. F. Casati, S. Ceri, S. Paraboschi, and G. Pozzi. Specification and implementation of exceptions in workflow management systems. *ACM Trans. Database Syst.*, 24(3):405–451, 1999.
5. W. Du, J. Davis, and M.C. Shan. Flexible specification of workflow compensation scopes. In *GROUP*, pages 309–316. ACM, 1997.
6. J. Eder and W. Liebhart. Workflow recovery. In *CoopIS*, pages 124–134, 1996.
7. J. Eder and W. Liebhart. Contributions to exception handling in workflow management. In O. Burkes, J. Eder, and S. Salza, editors, *Proceedings of the Sixth International Conference on Extending Database Technology, Valencia, Spain, March 1998*, pages 3–10, 1998.
8. W. Gaaloul, S. Bhiri, and C. Godart. Discovering workflow transactional behavior from event-based log. In *On the Move to Meaningful Internet Systems 2004: CoopIS, DOA, and ODBASE*, pages 3–18. Springer, October 2004.
9. M. Golani and S.S. Pinter. Generating a process model from a process audit log. In M. Weske W. van der Aalst, A. ter Hofstede, editor, *Lecture Notes on Computer Science, 2678*, pages 136–151. Springer Verlag, 2003. Proceedings of the Business Process Management International Conference, BPM 2003, Eindhoven, The Netherlands, June 26-27, 2003.
10. C. Hagen and G. Alonso. Exception handling in workflow management systems. *IEEE Trans. Software Eng.*, 26(10):943–958, 2000.
11. G.H. Hwang, Y.C. Lee, and B.Y. Wu. A new language to support flexible failure recovery for workflow management systems. In Jesús Favela and Dominique Decouchant, editors, *CRIWG*, volume 2806 of *Lecture Notes in Computer Science*, pages 135–150. Springer, 2003.
12. M. Kamath and K. Ramamritham. Failure handling and coordinated execution of concurrent workflows. In *ICDE*, pages 334–341. IEEE Computer Society, 1998.
13. M. Reichert and P. Dadam. Adept$_f$lex-supporting dynamic changes of workflows without losing control. *Journal of Intelligent Information Systems (JIIS)*, 10(2):93–129, March-April 1998.
14. S. Rinderle, M. Reichert, and Peter Dadam. Correctness criteria for dynamic changes in workflow systems - a survey. *Data Knowl. Eng.*, 50(1):9–34, 2004.
15. S. Sadiq, O. Marjanovic, and M. E. Orlowska. Managing Change and Time in Dynamic Workflow Processes. *International Journal of Cooperative Information Systems*, 9(1-2):93–116, 2000.
16. W.M.P van der Aalst et al. Advance workflow patterns. In O. Etzion and P. Scheuermann, editors, *Cooperative Information Systems, 8th International Conference, CoopIS 2000, Eilat, Israel,Proceedings*, Lecture Notes in Computer Science 1901, pages 18–29. Springer, 2000.
17. Worflow management coalition, the workflow reference model (wfmc-tc-1003), 1995.
18. workflow management coalition 1998. interface 5 - audit data specification. technical report wfmc-tc-1015 issue 1.1. workflow management coalition.
19. P. Wohed, W.M.P. van der Aalst, M. Dumas, and A.H.M ter Hostede. Analysis of web services composition languages: The case of bpel4ws. In Song et al., editor, *Conceptual Modeling - ER 2003 - 22nd international Conference on Conceptual Modeling, Chicago, IL USA, October, 2003,Proceedings*, Lecture Notes in Computer Science 2813, pages 200–215. Springer, 2003.
20. Specification: Web services description language (wsdl) version 2.0. http://www.w3.org/TR/wsdl.

Semi-automatic Generation of Web Services and BPEL Processes - A Model-Driven Approach

Rainer Anzböck[1] and Schahram Dustdar[2]

[1] D.A.T.A. Corporation,
Invalidenstrasse 5-7/10, 1030 Wien, Austria
ar@data.at
[2] Distributed Systems Group, Vienna University of Technology,
Argentinierstrasse 8/184-1, 1040 Wien, Austria
dustdar@infosys.tuwien.ac.at

Abstract. With the advent of Web services and orchestration specifications like BPEL it is possible to define workflows on an Internet-scale. In the health-care domain highly structured and well defined workflows have been specified in standard documents. To reduce the complexity of creating Web service orchestration specifications, we provide a model-driven design approach, consisting of manual and automatic transformations. Security and transaction requirements are covered additionally. The resulting Web services can be bound dynamically at run-time. Therefore, we gain the flexibility to integrate processes that are already established with specific business protocols. Parts of this approach should be applicable to other domains, too.

1 Introduction

1.1 Motivation

Healthcare workflows are distributed across several locations and executed by a large number of applications. Most scenarios in the past covered the inter-hospital execution of processes, such as exchanging patient information or diagnosis data. With the advent of Web services the Internet is capable of providing an infrastructure including a platform for patient social security cards and for supporting healthcare workflows. One way to reach this goal is to enable existing business protocols (HL7 [1], DICOM [2]) to be executed using Web services. Furthermore, the definition of Web service processes is a time-consuming and error-prone task. A semi-automatic, model-driven approach reduces the steps involved. Dynamic run-time behavior of a process enables a single Web service to bridge existing business protocols between business partners which further reduces the complexity of the solution. For our solution we focus on Web service standards that turn out to receive most support from the industry, for example BPEL [6], WSDL [5], WS-security [12], WS-transaction [13] and WS-policy [14].

1.2 Goals

In this paper we provide a model-driven approach for semi-automatic Web service descriptions with run-time binding and a Web service process. The goals we want to

W.M.P. van der Aalst et al. (Eds.): BPM 2005, LNCS 3649, pp. 64–79, 2005.

reach are, (i) to define a modeling process for Web service orchestration. The steps in the modeling process are supported by automatic transformations to reduce the effort that has to be put into the process; (ii) to specify the Web service orchestration in a way that it can be dynamically invoked by all applications that currently interact using established processes with specific business protocols; (iii) to integrate additional security and transaction properties of the orchestration, to satisfy requirements of real-world scenarios; and (iv) to complement this design-time process with a run-time perspective, to gain a better understanding of the execution of the orchestration.

Overall, this approach supports the implementation of Internet-scale healthcare workflows by reducing the complexity of creating Web service orchestration specifications. Although, we use an example from the healthcare domain, valuable parts of this model-driven approach should be applicable to other domains, too.

The paper is structured as follows. Section 2 introduces an example and the basic idea of our model-driven approach. Section 3 describes the modeling process in an overview and with each step in detail. Section 4 provides additional information about the run-time behavior of the Web service orchestration. Section 5 concludes the results and states topics of further research.

1.3 Related Work

Our previous work covers interorganizational workflow in the medical imaging domain [4]. The paper covers the separation of a workflow layer using WSDL and BPEL, and a domain layer using DICOM and HL7. Subsequent papers then focused on Web service modeling and the mapping between BPEL activities and DICOM and HL7 messages [9, 10]. Besides our work, Artemis [26], an EU supported project, develops Semantic Web services for the healthcare domain. Its main focus is semantic mediation of services, in contrast to our process modeling oriented approach.

Furthermore, one paper [9] compares classical workflow models for medical imaging with Biztalk. This work is related to the middleware paradigm in an intranet based environment. One paper on application integration [10] helps understanding the "large picture" of medical workflows and the IHE framework but does not focus on modeling Web services. Another paper covers a model-driven approach for Web service transactions that supports more sophisticated scenarios of interaction [23].

Finally, there is work related to the medical industry and Web services standards as referenced throughout this paper. However, the focus of our paper on modeling BPEL processes based on the IHE framework is, to the best of our knowledge, not covered in the literature so far.

2 Example Workflow

2.1 Overview

In the healthcare domain highly structured and well defined workflows have been specified through the HL7 and the DICOM protocol standards. Those standards have been extended with the IHE (Integrating the Healthcare Enterprise) [3] framework that defines scenarios and profiles for these standards and specifies the most common application roles and workflows in detail. This is the source for our modeling process. In other domains similar specific sources have to be identified or created. Most IHE roles used in the workflow are covered by HIS and RIS (Hospital and Radiology Information System) and PACS (Picture Archiving and Communication System) applications. They are comparable to ERP (Enterprise Resource Planning) or SCM (Supply Chain Management) applications. For a more detailed description of the domain and the capabilities of these applications refer to [9, 10].

2.2 Example

For our example we focus on a specific workflow within the IHE framework, the IHE *administrative workflow*. Figure 1 shows an overview of roles and transactions, the grey-shaded area, the *patient registration* IHE transaction, is where we dive into. The lighter-shaded area is also mentioned as it is part of the same workflow. An IHE transaction is comparable to a BPEL process (Appendix C, Table 1 [27]).

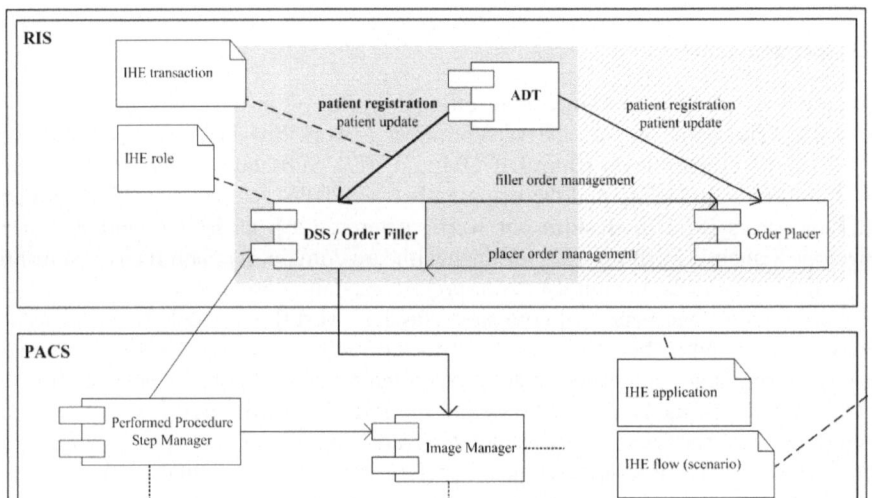

Fig. 1. IHE *administrative workflow* - focus of our example

The *patient registration* transaction is performed between two systems: *ADT* (Admission, Discharge and Transfer) and the *DSS* (Department System Scheduler). The *ADT* corresponds to an administration application that provides patient data to different subsystems. The *DSS* is responsible for scheduling medical examinations. The transaction transfers patient registration information from the *ADT* to the *DSS*. The *ADT* and the *DSS* are IHE roles, an application (IHE actor) can act as several IHE roles. There is a direct relation to the BPEL partner model (Appendix C, Table 1 [27]). As shown in Figure 1, several roles and transactions are involved in a specific implementation scenario. Our model-driven approach is appropriate for this environment in general. Next, we provide an overview of the source and result of our modeling approach (see Figure 2).

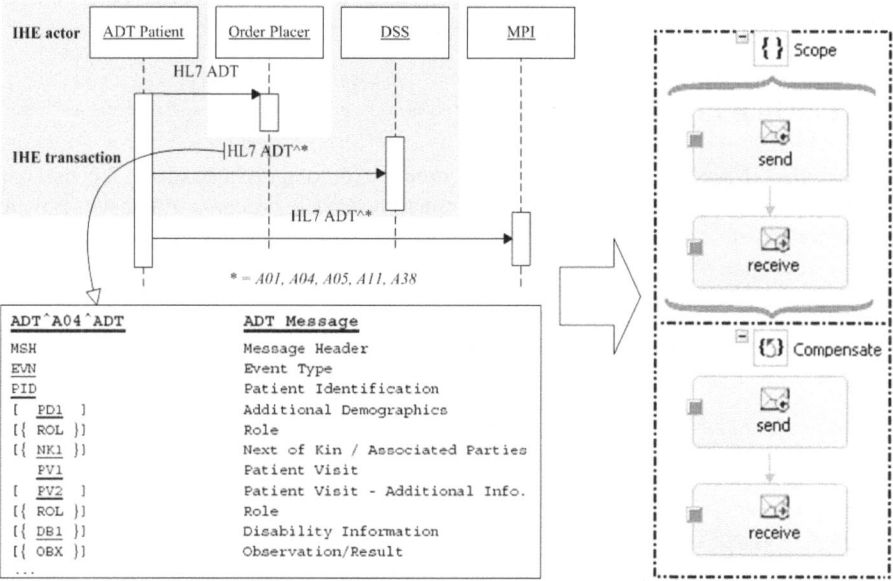

Fig. 2. source and result of our model-driven approach

The HL7, DICOM and IHE standard documents serve as the starting point for the modeling process. In our example, the two main sources are the IHE transaction UML sequence diagrams and (for our example) the HL7 message scheme definition (on the left side). The modeling result should be an executable BPEL process (on the right side). Of course, this basis cannot be mapped directly to a Web service orchestration. It provides process information, structure and data which have to be extended manually and transformed automatically in several steps. From the sequence diagram we derive the business process. This source has to be transformed and extended with orchestration flow constructs and security and transaction requirements manually. From the HL7 message we extract the structure and data for message correlation, business partner and communication port configuration. It also provides data to control the orchestration itself. In non-Internet environments, application providers implement the workflow according to the IHE framework sequence diagrams. They provide a native HL7 (over TCP/IP) interface business protocol. The outcome of our approach is to execute the processes and exchange the messages using Web services and BPEL orchestration. Section 3 describes the modeling process in detail.

2.3 Requirements of Further Workflows

Healthcare workflows have different requirements. A model-driven approach should be evaluated using several examples with different requirements. For example, the

IHE framework contains transactions using the DICOM protocol, where large amounts of medical image data (more than 100MB per transaction) have to be transferred and, therefore, be compressed. In [9, 10] we investigated the requirements of healthcare workflows in a Web service environment. Conclusions have been considered in our approach.

3   Modeling Process

In this chapter we provide the modeling process for our orchestration. We use our example, the IHE *patient registration* transaction (BPEL process). First, we provide an overview with a short description before we show the modeling steps in detail.

3.1   Process Overview

The modeling steps are described throughout the sections as shown in Figure 3.

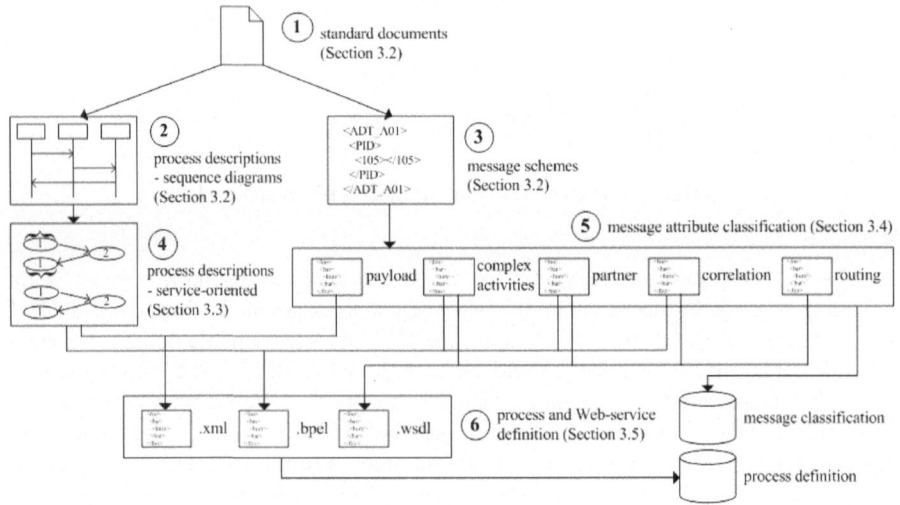

Fig. 3. Modeling process steps

The modeling process occurs at design-time. The run-time behavior of an executing BPEL process is shown in Section 4. The process starts with the available standard documents (Step 1) as shown in the introduction of the example. In Step 2 and 3 the UML sequence diagrams and HL7 message schemes are stored in a database or file-system. Step 4 converts the diagrams to a process oriented UML activity diagram and applies transaction and security concerns. Step 5, on the other hand, classifies the message attributes. The classification covers *structured activities* (for structured activities, like BPEL *switch* statements), the business *partner* and partner-link definitions (corresponding to the sections in the BPEL process definition), *correlation* attributes (used in BPEL correlation-sets) and *routing* attributes that define the destination of the messages and are used to configure the ports in the BPEL process. The *payload* contains the whole HL7 message that is sent as an attachment to the SOAP [25] message by the Web service. Step 6 merges the information of the process and the message attributes and generates three output files, a BPEL process description, a WSDL file that defines the communication end-points and an XML containing additional security and transaction properties using WS-policy.

3.2 Step 1-3: Digital Source Representation

We start with a digital representation of the IHE transaction and HL7 message. Figure 4 shows the source representation of the *patient registration* transaction.

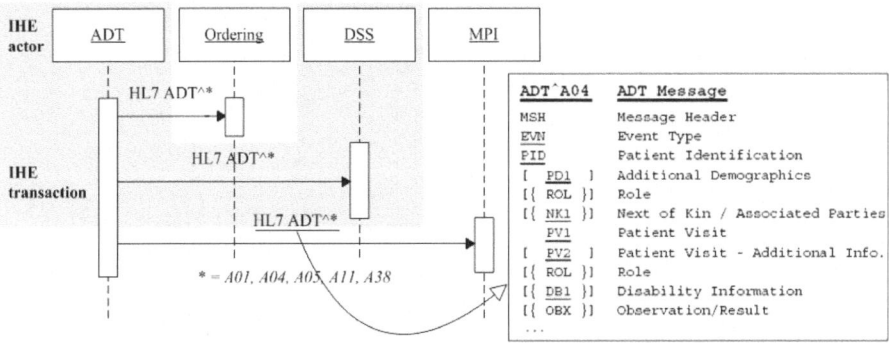

Fig. 4. IHE *patient registration* transaction and HL7 message

In general, from the sequence diagram we extract the process, from the message we extract a design-time configuration and run-time properties of the process. The information has to be digitized into a file-based or database storage.

The UML diagrams of the IHE framework are of proprietary file format (in our case Microsoft Visio). The diagrams contain several ambiguities and errors that have to be resolved. For example, the arrows represent more than one message. Furthermore, a sequence diagram is not appropriate for a BPEL process as it contains more than two business partners for which an interface should be defined. In the next section we show how an activity diagram and several adaptations solve these problems.

The HL7 message has a hierarchical format that consists of several modules (which are reused between messages) and each module consists of a set of attributes. The attributes are of specific (simple and complex) data types that can be represented in XML [11]. DICOM messages for comparison contain data and service descriptions, but nevertheless, can be broken down to data types and payload data (images, documents). Also some XML messages carry documents as attachments. However, to setup our process we are only interested in those parts of the messages that contain information to identify partners, configure and control our process and route messages. All other data resides in the HL7 message, which is sent as an attachment of the SOAP message.

3.3   Step 4: Service-Oriented Process Description

In this step we convert the sequence diagram into an activity diagram.

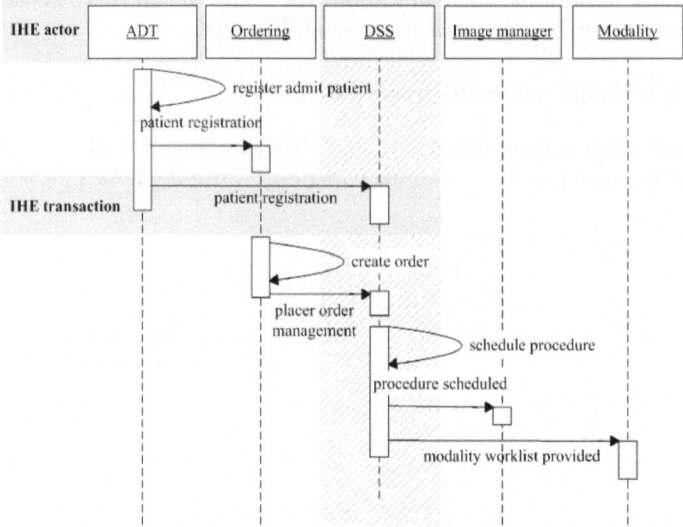

Fig. 5. IHE administrative flow

The following sub-steps are performed during the manual transformation:

- resolve errors in the standard document
- select partner to define a public process
- convert to an activity diagram and skip private activities
- apply security requirements
  (repeat the next steps for each IHE transaction)
- select a specific IHE transaction
- extend the diagram to represent different control flows
- extend the diagram to represent acknowledgement messages
- apply transaction requirements

Each sub-step is described in more detail throughout the rest of this section. In contrast to the introduction the sequence diagrams of the IHE standard are provided at two levels of detail. We start with the *coarse-grained level* to perform several sub-steps of the transformation for multiple transactions at once. Figure 5 shows the sequence diagram of the IHE *administrative workflow* which is the "large-picture" where the *patient registration* transaction (grey-shaded are) is performed. From here we start with the following changes.

Substep 1: Resolve errors in the standard document: In this diagram of the IHE framework we found, that a transaction has been drawn in the wrong direction (*mo-*

dality worklist provided transaction). Manually created sources always have to be reviewed in detail.

Substep 2: Select partner to define a public process: As we model executable BPEL processes, it is necessary to select an IHE actor (BPEL partner), whose public process has to be represented. We select the *DSS* actor and skip all activities that are not sent or received by this actor.

Substep 3: Convert into activity diagram and skip private activities: Compared to the sequence diagrams, each arrow (IHE transaction) is represented with two activities, one BPEL *invoke* and *receive*. For each IHE actor a lane is generated. Furthermore, internal activities, activities that are performed by an actor on itself, are skipped, as no BPEL process related activity is necessary. The resulting diagram is shown in Figure 6.

Substep 4: Apply security requirements: Next, security requirements between business partners are defined. In [8] we defined security zones and boundaries to represent groups of applications that trust each other. The organizational trust information might be modelled using WS-Trust and stored globally in a database for all modelled processes. However, this is out of the scope of this paper. Figure 6 shows the result after these four sub-steps.

In our case, the *DSS* actor is in the same zone as several other actors but in a different one as the *ADT* actor. Therefore, we require message encryption using WS-security when executing the *register patient* transaction.

Substep 5: select a specific IHE transaction: We select the *patient registration* transaction and now focus on the sequence diagram provided by IHE on the *fine-grained level* (diagram shown in Figure 4).

Substep 6: Extend the diagram to represent different control flows: In our case the transaction consists of sending one of three HL7 messages (*ADT_A01*, *ADT_A04* and *ADT_A05*). Which message is sent depends on the class of patient which is represented as the *PatientClass* attribute within the HL7 message. This decision can be modelled in BPEL using a *switch* structured activity. It is represented in the activity diagram accordingly.

Substep 7: Extend the diagram to represent acknowledgement messages: In HL7 each message sent is followed by receiving an acknowledgment message (*ACK_A01*, *ACK_A04*, *ACK_A05*). Therefore, the diagram has to be extended to represent this behavior. DICOM uses status messages to represent similar behavior. Figure 7 shows the resulting diagram. The outer frame corresponds to the original invoke-receive pairs (compare to Figure 6), which has been extended through the substeps 6 and 7.

Substep 8: Apply transaction requirements: Referring to the requirements stated in Section 3.2 we integrate transaction requirements using compensation activities. Atomic transactions are currently not considered, although, there are several operations in the DICOM standard suited for it. We currently refer to the work presented in [23] for an extended transaction modeling approach. Figure 8 shows the resulting diagram.

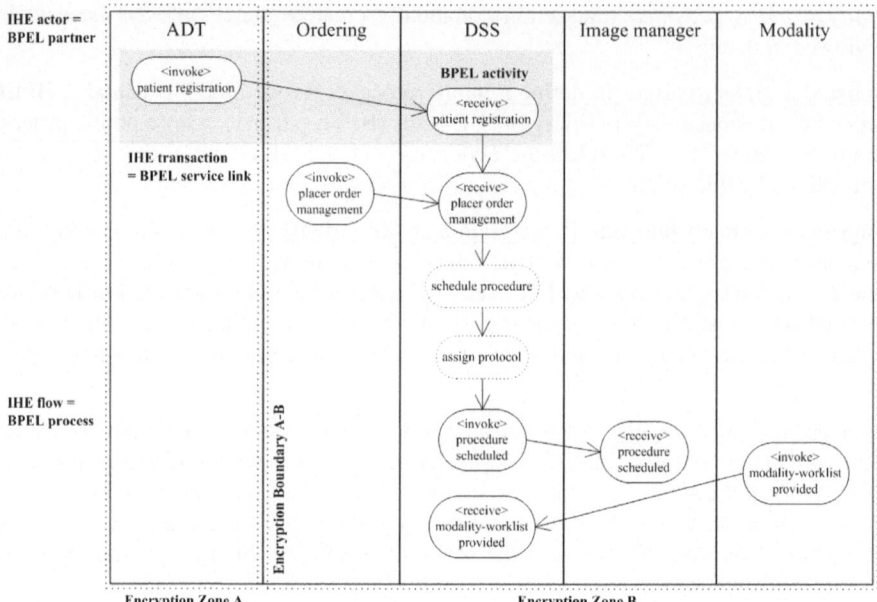

Fig. 6. Administrative flow - BPEL public process of *DSS* (Department System Scheduler)

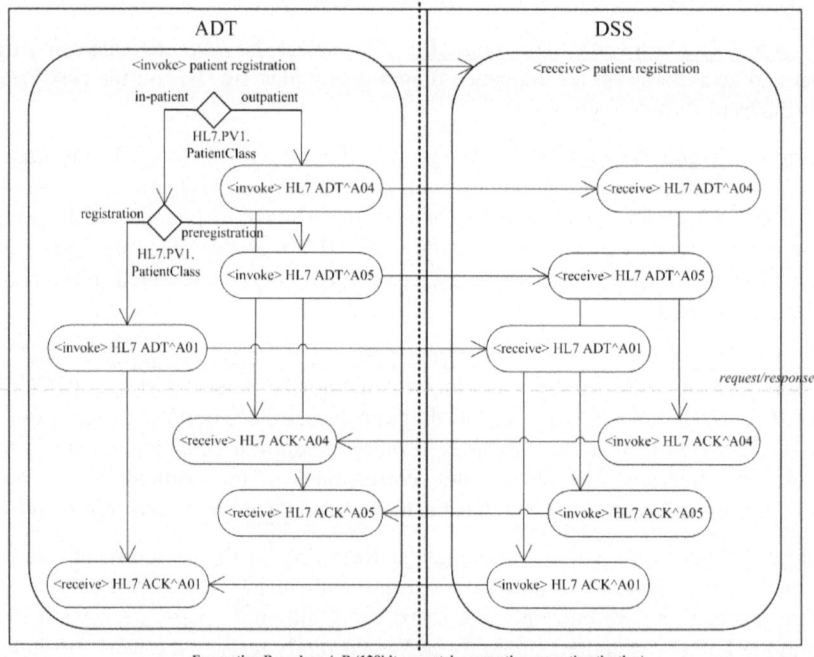

Fig. 7. *patient registration* transaction - public process

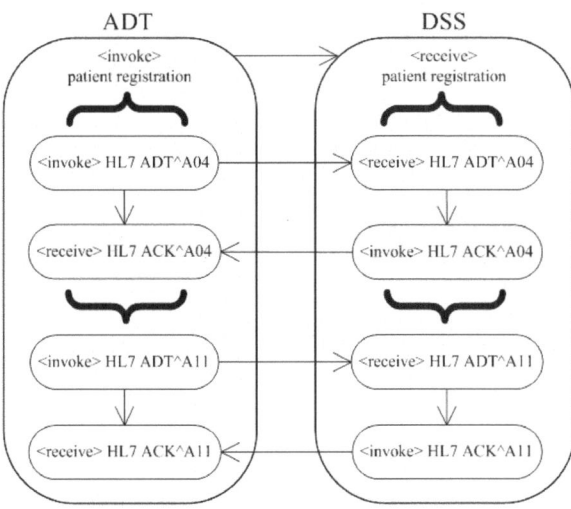

Fig. 8. Compensation-based transaction for the *patient registration* transaction

For HL7 negative acknowledge messages (*ACK_A11*) are generated in case of errors. Transactional behavior is directly expressed in the process definition, as it is part of the BPEL specification, while security parameters have to be configured for interfaces and are later on stored in the policy file (see Section 3.5.3).

3.4 Step 5: Message Attribute Classification

In the next step we turn to the HL7 message itself. Not all parts of the message are equally important for a process definition. We can distinguish the following five categories for message attributes (which are also valid for other business protocols):

- Class 1: attributes required for *binding* (WSDL interfaces binding information)
- Class 2: attributes required for *partner* definition (BPEL partner definitions)
- Class 3: attributes required for *complex activities* (complex BPEL activities)
- Class 4: attributes required for correlating the message (BPEL *correlation-sets*)
- Class 5: other message attributes (used by the underlying business protocol)

As stated in Section 3.2 the hierarchical structure of message attributes can be represented in XML. Related to our example we classify the HL7 ADT_A04 message as shown in Figure 9. The structure of all ADT messages is the same regarding attribute classification.

On the left, the original message structure is shown. Through an analysis of the attribute descriptions in the HL7 standard and the IHE framework the attributes listed on the right side have been selected for each class respectively. An XML scheme file for this structure can be found in Appendix A [27]. The result of this step is a classification file for each message used in the process. The file is required in the next step of

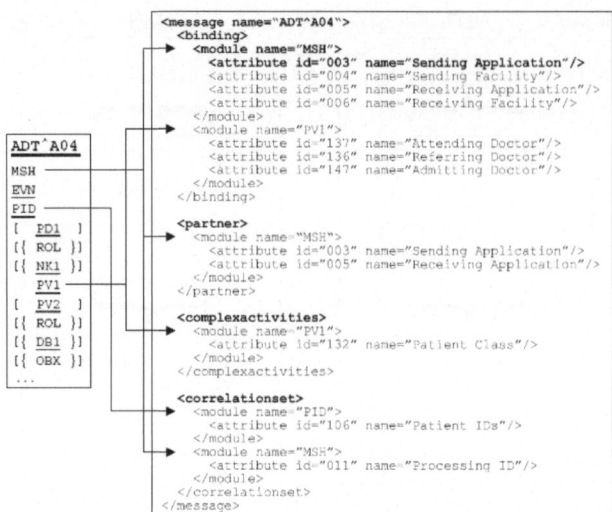

Fig. 9. Message attribute classification of HL7 ADT^A04

orchestration definition and during run-time execution (Section 4) for message parsing. It is stored together with the message scheme files in a database.

The Class 1-3 attributes are the same for every business partner. For Class 4 it is possible that the receiving partner requires different attributes for BPEL complex activities. Therefore the analysis has to take into account the processing of all partners involved in the transaction. In our case no extension to the definition is necessary. A further conclusion is that Class 1-4 attributes have to be part of the SOAP message and Class 5 attributes reside in the attachment. However, the security requirements (see previous section) are always defined for the whole SOAP message (the parameters and the attachments).

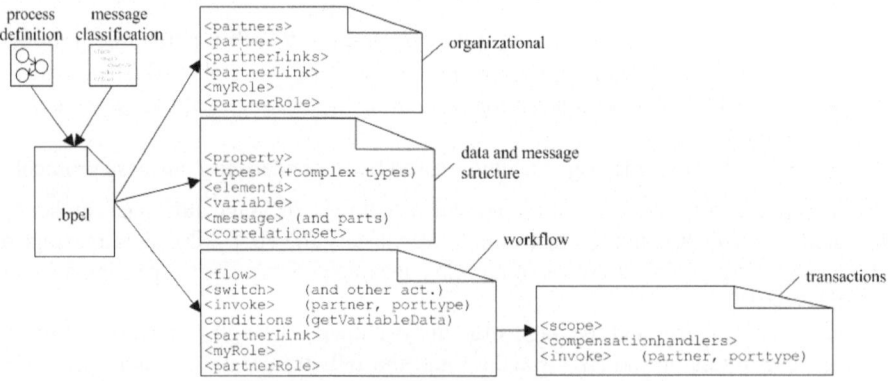

Fig. 10. BPEL process definition

3.5 Step 6: Orchestration Definition

The orchestration definition step is split into three sub-steps which are described in detail throughout this section. We present parts of the files for illustration, a complete listing can be found in Appendix B [27].

3.5.1 BPEL Process Definition
The first sub-step is the definition of the BPEL process. Figure 10 points out which parts of the model contribute to the content of the file and how the file is structured.

One part of the BPEL definition is created by converting the activity diagram into BPEL constructs. Several conversions are performed to create parts of the *flow* section of the BPEL file from the process (see Table 2 in Appendix C [27]). The second part is extracted from the message attribute classifications partner and binding sections. Therefore, the table also lists the mapping between message attributes and BPEL sections and tags. Conversions of IHE activity diagrams into BPEL flows have been covered in detail in an earlier paper (see [7]). There is also a paper that covers UML conversions in general [22]. Of special interest here are the components for the partner definition, which are directly derived from the IHE roles and transactions. Those names are generic and allow, together with a run-time generation of WSDL interfaces (see next section), a dynamic model of BPEL process execution (see Section 4). Finally, XPath [17] expressions are generated using a lookup in the message attribute classifications *complexactivities* section. In our example the value for the PatientClass variable is "HL7_A04_TYPE/PV1-132", if an A04 type message is sent by the application.

3.5.2 WSDL Interface Definition
The second part is the WSDL interface definition. Here we have to distinguish between design-time and run-time operations. During design-time the portTypes, which are required by the BPEL process, are defined. As Figure 11 shows, message attributes are used here.

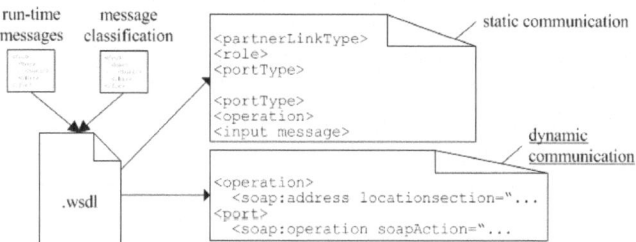

Fig. 11. WSDL process definition as design-time

Appendix C, Table 3 [27] contains the mapping between those elements and the WSDL content. We want to focus on the dynamic binding part in the WSDL definition, which contains the information about the communication endpoint. Selection of a specific endpoint can be performed dynamically during run-time by configuring the WSDL file. This can be done by the Web service or the BPEL engine, if it supports dynamic endpoint configuration. The dynamic parts of the WSDL file are the *binding* and *port* sections. The definition consists of a base URI and an extension that references the specific service.

- WSDL section: operation, element: soap:address, attribute: location
- WSDL section: port, soap:operation, attribute: soapAction

As each business partner always uses the same generic Web service, it is only necessary to store a mapping for one destination URI. The source for the URI mappings can be any value of the HL7 and DICOM message attributes that have been classified in the *bindings* section of the message classification document. During run-time the values are extracted from the messages. Then the mapped URI is calculated and the WSDL file is configured for the required endpoint. In the next step, BPEL processes can perform activities with the dynamically added business partner using the newly configured port.

3.5.3 Policy Definitions

The third part is the policy definition. Security and attachment requirements have to be converted to WS-security and proprietary constructs (see Figure 12).

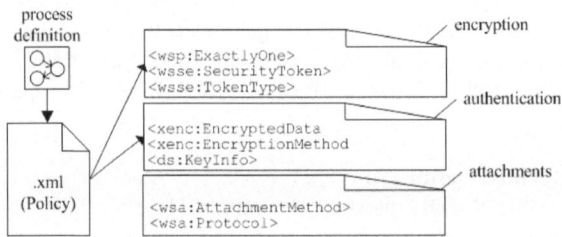

Fig. 12. WS-policy definition

For the WS-policy definition we use the WS-SecurityPolicy [21], WS-Encryption [19] and WS-Signature [20] standards. For attachments we defined a proprietary policy description. The attachment requirement is constant, all messages contain attachments. Currently, the DIME [15] standard is specified but is supposed to be superseded by MTOM [16] soon.

For the security part the process diagram has to be parsed and for all security boundaries the properties for encryption and authentication have to be applied (Appendix C, Table 4 [27]). The security credentials are provided at run-time (Section 4).

4   BPEL Process at Run-Time

We split the run-time activities into 3 phases. Figure 13 shows phase 1.

First, a component (that we call workflow engine) receives a HL7 message from an application using the business protocol. Then it has to lookup and cache the message classification and perform a classification of message attributes on the received message instance. According to the XSL scheme (Appendix A [27]), classification values are extracted and stored. The next phases 2 and 3 are shown in Figure 14.

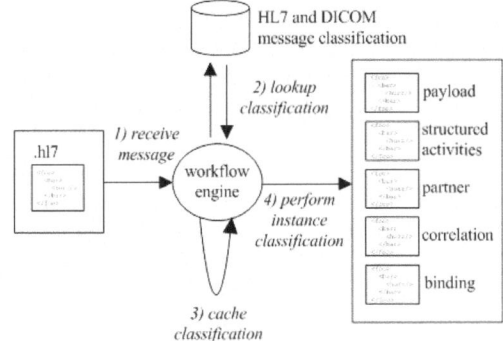

Fig. 13. Run-time phase 1: receiving a message

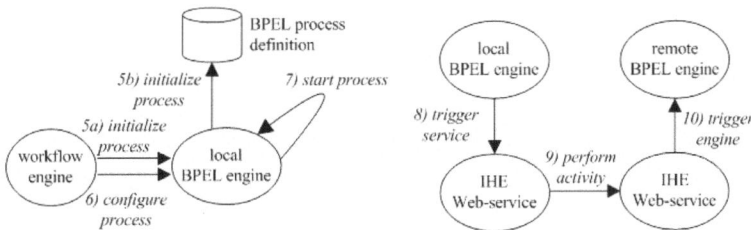

Fig. 14. Run-time phase 2 and 3: start and execute BPEL process

In Phase 2, the BPEL process is initialized (if it is a start message of a BPEL process) using its process definition and configured with the values of the message classification. For each initial and subsequent message the dynamic ports are bound according to the values of the *binding* section in the message classification. Next, depending on the capabilities of the BPEL engine, the security requirements are selected by the engine (in Phase 2) or the Web service (in Phase 3) and are used to configure secure ports. Biztalk Server 2004 [18] for example supports secure ports with an additional Web service adapter. The security credentials (for example asymmetric keys) depend on the communicating applications. Information from the *partner* section of the message classification is used to lookup credentials in a database (via UDDI [24] for example). The credentials are inserted into the SOAP message before the Web service calls the partner. The same steps occur on the receiving Web service. As the sending port is now identified by the initiating application, a lookup to a database can be performed to decode and validate the SOAP message, before it is passed to the BPEL engine for workflow processing. Additionally, both partners have to insert the business protocol message into the attachment part of the SOAP message.

78 R. Anzböck and S. Dustdar

5 Conclusions and Future Work

In this paper we presented a model-driven approach to define Web service orchestration in the healthcare domain. We were able to meet our goals, to define a design-time modeling process. Through semi-automatic modeling steps we produced results out of standard documents of the business protocols and applied security and transaction requirements. We created several artifacts: a BPEL file for the process definition, a WSDL file with a dynamic port configuration, a WS-policy file containing security and attachment requirements. We have also shown the run-time behavior of the suggested solution using the artifacts produced during modeling. The benefits stated initially, the reduction of complexity and required effort can be concluded from our work. Further, the modeling steps *public process definition* and *message classification* can be applied to other domains, too. We find it especially noticeable, that it is possible to execute several BPEL processes using one generic Web service. For future work we plan to extend our current prototype implementation to encompass a model-driven toolset. Currently, several steps need more standardization before we can proceed. Furthermore, atomic transaction requirements should be investigated and an executable example transaction should proof our approach.

References

1. HL7 Organization: Health Level 7, http://www.hl7.org (2000)
2. NEMA and Global Engineering Group: DICOM 3 Standard, http://www.nema.org (1998)
3. Radiological Society of North America: IHE Technical Framework 1.1, http://www.rsna.org /IHE/index.shtml (2003)
4. Anzböck, R., Dustdar, S.: Interorganizational Workflow in the Medical Imaging Domain. Proceedings of the 5th International Conference on Enterprise Information Systems (ICEIS), Angers, France, Kluwer Academic Publishers (2003)
5. W3C: Web services Description Language 1.1, http://www.w3.org/TR/wsdl.html (2001)
6. BEA Systems, IBM, Microsoft, SAP AG and Siebel Systems: Business Process Execution Language for Web services version 1.1, http://www-128.ibm.com/developerworks/library/specification/ws-bpel/ (2003)
7. Anzböck, R., Dustdar, S.: Modeling Medical Web services, BPM 2004 - Conference on Business Process Management (2004), Springer LNCS 3080, pp. 49 – 65.
8. Anzböck, R., Dustdar, S.: Modeling and Implementing Medical Web services. Data and Knowledge Engineering, Elsevier, forthcoming (2005)
9. Von Berg, et.al.: Business Process Integration for Distributed Applications in Radiology, Philips Research; Hamburg, Germany (2001)
10. From PACS to integrated EMR: Osman Ratiba, Michael Swiernik, J. Michael McCoy, Computerized Medical Imaging and Graphics 27 Pages 207–215 (2003)
11. HL7 Organization: HL7 XML encoding scheme: http://www.hl7.org/ (2003)
12. BEA, IBM, Microsoft: Web Services Security (WS-Security), www-106.ibm.com/develop-perworks/webservices/library/ws-secure/ (2002)
13. BEA, IBM, Microsoft: Web Services Transactions (WS-Transactions), http://www.ibm.com/developerworks/library/ws-transpec/ (2002)
14. BEA, IBM, Microsoft, SAP, Sonic, VeriSign: http://msdn.microsoft.com/library/default.asp?url=/library/en-us/dnglobspec/html/WS-policy.asp (2004)

15. Microsoft: Direct Internet Message Encapsulation (DIME), http://msdn.microsoft.com/library/en-us/dnglobspec/html/draft-nielsen-dime-02.txt (2002)

16. BEA, Canon, IBM, Microsoft: http://www.w3.org/TR/soap12-mtom (2005)

17. W3C: XPath, www.w3.org/TR/xpath (1999)

18. Microsoft Biztalk Server 2004: Microsoft Corporation, www.mirosoft.com (2004)

19. XML-Encryption: W3C Working Draft, "XML Encryption Syntax and Processing,", http://www.w3.org/TR/xmlenc-core/ (2002)

20. XML-Signature: W3C Proposed Recommendation, "XML Signature Syntax and Processing,", http://www.w3.org/TR/2001/PR-xmldsig-core-20010820 (2001)

21. Microsoft, IBM, Verisign, RSA Security: WS-SecurityPolicy, http://msdn.microsoft.com/webservices/default.aspx?pull=/library/en-us/dnglobspec/html/WS-Securitypolicy.asp (2002)

22. IBM: From UML to BPEL, http://www.ibm.com/developerworks/webservices/library/ws-uml2bpel/ (2003)

23. Schmit, B.A., Dustdar, S. (2005). Model-driven Development of Web service Transactions, XML4BPM 2005 - XML for Business Process Management Workshop, 11th GI Konferenz Business, Technologie, und Web (BTW 2005), 1 March 2005, Karlsruhe, Germany.

24. IBM/Microsoft/SAP, et.al.: UDDI 3.0.2, http://www.oasisopen.org/specs/index.php#uddiv3 (2005)

25. W3C: SOAP Version 1.2, http://www.w3.org/TR/soap12-part1/ (2003)

26. Dogac, A., et.al.: Artemis: Deploying semantically enriched Web services in the healthcare domain, Elsevier, Information Systems, (2005), Article in Press

27. Appendix see http://www.infosys.tuwien.ac.at/Staff/sd/papers/BPM2005Appendix.pdf

A Human-Oriented Tuning of Workflow Management Systems

Irene Vanderfeesten and Hajo A. Reijers

Eindhoven University of Technology, Department of Technology Management,
PO Box 513, NL-5600 MB Eindhoven, The Netherlands
{i.t.p.vanderfeesten, h.a.reijers}@tm.tue.nl

Abstract. Workflow Management Systems (WfMS's) offer a tremendous potential for organizations. Shorter lead times, less mistakes in work handoffs, and a better insight into process execution are some of the most notable advantages experienced in practice. At the same time, the introduction of these systems on the work floor undoubtedly brings great changes in the way that professionals work. If a WfMS's work coordination is experienced as too rigid or mechanistic, this may negatively affect employees' motivation, performance and satisfaction. In this paper, we propose a set of measures to "tune" functioning workflow systems and minimize such effects. The measures we propose do not require undue cost, time, or organizational changes, as they characteristically lie within the configuration options of a WfMS. We have asked an expert panel to select and validate the 6 most promising measures, which we present in this paper. From our evaluation of three commercial WfMS's, we conclude that it depends on the specific system to what level these general measures can be easily implemented.

1 Introduction

A workflow management system (WfMS) is a software system that supports the specification, execution, and control of business processes [19]. Commercial WfMS's have been around since the early nineties, while their conceptual predecessors range back even further, see e.g. [6]. They have become "one of the most successful genres of systems supporting cooperative working" [5]. The worldwide WfM market, estimated at \$213.6 million in 2002, is expected to redouble by 2008 [38]. Furthermore, WfM functionality has been embedded by many other contemporary systems, such as ERP, CRM, and call-center software. WfM technology, in other words, has become quite successful and widespread. The reason for this popularity is fourfold [33]:

- *The coordination of work becomes easier.* A WfMS liberates human actors from the efforts to coordinate their work ("what do I do have to do next?", "where is the *#& client file?", "who must check this proposal next?")
- *A higher quality of service is delivered.* The WfMS will ensure that the process is executed in correspondence with the intended procedure: important steps can no longer be forgotten, work will not get lost, and authorization policies are automatically enforced.

W.M.P. van der Aalst et al. (Eds.): BPM 2005, LNCS 3649, pp. 80–95, 2005.

- *The work is executed more efficiently.* Work items will only be allocated to workers by the WfMS if and when they are required to be executed.
- *The process becomes more flexible.* Ejecting the business control flow from traditional applications and moving it towards a WfMS simplifies the redesign of the process.

Recent successful implementations of WfMS's are e.g. reported within the banking, automotive and IT industry [2,3,25]. Despite its success, WfMS's have received their share of criticism as well, see e.g. [5,7]. Skeptical arguments are mainly raised by employees - the potential users - and work psychologists, who fear that workflow systems might lead to a mechanical approach to office work where man is seen as an exchangeable resource, e.g. like a machine, and not as a human being. In a study by Küng [24], an interviewee at an organization described the effects of a WfMS introduction within his organization as follows:

"Jobs became more monotonous. The system forces the employees to work strictly according to the process definition. Through the use of the workflow system, we now have some kind of 'chain production' in the office."

The image of a WfMS as a rigid system is also produced very glaringly in the well-known case study of a WfMS implementation in the UK print industry [1]. The respective system was not accepted by the end users, who invented various ways to work around the intended procedures. The previous examples illustrate that through the rigid structure of a workflow system there is a risk of creating similar problems as to mass production and assembly line work in the previous 19th and 20th century, e.g. boring work, decreasing performance, unsatisfied and unmotivated employees.

This paper proposes measures that can be taken to reconfigure an implemented WfMS so that it becomes more agreeable to the needs of performers working with such a system. An important driver in the creation and selection of these proposals was to come up with measures that have a wide applicability and are easy to implement. The proposals have emerged from the confrontation of mainly two perspectives. On the one hand, we have considered the general characteristics that positively influence the motivation, performance and job satisfaction of performers. On the other hand, we looked at the policies that WfMS's generally use for distributing and assigning work to performers. Even though such policies do not affect the work that has to be executed itself - as specified in an underlying workflow definition - they have a direct impact on the way people experience performing that work.

The paper is organized as follows. In the following section, we will give the theoretical background of the perspectives we mentioned, as used for generating the proposals. In Section 3, we will describe the various proposals, how they have been selected, and how they were validated by an expert panel. In Section 4, we present the evaluation of three current, commercially available WfMS's to determine to what extent these specific systems can be reconfigured in accordance with the presented general proposals. This paper ends with our conclusions and recommendations.

2   Background

In this section, we will present the theoretical background for the development of the "tuning" measures. In particular, we will introduce a frame of reference which captures the scope of the various measures. In addition, we will discuss the two theoretical models that have been most fruitful for developing the proposed tuning measures. They focus on small adaptations of an already implemented WfMS and as such improve on how the system meets human needs. The needs we focus on go beyond primary needs like e.g. physiological needs. The proposed measures, instead, focus on the improvement of a person's esteem (i.e. the need for a feeling of self-worth and for respect and admiration from others) and self-actualization (i.e. the need to make the most of one's life, i.e. the need to obtain self-fulfillment) [27].

Our first observation is that the way that people perceive the use of a WfMS is influenced by more than merely the characteristics of the technology itself. Consider, as an extreme example, an organizational policy enforcing that the WfMS can be accessed through one specific work station only. As a result, workers may have to move to and from, stand in line, etc. The irritation that this policy may cause is not due to a technical characteristic of the system, but will affect people's perception of using the WfMS nonetheless. Therefore, we see an implemented WfMS as being part of a more abstract, larger system, which involves organizational, process, human, and technical components. Inspired by the well-known reference model of the Workflow Management Coalition (WfMC) [37], we consider an abstract workflow system as shown in Figure 1. The abstract workflow system consists of four levels:

- **Organizational structure -** On this level the structure of the organization is defined. This includes for example, the division in departments or business units, hierarchical relations, functions, physical employees, geographical position, competencies, authorization and rules.
- **Roles -** This level contains the roles that can be performed by employees in the organization. To fulfil a certain role an employee has to meet the accompanying requirements (e.g. concerning competencies or the proper organizational unit).
- **Process automation layer -** The process automation layer can be divided into two parts: the distribution of work items and the automatic and computer supported execution of work items. This includes: the workflow enactment service, the automatic execution of work items, the computer applications an employee needs to perform an activity, the shared or individual worklists and the administration and monitoring tools.
- **Workflow definition -** The workflow definition is a static representation of the process. It consists of the process model, the resource classification and the relationships between those two.

In comparison with the reference model of the WfMC [37], this model explicitly adds the organizational context. Also, it more clearly separates the build-time from the run-time components of a WfMS.

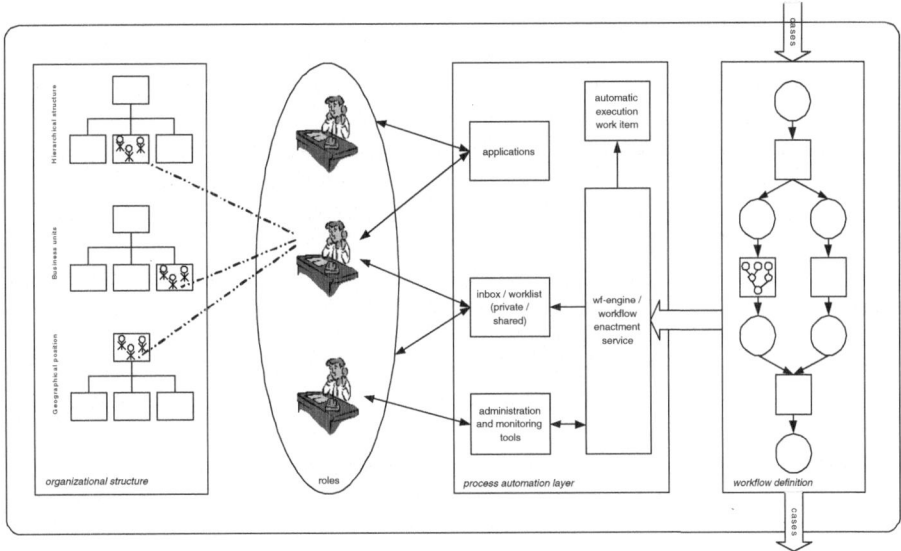

Fig. 1. Scope of measures

The presented abstract workflow model has been used to settle the scope and as a reference model to generate ideas on tuning measures, focusing on each of the separate layers as well as on their interactions. A wide variety of sources has been used for this purpose, e.g. [12,13,14,15,34]. The process of generating and evaluating these measures will be described in more detail in Section 3. While this process and the consultation of literature sources have rendered ideas for all parts of the abstract workflow model, two sources have been particularly fruitful to establish ideas that seemed both effective and easy to implement. Therefore, we will deal with these two sources in more detail in the remainder of this section.

2.1 JCM Model

In general, employees experience their work based on their perception of the work and their values[1]. Such experiences lead to job attitudes (i.e. evaluative statements or judgements concerning the work), such as job satisfaction and well-being. These attitudes affect an employee's motivation [34]. In turn, the behavior and especially the performance of an employee are influenced by his or her motivation.

Personality, well-being, job satisfaction and motivation of an employee are correlated to job performance [20,21,22]. Even though the correlation between

---

[1] In every day life these terms are not clearly defined. For the purpose of this paper we use the definitions as they are used in psychology. Perception is defined as a process by which individuals organize and interpret their sensory impressions in order to give meaning to their environment [34]. Values are defined as basic convictions that a specific mode of conduct or end-state of existence is personally or socially preferable to an opposite or converse mode of conduct or end-state of existence [34].

job satisfaction and performance is not always obvious [17,20,35], it is generally believed that a users' satisfaction and motivation is a very important part in making the implementation of automation systems a success [35].

In their overview of job design theory Holman et al. [16] show that, although many theories on job design exist, up till now little research has been done on implementing and applying job design theory in concrete automation systems. We believe a human-oriented design of the technical system indeed can contribute to the success of information systems, particularly by improving an employee's experience of the work he or she performs. Therefore we consider the important dimensions on which a job can be assessed in order to determine the degree to which a job is pleasant to the performer.

Based on theory of human needs (for instance [27]), Hackman and Oldham developed the Job Characteristics Model (JCM) [9,10]. Today this model is known as the dominant framework for defining task characteristics and understanding their relationship to employee motivation, performance and satisfaction. According to this theory a job can be characterized in terms of five core job dimensions [11,34]:

- **Skill variety** - the degree to which the job requires a variety of different activities so the worker can use a number of different skills and talent.
- **Task identity** - the degree to which the job requires completion of a whole and identifiable piece of work.
- **Task significance** - the degree to which the job has a substantial impact on the lives or work of other people.
- **Autonomy** - the degree to which the job provides substantial freedom, independence, and discretion to the individual in scheduling the work and in determining the procedures to be used in carrying it out.
- **Feedback** - the degree to which carrying out the work activities required by the job results in the individual obtaining direct and clear information about the effectiveness of his or her performance.

The higher a job scores on each of these characteristics, the better it is and the higher the motivation, performance and satisfaction of the person executing this job will be [34]. Therefore, it makes sense to design and improve working environments considering the impact on these characteristics.

The JCM-model has proven its validity, because it is used in many kinds of research on the quality of jobs (see for example [30]). Moreover, [8] shows that there is quite a strong correlation between the job characteristics and job satisfaction and a small correlation between the job characteristics and performance.

2.2   Assignment and Synchronization Policies

While the JCM model has been valuable to assess the effect of generated tuning measures, another model turned out to be the most fruitful source for tuning measures that seemed both effective and easy to implement. This model describes the policies that WFMS's generally use for distributing and assigning work to performers, as published by zur Muehlen in 2004. It is part of a paper on organizational management in workflow applications [28], as modified from the research of Hoffman et al [15]. The paper in question describes the assignment and synchronization policies in a very detailed way, listing a number of properties that can be used to characterize the various policies and their range of values. We will refer to this model as "AS-policies".

The AS-policies model consists of two parts. The first part, assignment policies, deals with the distribution of work to a shared work list, accessible by qualified employees. The second part, synchronization policies, explains how a work item that is placed on such a shared worklist, can be accessed by individual workflow participants (according to [28]). Although the author does not claim universal applicability of the AS-policies, we believe these policies are recognizable and general for most WfMS's. The AS-policies only denote possible configuration changes, but do not elaborate on the way to implement them.

In Figure 2 and Figure 3, the policies are depicted. For clarity, we will shortly discuss three policies (one assignment policy and two synchronization policies) and refer the reader to the original paper [28] for further information.

Property	Possible Values		
Planning of new work items	Net change	Re-planning	
Time of notification	Upon availability	Between availability and latest start time	At latest start time
Queuing of new work items	Queue	Pool	Combination
Activity execution	Individual	Collaborative	
Decison hierarchy	Final Assignment	Delegation possible	

Fig. 2. Assignment policies (from [28])

Property	Possible Values					
Coordination	Hierarchy		Group negotiation		Schedule	
	System	Manager	Market	Auction	FCFS	Other
Allocation mechanism	Fully automated		Partially automated		Manual	
Participant selection	Direct		Indirect			
	w/o substitution	w/ substitution	Role	Or. Pos.	Org. Unit	Other
Assignment specification	Static			Dynamic		
Assignment of work items	Push		Pull		Combination	
Participant autonomy	Rejection of assignment possible			Assignment is final		

Fig. 3. Synchronization policies (from [28])

A policy can be seen as an axis on which a certain variable can be varied. First, consider the *"planning of new work items"* policy from Figure 2. This variable can be set to a net change strategy or a re-planning strategy. In a net change strategy, the workflow system assigns available work items to certain people and places them in their worklists. The work items stay there until they are performed. However, if a re-planning strategy is implemented in the system the work items are assigned to worklists, but when they have not been picked up by the performer they can be recalled and together with the newly available work items they are re-distributed among the employees and their worklists. This might mean that a work item then is assigned to another employee.

The *"assignment of work items"* in Figure 3 describes the way in which work items are offered to an employee. When a push-mechanism is used the system determines who is going to work on what work item at what time. When a pull-manner is used the employees can decide themselves when they are going to work on which of the available work items.

Finally, *"participant autonomy"* describes to what degree the assignment of a work item is final. When rejection of assignment is possible an employee can reject to perform a work item that is assigned to him or her. When the assignment is final the employee has no choice and has to execute the work item.

The differences between the two parts of the AS-policies model are subtle and policy implementation decisions may affect each other. For example, when a work item is directly pushed from a shared worklist to a certain employee (this is a specific synchronization policy), then this precludes the effective use of the assignment policy of "pooling" where resources can choose freely between work items. Nonetheless, the AS-policies model can be seen as a fairly complete overview of configuration policies in WfMS's. Although the original paper [28] does not give much justification about its correctness and generalizability, it has been very useful as a steppingstone to think about a human-oriented configuration of WfMS's.

By showing the scope of the tuning measures on the one hand and the two theoretical models on the other, we have shown our framework for developing easy-to-implement, human-oriented "tuning" measures for WfMS's. In the next section, the development of these measures is elaborated.

3   Tuning Measures

In this section, we present the most promising set of measures to "tune" a workflow system in such a way that working in the system becomes more pleasant. We will illustrate how these ideas have been generated and clarify the validation process.

3.1   Method Description: Idea Generation, Selection, Validation

The generation of ideas to "tune" a workflow system in a human-oriented way has been a creative process. By mainly considering the various options to implement the AS-policies, we tried to identify those options that would affect people's job characteristics most positively. In addition to the AS-policies, we occasionally used additional sources for idea generation, as will become clear in this section. To illustrate the process of generating "tuning" measures, we will give two examples. In the explanation of the re-planning strategy in planning new work items, zur Muehlen [28] states:

"... while a re-planning strategy would re-allocate all work items that have not yet been started, possibly removing work items from some performer's worklists and placing them on other worklists".

We think an employee will not like the fact that work that is allocated to him/her suddenly is removed or changed. The re-planning may decrease the experienced level of self-determination or control with employees. Therefore, one of the ideas we generated is: "Do not re-plan work items by workflow enactment service".

As a second illustration, we adopted the idea from Hoffmann and Loser [14] that "meaningful decisions in the process should be made by the employees, even if they could be performed by the workflow system". In their paper they describe a case study at a transport company where parcels are received, checked and distributed. In this process two scenarios are considered: a workflow process design in which decisions about the received parcels are made by the system and the same process in which the decisions are made by employees. It turned out the latter situation performed best.

In this way 32 ideas for tuning measures have been generated. In Appendix A the complete list can be found. Next, we have critically assessed these ideas on ease of implementation. We defined ease of implementation as an intervention that does not take too much time nor too much money to be realized. 21 ideas survived this assessment and 11 were eliminated (see Appendix A). In particular, the ideas that caused changes in the abstract workflow system's layers of organizational structure or workflow definition were eliminated (see Figure 1). For example, the measure "Do not over-specify the content of an activity" is one of the eliminated ideas. This measure requires a different way to define and enact a workflow process with a workflow system. These kind of decisions are usually made during the design phase of a workflow implementation. When they need to be changed for a running system, probably the system has to be shut down and implemented from the start again. Typically, change projects affecting the organization or workflow definition are costly and time-consuming.

The remaining ideas from the assessment are particularly located in the process automation layer of the abstract workflow system model (the workflow engine, worklist and administration and monitoring tools), and its interfaces to the employees. These are mainly the parts that are described by the AS-policies.

In the next step, the 21 remaining ideas have been validated by a qualitative expert validation. The goal of this validation was to gain qualitative feedback on the proposed measures. Six experts (with diverse backgrounds, from both psychology and IT, practice and research) were willing to give their view on the ideas during an interview. All interviews were taken in May and June 2004. We used a face-to-face interview approach, providing the possibility to give more explanation where needed, except for one interview: respondent R5 answered the questions by e-mail.

The six experts can be divided into two categories: three of them are researchers (two females, one male) and the other three are people with practical experience in the area of workflow management. Respondent 1 (R1) and respondent 4 (R4) are workflow designers with a Dutch consultancy firm and a Dutch bank, respectively. R6 is a workflow project manager with a Dutch insurance and banking company. R2, R3 and R5 are researchers at respectively Delft University of Technology (area of information systems/business processes), Eindhoven University of Technology (organizational behavior) and Stevens Institute of Technology (process automation and workflow management). The background of the experts is summarized in Table 1. We have to note that R2 also has a lot of practical experience in workflow projects and that R3 is an expert on work psychology and organizational behavior. Her expertise in technical systems is less, which made it not very feasible to answer the question of ease of implementation for each idea.

During the interview the respondents were asked to indicate if they thought the proposed measure would have a positive impact on the employee, if it would be easy to implement and if they could rank a list of the top five ideas from the twenty-one presented ideas. The outcomes of the interviews can be found in Appendix B. Based on these expert rankings we selected the six most promising measures for further research.

Table 1. Background of interview respondents

Respondent	Gender	Category	Function	Company / Institute
R1	Male	Practice	Workflow designer	Dutch consultancy firm
R2	Female	Research	Researcher	Delft University of Technology
R3	Female	Research	Researcher	Eindhoven University of Technology
R4	Male	Practice	Workflow designer	Dutch bank
R5	Male	Research	Researcher	Stevens Institute of Technology
R6	Male	Practice	Workflow project manager	Dutch insurance & banking company

3.2 Six "Tuning" Measures and Their Aimed Effect on JCM Characteristics

Below we will shortly describe the most promising measures, selected by the experts, and we will explain their expected impact on the job characteristics from the JCM-model in Section 2. Primarily, these measures provide workers with more autonomy or, in other words, with more self-determination, while the WfMS stays in control of process coordination.

SH PULL - **Use a shared worklist, from which an employee can choose himself: pull-manner**. The first measure gives the worker more autonomy. By using a pull mechanism (in stead of a push-mechanism) the employee can decide for himself or herself when he or she starts which work item. The execution of work is not forced by the system and thus the employee has more freedom. Through this freedom the worker can also ensure that the work he or she is doing is alternating, which may positively affect skill variety.

TARGET - **Show an employee if he or she works hard enough, if he or she is satisfying the targets**. This measure is improving feedback to employees. In many settings, performers have to satisfy targets with respect to the amount of cases they have to process every hour or every day. It is good for an employee to know if he or she meets the requirements that are asked. This information should of course be private.

RESUB - When a work item has to be performed again after a (negative result of a) check, return it to the same employee to execute it again. The aim of this measure also is to improve feedback. Often the execution of important steps in a process is checked by, for instance, a supervisor. In such a case, the supervisor determines whether the step has been performed properly. If that is not the case, the step has to be redone. When an employee has made a mistake or error in executing an activity for a certain case, it can be very valuable to know what went wrong and why it went wrong. Therefore, the case should be sent back to the same employee that made the mistake, so he or she can learn from it.

TEAMBAT - Create 'team batches' of work items. A team of employees (having the same competences/role) can divide the work according to their own preferences. (Here we assume the allocation mechanism is manual, but is not necessarily controlled by a team leader or manager.) By creating "team batches", employees will experience more autonomy, skill variety and task significance. In "team batches", the work that is assigned to the team still has to be divided amongst the members of the team. By negotiating and discussing who should do what, employees can have more influence on the work they are supposed to perform and they can experience more task significance.

APPEAR - Give employees the opportunity to adjust the ordering of work items in their worklists to their own preferences: FIFO, earliest due date, random, etc. (Here we assume the assignment of work items is in a pull manner and the worklist is private.) This measure provides an employee with more autonomy. When there is a possibility to adjust the appearance of work items in the worklist the employee can create a better overview of the things he or she has to do according to his or her own preferences. This makes it easier to decide for oneself which work item should be performed next.

CASEMAN - Case management: let an employee work on the same case as much as possible. Finally, case management improves the task identity and task significance for employees. When employees work as much as possible on the same case they know the ins and outs of the case, they will get more involved with the customer's interests and they will feel more helpful.

The final step we conducted in this research is an evaluation of the measures in terms of current workflow technology. This evaluation will be described in section 4.

4   Evaluation

To test these theoretical ideas to tune a workflow system, we have evaluated the most promising measures using three commercially available WfMS's: Staffware

(Tibco), COSA (Transflow) and FLOWer (Pallas Athena). Based on the documentation of these systems [18,26,31,36] we have identified to what degree the six measures are supported by these systems. The goal of this validation is to determine whether the ideas are practically feasible with current workflow technology. In the first sub section, we will explain why we selected these systems. Next, we will present the results of this evaluation and this section will be concluded with a short discussion.

4.1 Selection of Systems

An important criterion for selecting the three WfMS's we mentioned in the introduction is their popularity. At the moment, Staffware and COSA both have a substantial market share in Europe. They are suited for production workflow, i.e. handling a large number of cases that all have to be processed in a similar, structured way. Furthermore, we felt FLOWer is an interesting system because of its rapid growth in popularity and its case handling paradigm [33], which provides more flexibility in the system. The differences between those systems with respect to the measures are discussed in the next section.

Table 2. Summary of the results of implementability of the six best ideas

Acronym	Staffware 9.0	FLOWer 3.0	COSA 4.2
SH PULL	+	-	+
TARGET	-	-	+/-
RESUB	-	+	+
TEAM BAT	-	-	-
APPEAR	+	-	+
CASEMAN	+/-	+/-	+/-

4.2 Outcome Evaluation

Based on documentation of the systems we identified whether an idea could be implemented or configured in the system. First we will give an overview of the evaluation results and next we will shortly discuss the outcome. Table 1 shows to which degree the actual WfMS is able to support a measure. This degree is expressed by the following symbols:

+   The idea can be directly supported by the WfMS itself.
+/-   The idea can be partly supported by the WfMS, some small adaptations to the WfMS have to be made or some "add-on's" have to be installed.
-      The idea can not be supported by the WfMS, or the underlying concept of the WfMS makes the facilitation of the idea not possible.

As turns out from Table 2, not all the ideas can be implemented or supported (yet) by the three contemporary WfMS's we considered. For instance, it is not possible in Staffware to send a work item back to the employee who previously worked on the case (cf. measure RESUB). This is a dynamic way to assign work items, while Staffware only provides a static way to assign work items, i.e. up front at design time it can be stated to which person (by this person's name) a work item has to be sent. In contrast, FLOWer and COSA support the resubmission measure.

Overall, we can conclude that COSA provides the best support for realizing these ideas. A remarkable result is the difference between the two production workflow systems, COSA and Staffware. Although they are based on the same concept, they do not have the same support for the measures.

Moreover, FLOWer - as the newer type of workflow management system - seems to provide less support. This is due to the difference in concept. The case handling paradigm [33] already provides a lot of flexibility and autonomy to users, but this is done in a way that makes some of the tuning measures impossible to realize. For instance, FLOWer provides a "Case Query" mechanism through which employees can search for cases that are available. This is rather similar to the idea of a shared worklist ("pool") with a pull-mechanism, but it shows cases in stead of work items. Therefore, an employee can not see what kind of work (which step in the process) has to be performed for a particular case. We considered the "Case Query" as different from a work items worklist. Thus, FLOWer can not support measure [SH PULL] because there is no worklist of work items. In this case FLOWer gets a somewhat negative result, but we should note that it may actually provide more flexibility and autonomy than we initially aimed for through the measures.

5   Conclusion

Although workflow systems are very successful in companies lately, many critiques are raised too. Especially when it concerns workers and their experience of the workflow system, the views are not necessarily positive. The schism around WfMS's is in our eyes accurately captured as follows [5]:

> "On the one hand, they are perhaps the most successful form of groupware technology in current use; but on the other, they have been subject to sustained and cogent critiques, particularly from perspective of the analysis of everyday working activities."

In this paper, we have looked for practical ways to make these type of systems more agreeable for those who have to use them in their everyday work. We have taken an approach that is uncharacteristic for much of the active workflow research, where substantial attention is devoted to make WfMS's more flexible (for an overview of the various approaches, see [23]). The driving idea is that the rigidity of WfMS's makes them unsuitable to deal with exceptional situations, in this way frustrating end users. However, this research direction is perhaps not the most effective way to go. Firstly, despite the broadness of the flexibility research, few research results make their way to commercially available WfMS's. This raises the question whether this type of research, aside from being intellectually satisfying, addresses organizational needs. Secondly, current research seems to indicate that the perceived usefulness of WfMS's by end users is not primarily determined by the flexibility it provides. On the basis of various case studies of workflow implementations and an extensive survey among end users, Poelmans [32] concludes that the provision of flexible features will likely not rule out the necessity of appropriating a WfMS in more thorough ways. A tentative conclusion from his research is that not the selection of the right WfMS, but the way it is configured and implemented is crucial in the success of a workflow implementation:

> "The most important factor is giving the end-users sufficient influence, after implementation, to have the system appropriated to their needs." (p.160)

This attention for reconfiguration possibilities is in line with earlier insights into the successfulness of IT technologies in general (see e.g. [29]).

The measures we have proposed in this paper are simple ways to reconfigure existing WfMS implementations to address the needs of end users. All of them are thought to positively affect the factors that make work enjoyable and satisfactory. The measures' validation by experts from both research and practice, IT and psychology, adds credibility to their usefulness and feasibility. Taking a general model of workflow policies as starting point, a wide applicability of the measures among WfMS's was aimed for. From the limited system evaluation we carried out, we can conclude that the specific brand of WfMS determines the ease of actually implementing a measure.

This also identifies the opportunities for further research. It would be valuable to broaden the scope of systems we considered to provide insight which measures can be used in what situations. Also, a more thorough evaluation of the various WfMS's will give a better insight into their reconfiguration capabilities. Additionally, it seems worthwhile to perform an actual validation in practice, i.e. an experiment with real workflow users in a realistic setting, to see if the measures really improve an employee's experience and satisfaction. Two recommendable designs to execute such a field study are an untreated control group design with pretest and posttest, and a nonequivalent dependent variables design [4]. A closer study of the other generated ideas may provide organizations with broader means to improve the efficiency of their existing operations with an eye for the human perspective.

To conclude, we have achieved our aims with this paper if it manages to inspire researchers and practitioners to look for those simple reconfiguration options that make working with a WfMS more enjoyable. In doing so, the organizational benefits of WfMS's can be exploited to their full potential.

References

1. J. Bowers, G. Button, and W. Sharrock. Workflow From Within and Without: Technology and Cooperative Work on the Print Industry Shopfloor. In: Proceedings of the Fourth European Conference on Computer-Supported Cooperative Work, 51-66, 1995.

2. C.T. Caine, T.W. Lauer, and E. Peacock. The T1-Auto Inc. production part testing (PPT) process: A workflow automation success story. Annals of Cases on Information Technology ,.5: 74-87, 2003.

3. J.L. Caro, A. Guevara, and A. Aguayo. Workflow: A solution for cooperative information system development. Business Process Management Journal. 9(2): 208-220, 2003.

4. T.D. Cook, D.T. Campbell. Quasi-experimentation: design and analysis issues for field settings. Rand McNally, 1979.

5. P. Dourish. Process descriptions as organizational accounting devices: the dual use of workflow technologies. In: C.A. Ellis and I. Zigurs (Eds.), Proceedings of the ACM 2001 Int. Conference on Supporting Group Work (pp. 52-60). New York, ACM Press, 2001.

6. C.A. Ellis. Information control nets: a mathematical model of office information flow. In: P.F. Roth and G.J. Nutt, Proceedings of the ACM Conference on Simulation, Measurement and Modeling of Computer Systems (pp. 225-240). New York: ACM Press, 1979

7. C.A. Ellis and J. Wainer. Goal-based Models of Collaboration. Collaborative Computing 1, 61-86, 1994.

8. Y. Fried. Meta-Analytical Comparison of the Job Diagnostic Survey and Job Chracteristcs Inventory as Correlates of Work Satisfaction and Performance. Journal of Applied Psychology, 76 (5), pages 690-697, 1991.

9. J.R. Hackman, G.R. Oldham. Development of the Job Diagnostic Survey. Journal of Applied Psychology, 60, pp. 159-170, 1975.

10. J.R. Hackman, G.R. Oldham. Motivation through the design of work: test of a theory. Organizational Behavior and Human Performance, 15, pp. 250-279, 1976.

11. J.R. Hackman, J.L. Suttle. Improving Life at Work: Behavioral Science Approaches to Organizational Change. Goodyear Publishing, 1977.

12. T. Herrmann, M. Hoffmann. Augmenting Self-Controlled Work Allocation in Workflow-Managmenet-Applications. Proceedings of HCI '99, pp. 288-292, 1999.

13. T. Herrmann. Evolving Workflows by User-driven Coordination. In: R. Reichwald, J. Schlichter (eds.): Verteiltes Arbeiten - Arbeiten der Zukunft. Tagingsband D-CSCW, pp. 103-114, 2000.

14. M. Hoffmann, K.-U. Loser. Mitarbeiter-orientierte Modellierung und Planung von Geschäftsprozessen bie der Einführung von Workflow-Management. Proceedings of EMISA-Fachgruppentreffens 1997, pp. 39-57. (In German)

15. M. Hoffmann, T. Löffeler, Y. Schmidt. Flexible Arbeitsverteilung mit Workflow-Management-Systemen. In: T. Herrmann, A.-W. Scheer and H. Weber, eds. Verbesserung von Geschäftsprozessen mit flexiblen Workflow-Management-Systemen, Physica, Heidelberg, Germany, 1999, pp. 135-159. (In German).

16. D. Holman, C. Clegg, P. Waterson. Navigating the territory of job design. Applied Ergonomics, 33. pp. 197-205, 2002.

17. M.T. Iaffaldano, P.M. Muchinsky. Job Satisfaction and Job Performance: A Meta-Analysis. Psychological Bulletin, 97(2), pp. 251-273, 1985.

18. IDS Scheer. ARIS Process Performance Manager (ARIS PPM). Presentation and whitepaper, http://www.ids-scheer.de/PPM/ , 2004.
19. S. Jablonski and C. Bussler. Workflow Management: Modeling Concepts, Architecture, and Implementation. Int. Thomson Computer Press, London, 1996.
20. T.A. Judge, C.J. Thoresen, J.E. Bono, G.K. Patton. The Job Satisfaction - Job Performance Relationship: A Qualitative and Quantitative Review. Psychological Bulletin, 127 (3), pages 376-407, 2001.
21. T.A. Judge, D. Heller, M.K. Mount. Five-Factor Model of Personality and Job Satisfaction: A Meta-Analysis. Journal of Applied Psychology, 87 (3), pages 530-541, 2002.
22. T.A. Judge, R. Ilies. Relationship of Personality to Performance Motivation.: A Meta-Analytical Review. Journal of Applied Psychology, 87 (4), pages 797-807, 2002.
23. M. Klein, C. Dellarocas, and A. Bernstein. Introduction to the Special Issue on Adaptive Workflow Systems. Computer Supported Cooperative Work, 9(3 4): 265-267, 2000.
24. P. Küng. The Effects of Workflow Systems on Organizations: A Qualitative Study. In: Wil M. P. van der Aalst, Jörg Desel, Andreas Oberweis (Eds.): Business Process Management, Models, Techniques, and Empirical Studies. Lecture Notes in Computer Science 1806 Springer 2000, p. 301-316.
25. P. Küng and C. Hagen. Increased performance through business process management: an experience report from a swiss bank. In: Neely, A. et al. (Eds.): Performance Measurement and Management - Public and Private, Cranfield, 1-8, 2004.
26. Ley GmbH. COSA 4, User's Guide, 2002 and Business-Process Designer's Guide, 2003.
27. A.H. Maslow. A theory of human motivation. Psychological review, 50, 1943, 370-96.
28. M. zur Muehlen. Organizational Management in Workflow Applications - Issues and Perspectives. Information Technology and Management Journal 5(3), pp. 271-291, 2004.
29. W. Orlikowski. The Duality of Technology: Rethinking the Concept of Technology in Organizations. Organization Science, 3(3): 398-427, 1992.
30. R.J. van Ouwerkerk, T.F. Meijman, G. Mulder. Industrial Psychological Task Analysis. Lemma, Utrecht, 1994. (In Dutch).
31. Pallas Athena. Administrator Guide, Designer's Guide, User Guide FLOWer 3.0. Pallas Athena, the Netherlands, 2004
32. S. Poelmans. Making Workflow Systems Work: An Investigation into the Importance of Task-appropriation Fit, End-user Support and other Technological Characteristics. Ph.D. thesis. Doctoral dissertation series Faculty of Economic and Applied Economic Sciences nr 161., Katholieke Universiteit Leuven, 2002.
33. H.A. Reijers, J.H.M. Rigter, and W.M.P. van der Aalst. The Case Handling Case. Int. Journal of Cooperative Information Systems, 12(3): 365-391, 2003.
34. S.P. Robbins. Organizational behavior. Prentice Hall, New Jersey, 2001.
35. F.E. Saal, P.A. Knight. Industrial/Organizational Psychology. Brooks/Cole Publishing Company, California, 1995.
36. Staffware. Staffware Process Suite, Using the Staffware Process Client, Issue 2, and Defining Staffware Procedures, Issue 2, 2002.
37. Workflow Management Coalition. WFMC Home Page: http://www.wfmc.org. The Workflow Reference Model (WFMC-TC-1003). 1995.
38. Wintergreen. Business process management (BPM) market opportunities, strategies, and forecasts, 2003 to 2008. Lexington: WinterGreen Research, 2003.

A   List of Tuning Measures

A list of all generated human-oriented tuning measures, including the eliminated ideas, can be found on http://is.tm.tue.nl/staff/ivanderfeesten/tuningmeasures.htm.

B   Results of Expert Validation

Table 3. Results of expert validation

	R1	R2	R3	R4	R5	R6	Total number of *		Number of selections
1	****			***			*******	(7)	2
2			*	****			*****	(5)	2
3		*	*				**	(2)	2
4								(0)	0
5			****				****	(4)	1
6		**	**		*	***	********	(8)	4
7	**						**	(2)	2
8	***	***	*****			*****	****************	(16)	4
9								(0)	0
10								(0)	0
11								(0)	0
12				*****	**		*******	(7)	2
13			**	****	*		*******	(7)	3
14	*****			*			******	(6)	2
15				***			***	(3)	1
16				***			***	(3)	1
17		****					****	(4)	2
18	*				**		***	(3)	2
19				*****			*****	(5)	1
20								(0)	0
21		*****	***		****		***********	(12)	3

The Price of Coordination in Resource Management

Kees van Hee, Alexander Serebrenik, Natalia Sidorova,
Marc Voorhoeve, and Jan van der Wal

Department of Mathematics and Computer Science,
Eindhoven University of Technology,
P.O. Box 513, 5600 MB Eindhoven, The Netherlands
{K.M.v.Hee, A.Serebrenik, N.Sidorova, M.Voorhoeve, Jan.v.d.Wal}@tue.nl

Abstract. We propose a resource management policy that grants or refuses requests for resources based only on the request made and the number of free resources. Computations at runtime are independent of the number of active cases. The policy requires little coordination and is therefore easy to implement in workflow management systems. This policy has been shown to be successful in avoiding deadlocks. In this paper we investigate its performance characteristics.

1   Introduction

Workflow nets [11,12,13,14], a special class of Petri nets, are frequently used to model business processes. In business processes, three elements are essential: *cases* to be processed, *tasks* to be performed on the cases and *resources* needed to perform these tasks. Typical examples of such resources are money, machinery and manpower. Traditionally, models of workflow nets emphasise the partial ordering of activities (i.e. executing a task for a case) in the process and abstract from the resources needed for them. The resources, however, cannot be ignored in many practical applications [1,3,5,6,10]. Resource-constrained workflow nets have been introduced in [15] for resources that are *durable* instead of *consumable*. Bad resource management may cause deadlocks, even if the workflow net is well-designed, i.e. the workflow net is *sound* (cf. [14]).

Assessment of business process models involves *correctness* and *efficiency*. Correctness requirements include *proper termination*, i.e. that given some minimal initial number of resources, in each reachable state of the business process it is possible to release all claimed resources and complete all open cases. Efficiency criteria are *quality of service* (e.g. the time between the arrival and the completion of cases) and *costs of operation* (e.g. the number of resource-hours needed).

We consider business processes where an arbitrary number of cases is handled and the resources belong to a single class. Cases are independent, i.e. they only communicate with the resource manager by claiming and releasing resources in various quantities. A resource manager, a human being or a software component in a workflow management system, may either grant or refuse these resource claims. In principle the resource manager may base its decisions on the *global state* of the process, i.e. the number of cases, the state of each case and the number of available resources. The first task of resource management is to ensure correctness of the resulting business process.

W.M.P. van der Aalst et al. (Eds.): BPM 2005, LNCS 3649, pp. 96–108, 2005.

This involves *scheduling*. Note that we are not dealing with a *static* scheduling problem in which all cases to be handled are known. Instead we are dealing with a *dynamic* scheduling problem, in which new cases arrive according to a random process and the routing and resource consumption of cases are largely unknown to the resource manager. The banker's algorithm of E.W. Dijkstra [4] ensures correctness in this way. This algorithm considers for each case only the maximal number of resources needed by it (credit limit) and the number of resources claimed and not yet returned so far (debt), plus the number of available resources.

The original paper [4] does not consider efficiency, but it leaves room for prioritising between cases that need resources, when they become available. Such a choice can be based on heuristics like FIFO (first in first out), SPT (shortest rest processing time) or EDD (earliest due date).

In [15], a property for cases is defined that we call *solidity*. When all cases are solid, the business process is guaranteed to terminate properly. Cases that are not solid can be *solidified* by setting a *threshold* for the number of resources that need to be available for each claim to be successful. In practise, solidification amounts to the following policy: *before committing resources to a case, make sure that there are enough resources available to allow its completion independently of other cases.* This is akin to a well known approach in production control [2]. The thresholds can be chosen in advance, when the business process is defined.

In this paper, we investigate resource scheduling based on solidification. Resources are granted to a task when the number of available resources exceeds a given threshold. To determine thresholds that perform well, an iterative, simulation-based approach is proposed. We illustrate our approach with a small example inspired by the building industry.

The sketched approach leads to a robust "uncoordinated" resource management. Note that the computation time needed at runtime by the resource manager is independent of the number of active cases. It is interesting to compare the performance of these robust resource managers w.r.t. more sophisticated ones, thus investigating the price of coordination (cost of robustness). For a very small (tandem queue) example, it is possible to compute an optimal global scheduler by Markov decision theory [9]. We compare the performance of the robust and optimal approaches.

The remainder of the paper is organised as follows. We describe our motivating example in Section 2. Basic notions from Petri net theory are defined in Section 3. Correctness and solidification are introduced in Section 4. Then, efficiency criteria are proposed and assessed by means of two examples in Section 5. Finally, we discuss the results presented.

2 Motivating Example

The business processes investigated are modelled as workflow nets. Our example model describes the business process of a building contractor, who constructs buildings of a similar nature in large numbers for various clients. Each building under construction represents a case. Each case is divided into tasks, that are performed by subcontractors and require a certain amount of time to complete. The resource considered is money; the contractor has an account at his bank with a fixed credit limit. At the onset of a task, an amount of money has to be paid to the subcontractor. At termination of a task, the client pays some amount of money (not necessarily the same as paid to the subcontractor). All payments pass through the contractor's bank account. So each construction task can be characterised by three parameters: payment to a subcontractor, duration and payment from the client. The net model for each task is shown in Figure 1. Horizontally, the control flow is depicted; the remaining edges describe the resource flow. Initially money is transferred to the account of the subcontractor and money is received at the end.

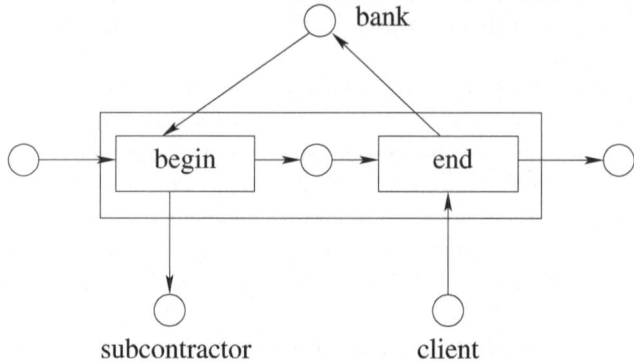

Fig. 1. A typical construction task

Construction starts by paying a subcontractor 2 mu (units of money, e.g. 20.000 euro) for the groundwork. This process takes twenty days and upon completion the client pays 1 mu. The next task is known as framing and includes building walls, windows and a roof. For this stage 4 mu is required by a subcontractor, and after thirty days the work is finished and 2 mu is paid by the client. Next, additional commissions (add-ons) of ten days long may be requested by the client. Each add-on requires 1 mu as initial payment to the subcontractor. The same amount is charged from the client when the task is finished. The independence of add-ons is modelled as a loop with 45% exit chance. Then, the internals of the building are installed (plumbing, heating, electricity). This task takes thirty days, requiring an initial sum of 1 mu for the subcontractor. This sum is payed by the client at termination of this task. Finally, when the construction is approved by the customer, she pays back the remaining 3 mu. The workflow net corresponding to this process description is presented in Figure 2. When considering correctness, we treat the tasks as transitions. However, when treating performance, the indicated timing delays must be observed and tasks are treated as subnets defined by Figure 1. For the sake of simplicity we abstract from the communication with subcontractors and clients, including money transfers.

The process described can deadlock and thus is not guaranteed to terminate properly. Suppose the credit limit equals 16 mu and the groundwork is started for 8 clients on days 1 to 8. On days 9 to 15, applications of 7 more clients are received, but these cases cannot start due to a lack of resources. On day 22, the groundwork for the first

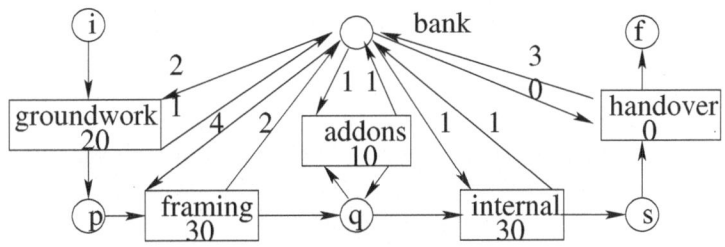

Fig. 2. Workflow of a building construction

two clients has terminated, 2 mu is available and the groundwork for client 9 is started. By immediately starting groundwork for new clients as soon a 2 mu becomes available rather than waiting for 4 mu and starting framing activities, we will arrive at a state where only 1 mu is available and 15 clients are in state p, which is a deadlock state. Such a deadlock can be achieved for larger credit limits as well.

As explained in the introduction, several deadlock avoidance policies can be implemented. Dijkstra's scheduling algorithm [4] will prevent groundwork for the 13-th client to start in the above scenario. This will allow framing to start for at least one client, eventually freeing the resources invested. Deadlock is likewise avoided by solidifying the process, setting a threshold of 3 mu for the groundwork task. Groundwork is started for a new client, investing 2 mu, only if 5 mu (the required 2 plus the threshold amount) are available.

However, deadlock avoidance alone does not guarantee a good performance. Assume that all cases have one add-on option. With the threshold 3 and 16 mu, it is possible to reach a global state where 12 cases are waiting in state p and for one case framing has started. After 70 days, 5 mu becomes available, which allows either to start groundwork for a new case or to start framing for a waiting case. If the groundwork option is chosen, 3 mu are left, so framing has to wait 20 more days, upon which framing can start for one case and we are back in the initial state. Thus, only one case in 90 days is completed. It is not difficult to find a different resource management scenario allowing the completion of 3 cases every 70 days. The reason for the bad performance of the sketched scenario is that the number (13) of active cases is too high. By increasing the minimum threshold of 3 for starting the groundwork of a new case, the number of active cases can be reduced.

3   Preliminaries

We adopt standard notation for sets, bags and transition systems. A *Petri net* is a tuple $N = \langle P, T, F^+, F^- \rangle$, where:

- P and T are two disjoint non-empty finite sets of *places* and *transitions* respectively, the set $P \sqcup T$ are the *nodes* of N;
- F^+ and F^- are mappings $(P \times T) \to \mathbb{N}$ that are *flow functions* from transitions to places and from places to transitions respectively.

We present nets with the usual graphical notation.

Markings are states (configurations) of a net. We denote the set of all markings reachable in net N from marking m as $\mathcal{R}(N,m)$. We will drop N and write $\mathcal{R}(m)$ when no ambiguities can arise. Given a transition $t \in T$, the *preset* $^\bullet t$ and the *postset* $t^\bullet$ of t are the *bags* of places where every $p \in P$ occurs $F^-(p,t)$ times in $^\bullet t$ and $F^+(p,t)$ times in $t^\bullet$. Analogously we write $^\bullet p, p^\bullet$ for pre- and postsets of places. We will say that a node n is a *source* node if and only if $^\bullet n = \emptyset$ and n is a *sink* node if and only if $n^\bullet = \emptyset$.

A transition $t \in T$ is *enabled* in marking m if and only if $^\bullet t \leq m$. An enabled transition t may *fire*. This results in a new marking m' defined by $m' \stackrel{\text{def}}{=} m - {}^\bullet t + t^\bullet$. We interpret a Petri net N as a transition system/process where markings play the role of states and firings of the enabled transitions define the transition relation, namely $m + {}^\bullet t \stackrel{t}{\longrightarrow} m + t^\bullet$. The notion of reachability for Petri nets is inherited from the transition systems. A net $N = \langle P, T, F^+, F^- \rangle$ is called a *state machine* if $^\bullet t$ and $t^\bullet$ are singleton bags for all $t \in T$.

Given a Petri net, a *place invariant* is a row vector $I : P \to \mathbb{Q}$ such that $I \cdot F = 0$.

In this paper we primarily focus upon the *Workflow Petri nets (WF-nets)* [11]. As the name suggests, WF-nets are used to model the processing of tasks in workflow processes. The initial and final nodes indicate respectively the initial and final states of processed cases.

Definition 1. *A Petri net N is a* Workflow net (WF-net) *if and only if :*

1. *N has two special places: i and f. The initial place i is a source place, i.e. $^\bullet i = \emptyset$, and the final place f is a sink place, i.e. $f^\bullet = \emptyset$.*
2. *For any node $n \in (P \cup T)$ there exists a path from i to n and a path from n to f.*

We extend the notion of WF-nets in order to include information about the use of resources into the model. A resource belongs to a type; we have one place per resource type in the net where the resources are located when they are free. We assume that resources are durable, i.e. they can neither be created nor destroyed, they are claimed during the handling procedure and then released again. By abstracting from the resource places we obtain the WF-net that we call *production net*.

Definition 2. *We will say that a WF-net $N = \langle P_p \cup P_r, T, F_p^+ \cup F_r^+, F_p^- \cup F_r^- \rangle$ with initial and final places $i, f \in P_p$ is a* Resource-Constrained Workflow net (RCWF-net) *with the set P_p of* production places *and the set P_r of* resource places *if and only if*

- $P_p \cap P_r = \emptyset$,
- F_p^+ *and* F_p^- *are mappings* $(P_p \times T) \to \mathbb{N}$,
- F_r^+ *and* F_r^- *are mappings* $(P_r \times T) \to \mathbb{N}$, *and*
- $N_p = \langle P_p, T, F_p^+, F_p^- \rangle$ *is a WF-net, which we call the* production net *of N.*

The processes that we consider can be modelled as WF nets with only one resource place, where the production net is a state machine (SM1WF-nets). In [15] it is shown that a business process modelled by an arbitrary workflow net can be converted to a state machine workflow net, provided that cases are independent.

Definition 3. *An RCWF-net $N = \langle P_p \cup P_r, T, F_p^+ \cup F_r^+, F_p^- \cup F_r^- \rangle$ is called a* state machine workflow net with one resource type (SM1WF-net) *if $P_r = \{r\}$ and the production net N_p of N is a state machine.*

Observe that the net in Figure 2 is indeed an SM1WF-net.

4   Correctness

As explained in the introduction the correctness criterion we consider is *proper termination*, also known as *soundness* in WF-nets. Proper termination is the property that every marking reachable from an initial marking can reach the corresponding final marking. Initial markings of the net have some tokens (say k) in the initial place and a number of resource tokens on each resource places. The corresponding final marking has k tokens in the final place; the resource places must contain the same number of tokens as initially. We assume that the number of resource available initially is sufficient.

Another correctness requirement that should be reflected by the definition is that resource tokens cannot be created during the processing, i.e. at any moment of time the number of available resources does not exceed the number of initially given resources. The definition of proper termination reads thus as follows:

Definition 4. *Let N be an RCWF-net.*
N is (k,r)-sound for some $k \in \mathbb{N}, r \in \mathbb{N}^{P_r}$ if and only if for all $m \in \mathcal{R}(k[i] + r)$ holds:
$m_r \leq r$ and $m \xrightarrow{*} (k[f] + r)$.
N is sound if and only if there exists $r \in \mathbb{N}^{P_r}$ such that it is (k,r')-sound for all $k, r' \in \mathbb{N}, r' \geq r$.

In [15], it is proved that for any sound SM1WF-net there exists a unique place invariant W such that $W(i) = W(f) = 0$, $W(r) = 1$ and for all $p \in P_p$, $W(p) \geq 0$. Given a place p we call $W(p)$ the *weight* of p.

Example 1. Recall the construction Petri net presented in Figure 2. Then, $W(i) = W(f)$ $= 0$, $W(p) = 1$, $W(q) = W(s) = 3$.

The discussion in Section 2 shows that the existence of a place invariant is necessary but not sufficient.

4.1   Solidity

The key ingredient in determining proper termination of SM1WF nets, called *solidity*, is the possibility that all resources claimed are eventually released. Important in the algorithm is the *path* concept. A path is a sequence of transitions such that the output state of a transition is the input state of the next transition. A path has an input and output state, resp. the input state of the first transition and the output state of the last transition. A path p is a *successor* of a path q if the input state of p equals the output state of q. If the weight of the input state of a path is less than the weight of its output state, the path is called a *consumption* path, if it is more than the weight of the output state the path is called a *production* path. Finally, we define the *resource need* of a path. This is the minimum number of resources needed for the execution of the path.

In our example net, the sequence (*framing, addons, internal*) is a path with input place p and output place s. Since p has weight 1 and s has weight 3, it is a consumption path. Its resource need is 5, since 4 free resources plus 1 resource occupied in the input place p are sufficient to fire the sequence, leading to s occupying 3 resources plus 2 free resource.

The above definition allows to formulate the necessary and sufficient condition for solidity: *Each consumption path produces enough resources to fulfil the resource need of at least one successive production path.*

Our example net does not satisfy this condition: the path $\sigma = (groundwork)$ (consisting of only one transition) is a consumption path, since its input place i has weight 0 and its output place p has weight 1. Its resource need equals 2. Any production path succeeding σ has input place p and thus must start with transition *framing* needing 4 free resources, so the resource need of such a production path is at least 5 (4 free ones and one occupied by p). So σ has no production successor with a resource need not exceeding 2.

In order to verify solidity, given an SM1WF-net with P_p production places, we introduce a matrix M, such that $M(p, p)$ is defined to be $W(p)$ for all $p \in P_p$ and $M(p, q)$ is defined as the sum of $W(q)$ and the minimal resource production of transitions from p to q. If there are no such transitions $M(p, q)$ is defined to be ω (denoting infinity). The condition above has considered paths rather than individual transitions. Therefore, we need to extend the definition of M to include paths of arbitrary length. To do so, we have introduced a binary operation $\circ$ such that for any $A, B : P_p \times P_p \to \mathbb{N}$, the product $A \circ B$ is defined as $C : P_p \times P_p \to \mathbb{N}$ where $C(p, q) = \min\{\max(A(p, r), B(r, q)) \mid r \in P_p\}$. We denote $A \circ A$ by A^2 etc. One can show that $M, M^2, M^4, \ldots$ converges to a fixpoint, which we call μ. Then, the intuitive condition for solidity stated above can be expressed as follows:

Corollary 1. *([15]) The SM1WF-net N is solid if and only if*

$$\forall x \in P_p : \min_y\{\mu(y, x) \mid W(y) < W(x)\} \geq \min_z\{\mu(x, z) \mid W(x) > W(z)\}.$$

In our running example, the following holds:

$$M = \begin{array}{c|ccccc} & i & p & q & s & f \\ \hline i & 0 & 2 & \omega & \omega & \omega \\ p & \omega & 1 & 5 & \omega & \omega \\ q & \omega & \omega & 3 & 4 & \omega \\ s & \omega & \omega & \omega & 3 & 3 \\ f & \omega & \omega & \omega & \omega & 0 \end{array} \qquad M^2 = \begin{array}{c|ccccc} & i & p & q & s & f \\ \hline i & 0 & 2 & 5 & \omega & \omega \\ p & \omega & 1 & 5 & 5 & \omega \\ q & \omega & \omega & 3 & 4 & 4 \\ s & \omega & \omega & \omega & 3 & 3 \\ f & \omega & \omega & \omega & \omega & 0 \end{array} \qquad M^4 = \begin{array}{c|ccccc} & i & p & q & s & f \\ \hline i & 0 & 2 & 5 & 5 & 5 \\ p & \omega & 1 & 5 & 5 & 5 \\ q & \omega & \omega & 3 & 4 & 4 \\ s & \omega & \omega & \omega & 3 & 3 \\ f & \omega & \omega & \omega & \omega & 0 \end{array}$$

M^4 is the fixpoint. Soundness condition is violated by p since

$$\min\{\mu(y, p) \mid W(y) < W(p)\} = 2 < 5 = \min\{\mu(p, z) \mid W(p) > W(z)\}.$$

However, it is possible to *solidify* SM1WF nets with a resource invariant. This is done by *thresholding* some of its transitions. A transition is thresholded by not firing it when the resources it needs are available *unless* some additional resources are available too. The amount of required extra resources is called the *threshold*. Thresholding replaces scheduling as by Dijkstra's algorithm ([4]; it is similar to the order acceptance strategy of ([2]). We have developed a method to find algorithmically minimal thresholds for a net to become solid. It should also be noted that any threshold exceeding the minimal one makes the net solid too.

The threshold solution proposed at the end of Section 2 has been obtained by the solidification technique.

5 Efficiency

5.1 Defining Efficiency

In order to estimate the quality of resource management, we need to define efficiency criteria. As mentioned in the introduction, we distinguish two kinds of criteria: those considering the quality of service, and those considering the cost of operation. Quality of service is focused towards minimising the throughput or cycle time of a case, i.e. the time between the arrival and the completion of the case. In our case minimising the throughput is equivalent to minimising the waiting time of a case, since the processing times of tasks are independent of the resource allocation. For a random stream of cases it is natural to consider as quality of service the average cycle time or waiting time:

$$\lim_{n \to \infty} \frac{\sum_{i=1}^{n} w(i)}{n},$$

where $w(i)$ is a waiting time of case i. An alternative for the minimising the average expected time is minimising the probability that a case has a waiting time larger than some given bound.

For the cost of operation we consider as criterion the average expected use of resources. However we need an additional condition for this criterion to eliminate the situation when the cost of operation is zero but no cases are handled. There are at least two approaches to deal with this: one can assume some minimum level of service as a boundary condition (like the average expected waiting time) or one can assign cost to resources and rewards for handling cases. Then the cost of operation is transformed into the value of the operation by subtracting the the reward of handled cases from the average cost of resource usage. We choose this last option; a negative cost of operation corresponds to making profit.

To evaluate the solidification technique, we computed an optimal scheduler described in Section 5.2, comparing it to the solidification approach. Unfortunately, optimisation is only feasible for very small examples, like the tandem queue described in the next subsection. To tackle our building example, only simulation techniques have been used (Section 5.3).

5.2 Tandem Queue Example

In order to determine what extra cost we incur in case of the solidification approach, we need to find an optimal resource allocation strategy. Such a strategy defines for each possible global state a decision for resource assignment. Finding an optimal strategy is, in general, an extremely difficult task. Therefore, we consider a *tandem queue example*.

The tandem queue example has two sequential tasks as shown on Figure 3. Like in Figure 2, numbers inside the task boxes denote the duration of the task. We assume exponential service times with mean service times equal to 1 for both tasks. [1] The arrival

---

[1] In order that the system satisfies the Markovian property, we need at least phase type service time distributions. However, since each extra phase adds an extra dimension to the state space, the number of phases has to be limited so that the Markov decision approach is still feasible. For random service times with a squared coefficient of variation larger than 0.5, two phases suffice to mimic the first two moments. Here we have chosen for exponential service times.

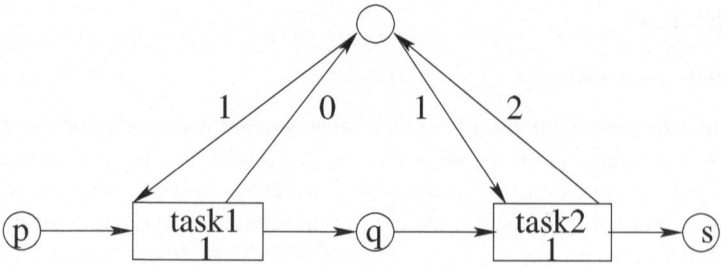

Fig. 3. Tandem queue example

process is Poisson with 4 arrivals per time unit on average. For this example the optimal strategy can be calculated by means of a Markov decision process theory [9,7], using techniques such as successive approximation.

To model this workflow system as a Markov decision process with continuous time we consider the two tasks as multi-server service stations, where the servers are the resources. The state of the Markov decision process is a quadruple $\langle q_1, t_1, q_2, t_2 \rangle$ where q_1 and q_2 are the number of waiting cases for the first and the second tasks and t_1 and t_2 are the number of cases treated in these service stations respectively. The possible actions are: admission of a newly arrived case, assigning resources to a waiting case in the queue q_1 and assigning resources to a waiting case in the queue q_2. Using the well-known uniformisation technique [7] we translate the model with continuous time into a Markovian decision process with discrete time. In order to determine the optimal strategy we need a finite state space, so we need to add some more restrictions. We assume $q_1 \leq 3$ and $q_2 \leq 3$. These assumptions imply that when a new case arrives to a queue but the queue is full, the case is lost. In order to achieve finiteness of the state space, t_1 and t_2 should be bounded as well. We set $t_1 \leq 4$ and $t_2 \leq 7$. Under these restrictions, the number of states space of the system is bounded.

The boundedness restrictions are not too severe if the loss of cases lost is penalised. High penalties imply that queue overflow is avoided as much as possible. Case loss in q_1 is punished by 5 cost units, case loss in q_2 by 10 cost units. We consider these penalties as opportunity costs.

Finally, to complete the specification of the Markov decision process, the performance measures have to be chosen. We consider two different measures, namely quality of service (QoS) and cost of operation (CoO). For the QoS measure, the waiting times at $q1$ and $q2$ plus the opportunity costs for queue overflow are added. For the CoO measure, we consider the average resource occupation of a case, i.e. the processing time of the first station plus the waiting time at $q2$ (where one resource is occupied by the case) plus twice the processing time of the second station (where two resources are occupied). As explained earlier, we subtract the reward for completed cases, which equals 7. After the subtraction, the cost of operation becomes negative for successful executions. Hence, minimising the cost of operation reflects the best possible scenario from "a case's point of view".

Using the standard successive approximation techniques (c.f. [7]), we derive an optimal strategy: it defines an action for each possible state. Since the state space consists of 640 states, the strategy becomes rather complex. Recall that the Bellman equation describes successive approximations $v_0, \ldots$ as follows:

$$v_0(s) = 0$$
$$v_{n+1}(s) = \min_{a \in A}\{c(s,a) + \sum_{s' \in S} P(s' \mid s,a)v_n(s')\},$$

where $s \in S$.

In the equations above, S is the set of all possible states, i.e. the set of quadruples $\langle q_1, t_1, q_2, t_2 \rangle$, A is the set of all possible actions: $\{reject, assign\ to\ q_1, assign\ to\ q_2\}$, $P(s' \mid s,a)$ is the probability to move from the state s to state s' by executing an action a, and $c(s,a)$ is the cost per time unit when the system is in state s and action a is taken. The value $v_n(s)$ is the total cost for running the system for n time units from state s under an optimal strategy. We consider the two cost functions QoS and CoO as described above.

We note that the average cost g per time unit under an optimal strategy satisfies:

$$\min_{s \in S}(v_{n+1}(s) - v_n(s)) \leq g \leq \max_{s \in S}(v_{n+1}(s) - v_n(s))$$

In case the process is recurrent, which means that from every state one can reach every other state, these two bounds converge to the same value. In this way one can compute the exact values of quality of service and cost of operation for the tandem queue. The optimal values found are

$$\text{QoS}: 1.73$$
$$\text{CoO}: -2.88.$$

Next, we apply our solidification approach. We start by observing that the place invariant W exists and satisfies $W(p) = W(s) = 0, W(q) = 1$. Therefore, the solidification approach is applicable. The powers of the matrix M are defined as follows:

$$M = \begin{array}{c|ccc} & p & q & s \\ \hline p & 0 & 1 & \omega \\ q & \omega & 1 & 2 \\ s & \omega & \omega & 0 \end{array} \qquad M^2 = \begin{array}{c|ccc} & p & q & s \\ \hline p & 0 & 1 & 2 \\ q & \omega & 1 & 2 \\ s & \omega & \omega & 0 \end{array}$$

The fixpoint μ is reached with M^2. This net is unsound since

$$\min\{\mu(y,q) \mid W(y) < W(q)\} = 1 < 2 = \min\{\mu(q,y) \mid W(y) < W(q)\}.$$

Solidification requires that the resource request of the first task is granted only if there is at least one more resource available. In this way one guarantees that at least one resource is available when a case arrives at place q in the net, so that the second transition can fire. Observe that if more than one additional resource is available, proper termination is guaranteed as well. Therefore, we have performed a number of simulation runs for different values of the parameter k—the number of additional resources— ranging from one to three. For each one of the values of k, two priority configurations were considered. The first uses FCFS (first come first served) priority for all tasks of all cases. The second uses SPT priority for tasks (i.e. the second task has priority over

Table 1. QoS and CoO for the tandem queue example

Strategy	Threshold	QoS	CoO
Optimal		**1.73**	**-2.88**
Solidification;	0	3.02	-2.13
FCFS	1	2.14	**-2.67**
	2	**2.09**	-2.72
	3	2.53	-2.44
Solidification;	0	2.23	-2.60
SPT+FCFS	1	**1.94**	**-2.80**
	2	2.09	-2.72
	3	2.53	-2.44

the first one) and FCFS for the same task of different cases. In both cases, "greedy" resource allocation takes place: if the set of tasks that can be started (i.e. for which the number of resources is at least the requested number plus the threshold) is nonempty, some task will start immediately.

Finally, for each one of the cases two values have been measured: cost of operation (CoO) and quality of service (QoS).

Table 1 represents our results for quality of service and cost of operation. The performance w.r.t. the optimal strategy is given at the top. Then come solidification strategies with various threshold values. The thresholds indicated represent the extra number of available resources required for entering the queue of the first task or station. Since the mathematical model has queues with finite capacity, no deadlocks are possible, so solidification is not required. The cases that are waiting in the queues are treated in FCFS order; when resources become available, the longest waiting case that can be served is selected. The third group of results stem from the solidification extended with the SPT priority rule. When resources become available, the second station has priority. Cases of that station are treated in FCFS order.

We observe that the simulation minima are obtained for QoS for a threshold value of 1 for the prioritised configuration and 2 without SPT priorities, and for CoO for threshold value 2, either with or without priorities. The exact values are 1.94 for QoS and -2.80 for CoO.

While comparing the theoretical results with the results of simulation, we observe that the relative error is quite small for CoO and somewhat larger for QoS. Probably, this difference is caused by the penalty for lost opportunities.

By examining scenarios where the solidification approach makes suboptimal decisions, it is possible to arrive at heuristics that improve upon the decisions made. It seems that information predicting the *future* availability of resources can be of some use here. Of course, improving performance in this way decreases the robustness, i.e. more information is needed and a less straightforward computation.

5.3 Simulation Results for the Construction Example

A simulation study in Arena [8] has been conducted to determine the optimal solidification of the example net from Figure 2. We assumed that the credit limit is 50 mu

Table 2. QoS and CoO for the construction example

Strategy	Threshold	QoS	CoO
Solidification;	1	deadlock	
FCFS	2	deadlock	
	3	106.05	17.59
	4	**43.48**	6.93
	5	51.19	**6.89**
Solidification;	1	deadlock	
SPT+FCFS	2	50.93	7.71
	3	45.66	6.77
	4	**38.79**	**5.97**
	5	51.15	6.49

and that the arrival of new customers is Poisson with the average time between arrivals being 8.5 days.

In Table 2, the simulation results for resource management in our building example are given. In the first series of simulations, resources are assigned to transitions based on the FCFS principle. This means that when the resources become available, the longest waiting case that has become enabled can continue. In the second series, this FCFS strategy is extended with the SPT priority rule, like in the tandem queue example. The first task (groundwork) gets lowest priority, next comes the framing task and the other two tasks have highest priority. Parameter of the simulation is again the threshold value for the groundwork task, which ranges from one to five. Thresholds one and zero yield deadlock in both cases. With a threshold of two, the simulation does not deadlock in combination with the SPT priority rule, although this is theoretically possible. Thresholds greater than three do solidify the net. For thresholds above five, both the CoO and QoS performance measure increase rapidly, so they have not been listed.

Summarising these results, we observe that the best quality of service is always achieved for threshold four: 43.48 in the FCFS case and 38.79 for SPT extension. Unlike this, the optimal threshold for the lowest cost of operation depends on the resource assignment policy. For FCFS, the minimum is obtained for threshold five (6.89), while for the SPT extension it happens for threshold four (5.97).

6 Conclusion

In this paper, we have presented a way, called *solidification*, to obtain a deadlock free scheduler that requires minimal coordination. The computation of this scheduler needed at runtime are independent of the number of active cases. This can be of importance in the implementation of workflow management systems.

We have studied the price of coordination in resource management, i.e. the difference in performance between the optimal (global) and the robust (local) scheduler based on the solidification approach. The performance criteria studied did correspond to quality of service and cost of operation respectively. Our experiments indicate that the performance loss due to a minimal coordination scheduler is not too high, but more convincing realistic case studies are sorely needed. The solidification scheduler can be significantly improved by extending it with an SPT priority rule.

For further research, it is interesting to apply our method to real-life resource scheduling problems. As it is computationally infeasible to compute an optimal scheduler for such processes, comparisons have to be made with heuristic schedulers used in practice. A second line of investigation is the improvement of our resource allocation strategy by adding more information without compromising robustness too much.

References

1. K. Barkaoui and L. Petrucci. Structural analysis of workflow nets with shared resources. In *Workflow management: Net-based Concepts, Models, Techniques and Tools (WFM'98)*, volume 98/7 of *Computing science reports*, pages 82–95. Eindhoven University of Technology, 1998.
2. J. Bertrand, J.C.Wortmann, and J.Wijngaard. *Production Control, A Structural and Design Oriented Approach*. Educatieve Partners, 1998. Second revised edition.
3. J. Colom. The resource allocation problem in flexible manufacturing systems. In W. van der Aalst and E. Best, editors, *Application and Theory of Petri Nets 2003, ICATPN'2003*, volume 2679 of *Lecture Notes in Computer Science*, pages 23–35. Springer-Verlag, 2003.
4. E. W. Dijkstra. *Selected Writings on Computing: A personal Perspective*. Texts and Monographs in Computer Science. Springer Verlag, 1982.
5. J. Ezpeleta. Flexible manufacturing systems. In C. Girault and R. Valk, editors, *Petri nets for systems engineering*. Springer-Verlag, 2003.
6. J. Ezpeleta, J. M. Colom, and J. Martínez. A Petri net based deadlock prevention policy for flexible manufacturing systems. *IEEE Transactions on Robotics and Automation*, 11(2):173–184, 1995.
7. E. Feinberg and A. Shwartz. *Handbook of Markov Decision Processes: Methods and Algorithms*. Kluwer, 2002.
8. W. Kelton, R. Sadowski, and D. Sadowski. *Simulation with Arena*. McGraw-Hill, 1998.
9. M. Puterman. *Markov decision processes: discrete stochastic dynamic programming*. Wiley, New York, 1994.
10. M. Silva and E. Turuel. Petri nets for the design and operation of manufacturing systems. *European Journal of Control*, 3(3):182–199, 1997.
11. W. M. P. van der Aalst. Verification of workflow nets. In P. Azéma and G. Balbo, editors, *Application and Theory of Petri Nets 1997, ICATPN'1997*, volume 1248 of *Lecture Notes in Computer Science*. Springer Verlag, 1997.
12. W. M. P. van der Aalst. The Application of Petri Nets to Workflow Management. *The Journal of Circuits, Systems and Computers*, 8(1):21–66, 1998.
13. W. M. P. van der Aalst. Workflow verification: Finding control-flow errors using Petri-net-based techniques. In W. M. P. van der Aalst, J. Desel, and A. Oberweis, editors, *Business Process Management: Models, Techniques, and Empirical Studies*, volume 1806 of *Lecture Notes in Computer Science*, pages 161–183. Springer-Verlag, 1999.
14. W. M. P. van der Aalst and K. M. van Hee. *Workflow Management: Models, Methods, and Systems*. MIT Press, 2002.
15. K. van Hee, A. Serebrenik, N. Sidorova, and M. Voorhoeve. Soundness of resource-constrained workflow nets. In G. Ciardo and P. Darondeau, editors, *Application and Theory of Petri Nets 2005, ICATPN'2005*, Lecture Notes in Computer Science. Springer Verlag, 2005. accepted.

sPAC (Web Services Performance Analysis Center): Performance Analysis and Estimation Tool of Web Services

Hyung Gi Song[1] and Kangsun Lee[2,*]

[1]R&D Institute, Netville Co., Ltd.,
161-7 Yeomni Mapo, Seoul,
121874 South Korea
joshuasong@netville.co.kr
[2] Dept. of Computer Engineering, Myongji University,
San 38-2 Namdong Yongin, Kyungki,
449728 South Korea
ksl@mju.ac.kr
(tel) +82-31-330-6444 (fax) +82-31-330-6432

Abstract. Web service is a promising technology to efficiently integrate disparate software components over various types of systems and to exchange various business artifacts among business organizations. As many web services are nowadays available on Internet, quality of services (QoS) becomes increasingly important to distinguish different service providers. Performance mainly characterizes QoS especially in mission critical services. However, performance analysis is a very difficult job, since it involves nondeterministic networks, frequent changes on workload intensity and unexpected usage patterns. In this work, we introduce sPAC (Web Services Performance Analysis Centre) and show how customers can verify timeliness of their web services semi-automatically. sPAC 1) graphically describes the workflow of web services, 2) automatically generates test codes for the web services and invokes them for performance tests using Java threads, 3) automatically generates a simulation model for the specified workflow model, and conducts extensive simulations for various load conditions and usage patterns, and 4) reports analysis and estimation results to help customers determine if the composed web services can meet the performance requirements.

Keywords: Process simulation, Quality of Service in business processes, Process verification and validation, Processes and service composition.

1 Introduction

Web services are expected to provide the ideal platform for integrating business artifacts disparate platforms, systems, and organization [1]. As many web services with similar functionalities are available on the Internet, QoS (Quality of Services) and the

---

* Corresponding author.

W.M.P. van der Aalst et al. (Eds.): BPM 2005, LNCS 3649, pp. 109–119, 2005.
© Springer-Verlag Berlin Heidelberg 2005

performance/cost will distinguish service providers from each other when a customer makes decisions of selecting suitable web services [2]. Therefore, QoS analysis becomes increasingly more important to provide accurate QoS information and to establish SLAs (Service Level Agreements) between service customers and service providers, accordingly.

QoS of web services is a combination of several qualities or properties of the comprising web services, for example, availability, security, reliability and performance [3,4]. These QoS properties are inherently dynamic and change in real time depending on how these services are actually performing. Among the dynamic QoS factors, performance may be the most difficult factor to access, since it involves nondeterministic network, abrupt changes on load intensity, and unexpected usage patterns.

In this work, we propose a simulation-based methodology to analyze and estimate the performance of web services. Our methodology tests web services by actually invoking the web services and analyses the resulting response time under low load conditions. Then, the given web services are automatically translated into a simulation model, while the test results are fed into our simulation engine as simulation parameters. The simulation model is used for extended performance analysis; we make virtual heavy load and estimate the expected performance of the web services without actually invoking them on networks and system resources to save time and cost. sPAC (Web Services Performance Analysis Center) is a performance analysis and estimation tool to support our methodology. We introduce useful facilities of sPAC and show how a customer uses sPAC to design, reengineer, verify their web services and finally produce a new business process (or web process) with guaranteed performance. The methodology of sPAC can be the foundation of SLA automation [2] (i.e., automatic SLA creation, SLA monitoring and control) if SLAs are mainly described by performance requirements as in the case of mission-critical services.

This paper is organized as follows. In Section 2, we present the related work in this area and highlight key features of our methodology. In Section 3, sPAC architecture is introduced with detailed explanation on key components. Section 4 demonstrates sPAC with an example. Section 5 concludes this paper with future works to achieve.

2   Related Research

Building web processes is an active area of research and development. Many research groups have developed flow languages for composing web services into web process, including WSFL [5], XLANG [6] and BPEL4WS [7]. One of the problems in the flow languages is that they do not have explicit supports to guarantee QoS in the composition of web services. Supporting QoS in web services is an active research area due to this reason, but is still at its infancy.

QoS issues have been addressed from the perspective of the Service providers of web services, and from the perspective of the Service consumers of these services [8]. Many research groups have been studying QoS of web services with the perspective of service providers. Shuping Ran [9] and Peter Farkas [10] researched a new web services discovery model, which enables QoS-based composition of web services by enhancing standard UDDI specification. Tao Yu [11] proposed a framework and algorithm for providing QoS information of web services by designing QoS broker module for web service servers. These research results have been advancing QoS-guaranteed composition of web services, but the specification of standard web services technology or web services servers have to be modified accordingly in order to enable the proposed methods. John A. Miller [4] researched on estimating performance of web services and developed SCET (Service Composition and Execution Tool) to support the perspective of service customers. SCET works without modifications on web services standards, and provides many convenient facilities to help service customers to estimate performance of the composite web services. SCET estimates the performance of web services by building and executing a simulation model. However, the simulation model and necessary parameters are not validated against the real data, and therefore it might be possible to produce wrong estimation data. Test-based analysis evaluates the performance of web services more accurately by actually executing web services in real world conditions [12]. However, it requires high cost and time, and cannot be used when a given test load exceeds resource capability of the test-host computer. Combining simulation-based analysis with test-based analysis might be the best way to produce accurate performance estimation and to save cost, at the same time.

3   sPAC (Web Services Performance Analysis Centre)

In this section, we present the methodology, performance metrics, and the architecture of sPAC (Web Services Performance Analysis Centre). sPAC works with the perspective of service consumers and mixes simulation and test-based analysis method to evaluate and estimate the performance of web services. Followings are the important features of sPAC:

- *Automatic generation of simulation models*: Customers specify how web services are formed into a new web process with UML (Unified Modeling Language)'s [13] activity diagram. Then, the activity diagram is automatically translated into a simulation model. The simulation model is equivalent to the activity diagram by following translation rules briefly explained in Section 3.1. This feature guarantees the validity of the simulation model and thereby increases the accuracy of performance evaluation.
- *Simulations with testing results*: A new web process is executed by actually invoking the web services with low load intensity. Then, the testing results are fed into the simulation model as simulation parameters to increase estimation accuracy in heavy load conditions.

Detailed explanations are found in Section 3.1

3.1   sPAC Methodology

sPAC is performed with the following steps:

Step 1: A customer defines how web services are formed into a new business process (or web process). We use UML's activity diagram to represent workflow of web services. With activity diagrams, web services are represented as *nodes*, while the execution path is represented as *links* with decorations to specify *fork*, *join*, *parallel* and *sequential* execution, and other various conditions.

Step 2: sPAC dynamically invokes the web services and executes them to get DRT (Dissected Response Time) and TRT (Traced Response Time) with low load intensity. DRT and TRT are our performance measurements and will be explained in Section 3.2.

Step 3: sPAC automatically generates a simulation model for the web process. Our simulation model is constructed based on Simjava [14], a process-based discrete event simulation package for Java. With Simjava, nodes and links of the activity diagram are represented as *entities* and *ports*, respectively. The execution order and types are exactly preserved as they are in the activity diagram. More details on the generation of the simulation model are found in Reference [15].

Step 4: Results from DRT and TRT tests produced in Step 3 are fed into the parameters of the simulation model generated in Step 3.

Step 5: A series of simulations are performed to estimate the performance of the web process under heavy and complex load conditions.

Step 6: sPAC analyzes the test and simulation results, and normalizes them as DRT, TRT and TPM (Transactions per Minute). Analysis results are reported to customers with text and graphical forms and used to decide if the web process satisfies the performance criteria

3.2 sPAC Performance Metrics

DRT (Dissected Response Time) and TRT (Traced Response Time) are the performance metrics of sPAC.

DRT divides response time into three factors: *Network Time* (N), *Messaging Time* (M) and *Service Time* (S). The response time, T, for a single web service, s, is defined in Equation 1.

$$T(s) = N(s) + M(s) + S(s) \tag{1}$$

Network Time is the amount of delay determined by bandwidth of network path between customers and the providers of web services, network traffic and performance of network equipments. *Messaging Time* is the amount of time taken by service providers to process SOAP messages. SOAP is an XML-based protocol. Therefore, the size of the exchanging message is usually bigger than other binary-based protocols, and the time to process SOAP messages is not negligible. *Service Time* is the amount of time for a web service to perform its designated task. It depends on efficiency of business logic, hardware capability, framework for web services and/or operating system of web services.

When a web process is commercialised with packaged software or in the form of web application, each web service is expected to experience heavy load intensity. TRT performance analysis creates virtual users with Java threads, lets them invoke web services simultaneously, and collects DRT for various load conditions. While TRT test can answer how the composed web process performs well in various user load, it costs time and system resources. Moreover, the maximum testable load always has a limit; the testable load is determined by physical memory size, operating system's memory management policy, network conditions, and/or framework for web services of the test host computer. We use simulation to tear this barrier away. A simulation model is automatically generated as explained in Section 3.1 and used for testing out heavy load conditions without actually invoking web services through physical resources and networks. To achieve better accuracy, the real test results are normalized and fed into the simulation model as simulation parameters. With the TRT tests and simulation results, software architects could be aware of the performance of the web process in various conditions and foresee if the web process performs well after deployment.

3.3 sPAC Architecture

Figure 1 shows the system architecture of sPAC. Two major components of sPAC, *Web Process Composer* and *Performance Evaluator*, are explained in Section 3.3.1 – Section 3.3.2.

3.3.1 Web Process Composer

Web Process Composer helps users to specify their web process with UML's activity diagram. *Web Process Specification Interface* is a set of graphical interfaces allowing users to compose web process conveniently. Users also determine desired values of parameters to control tests and simulations (for example, minimum load, maximum load, and test & simulation frequency). *WSDL Analyzer* fetches WSDL file from remote web services and analyses it to recognize the interface of web services.

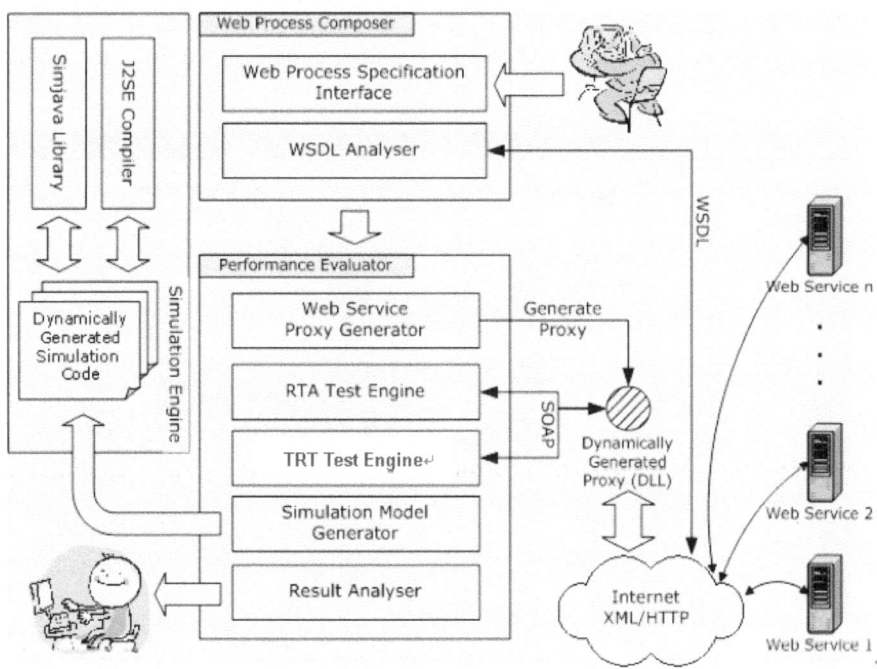

Fig. 1. System architecture of sPAC

3.3.2 Performance Evaluator

Performance Evaluator controls DRT and TRT test for the specified web process. *Web Service Proxy Generator* creates proxy classes of web services in runtime. With the help of Common Language Runtime (CLR) of the Microsoft .NET Framework, *Web Service Proxy Generator* creates corresponding source codes of proxy classes in C# language, compiles them as Dynamic Link Library, and loads the DLLs onto the memory so that *Test Engines* can communicate with each web service. *Test Engines* invoke web services, bind and execute them to perform DRT and TRT tests under fairly low load intensity. *Simulation Model Generator* creates a simulation model for the given web process, and interacts with *Simulation Engine* to perform extended TRT performance analysis under heavy load conditions. *Result Analyzer* summarizes the analysis and estimation results with graphs and text reports.

4   Example: My Travel Planner

In this section, we illustrate sPAC with an example of *My Travel Planner*. Suppose *My Travel Planner* will provide services for booking a flight, reserving an accommodation, renting a car, finding out the current currency rate between Korea and France, exchanging travel money, and processing credit cards. Also, we would like to create *My Travel Planner* just by integrating the existing web services available on the web.

As shown in Figure 2, users can search reusable web services in UDDI and graphically specify how they are formed into a new web process.

After the web process is composed, users set execution parameters (i.e., input and parameters for the composite web services) and environment parameters (i.e., maxi

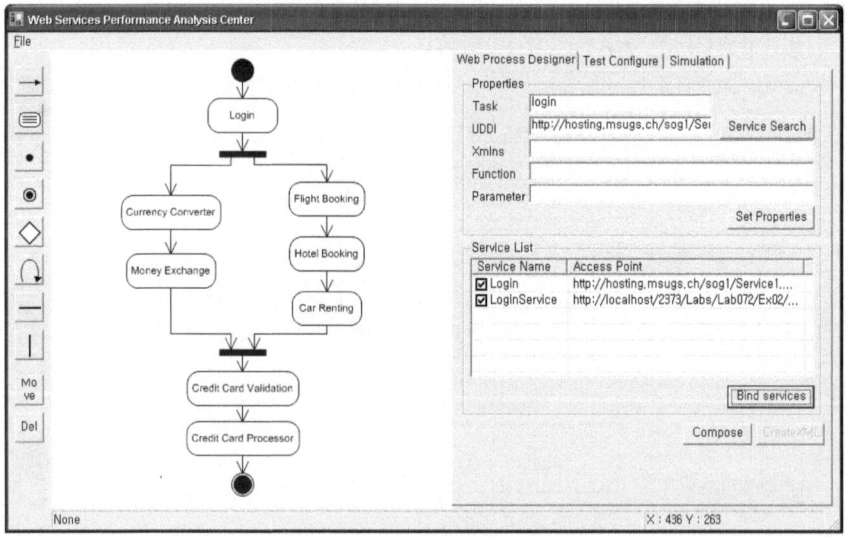

Fig. 2. Web Process Specification

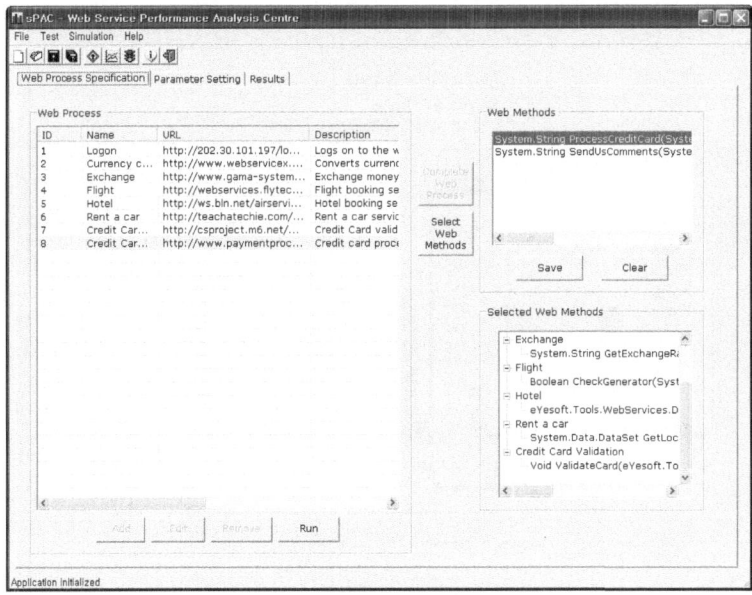

Fig. 3. Web Process Information

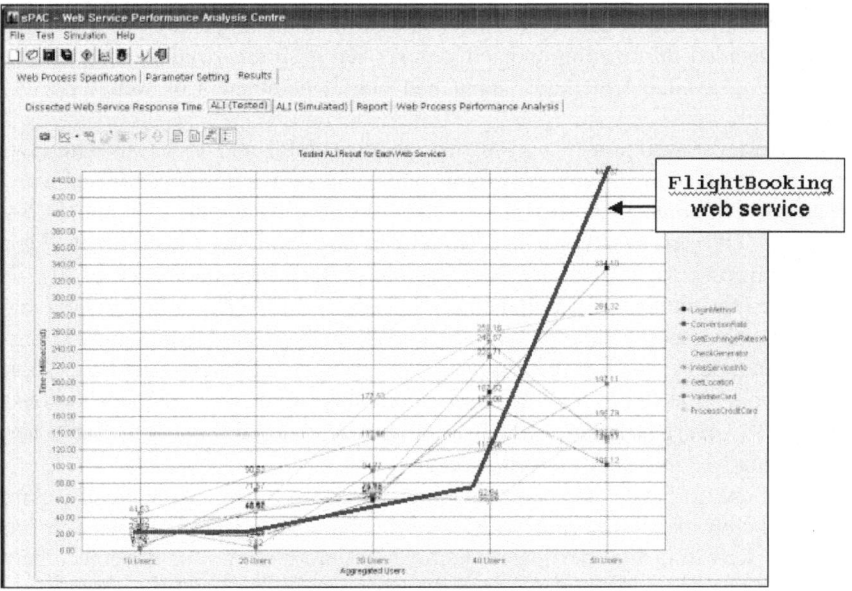

Fig. 4. TRT test results

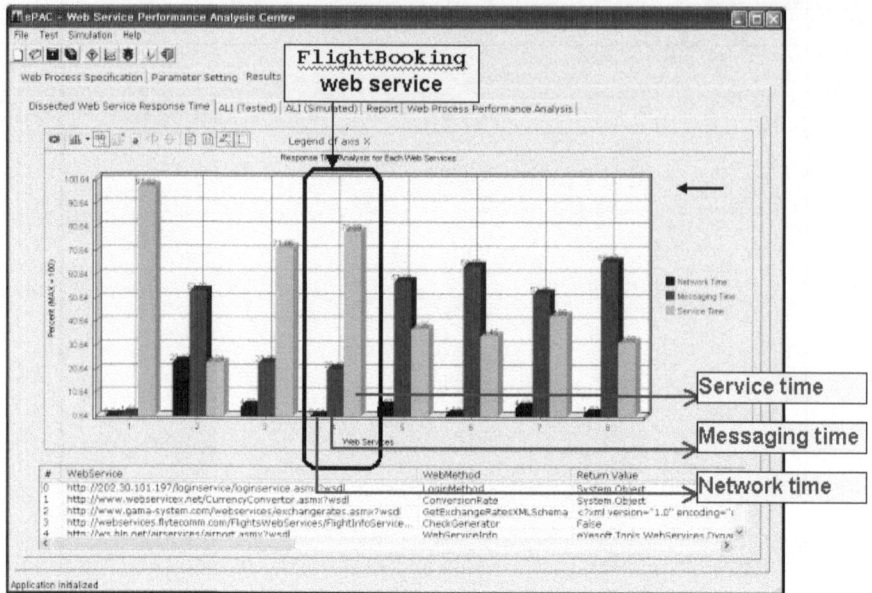

Fig. 5. DRT test results

mum load, minimum load, and testing and simulation frequency) with appropriate values. Detailed information on each web service is summarized as shown in Figure 3, including available methods, input and output parameters of web services, and return types.

For the given web process, sPAC first conducts DRT and TRT tests with low load intensity. In *My Travel Planner* example, DRT and TRT tests are conducted with 10 – 50 numbers of simultaneous users, and take 40 second to complete. Figure 4 - 5 show DRT and TRT test results for *My Travel Planner* example. As shown in Figure 4, FlightBooking web service becomes the performance bottleneck of *My Travel Planner* as the number of simultaneous requests increases. According to DRT tests in Figure 5, the response time of FlightBooking web service is mainly dominated by service time (78.89% of the response time) compared to messaging time (20.46 %) and network time (0.64%). This observation suggests us to reengineer the business logic of FlightBooking, or to increase hardware capability, or to find other service alternatives for better performance.

Then, sPAC creates the corresponding simulation model automatically, compiles the simulation model and runs a series of simulations to foresee if the web process performs well in heavy user load conditions. Figure 6 shows the difference between test-based analysis and simulation-based analysis under the low load intensity of 10 – 50 simultaneous requests. The graph shows our simulation-based analysis is accurate enough to estimate the performance of *My Travel Planner* under the heavy load intensity of 50 – 300 simultaneous requests. Figure 7 shows simulation-based TRT estimation for the web services. Under heavy load of 200 simultaneous requests, Flight-

`Booking` and `CreditcardValidation` are expected to be the performance

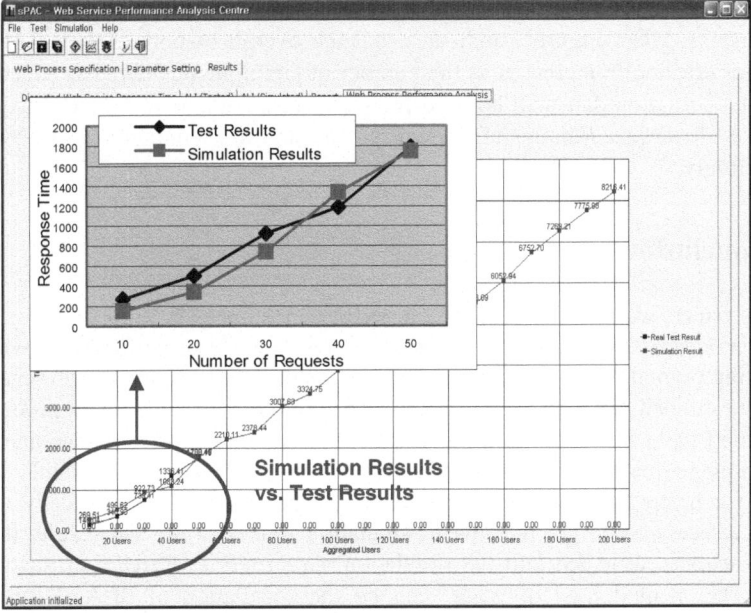

Fig. 6. Simulation Accuracy and Estimation Results

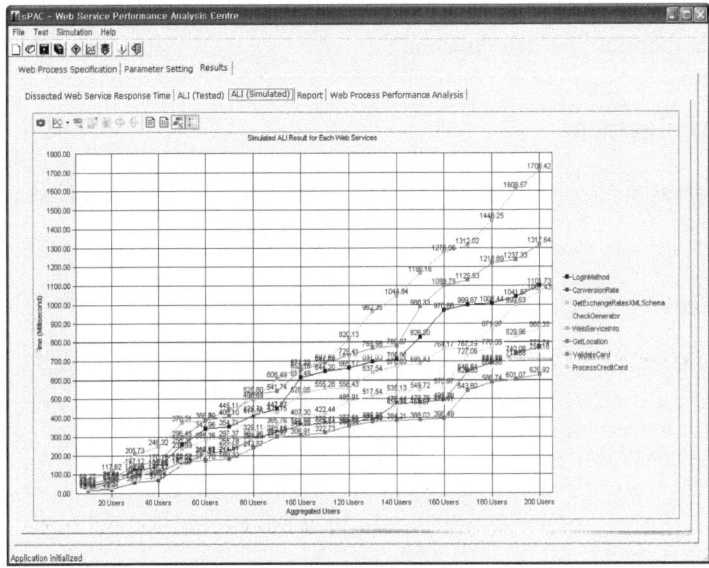

Fig. 7. Simulation-based TRT estimation

bottlenecks. Also, Figure 7 indicates that the overall response time of *My Travel Planner* drastically increases as the number of simultaneous requests exceeds 100. All the estimation data will be used for software architects to foresee the perform-ance of *My Travel Planner* after deployment, and to reengineer the web services, accordingly.

5 Conclusion

In this paper, we introduced sPAC, a performance analysis and estimation tool for web services. sPAC enables software architects to evaluate performance without re-modeling or modifying existing web service standard technology or web service serv-ers. We defined DRT and TRT as the performance metrics for web processes and embodied them in sPAC. In order to save time and cost, performance analysis has been done in dual mode: test-based mode for low load intensity and simulation-based mode for heavy load intensity. The given web process was automatically translated into a process-based discrete event simulation model, while the results from test-based analysis were fed into the simulation parameters to increase estimation accu-racy. The methodology of sPAC can be the foundation of SLA (Service Level Agreement) automation (i.e., automatic SLA creation, SLA monitoring and control) if SLAs are mainly described by performance requirements as in the case of mission-critical services.

We will extend our QoS analysis methods to other dynamic QoS properties, such as, availability, reliability and security for future work. Reference [16] is part of our achievement for this research direction.

Acknowledgement

This work was supported by grant No. R05-2004-000-11329-0 from Korea Research Foundation.

References

1. Preeda Rajasekaran, John A. Miller, Kunal Verma and Amit P. Sheth, *Enhancing Web Services Description and Discovery to Facilitate Composition,* In Proceedings of the First International Workshop on Semantic Web Services and Web Process Composi-tion (SWSWPC'04), San Diego, California, 2004, pp. 34-47
2. Li-jie Jin, Vijay Machiraju, Akhil Sahai, *Analysis on Service Level Agreement of Web ser-vices,* Software Technology Laboratory, HP Laboratories, Palo Alto, HPL-2002-180, Hewlett Packard Company, 2002, June
3. Liangzhao Zeng, Boualem Benatallah, Marlon Dumas, *Quality Driven Web Services Com-position,* In Proceedings of WWW2003, 2003, Budapest, Hungary, pp. 411 - 421

4. Gregory. Silver, A. Maduko, R. Jafri, and et. al, *Modeling and Simulation of Quality of Service for Composite Web Services*, Proceedings of the 7th World Multiconference on Systems, Cybernetics, and Informatics (SCI'03), Orlando, Florida, pp. 420-425, July 2003

5. Frank Leymann, *Web Services Flow Language (WSFL 1.0)*, http://www-3.ibm.com/ software/ solutions/Webservices/pdf/WSFL.pdf, 2001

6. Satish Thatte, *XLANG: Web services for business process design*, http://www.gotdotnet.com/team/ xml_wsspecs/xlang-c/default.htm, 2001

7. Tony Andrews, Francisco Curbera, and et. al, *Specification: Business Process Execution Language for Web Services Version 1.1*, http://www-128.ibm.com/developerworks/library/ws-bpel, 2003

8. Daniel A. Menasce, *QoS Issues in Web services*, IEEE Internet Computing, Nov. 2002, pp. 72-75

9. Shuping Ran, *A model for Web services discovery with QoS*, ACM SIGecom Exchanges, Volume 4, Issue 1, Spring, 2003, pp. 1-10

10. Peter Farkas, Hassan Charaf, *Web Services Planning Concepts*, Journal of WSCG, Vol.11, No.1, 2003, ISSN 1213-6972

11. Tao Yu, Kwei-Jay Lin, *The Design of QoS Broker Algorithms for QoS-Capable Web Services*, International Journal of Web Services, Vol. 1, No. 4, 2004, pp. 17-24

12. J.D. Meise, Srinaath Vasireddy, Ashish Babbar, Alex Mackman, *How to: Use ACT to Test Web Services Performance*, Microsoft Developer Network, Microsoft Corporation, 2004

13. Object Management Group, *UML (Unified Modeling Language) ᵀᴹ Resource Page,* http://www.uml.org/, January 2005

14. Fred Howell, Ross McNab, *Simjava Library,* http://www.dcs.ed.ac.uk/home/hase/simjava, 1996

15. Hyunggi Song, *sPAC: Web Services Performance Analysis Center*, Master Thesis, Department of Computer Engineering, MyongJi University, Korea, 2004

16. Heejung Chang, Hyungki Song, Kangsun Lee and et. al, *Simulation-Based Web Service Composition: Framework and Performance Analysis*, Lecture Notes in Computer Science, Springer Verlag, vol. 3398/2005, Feb. 2005, pp. 352-361

Specifying Web Workflow Services for Finding Partners in the Context of Loose Inter-organizational Workflow

Eric Andonoff, Lotfi Bouzguenda, and Chihab Hanachi

IRIT Laboratory,
University Toulouse 1, 1 Place Anatole France
31042 Toulouse Cedex, France
{andonoff, lotfi.bouzguenda, hanachi}@univ-tlse1.fr

Abstract. This paper deals with Web Workflow Services (W2S) description languages that help organizations to find partners in the context of loose Inter-Organizational Workflow (IOW). Loose IOW refers to occasional cooperation between organizations, free of structural constraints, where the partners involved and their number are not pre-defined. Such a dynamic and heterogeneous context requires the definition of a W2S description language allowing workflow service providers to publish their capabilities and workflow service requesters to express their needs. Current Web services languages do not permit to describe adequately workflow services (structure and behavior) by lack of expressive power and/or formal semantics. In this paper, we show how the appropriate combination of Petri Nets with Objects (PNO) and OWL-S allows the specification, validation and publication of workflow services. On the one hand, PNOs permit the formal and graphical specification of workflow services, their simulation and validation. On the other hand, OWL-S permits the publication of workflow services on the Web. OWL-S has also the advantage to include the ontology concept which can be used to solve semantic problems between IOW partners. Moreover, we provide rules and algorithms which automatically derive OWL-S specifications from PNOs ones. This work has been implemented.

1 Introduction

Inter-Organizational Workflow (IOW) is a key technology for helping the necessary cooperation of organizations facing the emergence of the open and dynamic worldwide economy. The different organizations involved in such cooperation need to put resources and skills in common, and coordinate their respective *business processes* in order to reach a common goal corresponding to a value-added service. In such a context, IOW is an adequate technology since it supports the cooperation of distributed and heterogeneous business processes running in different organizations [1].

A fundamental problem for IOW is the *coordination* of the different distributed and heterogeneous processes. By coordination, we mean all the work needed for putting all these processes together in order to provide the value-added service in an efficient way. This coordination can be investigated in the context of two different scenarios: loose IOW and tight IOW [2]. In this work, we focused on *loose IOW*

W.M.P. van der Aalst et al. (Eds.): BPM 2005, LNCS 3649, pp. 120–136, 2005.

which refers to occasional cooperations between organizations, free of structural constraints, where the organizations involved and their number are not pre-defined but are selected at run time in an opportunistic way.

Coordination in loose IOW raises several problems such as the finding of partners, the negotiation of the processes themselves between partners according to certain criteria (due time, precision, visibility of the evolution process, way of doing it...), and the synchronization of the distributed and concurrent execution of these different processes. In this work, we focused on the *finding of partners*. One possible way to select organizations is to sub-contract the research to a mediator, as it is presented in [3], thanks to a *matchmaker*. The aim of the matchmaker is to connect workflow service requesters to workflow service providers according to the following protocol: (i) a workflow service provider advertises the offered service to the matchmaker, (ii) the matchmaker stores the advertisement, (iii) a workflow service requester asks the matchmaker whether it knows providers offering the desired service, and finally (iv) the matchmaker matches the request against the stored advertisements and returns the result as a set of workflow service providers. For this protocol to work, in this highly heterogeneous environment, an ontology is required to assist semantic interoperability between partners, i.e. by allowing them to adopt a shared business view through a common terminology. In this way, the use of a matchmaker finally requires the definition of a workflow service description language allowing providers to publish their capabilities and requesters to express their needs, capabilities and needs being expressed in the terms of a common ontology.

As the Web provides many widely available facilities for inter-organizational communication, we propose to define a Web Workflow Service (W2S) description language. By *W2S description language*, we mean a language able to describe workflow services of which description and execution are accessible through the web. Such a language must permit the expression of three workflow complementary aspects, usually described through three different interacting models: the organizational, informational and process models. The organizational model structures the workflow actors and authorizes them, through the notion of role, to perform tasks making up the processes. The informational model defines the structure of the documents and data required and produced by the processes. The process model defines component tasks, their coordination as well as the required resources (information, actors).

The problem addressed in this paper is *"what language for W2S description: do we define a new language or do we chose an existing one?"*.

Most of the existing languages proposed in the context of Web services do not meet the previous requirements. Indeed, Web services languages, such as WSDL [4], do not allow the expression of the process concept as it is defined in the workflow, i.e. as a set of coordinated tasks. Regarding composition Web service languages, such as BPEL4WS [5] or WSFL [6], they neither completely describe the organizational and informational models, nor integrate semantics aspects through ontology. If we consider workflow technology, the proposed languages, such as YAWL [7] or XPDL [8], describe the three workflow models, and tools are provided to derive XML workflow specifications. Unfortunately, these XML specifications solve only syntactic conflicts between organizations, while in loose IOW, the heterogeneous context requires semantic conflicts solving mechanisms.

Conversely, languages proposed in the context of semantic Web services [9], and more particularly OWL-S [10], which is recommended by the World Wide Web consortium, seem to be appropriate for W2S. Indeed, first, OWL-S captures the concepts involved in the three workflow models, second, it allows the description of workflow services referencing ontology, and third, it enables them to be published in a Web accessible format.

However, OWL-S has two main drawbacks: first, it does not provide any graphical tool to specify workflow services, and, second, it lacks theoretical foundations with an operational semantics to simulate and validate the workflow services.

The aim of the paper is to compensate the previous drawbacks by proposing a solution based on the three following principles:

- (i) The use of a graphical and formal language, namely Petri Nets with Objects (PNO), to specify and validate workflow services. PNO [11] are used as a graphical tool to help a designer to define a workflow service; they also provide formal and executable specifications to analyze, simulate, check and validate the described workflow service behavior.
- (ii) The automatic derivation of the previous workflow services specifications onto OWL-S specifications.
- (iii) The publication of the workflow services by means of OWL-S.

Thus, PNO can be seen as a graphical tool for specifying OWL-S services and as a formalism providing an operational semantics to OWL-S.

The remainder of this paper is organized as follows. Section 2 briefly introduces the PNO formalism and explains why this formalism is convenient for workflow service specification and validation. Section 3 presents OWL-S and explains the reasons of this choice for publication. Section 4 gives an operational semantics to OWL-S by formalizing its service profile and service process using PNO. This section first presents our approach, and then specifies the rules and algorithms we propose to derive a PNO onto OWL-S service profile and service process specifications. Section 5 describes some aspects of the implementation of this work. Section 6 briefly compares our proposition to related works and concludes the paper.

2   Petri Nets with Objects: A Convenient Language for Workflow Service Specification and Validation

This section is dedicated to the presentation of the Petri Nets with Objects formalism and gives the reasons of the choice of this language for workflow service description and validation. However, the comparison of this language with others closed languages, such as YAWL [7] or UML Activity diagrams for instance, is briefly discussed in section 6.

2.1 What Are Petri Nets with Objects?

Petri Nets with Objects (PNO) [11] are a formalism combining coherently Petri nets (PN) technology and the Object-Oriented (OO) approach. While PN are very suitable to express the dynamic behavior of a system, the OO approach permits the modeling and the structuring of its active (actor) and passive (information) entities. In a conventional PN, tokens are atomic, whereas they are objects in a PNO. As any PN, a PNO is made up of places, arcs and transitions, but in a PNO, they are labeled with inscriptions referring to the handled objects. More precisely, a PNO features the following additional characteristics:

- Places are typed. The type of a place is a (list of) type of an (list of) object(s). A token is a value matching the type of a place such as a (list of) constant (e.g. 2 or 'hello'), an instance of an object class, or a reference towards such an instance. The value of a place is a set of tokens it contains.
- Arcs are labeled with parameters. Each arc is labeled with a (list of) variable of the same type, as the place the arc is connected to. The variables on the arcs surrounding a transition serve as formal parameters of that transition and define the flow of tokens from input to output places. Arcs from places to a transition determine the possible condition of the transition: a transition may occur (or is possible) if there exists a binding of its input variables with tokens lying in its input places.
- Each transition is a complex structure made up of three components: a precondition, an action and emission rules. A transition may be guarded by a precondition, i.e. a side-effect free Boolean expression involving input variables. In this case, the transition is only permitted by a binding if this binding evaluates the precondition to be true. Passing a transition through depends on the precondition, on the location of tokens and also on their value. Most transitions also include an action, which consists in a piece of code in which transitions' variables may appear and object methods be invoked. This action is executed at each occurrence of the transition and it processes the values of tokens. Finally, a transition may include a set of emission rules i.e. side-effect free Boolean expressions that determine the output arcs that are actually activated after the execution of the action.

Figure 1 gives an example of a PNO describing a simple task providing the references of a flight, given the departure and arrival airports, the traveling date and the agency in charge of finding the flight. This PNO is composed of a transition, four input places and two output places. Each place is typed with one of the four following object classes: *Airport, Date, TravelAgency* and *Result*. Each input place contains a token of which value is indicated by a comment linked to it by an arrow. From left to right, the first two input places called *DepartureAirport* and *ArrivalAirport* contain one token corresponding to an Airport. The object class *Airport* has two attributes *{Name,City}* and we can read that the flight requested is between Tunis and Toulouse. With the same principle, we can deduce from the *DepartureDate* and *Agency* input places that the travel date is 01/01/2005 and the travel agency in charge of finding the flight is *Bravo Agency*. Let us also remark that the *TravelAgency* object class, in addition to three attributes *{Name,Phone,Address}*, features a method *{GetFlightDetails}* as well. Now let us consider the transition *GetDesiredFlightDetails*. It has a precondition *{(DD.DateD>Date())* *and (DA.Name<>AA.Name)}* which indicates that the departure date must be in the future and departure and arrival airports are different. Let us notice that this precondition is expressed with the formal parameter of the input arcs (*DD, DA* and *AA*). If this precondition is satisfied, the action is executed and the object *Travel Agency* is asked to execute the *GetFlightDetails* method. According to the result R, returned by this method, the emission rules will direct the process through one path or another. If a flight is found, the result R is not null and then a token is put in the *Success* output place. In the other case, a token is put in the *Fail* output place.

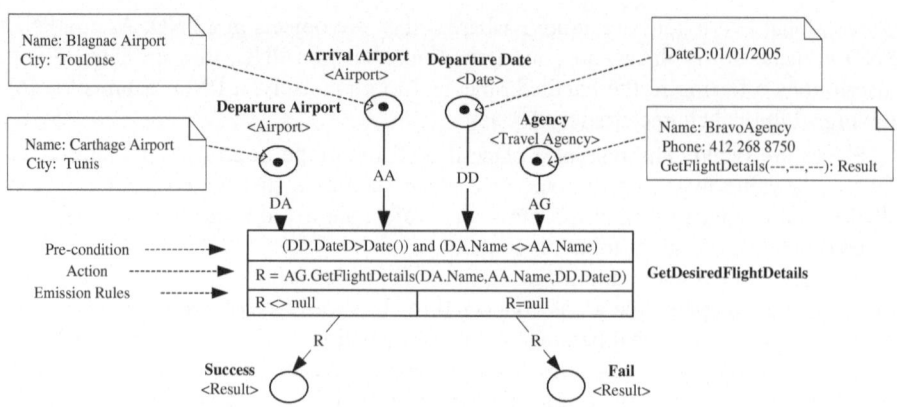

Fig. 1. Example of a PNO

2.2 Motivations for Using Petri Nets with Objects

Petri nets are widely used for workflow service specification [12]. Several good reasons justify their use:

- *An appropriate expressive power* that permits the description of the different tasks involved in a workflow service and their coordination.
- *A graphical representation* that eases the workflow service specification.
- *An operational semantics* making an easy mapping from specification to implementation possible.
- *Theoretical foundations* permitting analysis and validation of behavioral properties and simulation facilities.

Unfortunately, conventional Petri nets focus on the process definition and do not perfectly capture the organizational and the informational dimensions of a workflow. As mentioned previously, Petri nets with Objects extend Petri nets by integrating high-level data structure represented as objects, and, therefore provide the possibility to integrate in a coherent way the two dimensions missing in Petri nets. Thus, using PNO, actors of the organizational model are directly represented as objects and they may be invoked through methods in the action part of a transition. In the same way, data and documents of the informational model are also represented by objects flowing in the PNOs and transformed by transitions. In the previous example (cf. figure 1), the object *Agency* refers to an actor of the organizational model while the *DepartureAirport, ArrivalAirport* and *DepartureDate* objects are data of the informational model.

To summarize, PNOs are convenient for workflow service specification because they really are a strong link between the three workflow models since they permit the description, in a same representation, of actors of the organizational model, data and documents of the informational model, and tasks of the process model. Moreover, we use PNO as a graphical tool to specify a workflow service, and as a formal tool to define executable specifications in order to analyze, simulate, check and validate a workflow service.

3   OWL-S: A Semantic Web Service Language for W2S Publication

3.1   Brief Overview of OWL-S

OWL-S is a semantic markup language that enables the description of Web services in order to be selected, invoked and composed [10]. OWL-S refers to an ontology of services that defines and structures the concepts for handling Web services. The resulting conceptual model is defined through a hierarchy of classes that may be variably refined according to the business domain considered. The essential properties of a service are described by the three following classes: ServiceProfile, ServiceModel and ServiceGrounding.

The *ServiceProfile* provides all the necessary information for a service to be found and possibly selected. The Service Profile is described by three groups of attributes. The first group describes the identity of the service with attributes such as *serviceName*, *textDescription* or *contactInformation* defining respectively the identity of the service, a natural-language description of it, and the organization providing it. The second group gathers attributes to classify a service (e.g. *serviceCategory, serviceParameter*) or to evaluate or compare it to others having the same capabilities (eg. *qualityRating*). The third group expresses the functional capabilities of the service with four attributes that are *inputs, outputs, preconditions* and *effects*. These attributes respectively define the required entries for starting the service, the results the service is able to produce, the constraints that must be satisfied by the inputs, and the output properties guaranteed after the service execution.

In OWL-S, services are viewed as processes. So, the *ServiceModel* describes the service in terms of a process model composed of two specifications: a service process and a process control. The *ServiceProcess* defines the structure of the process using three types of processes: atomic, simple and composite processes. Atomic processes correspond to operations that the service can directly execute; they have no subprocesses. Simple processes correspond to abstractions of atomic processes and are not directly invocable. Composite processes are collections of processes coordinated by control constructs including sequence, loops, conditionals and concurrency. Four attributes are defined for these processes: *inputs, outputs, preconditions* and *effects* having the same semantic as the functional capabilities of the ServiceProfile. Regarding the *ProcessControl*, OWL-S informally represents all the useful attributes for monitoring the execution of the service, notably its possible states at run-time (e.g. ready, ongoing, suspended, aborted…).

The *ServiceGrounding* defines how to access to the service by specifying the communication protocols and messages, and the port numbers to be used.

3.2   Motivations for Using OWL-S

OWL-S is appropriate to workflow service publication for two main reasons. The *first reason* that led us to choose OWL-S is that OWL-S has an adequate expressive power to describe workflow services as illustrated in figure 2. Indeed, it is possible to describe, using the service profile and the service model of OWL-S, the three different

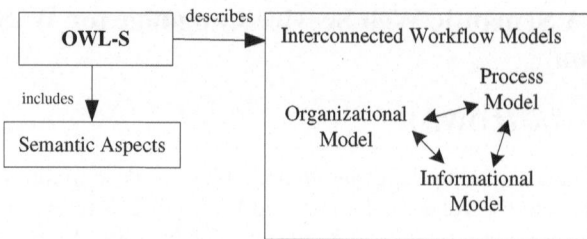

Fig. 2. Adequate Expressive Power of OWL-S for Workflow Service Publication

interrelated models of a workflow i.e. the organizational, informational and process models. Regarding the process, there is a direct mapping between the service model of OWL-S and the process model of a workflow. In OWL-S, the described service is broken down into tasks and their coordination is specified using control constructs such as sequence, loops, conditional, concurrency. In the process model of a workflow, the process is also broken down into tasks and workflow patterns are used to coordinate them. Even if OWL-S does not include all the PNO patterns, it provides the control constructs necessary to describe the majority of workflow process models since it allows the modeling of sequence, loops, conditional and concurrency. Regarding the informational and organizational aspects, OWL-S, through the service profile and the service process, gives a support for the description of actors, information (data or documents) and their availability as required in a workflow. This is possible thanks to the set of inputs (actors, data and documents), preconditions (actors able to play specific roles, empty documents), outputs (data and documents) and effects (documents well filled, compliant with a specific norm).

Moreover, OWL-S, which is a semantic Web service language, enriches Web services description based on WSDL with semantic information about the properties (ServiceProfile) and the structure of the service (ServiceModel). Moreover, this semantic information is based on an ontology, extensible according to the domain and described with a well defined mark up language. Ontology makes possible, in the context of loose IOW in which several heterogeneous organizations cooperate, to solve semantic conflicts between these organizations by defining a shared business view based on a common vocabulary. Moreover, OWL-S ontology has a first-order logic representation [13] that permits deduction and eases the implementation of matchmaking mechanisms useful to compare workflow services. Such mechanisms are very important when selecting partners.

The *second reason* that led us to choose OWL-S is that OWL-S is recommended by the World Wide Web consortium, which is not the case for WSMO [14], another interesting semantic Web language, still in the process of specification (v0.1).

Besides, as we will show later, there is an easy mapping of PNO onto OWL-S concepts. Roughly speaking, the OWL-S service profile can be derived from the input and output places of a PNO, the OWL-S service process can be built from the places and transitions of a PNO, and all the OWL-S control constructs have a corresponding PNO pattern.

4 Formalizing Service Profile and Service Process Using PNO

4.1 Our Approach

The idea was to use PNO as a graphic tool to specify, simulate and validate a work-flow service and then to deduce the corresponding OWL-S service profile and service process automatically.

The design process of a PNO workflow service, i.e. a workflow service designed using PNO, is a hierarchical construction as it is the case for a OWL-S service process specification. The designer first specifies a unique transition with input and output places. If this transition does not correspond to an atomic task (immediately executable), the designer refines it, using only the PNO patterns which have a corresponding control construct in OWL-S, that is *sequence*, *split*, *split-join*, *choice*, *iterate*, *unordered*, *repeat-until*, *repeat-while*, and *if-then-else*. The result is the definition of other transitions expanding the previous one, and having input places and producing output places. The so-defined transitions can themselves be refined if necessary. This top-down decomposition approach is repeated until we obtain only atomic transitions.

Figure 3a below illustrates this hierarchical design process. The first transition, named T, is defined first. It includes one input and two outputs with their corresponding conditions (respectively preconditions and post-conditions). This transition is refined using the *sequence* pattern. Three new transitions, named T1, T2 and T3, are defined and replace the previous one. Among these three transitions, T1 and T3 are

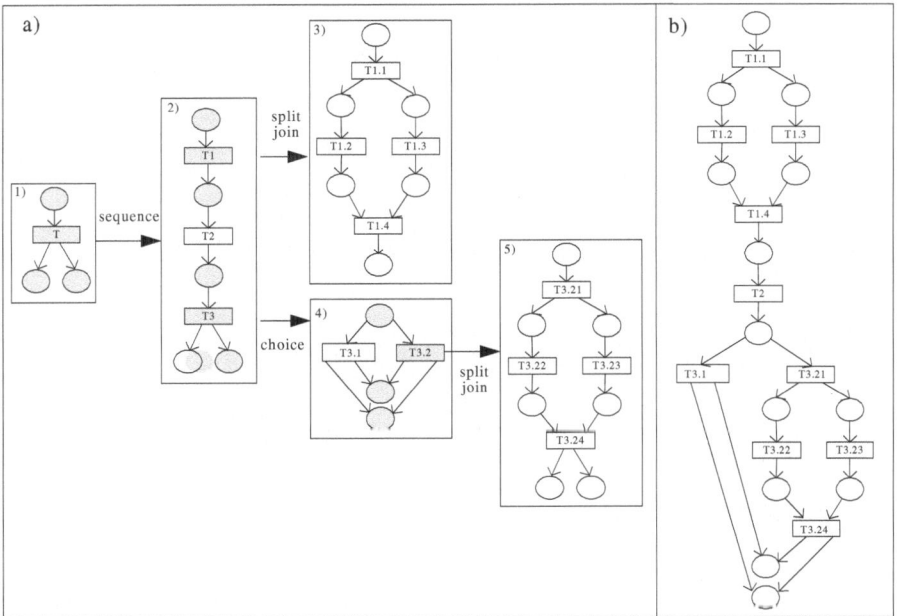

Fig. 3. Hierarchical Design of a Workflow Service and Final PNO

refined while T2 is atomic. T1 is refined using the *split-join* pattern while T3 is refined using the choice pattern. These transitions replace the refined ones. Finally, a new transition of T3, named T3.2, is in turn refined using the *split-join* pattern.

The final result is a global PNO describing a workflow service (cf. figure 3b). This PNO can be represented as a tree where non-terminal nodes are the refined transitions and terminal nodes (leaves) are atomic transitions. Every node (terminal or non-terminal) of the PNO tree include a data structure which indicates the name of the transition, the PNO pattern used when refining the transition (null for atomic transitions), and its corresponding inputs, outputs, preconditions and post-conditions. More precisely, for each input place of each transition we have a couple (InputName, Pre-Condition) where InputName is the name of the considered input place and PreCondition is the precondition of the transition in case this input place is involved in the precondition. In a similar way, for each output place of each transition, we have a couple (OutputName, EmRule) where OutputName is the name of the considered output place and EmRule is the emission rule associated to the considered output place. Figure 4 below visualizes the tree corresponding to the previous PNO.

Finally, before deriving the OWL-S service profile and service process specifications, the designer can use one of the PNO analysis techniques to simulate, check and validate the corresponding workflow service. Validation concerns a certain number of properties such as Ending (does a process effectively end?), Liveness (is a given task (transition) always possible?), Boundedness (is the number of possible configurations of a process finite?), Reachability (is there an evolution in the process leading to a given configuration (desired or not)?), and Quasi-Liveness (does a configuration exist where a given task is possible?).

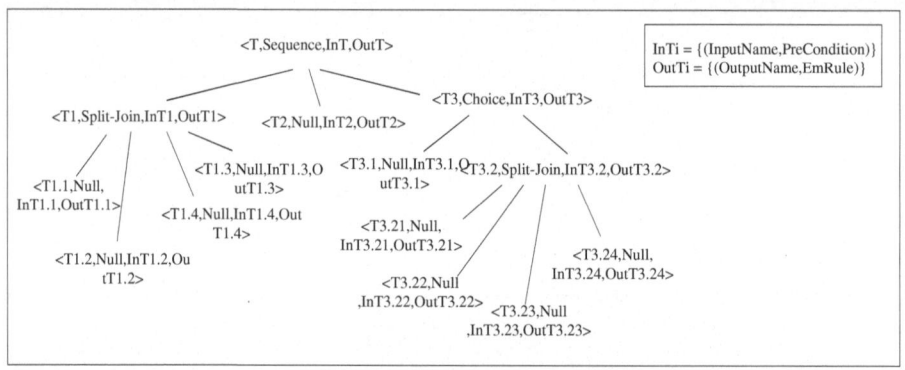

Fig. 4. PNO Tree

4.2 From PNO to OWL-S Service Profile

4.2.1 Principle and Rules

The starting point of the derivation is the global PNO. We considered three types of places: *begin* places which are exclusively input places, *intermediate* places which are input and output places, and finally *end* places which are exclusively output places. In this step, only *begin* and *end* places are to be considered. Table 1 below summarizes the different derivation rules. In this table, the variable I represents the set of input places of the global PNO, while O represents its set of output places.

Table 1. Mapping PNO with OWL-S Service Profile

PNO	OWL-S Service Profile
Begin place b $\in$ B, B=I-(I$\cap$O)	Parameter Name of an Input
End place e $\in$ E, E=O-(O$\cap$I)	Parameter Name of an Output
Precondition associated to a Begin place b $\in$ B	Parameter Name of a Precondition
Emission rule associated to an End place e $\in$ E	Parameter Name of an Effect

4.2.2 Algorithm

Algorithm 1 below explains how we deduce a OWL-S service profile from a PNO. The algorithm has two inputs. Its first input is the PNO itself, which is defined as a 9-uplet (C,P, T, V, PreCond, A, EmR, Pre, Post) as follows [11]:

- C is a set of object classes (which correspond to the ontology classes[1]),
- P is a set of places, typed by a function P$\rightarrow$C*,
- T is a set of transitions, each transition being identified by a name,
- V is a set of object variables, typed by a type function V$\rightarrow$C,
- PreCond is a set of preconditions, each one being necessary to trigger a transition,
- A is a set of actions, each action being triggered by a transition,
- EmR is a set of emission rules, each one corresponding to a logical expression
- Pre is the forward incidence function: PxT$\rightarrow$MultiSet(V*); Pre associates a multi-set of object variables to a (place, transition) couple,
- Post is the backward incidence function: PxTxEmR$\rightarrow$MultiSet(V*); Post associates a multi-set of object variables to a (place, transition, emission rule) triplet.

The second input of the algorithm is the urlProcess (file name of the service process) which is necessary to the OWL-S specification. Its unique output is the generated service profile.

The algorithm implements the rules presented in Table 1. Its first step produces the general tags of the OWL-S service profile along with its non-functional attributes while its second step generates its functional attributes (i.e. input, output, precondition and effect).

```
ALGORITHM From_PNO_To_OWL-S_ServiceProfile
INPUT PNO, urlProcess
OUTPUT SP  % is the generated service profile
```

---

[1] There is a close correspondence between PNO object classes and OWL-S ontology classes (which define the concepts used by a workflow service). Because of space limitation, this paper does not give the algorithm to derive OWL-S ontology classes of a service from PNO object classes.

```
BEGIN
   % B is the set of begin places and ComputeBeginPlaces
   % calculates it from P, Pre and Post; E is the set of end
   % places and ComputeEndPlaces calculates it from P, Pre and
   % Post
   B ← ComputeBeginPlaces(P,Pre,Post)
   E ← ComputeEndPlaces(P,Pre,Post)

   % Step 1 generates general information of the service
   % such as rdf files along with its non functional
   % attributes serviceName, textDescription, contactInforma
   % tion, serviceParameter, qualityRating, and serviceCategory
   SP ← Generate_Profile_GeneralInformation()

   SP ← SP + Generate_Profile_Non_Functional_Attributes()
   % Step 2 generates functional attributes of the service
   % i.e. inputs, outputs, preconditions and effects
   For each begin place b ∈ B Do
    SP ← SP + Gener ate_Profile_HasInput(urlProcess+"#"+b.Name)
   End for
   For each end place e ∈ E Do
    SP ← SP + Generate_Profile_HasOutput(urlProcess+"#"+e.Name)
   End for
   For each precondition pco ∈ PreCond associated to b ∈ B Do
    SP ← SP + Generate_Profile_HasPrecondition
    (urlProcess+"#"+pco.Name)
   End for
   For each emission rule er ∈ EmR associated to e ∈ E Do
    SP ← SP + Generate_Profile_HasEffect
       (urlProcess+"#"+er.Name)
   End for
END
```

Algorithm 1: From PNO to OWL-S Service Profile

4.3   From PNO to OWL-S Service Process

4.3.1   Principle and Rules

As shown previously in section 4.1, the result of the decomposition process used for designing a PNO workflow service is a PNO tree. This tree is the starting point to derive the OWL-S service process. We consider two types of nodes: *terminal* nodes and *non-terminal* nodes. Table 2 below summarizes the different derivation rules. In this table, N is the set of the PNO tree nodes (terminal and non terminal) and T is the set of its terminal nodes.

Table 2. Mapping PNO tree with OWL-S Service Process

PNO tree	OWL-S Service Process
Name of a Node $n \in N$	Name of a Process
(InputName, PreCondition) of a node	Input of a Process
$n \in N$	Precondition associated to the Input
(OutputName, EmRule) of a node	Output of a Process
$n \in N$	Effect associated to the Output
Terminal Node $t \in T$	Atomic Process
Non Terminal Node $n \in$ N-T	Composite Process

4.3.2 Algorithm

Algorithm 2 below explains how we deduce, from the PNO tree, the corresponding OWL-S service process. The algorithm has two inputs. Its first input is the PNO tree itself whose nodes have the following data structure:

- Name is the name of the node (corresponding to the name of a transition),
- Pattern is the PNO pattern of the node,
- In is a set of (InputName, PreCondition) couples which correspond to the input places of the PNO and their corresponding preconditions,
- Out is a set of (OutputName, EmRule) couples which correspond to the output places of the PNO and their corresponding emission rules.

We also have a set of functions permitting the handling of a tree:
- Depth(t) returns the depth of a tree t,
- Root(t) returns the root of a tree t,
- ListOfChildren(n) returns the children of a node n (non-terminal or terminal nodes),
- ListOfParents(t) returns the non-terminal nodes of a tree t,
- ListOfLeaves(t) returns the terminal nodes (leaves) of a tree t.

The second input of the algorithm is the urlOntology (file name of the ontology) which is necessary to the OWL-S specification. Its unique output is the generated service process. The algorithm implements the rules presented in Table 2. Its first step produces the general information of the OWL-S service process. Its second and the third steps generate a definition of a top level process as a composite process (for the root) along with composite processes for non terminal nodes. The last step of the algorithm generates the atomic processes.

```
ALGORITHM From_PNO_To_OWL-S_ServiceProcess
INPUT PNOTree, urlOntology
OUTPUT SP    % is the generated service process

BEGIN
    % Step 1 generates general information of the service
    % such as rdf files, instance definition process model
    SP ← Generate_Process_GeneralInformation()
    % Step 2 and step 3 only for a PNOTree whose depth > 1
```

```
If (Depth(PNOTree)>1)
  % Step 2 generates a definition of top level process as a
  % composite process for the root of PNOTree
    SP ←SP +Generate_Definition_TopLevel_Process(root
    (PNOTree).Name,root(PNOTree).pattern)
    For each cn ∈ ListOfChildren(root(PNOTree)) Do
      SP ← SP + Generate_Component_Process(cn.Name)
    End For
  % Step 3 generates a composite process from the non
  % terminal nodes expect the root
    For each node n ∈ ListOfParents(PNOTree)-root(PNOTree) Do
      % ControlConstruct returns the OWL-S control construct
      % corresponding to a PNO pattern
      % n.In allows to deduce OWL-S inputs and preconditions
      % n.Out allows to deduce OWL-S outputs and effects
          SP ← SP + Generate_Composite_Process(n.Name,
                ControlConstruct(n.Pattern), n.In, n.Out)
      For each cn ∈ ListOfChildren(n) Do
        SP ← SP + Generate_Component_Process(cn.Name)
      End For
      % ClassOf is a function which returns the corresponding
      % class of an object
      For each prein ∈ n.In Do
       SP ← SP + Generate_Input_Precondition(prein.Input
         Name, prein.PreCondition, ClassOf(prein), urlOntology)
      End For
      For each preout ∈ n.Out Do
       SP ← SP + Generate_Output_Effect(preout.OutputName,
             ClassOf(preout), preout.EmRule, urlOntology)
      End For
    End For
End If
% Step 4 generates Atomic Process from terminal nodes
For each node n ∈ ListOfLeaves(PNOTree) Do
  % SP is completed with an atomic process
  SP ← SP + Generate_Atomic_Process(n.Name,n.In, n.Out)
  For each prein ∈ n.In Do
   SP ← SP + Generate_Input_Precondition(prein.Input
       Name, prein.PreCondition, ClassOf(prein), urlOntology)
  End For
  For each preout ∈ n.Out Do
      SP ← SP + Generate_Output_Effect(preout.OutputName,
         ClassOf(preout), preout.EmRule, urlOntology)
  End For
End For
END
```

Algorithm 2: From PNO to OWL-S Service Process

5   Implementation

This work has been implemented as part of the MatchFlow project [3], whose objective is to connect workflow service requesters to workflow service providers, offers and requests being specified using PNOs and stored by a matchmaker in the OWL-S format. In the current version of MatchFlow, the matchmaker compares the offers' and requests' service profiles: it establishes flexible comparisons (exact, plug in, relaxed) based on an ontology.

MatchFlow has been implemented with the Madkit platform [15], which permits the development of distributed applications using multi-agent principles. Indeed, the *agent* technology provides *natural abstractions* to deal with autonomy, distribution, heterogeneity and coordination which are inherent to loose IOW. Moreover, Madkit is based on an organizational paradigm that provides high-level concepts to describe loose IOW coordination, thanks to notions such as role, agent and group. A group is an interaction space, governed by coordination laws, where each agent (representing an organization in our case) can enter, under certain conditions, to play a specific role.

Figure 5 above shows some screenshots of MatchFlow. The middle window (number 1) represents the matchmaker implemented as an agent. The top left Window (number 2) corresponds to an agent implementing a workflow service requester.

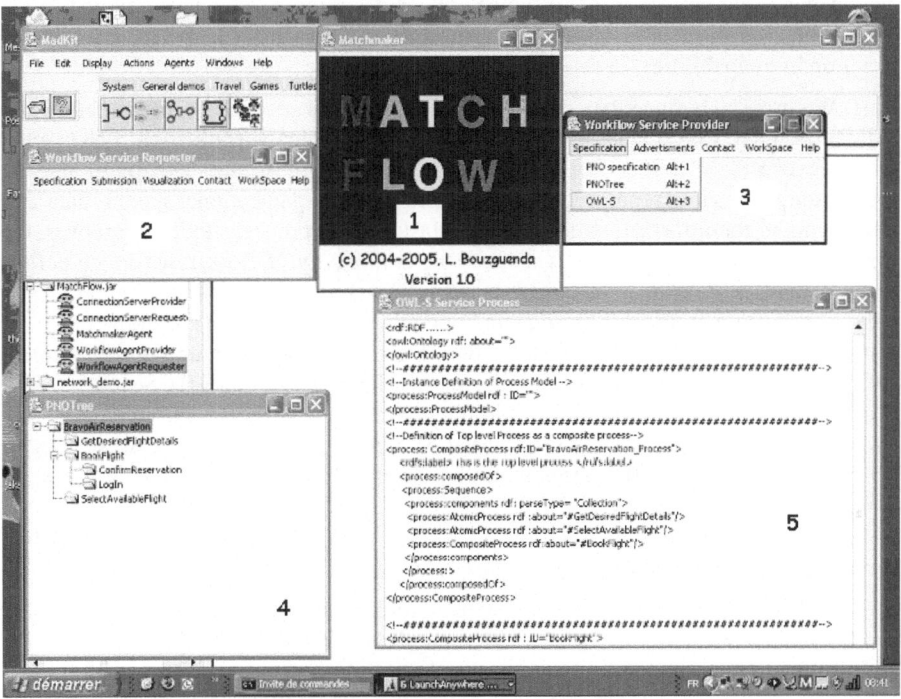

Fig. 5. Overview of the Implementation

Through this interface, the requester can i) specify a workflow service request (*Specification menu*), ii) advertise this request to the matchmaker (*Submission menu*), iii) visualize the providers offering services corresponding to this request (*Visualization menu*), iv) establish peer-to-peer connections with one of these providers (*Contact menu*), and, v) launch the execution of the selected service (*WorkSpace menu*). In a symmetric way, the top right window (number 3) represents an agent playing the role of a workflow service provider and a set of menus enables it to manage its offered services. As shown by window 3, the *Specification menu* includes three commands to support the specification and derivation process. The first command permits the specification of a PNO, the second one visualizes its corresponding PNO Tree (as shown by window 4), and the third one derives the corresponding OWL-S specification (as shown by window 5).

The example partially shown in windows 4 and 5 is based on the well-known *BravoAirReservation* case study proposed by the OWL-S Coalition. This example has been implemented. Windows 4 and 5 respectively give the *BravoAirReservation* PNO Tree and a partial view of its OWL-S service process. We do not give here the corresponding PNO, but figure 1 in section 2 is an extract of this net, restricted to the GetDesiredFlightDetails transition.

6   Discussion and Conclusion

This paper deals with Web Workflow Services description languages devoted to organizations involved in a loose IOW in order to help them describe their workflow needs and/or capabilities. In this paper:

- (i) We use PNOs since they permit workflow services specification, simulation and validation. First, PNOs are convenient to workflow service specification: PNOs are a glue between the different workflow models since they ease the description and the interaction, in a same representation, of actors of the organizational model, data and documents of the informational model, and tasks of the process model. Moreover, the graphical representation of PNOs reduces the complexity of workflow service definition. Second, PNOs are convenient to workflow services analysis, simulation and validation since they describe formal and operational (executable) specifications.
- (ii) We use OWL-S for W2S publication. First, OWL-S has an appropriate expressive power for workflow service description (as shown in section 3.2). Second, OWL-S includes ontology that eases semantic interoperability. Using an ontology, it is possible to build a shared business view based on a common vocabulary to solve semantic conflicts between IOW partners, and ease matchmaking mechanisms useful to compare workflow services.
- (iii) We provide rules and algorithms to derive PNO specifications onto OWL-S service profile and service process specifications.

Thus, PNO can be seen both as a graphical tool for specifying OWL-S services and as a formalism providing an operational semantics to OWL-S.

We found in the literature some works about specification of Web workflow services. Some of them concern workflow technology while others concern Web service technology.

Regarding workflow technology, the main proposition is YAWL [7]. YAWL permits a graphical specification of workflow services, validates them using a Petri Net representation, and provides tools to derive XML specifications. However, this proposition has two drawbacks. First, YAWL mainly focuses on the modeling of the process model and pay less attention to the informational and organizational models, and their interactions. So, we strongly believe that the PNO formalism is a better glue between the three workflow models than the YAWL one. Second, YAWL derive XML specifications, which solve only syntactic conflicts between organizations, while in loose IOW, the heterogeneous context requires semantics conflicts solving mechanisms.

Regarding Web services technology, [16] describes the OWL-S Editor Tool for visual modeling of OWL-S services. This tool permits the description of semantic Web services using standard UML Activity Diagrams and derives the corresponding OWL-S specifications. However, the UML Activity Diagrams notation (versus formalism) is not adequate for workflow services specification and does not provide a well-founded and operational semantics to validate the described services. Besides, OWL-S has a first-order logic representation [13], which is insufficient to capture and validate the behavior of business processes. [17] gives an operational semantics to DAML-S [18] using conventional Petri Nets. Unfortunately, DAML-S was discontinued in favor of OWL-S some while ago. Moreover, we believe that conventional Petri nets used in [17] are less adequate to workflow services specification than PNO as shown in section 2.2. Finally, [19] derives BPEL4WS specifications onto PNML specifications. The drawback of this proposition is to only consider syntactic aspects both at the specification and execution levels. Indeed, BPEL4WS [5], which is a composition Web service language, does not include ontology mechanism. PNML [20] is an XML specification which solves only syntactic conflicts between organizations. As seen before, such drawbacks are restrictive.

As future works, we plan to complete this work refining the OWL-S ontology in order to integrate some workflow processes particularities. Indeed, the analyze and simulation of a PNO enable the deduction of process properties (ending, reachability, quasi-liveness, …) and some performance evaluations (average throughput time, average waiting time, occupation rates of resources, …) which are not taken into account in the current OWL-S ontology. This information is however relevant and useful to workflow requesters to compare and select providers. Hence, we propose to refine the Service Model class by adding a specific sub-class describing these process properties.

References

1. van der Aalst, W.: Inter-Organizational Workflows: An Approach Based on Message Sequence Charts and Petri Nets. Int. Journal on Systems Analysis, Modeling and Simulation 34(3) (1999) 335–367
2. Divitini, M., Hanachi, C., Sibertin-Blanc, C.: Inter Organizational Workflows for Enterprise Coordination. In: Omicini, A., Zambonelli, F., Klusch, M., Tolksdorf, R. (eds): Coordination of Internet Agents, Springer-Verlarg, Berlin Heidelberg New-York (2001) 46–77

3. Andonoff, E., Bouzguenda, L., Hanachi, C., Sibertin-Blanc, C.: Finding Partners in the Coordination of Loose Inter-Organizational Workflow. 6th Int. Conference on the Design of Cooperative Systems (2004) Hyères (France) 147–162
4. World Wilde Web Coalition: the Web Service Description Language. Documentation available at: http://xml.coverpages.org/wsdl.html
5. BEA, IBM, Microsoft: Business Process Execution Language for Web Services. Documentation available at: http://xml.coverpages.org/bpel4ws.html
6. IBM: Web Services Flow Language. Documentation available at: http://xml.coverpages.org/wsfl.html
7. van der Aalst, W., Alderd, L., Dumas, M., ter Hofstede, A.: Design and Implementation of the YAWL System. 16th Int. Conference on Advanced Information System Engineering (2004) Riga (Latvia) 142–159
8. Workflow Management Coalition: XML Process Definition Language. Documentation available at: http://xml.coverpages.org/XPDL20010522.pdf
9. McIlraith, S., Son, TC., Zeng, H.: Semantic Web Services. Int. Journal on Intelligent Systems 16(2) (2001) 46–53
10. OWL Services Coalition: Ontology Web Language for Services Version 1.0. Documentation available at: http://xml.coverpages.org/ni2004-01-08-a.html
11. Sibertin-Blanc, C.: High Level Petri Nets with Data Structure. 6th Int. Workshop on Petri Nets and Applications (1985) Espoo (Finland)
12. van der Aalst, W.: The application of Petri Nets to Workflow Management. Int. Journal on Circuits, Systems and Computers 8(1) (1998) 21–66
13. Berardi, D., Gruninger, M., Hull, R., McIlraith, S.: Flows: A First-Order Logic Ontology for Web Services. Available at: www.wsmo.org/papers/presentations/FLOWS-WSMO-06-30-04.ppt
14. Lara, R., Roman, D., Polleres, A., Fensel, D.: A Conceptual Comparison of WSMO and OWL-S. 2nd International Conference on Web Services Europe (2004) Erfurt (Germany) 254–269
15. Ferber, J., Gutknecht, O.: TheMadKit Project: a Multi-Agent Development Kit. Documentation available at: http://www.madkit.org
16. Scicluna, J., Abela, C., Montebello, M.: Visual Modeling of OWL-S Services. Available at: http://www.daml.org/services/owl-s/pub-archive.html
17. Narayanan, S., McIlraith, S.: Simulation, Verification and Automated Composition of Web Services. 11th Int. World Wild Web Conference (2002) Honolulu (Hawaii) 77–88
18. Ankolekar, A., Burstein, M., Hobbs, J., Lassila, O., Martin, D., McDermott, D., McIlraith, S., Narayanan, S., Paolucci, M., Payne, T., Sycara, K.: DAML-S: Web Service Description for the Semantic Web. 6th Int. Semantic Web Conference (2002) Sardinia (Italy) 348–363
19. Vidal, JM., Buhler, P., Stahl, C.: Multi Agent Systems with Workflows. Int. Journal on Internet Computing 8(1) (2004) 76–82
20. Billington, J., Christensen, S., van Hee, K., Kindler, E., Kummer, O., Petrucci, L., Post, R., Stehno, C., Weber, M.: The Petri Net Markup Language: Concept, Technology and Tools. 23rd Int. Conference on Applications and Theory of Petri Nets (2003) Eindhoven (The Netherlands) 483–505

An Intuitive Formal Approach to Dynamic Workflow Modeling and Analysis

Jiacun Wang[1], Daniela Rosca[1], William Tepfenhart[1], Allen Milewski[1],
and Michael Stoute[2]

[1] Department of Software Engineering,
Monmouth University,
West Long Branch, NJ 07762, USA
{jwang, drosca, btepfenh, amilewsk}@monmouth.edu
[2] Intellipro, Inc.
255 Old New Brunswick Road,
Piscataway, NJ 08854, USA
jason@intellipro.com

Abstract. The increasing dynamics and the continuous changes of business processes raise a challenge to the research and implementation of workflows. The significance of applying formal approaches to the modeling and analysis of workflows has been well recognized and many such approaches have been proposed. However, these approaches require users to master considerable knowledge of the particular formalisms, which impacts the application of these approaches on a larger scale. This paper presents a new formal, yet intuitive approach for the modeling and analysis of workflows, which attempts to overcome the above problem. In addition to the abilities of supporting workflow validation and enactment, this new approach possesses the distinguishing feature of allowing users who are not proficient in formal methods to build up and dynamically modify the workflow models that address their business needs.

1 Introduction

Although workflow is an old concept [13] its research and implementation are gaining momentum due to the increasing dynamics and the continuous changes of the market places. The business environment today is undergoing rapid and constant changes. The way companies do business, including the business processes and their underlying business rules, ought to adapt to these changes flexibly with minimum interruption to ongoing operations [3,5]. This flexibility becomes of a paramount importance in applications such as an incident command system (ICS). An ICS would support the activities necessary for the allocation of people, resources and services in the event of a major natural or terrorist incident. An ICS would need to deal with frequent changes of the course of actions dictated by incoming events, a predominantly volunteer-based workforce, the need to integrate various software tools and organizations, a highly distributed workflow management.

Dealing with these issues generates many challenges for a workflow management system. The need of making many ad-hoc changes calls for an on-the-fly verification of the correctness of the modified workflow. This cannot be achieved without an underlying

W.M.P. van der Aalst et al. (Eds.): BPM 2005, LNCS 3649, pp. 137–152, 2005.

formal approach of the workflow, which does not leave any scope for ambiguity and sets the ground for analysis. Yet, since our main users will be volunteers from various backgrounds, with little computer experience, we need to provide a tool with highly intuitive features for the description and modification of the workflows.

A number of formal modeling techniques have been proposed in the past decades [1, 6, 8, 10, 11, 12]. Van der Aalst [9] identifies three reasons for using Petri Nets in workflow modeling. Firstly, Petri Nets possess formal semantics despite their graphical nature. Secondly, instead of being purely event-based, Petri Nets can explicitly model states, and lastly it is a theoretical proven analysis technique. Other than Petri Nets, techniques such as state charts have also been proposed for modeling WFMS [4]. Although state charts can model the behavior of workflows, they have to be supplemented with logical specification for supporting analysis. Singh et al [7] use event algebra to model the inter-task dependencies and temporal logic. Attia et al [2] have used computational tree logic to model tasks by providing their states together with significant events corresponding to the state transitions (start, commit, rollback etc) that may be forcible, rejectable, or delayable.

As indicated in [10], it is desirable that a business process model can be understood by the various stakeholders involved in an as straightforward manner as possible. Unfortunately, a common major drawback that all the above formal approaches suffer is that only users who have the expertise in these particular formal methods can build their workflows and dynamically change the business rules within the workflows. For example, in order to add a new task to a Petri-net based workflow, one must manipulate the model in terms of transitions, places, arcs and tokens, which can be done correctly and efficiently only by a person with a good understanding of Petri-nets. This significantly affects the application of these approaches on a large scale. This paper attempts to define a new formalism for the modeling and analysis of workflows, which, in addition to the abilities of supporting workflow validation and enactment, possesses the distinguishing feature of allowing users who are not proficient in formal methods to build up and dynamically modify the workflows that address their business needs.

The paper is organized as follows: Section 2 presents the new workflow formalism (WIFA – Workflows Intuitive Formal Approach), its state transition rules and its modeling power. Section 3 introduces well-formed workflows and how to build up a well-formed workflow. Section 4 gives a brief description of our tool for workflows modeling and analysis. Section 5 presents conclusions and ideas for the continuation of this work.

2   The WIFA Workflow Model

In general, a workflow consists of processes and activities, which are represented by well-defined *tasks*. The entities that execute these tasks are humans, application programs or database management systems. These tasks are related and dependent on one another based on business policies and rules [4]. In this section, we introduce the WIFA workflow model which captures tasks and relations among them in a workflow. We also define a set of state transition rules to facilitate the analysis of the dynamic behavior of a workflow.

2.1 WIFA Workflow Model Definitions

The control dependencies among tasks contain the order in which they can execute. Two tasks are said to have *precedence constraints* if they are constrained to execute in some order. As a convention, we use a partial-order relation <, called a *precedence relation*, over the set of tasks to specify the precedence constraints among tasks. A task T_i is a *predecessor* of another task T_j (and T_j a *successor* of T_i) if T_j cannot begin execution until the execution of T_i completes. A short-hand notation for this fact is $T_i < T_j$. T_i is an immediate predecessor of T_j (and T_j an immediate successor of T_i) if $T_i < T_j$ and there is no other task T_k such that $T_i < T_k < T_j$. We denote this fact with notation $p_{ij} = 1$. Naturally, the fact that T_i is *not* an immediate predecessor of T_j is denoted by $p_{ij} = 0$. Two tasks are independent when neither $T_i < T_j$ nor $T_j < T_i$. A classic way to represent the precedence constraints among tasks in a set T is by a directed graph $G = (T, <)$, in which each vertex represents a task in T, and there is a directed edge from vertex T_i to vertex T_j if T_i is an immediate predecessor of T_j. The graph is called a *precedence graph*.

Definition 1 (preset of a task): The preset of a task T_k, denoted by $*T_k$, is

$$*T_k = \{T_i \mid p_{ik} = 1\}.$$

Definition 2 (postset of a task): The postset of a task T_k, denoted by T_k*, is

$$T_k* = \{T_i \mid p_{ki} = 1\}.$$

Basically, the preset of a task is the set of all tasks that are immediate predecessors of the task, while the postset of a task is the set of all tasks that are immediate successors of the tasks. If $|T_k*| \geq 1$, then the execution of T_k might trigger multiple tasks. Suppose $\{T_i, T_j\} \subseteq T_k*$. There are two possibilities: (1) T_i and T_j can be executed simultaneously, and (2) only one of them can be executed, and the execution of one will disable the other due to the conflict between them. We denote the former case by $c_{ij} = c_{ji} = 0$, and the latter case by $c_{ij} = c_{ji} = 1$.

If $|*T_k| \geq 1$, then based on the aforementioned classic precedence model, the execution of T_k won't start until *all* of its immediate predecessors are executed. This precedence constraint is also called *AND precedence constraint*. An extension to this classic precedence model is to allow a task to be executed when *some* of its immediate predecessors are executed. This loosens the precedence constraints to some extent, and the loosened precedence constraint is also called *OR precedence constraint*. Obviously, the *OR* precedence model provides more flexibility than the classic *AND* precedence model in describing the dependencies among tasks. So in this paper, the *OR* precedence model is adopted. The *AND* precedence model can be viewed as a special case of the *OR* precedence model.

Suppose $*T_k = \{T_{k1}, T_{k2}, \ldots T_{kn}\}$, $n \geq 1$. Define $A(T_k) = \{A_1, A_2, \ldots A_h\}$, $h \geq 1$ such that

1) $A_i \subseteq *T_k$, $i = 1, 2, \ldots, h$, i.e. $A(T_k)$ is a set of subsets of $*T_k$.
2) $A_i \neq A_j$, $\forall i \neq j$, $i, j \in \{1, 2, \ldots, h\}$, i.e. these subsets are all different.

3) T_k is executable if and only if all tasks in any $A_i \in A(T_k)$ are executed. In other words, T_k can be triggered by any subset in $A(T_k)$, but only after all tasks in that subset are executed.

The set $A(T_k)$ is used to specify the pre-condition set for T_k to become executable.

The state of a workflow can be described as an array whose elements are the states of all individual tasks in the workflow. Denote by S a state of a workflow, then $S = (S(T_1), S(T_2), ..., S(T_m))$.

Now we are ready to formally define our WIFA workflow model.

Definition 3 (workflow): A workflow is $WF = (T, P, C, A, S_0)$, where

1) $T = \{T_1, T_2, ..., T_m\}$ is a set of *tasks*, $m \geq 1$.

2) $P = (p_{ij})_{m \times m}$ is the *precedence matrix* of the task set. If T_i is the direct predecessor of T_j, then $p_{ij} = 1$; otherwise, $p_{ij} = 0$.

3) $C = (c_{ij})_{m \times m}$ is the *conflict matrix* of the task set. $c_{ij} \in \{0, 1\}$ for $i = 1, 2, ...m$ and $j = 1, 2, ... m$.

4) $A = (A(T_1), A(T_2), ..., A(T_m))$ defines *pre-condition set* for each task. $\forall T_k \in T$, $A(T_k): *T_k \rightarrow 2^{*T_k}$. Let set $A' \in A(T_k)$. Then $T_i \in A'$ implies $p_{ik} = 1$.

5) $S_0 \in \{0, 1, 2, 3\}^m$ is the *initial state* of the workflow.

Definition 4 (state values): Denote a state of the WF by $S = (S(T_1), S(T_2), ..., S(T_m))$, where $S(T_i) \in \{0, 1, 2, 3\}$.

1) $S(T_i) = 0$ means T_i is *not executable* at state S and *not executed previously*.

2) $S(T_i) = 1$ means T_i is *executable* at state S and *not executed previously*.

3) $S(T_i) = 2$ means T_i is *not executable* at state S and *executed previously*.

4) $S(T_i) = 3$ means T_i is *executable* at state S and *executed previously*.

By the definition of state values, at any state, only those tasks whose values are either 1 or 3 can be selected for execution. Suppose task T_i at state S_a is selected for execution, and the new state resulted from the execution of T_i is S_b, then the execution of T_i is denoted by $S_a(T_i)S_b$.

Now we can have a more accurate explanation on the conflict matrix C and the precondition set A of a task. Let tasks T_i, T_j and $T_k \in T$ with $p_{ki} = p_{kj} = 1$. Suppose there are three states S_a, S_b and S_c such that either $S_a(T_i) = S_a(T_j) = 1$ or $S_a(T_i) = S_a(T_j) = 3$, and $S_a(T_i)S_b$ and $S_a(T_j)S_c$.

1) If $S_a(T_i) = S_a(T_j) = 1$, then $c_{ij} = c_{ji} = 1$ implies $S_b(T_j) = S_c(T_i) = 0$, and $c_{ij} = c_{ji} = 0$ results in $S_b(T_j) = S_c(T_i) = 1$.

2) If $S_a(T_i) = S_a(T_j) = 3$, then $c_{ij} = c_{ji} = 1$ implies $S_b(T_j) = S_c(T_i) = 2$, and $c_{ij} = c_{ji} = 0$ results in $S_b(T_j) = S_c(T_i) = 3$.

On the other hand, suppose $A(T_k) = \{A_1, A_2, ... A_h\}$, $h \geq 1$. Then $S_a(T_k) \in \{1, 3\}$ if $\exists A_i \in A(T_k)$ such that $S_a(T_j) = 2$ for $\forall T_j \in A_i$.

Definition 5: (initial state) At the initial state S_0, for any task $T_i \in T$, if there is no T_j such that $p_{ji} = 1$, then $S_0(T_i) = 1$; otherwise $S_0(T_i) = 0$.

Note that tasks that have no predecessor do not need to wait for any other task to execute first. In other words, these tasks are executable immediately. We assume that there is always such kind of tasks in a workflow. They are the initial triggers or "starting" tasks of workflows. In Definition 1 there is no restriction on the preset and post-sets of tasks. Therefore, there may be multiple tasks whose presets are empty, and there may be multiple tasks whose postsets are empty. In other words, this formalism supports multiple "starting" tasks and "ending" tasks in a workflow.

2.2 State Transition Rules

The dynamics of a workflow can be captured by state transitions. Of course, state transitions should be guided by a set of state transition rules. In this subsection, we define the rules.

Definition 6: (state transition rules) If $S_a(T_i)S_b$, then $\forall\ T_j \in T$,

1) If $T_j = T_i$ then $S_b(T_j) = 2$;

2) If $T_j \neq T_i$ then the state value of T_j at new state S_b depends on its state value at state S_a. We consider four cases:

   Case A $- S_a(T_j) = 0$:
   If $p_{ij} = 1$ and $\exists A' \in A(T_j)$ such that $S_b(T_k) = 2$ for any $T_k \in A'$, then $S_b(T_j) = 1$; otherwise $S_b(T_j) = 0$.

   Case B $- S_a(T_j) = 1$
   If $c_{ij} = 0$ then $S_b(T_j) = 1$; otherwise $S_b(T_j) = 0$.

   Case C $- S_a(T_j) = 2$
   If $p_{ij} = 1$ and $\exists A' \in A(T_j)$ such that $S_b(T_k) = 2$ for any $T_k \in A'$, then $S_b(T_j) = 3$; otherwise $S_b(T_j) = 2$.

   Case D $- S_a(T_j) = 3$
   If $c_{ij} = 0$ then $S_b(T_j) = 3$; otherwise $S_b(T_j) = 2$.

According to the above state transition rules, a task's state value at a given state other than the initial state is 0 iff one of the following holds:

1) Its state value is 0 in the previous state, and it is not the successor of the task which is just executed.

2) Its state value is 0 in the previous state, and it is the successor of the task which is just executed, but for each of its precondition sets there is at least one task that is not executed.

3) Its state value is 1 in the previous state but it conflicts with the task which is just executed.

A task's state value at a given state other than the initial state is 1 iff one of the following holds:

1) Its state value is 0 in the previous state, it is the successor of the task which is just executed, and in at least one of its precondition sets all tasks are executed.

2) Its state value is 1 in the previous state and it does not conflict with the task which is just executed.

A task's state value at a given state other than the initial state is 2 if and only if one of the following holds:

1) It is just executed.
2) Its state value is 2 in the previous state, and it is not the successor of the task which is just executed.
3) Its state value is 2 in the previous state, and it is the successor of the task which is just executed, but for each of its precondition sets there is at least one task that is not executed.
4) Its state value is 3 in the previous state but it conflicts with the task which is just executed.

A task's state value at a given state other than the initial state is 3 if and only if one of the following holds:

1) Its state value is 2 in the previous state, it is the successor of the task which is just executed, and there is at least one of its precondition sets in which every task is executed.
2) Its state value is 3 in the previous state and it does not conflict with the task which is just executed.

Note that a state value can increment from 0 to 1, from 1 to 2 or from 2 to 3; it can also decrement from 1 to 0 or from 3 to 2. But it cannot decrement from 2 to 1. Fig. 1 illustrates possible state value changes for a given task when a workflow changes from one state to another state due to the execution of some task.

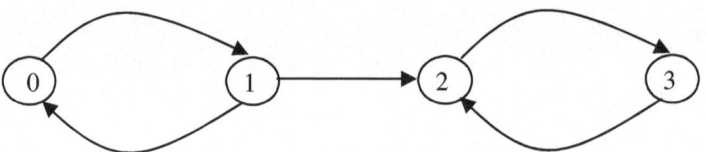

Fig. 1. State transition of an individual task

2.3 Example

We now illustrate how to apply the WIFA approach to workflow modeling and analysis through an example. Assume that we have a workflow with eight tasks, namely T_1, T_2, ... T_8. Its specification is as follows:

- T_1 is the direct predecessor of T_2 and T_3, T_2 is the immediate predecessor of T_4, T_4 is the immediate predecessor of T_5, T_5 is the immediate predecessor of T_6 and T_7, T_6 is the second immediate predecessor of T_2, T_3 is the immediate predecessor of T_7, and T_7 is the second immediate predecessor of T_8. See Fig. 2.

- T_6 and T_7 conflict with each other. In other words, after T_5 is executed, if T_6 is selected for execution, then the execution of T_6 will make T_7 not executable and vice versa.

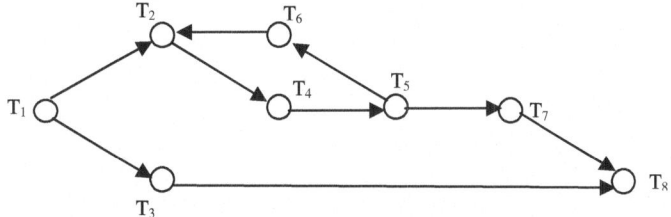

Fig. 2. Precedence graph of an eight-task workflow

- T_1 is executable when the workflow is started. T_2 and T_3 become executable when T_1 is executed. T_2 also becomes executable when T_6 is executed. T_4 becomes executable when T_2 is executed. T_5 becomes executable when T_4 is executed. T_6 and T_7 become executable when T_5 is executed. T_4 becomes executable when T_2 is executed. T_5 becomes executable when T_4 is executed. And T_8 becomes executable when both T_3 and T_7 are executed.

This workflow is formulated in the WIFA framework as:

$T = \{T_1, T_2, T_3, T_4, T_5, T_6, T_7, T_8\}$,

$$
P = \begin{bmatrix}
0 & 1 & 1 & 0 & 0 & 0 & 0 & 0 \\
0 & 0 & 0 & 1 & 0 & 0 & 0 & 0 \\
0 & 0 & 0 & 0 & 0 & 0 & 0 & 1 \\
0 & 0 & 0 & 0 & 1 & 0 & 0 & 0 \\
0 & 0 & 0 & 0 & 0 & 1 & 1 & 0 \\
0 & 1 & 0 & 0 & 0 & 0 & 0 & 0 \\
0 & 0 & 0 & 0 & 0 & 0 & 0 & 1 \\
0 & 0 & 0 & 0 & 0 & 0 & 0 & 0
\end{bmatrix}, \quad
C = \begin{bmatrix}
0 & 0 & 0 & 0 & 0 & 0 & 0 & 0 \\
0 & 0 & 0 & 0 & 0 & 0 & 0 & 0 \\
0 & 0 & 0 & 0 & 0 & 0 & 0 & 0 \\
0 & 0 & 0 & 0 & 0 & 0 & 0 & 0 \\
0 & 0 & 0 & 0 & 0 & 0 & 0 & 0 \\
0 & 0 & 0 & 0 & 0 & 0 & 1 & 0 \\
0 & 0 & 0 & 0 & 0 & 1 & 0 & 0 \\
0 & 0 & 0 & 0 & 0 & 0 & 0 & 0
\end{bmatrix}
$$

$A(T_1) = \emptyset, A(T_2) = \{\{T_1\}, \{T_6\}\}, A(T_3) = \{\{T_1\}\}$,

$A(T_4) = \{\{T_2\}\}, A(T_5) = \{\{T_4\}\}$,

$A(T_6) = A(T_7) = \{\{T_5\}\}, A(T_8) = \{\{T_3, T_7\}\}$.

$S_0 = (1, 0, 0, 0, 0, 0, 0, 0)$.

Now let us examine the execution of this workflow. At S_0, T_1 is the only executable task. Let $S_0(T_1)S_1$, then based on the state transition rule, we have

$S_1(T_1) = 2$ (Rule 1)

$S_1(T_2) = S_1(T_3) = 1$ (Rule 2A)

$S_1(T_4) = S_1(T_5) = S_1(T_6) = S_1(T_7) = S_1(T_8) = 0$ (Rule 2A)

So $S_1 = (2, 1, 1, 0, 0, 0, 0, 0)$.

At S_1, T_2, T_3 are executable, because their state values are 1. Let $S_1(T_2)S_2$, then based on the state transition rule, we have

$S_2(T_1) = 2$ (Rule 2C)

$S_2(T_2) = 2$ (Rule 1)

$S_2(T_3) = 1$ (Rule 2B)

$S_2(T_4) = 1$ (Rule 2A)

$S_2(T_5) = S_2(T_6) = S_2(T_7) = S_2(T_8) = 0$ (Rule 2A)

So $S_2 = (2, 2, 1, 1, 0, 0, 0, 0)$.

At S_2, T_3 and T_4 are executable, because their state values are 1. Let $S_2(T_3)S_3$, then based on the state transition rule, we have

$S_3(T_1) = S_3(T_2) = 2$ (Rule 2C)

$S_3(T_3) = 2$ (Rule 1)

$S_3(T_4) = 1$ (Rule 2B)

$S_2(T_5) = S_2(T_6) = S_2(T_7) = S_2(T_8) = 0$ (Rule 2A)

So $S_3 = (2, 2, 2, 1, 0, 0, 0, 0)$. Notice that T_6 is not executable now because neither T_4 nor T_5 is executed.

At S_3, only T_4 is executable, because it is the only task with state value 1 or 3. Let $S_3(T_4)S_4$, then it follows from the state transition rules that $S_4 = (2, 2, 2, 2, 1, 0, 0, 0)$. At S_4, only T_5 is executable. Let $S_4(T_5)S_5$, then it follows from the state transition rules that $S_5 = (2, 2, 2, 2, 2, 1, 1, 0)$.

At S_5, T_6 and T_7 are executable, because their state values are 1. The execution T_6 causes the workflow to proceed along the T_2-T_4-T_5-T_6-T_2 loop. Let $S_5(T_6)S_6$, then based on the state transition rule, we have

$S_6(T_1) = S_5(T_3) = S_5(T_4) = S_5(T_5) = 2$ (Rule 2C)

$S_6(T_2) = 3$ (Rule 2C)

$S_6(T_6) = 2$ (Rule 1)

$S_6(T_7) = S_6(T_8) = 0$ (Rule 2A)

So $S_6 = (2, 3, 2, 2, 2, 2, 0, 0)$. Notice that T_7 becomes not executable now because T_6 and T_7 are in conflict.

Task T_2 will execute at S_6, which results in $S_7 = (2, 2, 2, 3, 2, 2, 0, 0)$. Then task T_4 will execute at S_7, which results in $S_8 = (2, 2, 2, 2, 3, 2, 0, 0)$. Then task T_5 will execute at S_8, which results in $S_9 = (2, 2, 2, 2, 2, 3, 1, 0)$. Let $S_9(T_7)S_{10}$, then based on the state transition rule, we have

$S_{10}(T_1) = S_5(T_2) = S_5(T_3) = S_5(T_4) = S_5(T_5) = 2$ (Rule 2C)

$S_{10}(T_6) = 0$ (Rule 2A)

$S_{10}(T_7) = 2$ (Rule 1)

$S_{10}(T_8) = 1$ (Rule 2A)

So $S_{10} = (2, 2, 2, 2, 2, 2, 2, 1)$. Notice that T_8 is executable now because both T_3 and T_7 are executed. The execution of T_8 results in $S_{11} = (2, 2, 2, 2, 2, 2, 2, 2)$. At this state, no more tasks are executable.

The above analysis only traces one execution path. The entire state transition graph of this workflow is depicted in Fig. 3, which contains 22 states in total. The workflow may either stop at state $(2, 2, 2, 2, 2, 2, 2, 2)$ if the workflow is looped or at state $(2, 2, 2, 2, 2, 0, 2, 2)$ if it doesn't go through the loop.

Discussion: This example shows that our formal workflow model can be directly formulated from the users' specification of the workflow. The importance of this fact lies in that the proposed approach supports automated formulation from users' workflow description to a formal model of the workflow. More discussion will be provided in next section.

2.4 WIFA Modeling Power

The characteristics exhibited by the task executions of workflows such as concurrency, decision making, synchronization and loops are modeled very effectively with the WIFA model. These characteristics are represented using a set of simple constructs:

1) *Sequential execution*: In the example, tasks T_1 and T_2 are executed sequentially. This relationship is specified by $p_{12} = 1$ in the precedence matrix. Such precedence constraints are typical of execution tasks in a workflow. Also, this construct models the causal relationships among activities.

2) *Conflict*: In the example, tasks T_6 and T_7 conflict with each other. This is specified by $c_{67} = c_{76} = 1$ in the conflict matrix. Such a situation will arise, for example, when a user has to choose among multiple possible actions.

3) *Concurrency*: In example, tasks T_2 and T_3 are concurrent. Concurrency is an important attribute of a workflow. A sufficient condition for two tasks to be concurrent is that they are successors of some other task, and they are not in conflict.

4) *Synchronization*: Oftentimes, a task in a workflow has to wait for execution results of two or more other tasks before it can be executed. The resulting synchronization of tasks can be captured by the pre-condition set of a task. In the example, $A(T_8) = \{T_3, T_7\}$, means T_3 is synchronized with T_7 for T_8.

5) *Loop*: Loop is a common characteristic within a workflow structure where some tasks are executed repeatedly. As an example shown in Fig. 2, tasks T_2, T_1, T_5 and T_6 could be executed again and again.

6) *Mutual exclusion*: Mutual exclusion is defined as following.

Definition 7 (mutual exclusion) Two tasks T_i and T_j are said to be mutual exclusive based on the following recursive definition:

1) T_i and T_j are mutual exclusive if $c_{ij} = 1$.

2) If T_i and T_j are mutual exclusive and $*T_k = \{T_i\}$, then so are T_k and T_j.

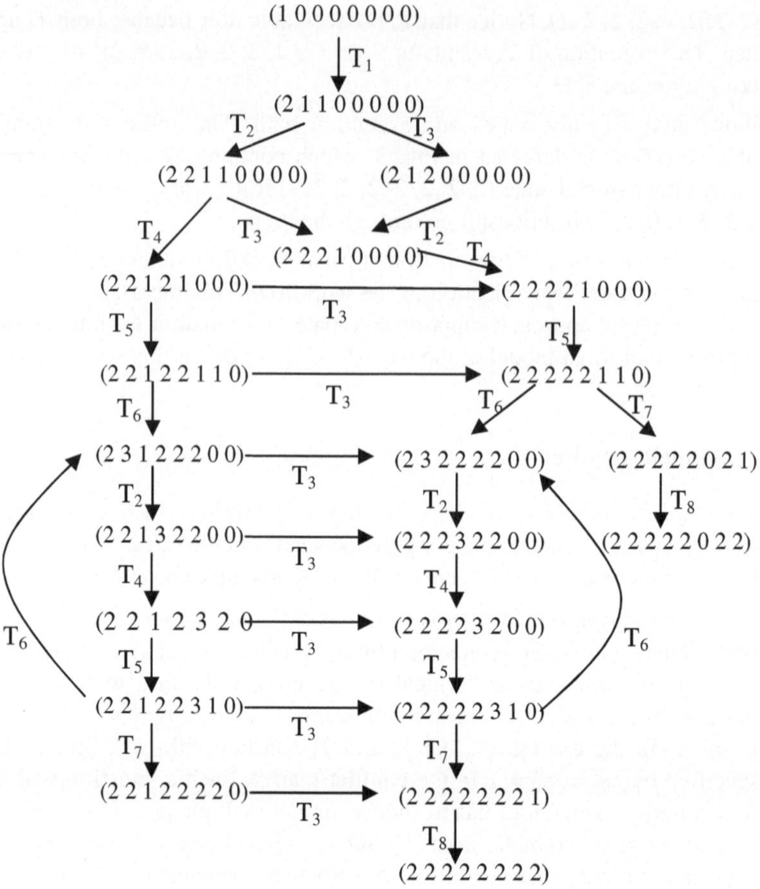

(1 0 0 0 0 0 0 0)
T_1
(2 1 1 0 0 0 0 0)
T_2
T_3
(2 2 1 1 0 0 0 0)
(2 1 2 0 0 0 0 0)
T_4
T_3
T_2
(2 2 2 1 0 0 0 0)
T_4
(2 2 1 2 1 0 0 0)
(2 2 2 2 1 0 0 0)
T_3
T_5
T_5
(2 2 1 2 2 1 1 0)
(2 2 2 2 2 1 1 0)
T_3
T_6
T_6
T_7
(2 3 1 2 2 2 0 0)
(2 3 2 2 2 2 0 0)
(2 2 2 2 2 0 2 1)
T_3
T_2
T_2
T_8
(2 2 1 3 2 2 0 0)
(2 2 2 3 2 2 0 0)
(2 2 2 2 2 0 2 2)
T_3
T_4
T_4
(2 2 1 2 3 2 0
(2 2 2 2 3 2 0 0)
T_6
T_3
T_6
T_5
T_5
(2 2 1 2 2 3 1 0)
(2 2 2 2 2 3 1 0)
T_3
T_7
T_7
(2 2 1 2 2 2 2 0)
(2 2 2 2 2 2 2 1)
T_3
T_8
(2 2 2 2 2 2 2 2)

Fig. 3. State transition graph of the example workflow

T_2
T_4
T_5
T_1
T_7
T_3
T_6

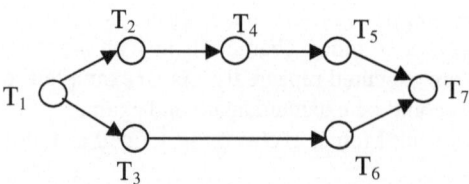

Fig. 4. A precedence graph where T_2 and T_3 are in conflict

According to this definition, any two mutual exclusive tasks are rooted from two conflicting tasks. For example, assume that tasks T_2 and T_3 in Fig.4 are in conflict. Then any task from set $\{T_2, T_4, T_5\}$ and any task from set $\{T_3, T_6\}$ are mutual exclusive. So, when T_2 and T_3 are triggered by T_1, either the branch T_2-T_4-T_5 or the branch T_3-T_6 will be chosen to execute.

3 Well-Formed Workflows

In this section, we introduce *well-formed workflows* which have no dangling tasks and are guaranteed to finish. We particularly discuss *confusion-free workflows*, which are a class of well-formed workflows and have some distinguishing properties. We introduce how to build confusion-free workflows, and how to ensure a workflow remains confusion-free when it needs to be changed.

3.1 Well-Formed Workflow Definitions

Definition 8 (reachable set): A state S of a workflow is *reachable* from the initial state if and only if there is a sequence of tasks that are executable sequentially from the initial state and the execution of these tasks leads the workflow to state S. The set of all reachable states, including the initial state, is called the reachable set. It is denoted by R.

Definition 9 (well-formed workflow): A workflow is *well-formed* if and only if the following two *behavior conditions* are met:

1) $\forall T_i \in T, \exists S \in R$ such that $S(T_i) = 1$. (i.e. there is no dangling task.)

2) $\exists S \in R$ such that $S(T_i) \in \{0, 2\}$ for $\forall T_i \in T$. (i.e. there is at least one ending state.)

The example workflow given in Section 2.3 is well-formed, because every task in this workflow is executable, and there are two ending states. In general, the validation of a workflow being well-formed requires the reachability analysis of the workflow. Below we introduce *confusion-free* workflows, which are a class of well-formed workflows with some restrictions imposed on their structure.

Definition 10 (confusion-free workflow): A well-formed workflow is *confusion-free* if and only if the following two *structural conditions* are met:

1) $\forall T_k \in T$ with $|T_k{}^*| \geq 3$, if $\exists T_i, T_j \in T_k{}^*$ such that $c_{ij} = 1$ (or $c_{ij} = 0$), then for $\forall T_a, T_b \subset T_k{}^*$ $c_{ab} = 1$ (or $c_{ab} = 0$) (i.e., either all tasks triggered by the same task are in conflict, or no pair of them are in conflict.)

2) $\forall T_k \in T$ with $^*T_k = \{T_{k1}, T_{k2}, ..., T_{kn}\}, n \geq 2$, either
$$A(T_k) = \{\{ T_{k1}, T_{k2}, ..., T_{kn}\}\}, \tag{1}$$
or
$$A(T_k) = \{\{T_{k1}\}, \{T_{k2}\}, ..., \{T_{kn}\}\} \tag{2}$$
(i.e., T_k becomes executable either when all of its predecessor tasks are executed, or when any one of them is executed.)

Based on this definition, the example workflow in Section 2.3 is also confusion-free. As will be described next in Theorem 1, it is easy to construct and validate a confusion-free workflow.

From the perspective of triggering condition and relation among triggered tasks, tasks in a confusion-free well-formed workflow can be classified into four types:

1) *And-In-Parallel-Out* A task belongs to this class iff it is not executable until all its direct predecessor tasks are executed, and after it is executed, all its direct successor tasks can be executed in parallel.

2) *And-In-Conflict-Out* A task belongs to this class iff it is not executable until all its direct predecessor tasks are executed, and after it is executed, only one of its direct successor tasks can be executed.

3) *Or-In-Parallel-Out* A task belongs to this class iff it is executable as long as one of its direct predecessor tasks is executed, and after it is executed, all its direct successor tasks can be executed in parallel.

4) *Or-In-Conflict-Out* A task belongs to this class iff it is executable as long as one of its direct predecessor tasks is executed, and after it is executed, only one of its direct successor tasks can be executed.

Without loss of generality, a task with only one direct predecessor is treated as an "And-In" task, and a task with only one direct successor treated as a "Parallel-Out" task. Denote by set T_{AP} for all And-In-Parallel-Out tasks, T_{AC} for all And-In-Conflict-Out tasks, T_{OP} for all Or-In-Parallel-Out tasks, and T_{OC} for all Or-In-Conflict-Out tasks. Then in the example workflow, we have $T_{AP} = \{T_1, T_3, T_4, T_6, T_7, T_8\}$, $T_{AC} = \{T_5\}$, $T_{OP} = \{T_2\}$, and $T_{OC} = \varnothing$.

3.2 Build a Well-Formed Workflow

Theorem 1: Given a confusion-free, well-formed workflow $WF = (T, P, C, A, S_0)$, by adding a new task T_k to it, the obtained new workflow is denoted by $WF' = (T', P', C', A', S_0')$. Then WF' is also a confusion-free workflow if it matches one of the following cases:

1) $^*T_k = T_k^* = \varnothing$, i.e., $p'_{ki} = p'_{ik} = 0$ for all $T_i \in T' \setminus \{T_k\}$.

2) $^*T_k = \varnothing$, $T_k^* \neq \varnothing$, and $\forall T_i \in T_k^*$, if $A(T_i)$ is defined in the form of (1) in Definition 10, then $A'(T_i)$ is also defined in the form of (1) by adding T_k to the only set. If $A(T_i)$ is defined in the form of (2), then $A'(T_i)$ is also defined in the form of (2) by adding $\{T_k\}$ to $A(T_i)$.

3) $^*T_k \neq \varnothing$, $T_k^* = \varnothing$. If $A(T_k)$ is defined in the form of (1) in Definition 10, then there exists a S_a in WF such that all tasks in $^*T_k$ have state value of 2; If $A(T_k)$ is defined in the form of (2) in Definition 10, then there exists a S_a in WF such that at least one task in $^*T_k$ has state value of 2. In addition, $\exists T_i \in {}^*T_k$, if T_i triggers two or more conflicting tasks, then T_k conflicts with each of these tasks, otherwise, $c_{kj} = 0$ for any $T_j \in T_i^*$.

4) $^*T_k \neq \varnothing$, $T_k^* \neq \varnothing$, with all other conditions appear in 2) and 3). Besides, $\forall T_i \in T_k^*$, if T_i is also a predecessor of T_k (i.e., T_k introduces a loop), then $A(T_i)$ can only be in the form of (2) in Definition 10.

Proof:

Case 1): T_k is an isolated task. Based on Definition 3, T_k will not be in any other task's pre-condition set, so it has no impact to the original workflow *WF*, and the two structural conditions of confusion-free workflows are all met in *WF'*. Because T_k has no predecessors, so it is executable in S'_0. Since *WF* is well-formed, there must be an ending state $S_q \in R(WF)$, then state $S'_q = S_q \cup \{S(T_k) = 2\}$ is an ending state of *WF'*. Therefore, *WF'* is confusion-free.

Case 2) In this case, T_k has no predecessors, so it is executable in S'_0. We need to make sure that all tasks that are successors to T_k are still executable after adding in T_k. $\forall T_i \in T_k^*$, if $A'(T_i)$ is defined in the form of (1) by adding T_k to the only set, then that *WF* is confusion-free indicates that there is a state S_a in *WF* such that all tasks in $*T_i$ have state value of 2. Because T_k is unconditionally executable, so there must be a corresponding state S_a' in *WF'* such that $S_a' = S_a \cup \{S_a'(T_i) = 2\}$. Thus T_i is still executable in *WF'*. If $A'(T_i)$ is defined in the form of (2) by adding $\{T_k\}$ to $A'(T_i)$, then the execution of any task in $*T_i$ in *WF'* can still trigger T_i as it does in *WF*, and T_k is just an additional task to trigger T_i. Thus T_i is still executable in *WF'*. Since *WF* is well-formed, there must be an ending state $S_q \in R(WF)$, then state $S_q' = S_q \cup \{S(T_k) = 2\}$ is an ending state of *WF'*. In addition, $A'(T_i)$ is defined in one of the two desired forms. Therefore, *WF'* is also confusion-free.

Case 3) In this case, T_k has no successors. The other conditions guarantee already that task T_k is executable, and the two structural conditions of confusion-free workflows are also met. We only need to prove that the introduction of T_k won't cause other tasks to become non-executable. It is easy to understand that the state transition behavior of *WF'* from any state S' in which $S'(T_k) = 0$ is not affected due to the introduction of T_k. Suppose that at state S_a' we have $S_a'(T_k) = 1$ and T_k is triggered by T_i ($T_i \in *T_k$). If all tasks triggered by T_i are able to execute in parallel with T_k ($c_{kj} = 0$ for any $T_j \in T_i^*$), then T_k has no impact to the execution of other triggered tasks. The other possibility is that T_k is in conflict with any other task triggered by T_i. In this case, if T_k is not chosen for execution, the state transition behavior from S' will be just like the case in state $S = S' \setminus \{ S'(T_k) = 1\}$ of *WF*. All these suggests that *WF'* is also a confusion-free workflow.

Case 4) This case is a combination of *Case* 2 and *Case* 3. The *WF'* can be proved confusion-free by jointly applying the reasoning for these two cases if T_k does not introduce a loop to the workflow, In case T_k introduces a loop, since we already restrict that $\forall T_i \in T_k^*$, if T_i is also a predecessor of T_k, then $A(T_i)$ can only be in the form of (2) in Definition 10, T_i can be triggered as it is without T_k in place. Adding T_k simply introduces one more trigger to T_i. So the loop does not cause any task un executable.

The theorem is proved. □

Theorem 1 can serve as a rule in building a confusion-free workflow. At the beginning, the task set is empty. When the first task is introduced, the workflow is well-formed, because this single task has no predecessors and successors and it is executable. Then we add a second task. This second task can either be an isolated one (Case 1 of Theorem 1), or be a successor of the first task (Case 2 of Theorem 1), or be a predecessor of the first task (Case 3 of Theorem 1), or even be both a predecessor and successor to the first task (Case 4 of Theorem 1). Since the first task is the only possible successor or predecessor to the second task, the new workflow (with these two tasks) is still confusion-free. When we continue to introduce more tasks to the workflow, as long as we make sure each new task is added in such a way that it satisfies the conditions defined in one of the four cases, then the new workflow is guaranteed to be confusion-free.

4 Tool Support

We are currently in the process of developing a visual tool to automate the workflow editing and enactment. In this section we briefly introduce the tool.

The tool has three components: an editor, a simulator and a validator. The editor enables users to create a workflow with an easy to use drag and drop interface. As shown in Fig. 5, the editor has a Tool Box which contains all the objects available for dragging and dropping into the Working area, with each of the four types of tasks represented by a unique icon. Connections add the directional flow from one task to the next. Every connection must have one start task and one end task. When a user is

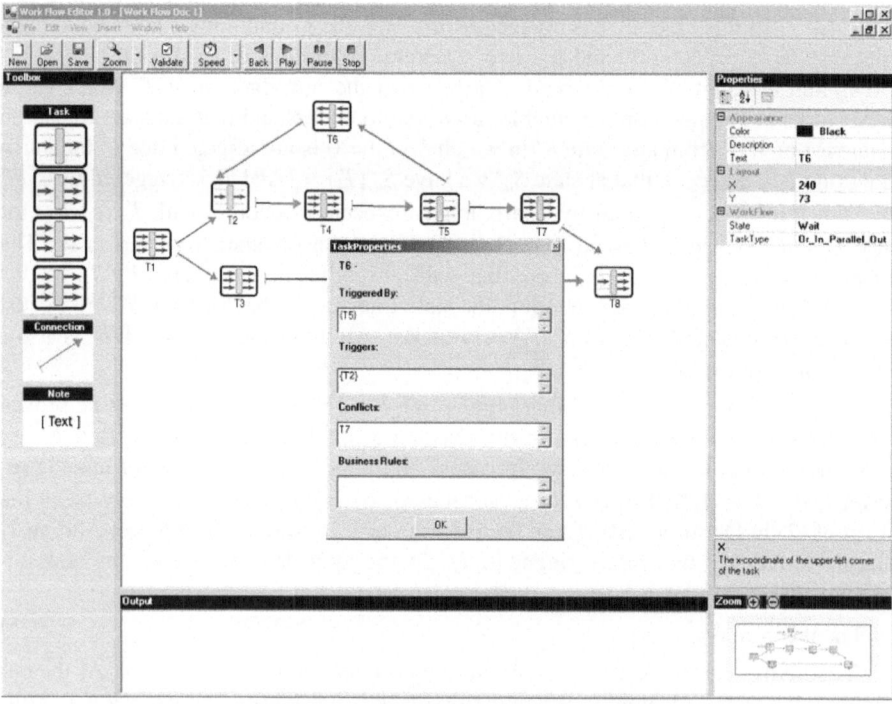

Fig. 5. Screenshot of the workflow editor

adding a connection, the working area will change into "connection start mode". The user will then select the start task by clicking on an existing task in the working area. Once a start task has been selected, the working area will change to "connection end mode". A phantom connection will follow the user until an end task is selected or the connection adding has been cancelled. Once the connection has been established, the connection is drawn and the working area returns to "normal mode".

Each object in the working area has general properties such as position, text, description among others. The general Properties area allows the user to see at a glance and change the properties of an object. This Properties area will be populated with the currently active object. Specially, the properties of each task can be seen by right-clicking a task and selecting the Task Properties. The task properties will show the tasks that the current task is triggered by, tasks that this task triggers, any conflicts with this task, and any business rules associated with the task.

A complete workflow can be saved as either an XML file or an image.

The simulator allows users to simulate the execution of a workflow. The users can set the simulation speed with the Speed command, and have options on Play, Back, Pause, and Stop.

The validator allows users to verify if their workflows are well-formed. The users can perform the validation at any stage of workflow construction.

5   Concluding Remarks

In this paper we presented a new formal, yet intuitive, approach for the modeling and analysis of workflows. We introduced our representation of tasks, relations among tasks, state transition rules, and the expressive power of this framework that enables the creation and enactment of workflows. We have showed our definition of well-formed workflows and how to build them, such that whenever a new task is added, it will not alter the well-formedness property of the workflow.

We are currently developing theorems on deleting a task from a well-formed workflow and changing some business rules in a well-formed workflow such that the modified workflow is still well-formed. Meanwhile, we are designing and implementing a visual tool to automate the workflow editing and enactment. The tool will allow the recording of an audit log that will permit the analysis and improvement of current workflows. We will also be working on extending our approach to the inter-organizational workflow modeling and analysis, to be able to represent the interactions between different people and organizations that need to work together for achieving different business goals.

References

1. N. R. Adam, V. Atluri and W. Huang, "Modeling and Analysis of Workflows Using Petri Nets", *Journal of Intelligent Information Systems*, pp. 131-158, March 1998.
2. P. C. Attie, M. P. Singh, A. Sheth and M. Rusibkiewicz, "Specifying Interdatabase Dependencies," *Proceedings 19th International Conference on Very Large Database*, pp.134-145, 1993.

3. P. Dourish, " Process Descriptions as Organizational Accounting Devices: The Dual use of Workflow Technologies", Paper presented at GROUP'01, (ACM), Sept. 30-Oct. 3, 2001, Boulder, Colorado, USA

4. P. Lawrence, editor, "Workflow Handbook 1997, Workflow Management Coalition", John Wiley and Sons, New York, 1997.

5. D.C. Marinescu, Internet-Based Workflow Management: Towards a Semantic Web, Wiley Series on Parallel and Distributed Computing, vol. 40, Wiley-Interscience, NY, 2002

6. D. Rosca, S. Greenspan, C. Wild, "Enterprise Modeling and Decision-Support for Automating the Business Rules Lifecycle", *Automated Software Engineering Journal,* Kluwer Academic Publishers, vol.9, pp.361-404, 2002.

7. M.P. Singh, G. Meredith, C. Tomlinson, and P.C. Attie, "An Event Algebra for Specifying and Scheduling Workflows," *Proceedings 4th International Conference on Database System for Advance Application*, pp. 53-60, 1995.

8. W.M.P. van der Aalst, "Verification of Workflow Nets", *Proceedings of Application and Theory of Petri Nets*, Volume 1248 of Lecture Notes in Computer Science, pp. 407-426, 1997.

9. W.M.P. van der Aalst, "Three Good Reasons for Using a Petri Net-Based Workflow Management System", *Proceedings of the International Working Conference on Information and Process Integration in Enterprises* (IPIC'96), pp. 179–201, Nov 1996.

10. W.M.P. van der Aalst, A.H.M. ter Hofstede, and M. Weske, "Business Process Management: A Survey." *International Conference on Business Process Management (BPM 2003)*, volume 2678 of *Lecture Notes in Computer Science*, pages 1-12. Springer-Verlag, Berlin, 2003.

11. J. Wang, Timed Petri Nets: Theory and Application, Kluwer Academic Publishers, 1998, ISBN: 0-7923-8270-6.

12. D. Wodtke and G. Weikum, "A Formal Foundation for Distributed Workflow Execution Based State Charts," *Proceedings 18th International Conference on Database theory*, 1997.

13. M.D. Zisman, "Representation, Specification and Automation of Office Procedures", *PhD thesis*, University of Pennsylvania, Warton School of Business, 1977.

Using the π-Calculus for Formalizing Workflow Patterns[⋆]

Frank Puhlmann and Mathias Weske

Hasso-Plattner-Institute for IT Systems Engineering,
at the University of Potsdam,
D-14482 Potsdam, Germany
{puhlmann, weske}@hpi.uni-potsdam.de

Abstract. This paper discusses the application of a general process theory – the π-calculus – for describing the behavioral perspective of workflow. The π-calculus is a process algebra that describes mobile systems. Mobile systems are made up of components that communicate and change their structure as a result of communication. The ideas behind mobility, communication and change can also enrich the workflow domain, where flexibility and reaction to change are main drivers. However, it has not yet been evaluated whether the π-calculus is actually appropriate to represent the behavioral patterns of workflow.

This paper investigates the issue and introduces a collection of workflow patterns formalizations, each with a sound formal definition and execution semantics. The formalizations can be used as a foundation for pattern-based workflow execution, reasoning, and simulation as well as a basis for future research on theoretical aspects of workflow.

1 Introduction

Recently, the π-calculus has been discussed as a formal foundation for workflow [1,2]. The advocates of the so called Third Wave claim that the π-calculus is a natural foundation for workflow as it is based on communication and change. Indeed, communication is required for inter-organizational workflow and service oriented languages like BPML, XLang, or BPEL4WS [3,4,5]. The ability to dynamically change workflows on demand is already an important topic in workflow research [6,7,8]. Despite these discussions, no formal and reasonably complete investigation of the π-calculus regarding the workflow domain has been made. This paper takes a first step by analyzing the capabilities of the π-calculus regarding workflow patterns [9]. It introduces a collection of workflow patterns formalizations, each with an unambiguous formal definition and execution semantics.

The formalizations can be used in two major directions. First, they build a foundation for pattern-based workflow execution, reasoning, and simulation, which is based upon the execution semantics and proving capabilities of a formal

---

[⋆] The work reported in this paper has been supported by the German Ministry of Research and Education (BMBF) by the PESOA project.

W.M.P. van der Aalst et al. (Eds.): BPM 2005, LNCS 3649, pp. 153–168, 2005.

algebra. Second, the formalizations show that the π-calculus is indeed a base for a precise definition of behavioral workflow requirements. At the same time it might open the door for future research, i.e. integrating other workflow perspectives like organizational, operational, or informational [10,11].

The remainder of this paper is organized as follows. Section 2 discusses related work. A brief introduction to the π-calculus is given in Section 3. Section 4 contains the formal definitions of workflow patterns; the main concepts are illustrated by examples. This paper is concluded with an outlook and directions for future work.

2   Related Work

Another approach of giving a detailed representation of the workflow patterns has been made with YAWL [12]. Starting as an endeavor as a workflow language of high expressiveness, YAWL has received considerable attention recently. The focus of YAWL is the convenient representation of all workflow patterns, as well as tool support and interfacing to various workflow tools. In the context of YAWL, a detailed representation of workflow patterns has been proposed [12]. As such, it is an important area of related work. However, the work presented in this paper aims at providing a broader exploitation and areas for future work, since the concepts provided by π-calculus allow for further representation, analysis, and reasoning, such as compliance of multiple processes.

From the context of process algebra, there has little been done for workflow purposes up to now. A Ph.D. thesis by Twan Basten researches basic process algebra and Petri nets [13]. A more practical approach of using CCS [14] to formalize web service choreography can be found in [15]. The only approach known to the authors on the use of the π-calculus for workflow definitions is from Yang Dong and Zhang Shen-Sheng and centers on basic control flow constructs and the definition of activities [16]. An approach close to process algebra is the logic based modeling and analysis of workflows by the use of concurrent transaction logic [17]. However, the expressiveness of this approach regarding to the workflow patterns has still to be investigated. Further approaches regarding the formalization of workflow patterns might include procedural techniques, which combine imperative, object–oriented and concurrent programming, logic–based attempts as well as graphgrammar– and net–based ones. Some approaches could be combined like the event–based and the process algebra has been used together in this paper.

3   The π-Calculus

The π-calculus is a modern process algebra that describes mobile systems in a broader sense [18]. The calculus is based on the concept of mobility, which includes communication and change. Communication takes place between different π-calculus processes. The structure of the processes changes over time by communication e.g., a process can dynamically include other processes which he received through communication. The communication itself is based on the concept of names. A name is a collective term for previous existing concepts like links, pointers, references, identifiers, etc., each of which has a scope. Assuming a name represents the reference to a process that currently processes a workflow activity, the scope of the name includes at that time only the active process. As soon as the process has finished, the scope is extruded to the process that handles the next workflow activity. Based on the flexibility of the π-calculus, which has only been sketched, many different possibilities arise to formalize the workflow patterns. We adopt an event, condition, action (ECA) approach, where each activity of a workflow is mapped conceptually to an independent π-calculus process. Those processes use events in the form of communication to coordinate the behavior of a workflow. Several processes together form a pattern of behavior, which represents a workflow pattern.

Syntax

As several different notations of the π-calculus exist [18,19,20,21], the one used throughout this paper is outlined. Details can be found in [22].

Basically, the π-calculus consists of processes and names, where names define links. The processes are defined through:

$$P \quad ::= \quad M \mid P|P \mid \mathsf{v}zP \mid !P \;.$$

The composition $P|P$ is the concurrent execution of P and P, $\mathsf{v}zP$ is the restriction of the scope of the name z to P, which is also used to generate a unique, fresh name z and $!P$ is the replication operator that satisfies the equation $!P = P \mid !P$. M contains the summations of the calculus:

$$M \quad ::= \quad \mathbf{0} \mid \pi.P \mid M + M$$

where $\mathbf{0}$ is inaction, a process that can do nothing, $M + M$ is the exclusive choice between M and M', and the prefix $\pi.P$ is defined by:

$$\pi \quad ::= \quad \overline{x}\,\langle y\rangle \mid x(z) \mid \tau \mid [x = y]\pi \;.$$

The output prefix $\overline{x}\,\langle y\rangle\,.P$ sends the name y over the name x and then continues as P. The input prefix $x(z)$ receives any name over x and then continues as P with z replaced by the received name (written as $\{^{name}/_z\}$). The unobservable prefix $\tau.P$ expresses an internal action of the process, and the match prefix $[x = y]\pi.P$ behaves as $\pi.P$, if x is equal to y.

Throughout this paper, upper case letters are used for process identifiers and lower case letters for names. Some additional process identifiers and names that represent special functions are introduced later on. Furthermore defined processes from the original paper on the π-calculus are used for parametric recursion, that is $A(y_1, ..., y_n)$ [18]. For the definitions given in this paper, defined processes are more applicable than the recent form with recursive definitions

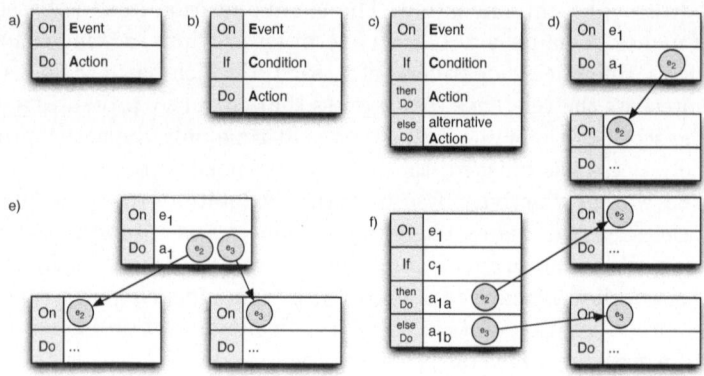

Fig. 1. The EA (a), ECA (b), and ECAA (c) notation for business rules and sequential (d), parallel (e), and optional (f) control flow

$K \triangleq (\widetilde{x}).P$ and constant applications $K\lfloor\widetilde{a}\rfloor$ where $\widetilde{x}$ and $\widetilde{a}$ represent sets of names [22].

We use the abbreviation $\sum_1^m(M)$ to denote the summation of m choices; e.g. $\sum_1^3(M_i) = M_1 + M_2 + M_3$. The abbreviation $\prod_1^m(P)$ is used to denote the composition of m parallel copies of P, e.g. $\prod_1^3(P) = P \mid P \mid P$. Also, $\{\pi\}_1^m$ denotes m subsequent executions of π, e.g. $\{\pi\}_1^3 = \pi.\pi.\pi$. All abbreviations could be used with an indexing variable, e.g. $\prod_{i=1}^3(d_i(x)) = d_1(x) \mid d_2(x) \mid d_3(x)$. Round brackets are used to define the ordering of a process definition. Given $\tau.P$ for instance, P might be expanded to $M+M'$ by using the summation rule from the π-calculus grammar. To avoid ambiguity, round brackets are put around the expanded symbol, e.g. $\tau.(M + M')$ instead of $\tau.M + M'$.

4   Pattern Representation

The formalization of the workflow patterns in the π-calculus starts with a mapping from activities to π-calculus processes[1]. Let every activity be an independent process. Each process has pre– and postconditions. A precondition for a process B could be that it should only start working after a process A has finished. A postcondition for process B could state that B has completed execution and then signals this to other processes.

The core idea is based on the ECA approach that originates from active database systems. ECA means *Event, Condition, Action* [23]. The event component specifies when a rule must be evaluated. After the rule has been evaluated, the conditional component must be checked and if it matches, the action component is executed. This approach has been adapted to specify control flow between

---

[1] We abbreviate the term π-*calculus process* to *process* in the following.

different activities in a workflow [24]. The adapted paradigm is called EC^mA^n. It allows m conditions and n actions. In the workflow domain ECAA, ECA and EA rules are most common (see figure 1). The figure also shows sequential, parallel, and optional control flow.

We can now map the ECA approach to process definitions. The preconditions of the processes comply to the event and conditional part of the ECA rules. Every process that has no event part represents an initially starting activity, as the process has no further dependencies. The events are modeled in the process definitions as input prefixes. After the input prefixes have been triggered (that is, the event has occurred) an optional condition has to be checked. This is modeled by a match prefix. It can be used to model global constraints like testing a cancellation flag. The action part is divided into two parts. First, the functional perspective of the activity is represented as an unobservable action. Second, the process can trigger other processes by output prefixes. Output prefixes represent postconditions. If a process does not trigger other processes it represents a final workflow activity. The complete process definition for a basic activity is:

$$x.[a = b].\tau.\overline{y}.\mathbf{0} \ . \tag{1}$$

A process receives a trigger x mapping to an event, makes a comparison $[a = b]$ mapping to a condition, does some internal work τ and finally triggers another process with $\overline{y}$ as the resulting action. This notation can be generalized to:

$$\{x_i\}_{i=1}^m.\{[a = b]\}_1^n.\tau.\{\overline{y_i}\}_{i=1}^o.\mathbf{0} \ . \tag{2}$$

A generic process can have m incoming triggers, n conditions, and o outgoing triggers. A process that represents an activity must have a functional part represented by τ. Note that it is explicitly allowed to have zero incoming triggers, conditions or outgoing triggers. The consequences have been discussed earlier. If a process representing an activity can be triggered more than once, the replication operator must be used.

The description given applies only to basic control flow structures. Advanced structures require slightly different approaches. Additionally to the processes that represent the workflow activities, system and helper processes are required. Those processes do not belong directly to the workflow, but are needed for reasoning and execution control.

The patterns given in the next paragraphs can be seen as small pieces of a workflow definitions. The postconditions of the processes that link to other processes are indicated by a process identifier with an apostrophe, like the process A has process A' as a postcondition. The process A' might represent any other workflow pattern.

4.1 Basic Control Flow Patterns

The basic control flow patterns capture elementary aspects of workflow control flow. They are structured like shown in equation 2.

Sequence. A sequence between two processes A and B is achieved by A sending a name over b to process B, which executes τ_B and afterward activates the continuation as B':

$$A = \tau_A.\bar{b}.0$$
$$B = b.\tau_B.B'$$

As can be seen, the actual process definition transmits no name, because the name is irrelevant for triggering another process. We abbreviate $\bar{b}\langle x\rangle \mid b(x).B'$ to $\bar{b} \mid b.B'$, when the argument count is zero. As explained earlier, this is called triggering.

Parallel Split. To achieve a parallel split from a process A to two processes B and C, A triggers two names b and c at the processes B and C.

$$A = \tau_A.(\bar{b}.0 \mid \bar{c}.0)$$
$$B = b.\tau_B.B'$$
$$C = c.\tau_C.C'$$

Synchronization. The synchronization between two processes B and C at another process D is represented by B and C each triggering the names d_1 or d_2 at D. The process D waits on those two names until it can continue as D'.

$$B = \tau_B.\bar{d_1}.0$$
$$C = \tau_C.\bar{d_2}.0$$
$$D = d_1.d_2.\tau_D.D'$$

Exclusive Choice. The exclusive choice between two alternative processes B or C after A is modeled by the π-calculus summation operator. Thereby A triggers either b or c.

$$A = \tau_A.(\bar{b}.0 + \bar{c}.0)$$
$$B = b.\tau_B.B'$$
$$C = c.\tau_C.C'$$

Simple Merge. The simple merge of two control flows from either processes B or C in D is achieved by B and C triggering a name d. Per definition of this pattern, B and C will never be executed in parallel, so D only needs to wait on one incoming name d. If B and C should be executable in parallel, the synchronizing merge pattern applies.

$$B = \tau_B.\bar{d}.0$$
$$C = \tau_C.\bar{d}.0$$
$$D = d.\tau_D.D'$$

4.2 Advanced Branching and Synchronization Patterns

This section covers advanced branching and synchronization patterns. They require advanced concepts and map only partly to equation 2. One pattern, the synchronizing merge, needs to know the number of incoming flows that depend on preceding multi–choices. However, this is only important at the execution level. At the design level considered here, all possibilities must be captured.

Multi–choice. The choice between processes B or C or B and C after A is modeled by A having three possibilities of execution. Either A triggers B or C or both, B and C.

$$A = (\mathbf{v}exec)\tau_A.(A_1 \mid A_2)$$
$$A_1 = \overline{exec}\,\langle b\rangle\,.0+$$
$$\overline{exec}\,\langle c\rangle\,.0+$$
$$\overline{exec}\,\langle b\rangle\,.\overline{exec}\,\langle c\rangle\,.0)$$
$$A_2 = !exec(x).\overline{x}.0$$
$$B = b.\tau_B.B'$$
$$C = c.\tau_C.C'$$

Note that this pattern uses the concept of an executor ($exec$) represented by process A_2. An executor receives a name and afterward triggers that name. The executor always immediately responds and decouples the triggering of the subject received in a parallel thread, thus not blocking the original caller. The executor workaround is needed, because a process $\overline{b}.0 + \overline{c}.0 + (\overline{b}.0 \mid \overline{c}.0)$ cannot be derived from the π-calculus grammar given. If we just specify $\overline{b}.\overline{c}.0$ to denote that both names, b and c should be triggered, the semantic is incorrect. For example image a more complicated construct. The preconditions of process B are extended so that he has to wait additionally on a name b_1. This could be written as $B = b_1.b.\tau_B.B'$. The name b_1 has not yet been triggered, so the process $\overline{b}.\overline{c}.0$ could not yet trigger the name c. Process C that only has c as a precondition cannot start execution. This is clearly not the intention of the multi–choice pattern.

Synchronizing Merge. The triggers for activating a process D can either come from B or C as well as from B and C. If B and C are executed in parallel, D has to wait on d_1 and d_2, otherwise only for d_1 or d_2.

$$B = \tau_B.\overline{d_1}.0$$
$$C = \tau_C.\overline{d_2}.0$$
$$D = d_1.\tau_D.D'+$$
$$d_2.\tau_D.D'+$$
$$d_1.d_2.\tau_D.D'$$

This pattern has no synchronization problem. Even if process C is able to signal d_2 earlier then B can signal d_1, the process C is blocked until B has signaled d_1. This confirms with the reduction rules of the π-calculus. Note that the semantics of this pattern does not describe how a runtime actually decides which summation of D is chosen.

Multi-merge. Process D can be triggered arbitrary times by incoming triggers from B or C. Each time D gets triggered, a new copy of D is created by replication.

$$B = \tau_B.\bar{d}.0$$
$$C = \tau_C.\bar{d}.0$$
$$D = !d.\tau_D.D'$$

Note that by using the replication operator to create multiple copies of a process D, all processes that are triggered by D must also support replication and so on. This also refers to all other patterns that create multiple copies by replication.

Discriminator. The discriminator pattern activates τ_D by triggering D_2 if process D_1 receives either d_1, d_2 or d_3. After D_2 has activated τ_D it waits for the triggers h from the remaining incoming branches of D_1. Finally D_2 resets the discriminator by using recursion.

$$A = \tau_A.\bar{d_1}.0$$
$$B = \tau_B.\bar{d_2}.0$$
$$C = \tau_C.\bar{d_3}.0$$
$$D = (\mathbf{v}h, exec)(D_1 \mid D_2)$$
$$D_1 = d_1.\bar{h}.0 \mid d_2.\bar{h}.0 \mid d_3.\bar{h}.0$$
$$D_2 = h.\overline{exec}.h.h.D \mid exec.\tau_D.D'$$

The process definitions A, B and C are trivial. The process definition D that represents the discriminator is split into two parts D_1 and D_2 with two fresh names h and $exec$. D_1 waits in parallel for all incoming triggers d_1, d_2 and d_3. If a trigger is received, D_1 anonymizes the trigger by signaling h to D_2. Afterward process D_1 waits for the remaining triggers. If another process signals a name that D_1 has already received, the signaling process is blocked. The process D_2 waits for an incoming name h and afterward executes τ_D in parallel, achieved through an internal trigger $exec$. This is needed due to the decoupling of the subsequent actions represented by D'. Afterward it waits for the remaining triggers from D_1 and then resets itself by the use of recursion. Note that all processes that are called by D' must have the capability of multiple execution.

A generic discriminator with m incoming control triggers is defined by:

$$D = (\mathbf{v}h, r)((\prod_{i=1}^{m} d_i.\bar{h}.0) \mid h.\bar{r}.\{h\}_1^{m-1}.D \mid r.\tau_D.D').$$

The generic discriminator uses the product operator $\prod$ from 1 to m to denote m different incoming triggers. After receiving the first h trigger, it uses the sequence operators $\{\}$ to wait on $m-1$ anonymized incoming triggers. Those operators are just notational sugar; they have to be expanded before the process can be executed.

Example: Discriminator. To illustrate the discriminator, one possible evolution[2] of the system defined by A, B, C, D_1 and D_2 is given:

$$DISC = A \mid B \mid C \mid (\mathbf{v}h, exec)(D_1 \mid D_2) \ .$$

The processes are defined initially as given in the discriminator paragraph. The evolution of $DISC$ begins with either A, B or C signaling a name to D_1. We start with A signaling name d_1 to process D_1:

$$DISC \longrightarrow DISC_1 \stackrel{def}{=} B \mid C \mid (\mathbf{v}h, exec)(D_{1_1} \mid D_2) \ .$$

The process A has vanished as no more prefixes other than $\mathbf{0}$ exist after signaling the name d_1. The process D_1 has evolved to D_{11} and is defined by $D_{11} \stackrel{def}{=} \overline{h}.\mathbf{0} \mid d_2.\overline{h}.\mathbf{0} \mid d_3.\overline{h}.\mathbf{0}$. Immediately after, a communication between D_{11} and D_2 is possible:

$$DISC_1 \longrightarrow DISC_2 \stackrel{def}{=} B \mid C \mid (\mathbf{v}h, r)(D_{12} \mid D_{21}) \ .$$

D_{11} signals the name h to D_2 and evolves to $D_{12} \stackrel{def}{=} d_2.\overline{h}.\mathbf{0} \mid d_3.\overline{h}.\mathbf{0}$. The left hand component has vanished as it reached inaction. The process D_2 evolves to $D_{21} \stackrel{def}{=} \overline{exec}.h.h.D \mid r.\tau_D.D'$. Now $exec$ can be triggered inside D_{21}:

$$DISC_2 \longrightarrow DISC_3 \stackrel{def}{=} B \mid C \mid (\mathbf{v}h, exec)(D_{12} \mid D_{22}) \ .$$

D_{22} is given by $D_{22} \stackrel{def}{=} h.h.D \mid D'$. Note that the right hand side of D_{22} now only consists of D'. We can assume D' to be $\mathbf{0}$ in our example. So the right hand side of D_{22} vanishes. Now process B can trigger d_2 and D_{12} can trigger h:

$$DISC_2 \longrightarrow DISC_3 \stackrel{def}{=} C \mid (\mathbf{v}h, exec)(D_{13} \mid D_{23}) \ .$$

Process B vanishes after triggering d_2. D_{12} evolves to $D_{13} \stackrel{def}{=} d_3.\overline{h}.\mathbf{0}$. Process D_{23} is given by $D_{23} \stackrel{def}{=} h.D$. Finally process C can trigger d_3:

$$DISC_3 \longrightarrow DISC_4 \stackrel{def}{=} D \ .$$

Process C vanishes after triggering d_3. D_{13} vanishes after receiving d_3 and triggering h. The only process now left is D which resets the discriminator through recursion. To make the discriminator work another time, we need new processes that trigger d_1, d_2 and d_3 again. A, B and C could also declared replicative, e.g. $A = !\tau_A.\overline{d_1}.\mathbf{0}$, etc. We could further trace other evolutions of the system described, e.g. starting with d_2 or d_3.

---

[2] We use the π-calculus semantics of reduction for this example, see [22].

N–out–of–M–Join. The n–out–of–m join generalizes the discriminator by executing the activity τ_D after n out of m triggers have arrived at process D. After the remaining triggers have been received, D resets itself by recursion.

$$D = (\mathbf{v}h, r)((\prod_{i=1}^{m} d_i.\bar{h}.\mathbf{0}) \mid \{h\}_1^n.\bar{r}.\{h\}_{n+1}^m.D \mid r.\tau_D.D')$$

The n–out–of–m–join simply expands the middle expression of the generic discriminator by waiting for m incoming triggers in a sequence. The remaining triggers are then counted from $n + 1$ to m.

4.3   Structural Patterns

Structural patterns show restrictions on workflow languages, as for instance that arbitrary loop are not allowed or that only one final node should be present. The π-calculus easily handles both of the following patterns.

Arbitrary Cycles. Arbitrary cycles are inherently given by the event based approach. The only thing that must be taken care of is the re–instantiation of processes that execute repeatedly.

$$A =\,!a.\tau_A.\bar{b}.\mathbf{0}$$
$$B =\,!b.\tau_B.\bar{c}.\mathbf{0}$$
$$C =\,!c.\tau_C(\bar{a}.\mathbf{0} + \bar{d}.\mathbf{0})$$
$$D = d.\tau_D.D'$$

The re–instantiation is modeled using the replication operator for all processes that could be executed more then once (A, B, C). Process C must decide if the loop is called another time by triggering a or to continue by triggering d. If arbitrary cycles are allowed in a workflow definition, the formal reasoning will be much more difficult.

Implicit Termination. The implicit termination pattern terminates a sub–process if no other activities can be made active. The π-calculus contains the special symbol $\mathbf{0}$ for this purpose. As $\mathbf{0}$ is the only final termination symbol of the π-calculus grammar, each (sub)–process must finally have an implicit termination.

4.4   Multiple Instance Patterns

Multiple instance patterns create several copies of workflow activities. A trivial pattern uses no synchronization whereas more advanced patterns synchronize the created copies afterward.

Multiple Instances without Synchronization. Any amount of multiple copies of a process B can easily spawn from a process A by replication.

$$A = \tau_A.!\bar{b}.\mathbf{0}$$
$$B =\,!b.\tau_B.B'$$

A recursive definition for A could be $A = \tau_A.A_1$ with $A_1 = \bar{b}.A_1 + \mathbf{0}$. This notation explicitly states that A_1 can spawn of new copies of B or stop execution.

Multiple Instances with a priori Design Time Knowledge. When the number of copies of B is known at design time and the copies have to be synchronized before the execution of τ_C, the following pattern is used (the example shows three copies of B).

$$A = \tau_A.\overline{b}.\overline{b}.\overline{b}.0$$
$$B = !b.\tau_B.\overline{c}.0$$
$$C = c.c.c.\tau_C.C'$$

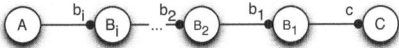

For n design time copies, the pattern is as follows:

$$A \mid B \mid C \equiv \tau_A.\{\overline{b}\}_1^n.0 \mid !b.\tau_B.\overline{c}.0 \mid \{c\}_1^n.\tau_C.C' \ .$$

Multiple Instances with a priori Runtime Knowledge. This pattern is runtime dependent like the synchronizing merge. At design time it can be modeled that A can spawn of an unknown number of processes B and only after A has finished creating the processes, τ_B gets activated by receiving a *start* trigger each. After all copies of B have finished, the name initially passed to A is triggered. The pattern needs a fresh name *start* private to A and B to work: $(\mathbf{v}start)(A \mid B)$. Note that this pattern uses defined processes for recursion.

$$A = (\mathbf{v}run)\tau_A.A_1(c) \mid run.!\overline{start}.0$$
$$A_1(x) = (\mathbf{v}y)\overline{b}\langle y\rangle.y\langle x\rangle.A_1(y) + \overline{run}.\overline{x}.0$$
$$B = !b(y).y(x).start.\tau_B.y.\overline{x}.0$$
$$C = c.\tau_C.C'$$

This pattern works like a dynamic linked list:

Initially A holds the name of the next process, that is c. An arbitrary number of processes B can be inserted between A and C using recursion. The created copies are started by triggering *run* in A which results in triggering *start* in all copies of B. Each copy of B triggers his predecessor after finishing τ_B. The initial predecessor is passed as a parameter to A; it is the name of the trigger that is activated after all copies of B have successfully executed τ_B. This pattern is a special case of the multiple instances without a priori runtime knowledge; an example is given later on.

Multiple Instances without a priori Runtime Knowledge. This pattern is much the same as the preceding one, with the difference that copies of B could be created all the time and start immediately.

$$A = \tau_A.A_1(c)$$
$$A_1(x) = (\mathbf{v}y)\overline{b}\langle y\rangle.y\langle x\rangle.A_1(y) + \overline{x}.0$$
$$B = !b(y).y(x).\tau_B.y.\overline{x}.0$$
$$C = c.\tau_C.C'$$

The only difference is the removal of the *start* and *run* triggers as well as the depending process parts.

Example: Multiple Instances without a priori Runtime Knowledge. We derive a trace of the multiple instances without a priori runtime knowledge pattern. The example shows how the recursive structure of the processes is build up while creating instances and how it is broken down while completing.

The process A is initialized with the link to the process that should be executed after all copies of B have been completed. That is c in our case:

$$A = \tau_A.A_1(c) .$$

Process A calls process A_1 with c as a parameter. A_1 has the choice between creating a copy of the process B or call the final process, that is c:

$$A_1(c) = (\mathbf{v}y)\overline{b}\,\langle y \rangle\,.y\,\langle c \rangle\,.A_1(y) + \overline{c}.\mathbf{0} .$$

We suppose process A_1 to create a new copy of process B. Therefore the left part of A_1 is executed: $(\mathbf{v}y_1)\overline{b}\,\langle y_1 \rangle\,.y_1\,\langle c \rangle\,.A_1(y_1)$. First, a fresh name y_1 is generated. We enumerate y with a subscript to mark different fresh names. The name y_1 is sent to process B which creates a new copy of itself through replication. Afterward, A_1 sends the name of the predecessor (that is c) to the new copy of process B. A_1 then calls itself with the fresh name y_1 as a parameter. Thereby the fresh name y_1 acts as the new predecessor. The processes A_1 and B now look like:

$$A_1(y_1) = (\mathbf{v}y_2)\overline{b}\,\langle y_2 \rangle\,.y_2\,\langle y_1 \rangle\,.A_1(y_2) + \overline{y_1}.\mathbf{0}$$
$$B =!b(y).y(x).\tau_B.y.\overline{x}.\mathbf{0} \mid \underbrace{\tau_B.y_1.\overline{c}.\mathbf{0}}_{\text{1st copy}} .$$

The process A_1 now has again the choice between creating a new copy of the process B or call the previous created process by y_1. Note that the 1st copy of B is already executing τ_B. We choose to create yet another copy of B:

$$A_1(y_2) = (\mathbf{v}y_3)\overline{b}\,\langle y_3 \rangle\,.y_3\,\langle y_2 \rangle\,.A_1(y_3) + \overline{y_2}.\mathbf{0}$$
$$B =!b(y).y(x).\tau_B.y.\overline{x}.\mathbf{0} \mid \underbrace{\tau_B.y_1.\overline{c}.\mathbf{0}}_{\text{1st copy}} \mid \underbrace{\tau_B.y_2.\overline{y_1}.\mathbf{0}}_{\text{2nd copy}} .$$

This is continued until A_1 decides to call the previous fresh name that was created; that is the parameter of A_1. In our example this is y_2. We suppose the τ_B of the copies of B to have been finished by now. So a communication between A_1 and B by y_2 could take place; otherwise we had to wait until the τ_B of the second copy has finished:

$$A_1(y_2) = \mathbf{0}$$
$$B =!b(y).y(x).\tau_B.y.\overline{x}.\mathbf{0} \mid \underbrace{y_1.\overline{c}.\mathbf{0}}_{\text{1st copy}} \mid \underbrace{\overline{y_1}.\mathbf{0}}_{\text{2nd copy}} .$$

Now the second copy of B has reduced to $\overline{y_1}.0$. A communication between the second and first copy by y_1 is now possible:

$$A_1(y_2) = 0$$
$$B = !b(y).y(x).\tau_B.y.\overline{x}.0 \mid \underbrace{\overline{c}.0}_{\text{1st copy}} \mid \underbrace{0}_{\text{2nd copy}} \ .$$

The second copy of B has reached inaction. The first copy can trigger the name c that references to the process that should be executed after all copies of B have finished. Thereafter the first copy reaches inaction. If we suppose A_1 as the only source of names b than no further communication is possible. The multiple instances without a priori runtime knowledge pattern is completed.

4.5   State Based Patterns

State based patterns capture implicit behavior of processes that is not based on the current case rather than the environment or other parts of the process. Some of the following patterns require the existence of an external process that represents the environment. This process is used as a source for external events. We denote the environmental process with the special process identifier $\mathcal{E}$. The names that are triggered from within $\mathcal{E}$ are marked with a subscripted env, as for instance a_{env} denotes an environmental trigger.

Deferred Choice. A deferred choice is much like the exclusive choice with the distinction that the choice if τ_B or τ_C get executed is not made explicit in A rather than by the environment. The environment is modeled as an external process $\mathcal{E}$ that signals either the name b_{env} or c_{env} but not both. The moment of choice is thereby as late as possible. Afterward the successful process signals the name $kill$ to the other process which leads to the empty process 0. B and C must share a fresh name $kill$: $(\mathbf{v}kill)(B \mid C)$

$$A = \tau_A.(\overline{b}.0 \mid \overline{c}.0)$$
$$B = b.(b_{env}.\overline{kill}.\tau_B.B' + kill.0)$$
$$C = c.(c_{env}.\overline{kill}.\tau_C.C' + kill.0)$$

Interleaved Parallel Routing. The interleaved parallel routing or unordered set is achieved by non–determinism in the π calculus. A, B and C share two fresh names $(\mathbf{v}x, y)(A \mid B \mid C)$ of which x is used to trigger B and C in any order. The name y is used to signal the complete execution of the triggered process. After all activities have been executed, the control is again at A.

$$A = \tau_A.\overline{x}.y.\overline{x}.y. A'$$
$$B = x.\tau_B.\overline{y}.0$$
$$C = x.\tau_C.\overline{y}.0$$

Milestone. A milestone is a test for a process A, if another parallel process B is in a given state. Thereby the two parallel processes share a private name *check* which returns either true (represented by the special name $\top$) if the condition holds or false ($\bot$) if not. A process definition $M(x)$ is used as a memory cell that keeps the condition. It is called by a private name m with $(\mathbf{v}m)(B)$:

$$A = check(x).([x = \top]\tau_{A1}.A' + [x = \bot]\tau_{A2}.A'')$$
$$B = M(\bot) \mid b.\overline{m}\langle\top\rangle.\tau_B.\overline{m}\langle\bot\rangle.B'$$
$$M(x) = m(x).M(x) + \overline{check}\langle x\rangle.M(x) \ .$$

4.6   Cancellation Patterns

The cancellation pattern describe the withdrawal of one or more processes that represent workflow activities.

Cancel Activity. The cancel activity pattern allows a process, that is waiting to get triggered, to be canceled. This pattern is modeled by the optional reception of a *cancel* trigger from an external environment process $\mathcal{E}$ with $(\mathbf{v}cancel)(A \mid \mathcal{E})$:

$$A \mid \mathcal{E} \equiv a.\tau_A.A' + cancel.\mathbf{0} \mid !\tau_{\mathcal{E}}.\overline{cancel}.\mathbf{0} \ .$$

Note that currently executed activities represented by τ could not be canceled due to the unobservability of τ.

Cancel Case. The cancel case pattern cancels a whole workflow instance. This is equal to Cancel Activity with the exception that all remaining processes receive a global cancel trigger.

5   Conclusion

In this paper, we introduced a formal semantics for workflow patterns, which is based on the π-calculus. All of the documented workflow patterns from [9] have been formalized with concise and unambiguous expressions. Based on the execution semantics of the π-calculus, the behavior of each workflow pattern has been defined precisely.

However, this paper is not to be understood as *the* formal semantics of the workflow patterns. Other notations, like Workflow Nets [25] or YAWL [12] use different approaches from Petri nets to transition systems to realize a formal specified behavior for some or all of the workflow patterns. Rather, this paper can be seen as a foundation for using modern process algebra in the workflow domain. The π-calculus supports mobility, communication and change. While it has not yet been shown how mobility can actually enrich the workflow domain, requirements like flexibility and reaction to change are ever more challenging [1]. Since the π-calculus was designed to model such highly dynamic systems, it might offer new ways to face the challenges in the workflow domain. As a starting point, this paper showed that the π-calculus is indeed able to handle all of the behavioral workflow requirements given by workflow patterns.

Based on the formalizations presented in this paper, further research has to be made. The π-calculus could be used as a formal foundation for graphical notations. Furthermore, formalized workflows can opening the door for reasoning on workflow process structures.

References

1. Smith, H., Fingar, P.: Business Process Management – The Third Wave. Meghan-Kiffer Press, Tampa (2002)
2. van der Aalst, W.M.P.: Pi calculus versus petri nets: Let us eat "humble pie" rather than further inflate the "pi hype". (http://is.tm.tue.nl/research/patterns/download/pi-hype.pdf (May 31, 2005))
3. BPMI.org: Business Process Modeling Language. Technical report (2002)
4. Microsoft: XLang Web Services for Business Process Design. (2001)
5. BEA Systems, IBM, Microsoft, SAP, Siebel Systems: Business Process Execution Language for Web Services Version 1.1. (2003)
6. van der Aalst, W.: Flexible Workflow Management Systems: An Approach based on Generic Process Models. In Bench-Capon, T., Soda, G., Tjoa, A., eds.: Database and Expert Systems Applications: 10th International Conference, DEXA'99, volume 1677 of LNCS, Berlin, Springer (1999) 186–195
7. van der Aalst, W.M.P.: Exterminating the Dynamic Change Bug: A Concrete Approach to Support Workflow Change. Information System Frontiers **3** (2001) 297–317
8. Rinderle, S., Reichert, M., Dadam, P.: Evaluation of Correctness Criteria for Dynamic Workflow Changes. In van der Aalst, W.e.a., ed.: Business Process Management 2003, volume 2678 of LNCS, Berlin, Springer (2003) 41–57
9. van der Aalst, W.M.P., ter Hofstede, A.H.M., Kiepuszewski, B., Barros, A.: Workflow patterns. Distributed and Parallel Databases **14** (2003) 5–51
10. Curtis, B., Kellner, M.I., Over, J.: Process Modeling. Communications of the ACM **35** (1992) 75–90
11. Weske, M.: Workflow Management Systems: Formal Foundation, Conceptual Design, Implementation Aspects. Habilitationsschrift, Fachbereich Mathematik und Informatik, Universität Münster, Münster (2000)
12. van der Aalst, W.M.P., ter Hofstede, A.H.M.: YAWL: Yet Another Workflow Language (Revised version. Technical Report FIT-TR-2003-04, Queensland University of Technology, Brisbane (2003)
13. Basten, T.: In Terms of Nets: System Design with Petri Nets and Process Algebra. PhD thesis, Eindhoven University of Technology, Eindhoven, The Netherlands (1998)
14. Milner, R.: Communication and Concurrency. Prentice Hall, New York (1989)
15. Brogi, A., Canal, C., E.Pimentel, Vallecillo, A.: Formalizing Web Service Choreographies. In: Proceedings of First International Workshop on Web Services and Formal Methods. Electronic Notes in Theoretical Computer Science, Elsevier (2004)
16. Dong, Y., Shen-Sheng, Z.: Approach for workflow modeling using π-calculus. Journal of Zhejiang University Science **4** (2003) 643–650

17. Davulcu, H., Kifer, M., Ramakrishnan, C.R., Ramakrishnan, I.V.: Logic Based Modeling and Analysis of Workflows. In: Proceedings of the seventeenth ACM SIGACT-SIGMOD-SIGART symposium on Principles of database systems, ACM Press (1998) 25–33

18. Milner, R., Parrow, J., Walker, D.: A calculus of mobile processes, Part I/II. Information and Computation **100** (1992) 1–77

19. Milner, R.: The polyadic π–Calculus: A tutorial. In Bauer, F.L., Brauer, W., Schwichtenberg, H., eds.: Logic and Algebra of Specification, Berlin, Springer-Verlag (1993) 203–246

20. Milner, R.: Communicating and Mobile Systems: The π-calculus. Cambridge University Press, Cambridge (1999)

21. Parrow, J.: An Introduction to the π–Calculus. In Bergstra, J.A., Ponse, A., Smolka, S.A., eds.: Handbook of Process Algebra, Elsevier (2001) 479–543

22. Sangiorgi, D., Walker, D.: The π-calculus: A Theory of Mobile Processes. Paperback edn. Cambridge University Press, Cambridge (2003)

23. Dayal, U., Hsu, M., Ladin, R.: Organizing long-running activities with triggers and transactions. In: Proceedings of the 1990 ACM SIGMOD international conference on Management of data, New York, ACM Press (1990) 204–214

24. Knolmayer, G., Endl, R., Pfahrer, M.: Modeling Processes and Workflows by Business Rules. In Aalst, W.v.d., Desel, J., Oberweis, A., eds.: Business Process Management: Models, Techniques, and Empirical Studies, volume 1806 of LNCS, Berlin, Springer-Verlag (2000) 16–29

25. van der Aalst, W., van Hee, K.: Workflow Management. MIT Press (2002)

Mining Workflow Recovery from Event Based Logs

Walid Gaaloul and Claude Godart

LORIA - INRIA - CNRS - UMR 7503
BP 239, F-54506 Vandœuvre-lès-Nancy Cedex, France
{gaaloul, godart}@loria.fr

Abstract. Handling workflow transactional behavior remains a main problem to ensure a correct and reliable execution. It is obvious that the discovery, and the explanation of this behavior, would enable to better understand and control workflow recovery. Unfortunately, previous workflow mining works have concentrated their efforts on control flow aspects. Although powerful, these proposals are found lacking in functionalities and performance when used to discover workflow transactional behavior.

In this paper, we describe mining techniques, which are able to discover a workflow model, and to improve its transactional behavior from event based logs. First, we propose an algorithm to discover workflow patterns. Then, we propose techniques to discover activities transactional dependencies that allow us to mine workflow recovery techniques. Finally, based on this mining step, we use a set of rules to improve workflow design.

Keywords: workflow mining, workflow recovery, transactional workflows, workflow patterns.

1   Introduction

Workflow Management Systems (WfMSs) are being increasingly used by many companies to improve the efficiencies of their processes and reduce costs. However, due to the overall complexity of workflows specification, deployment of a process without validation may lead to undesirable execution behavior that compromises process goals. WfMSs are expected to recognize and handle errors to support reliable and consistent execution of workflows, but as has been pointed out in [1], up to now, most WfMSs lack such functionalities. The introduction of some kind of transactions in WfMSs is unavoidable to guarantee reliable and consistent workflow executions. In contrast to advanced transaction models, **transactional workflows** focus on issues of consistency from the business point of view rather than from the database point of view. They are applied in case of failures and define mechanisms supporting the automation of failure handling during runtime. Basically they specify activities transactional interactions in order to resume correctly workflow execution.

The main problem at this stage is how to ensure that the specified workflow model guaranties reliable executions. Generally, formal previous approaches develop, using their workflow modelling formalisms, a set of techniques to analyze and check model correctness [2,3,4]. Although powerful, these approaches may fail, in some cases, to

W.M.P. van der Aalst et al. (Eds.): BPM 2005, LNCS 3649, pp. 169–185, 2005.

ensure workflow reliable executions. Besides, it is neither possible nor intended by most workflow designers to model all failures; the process description will in any case become complex very soon - especially if the original process is more complex [5]. Furthermore, workflow errors and exceptions are commonly not detected until the workflow model is performed. By gathering and analyzing information about workflow processing as they took place in run time, we can first identify these errors and related recovery techniques and second propose solutions to improve them.

The work described in this paper presents a new **workflow mining** [6] technique to discover workflow transactional behavior from logs and to be alerted of design gaps especially about failure handling and recovery techniques. This kind of analysis is very useful in showing cause effect relationships. In the next section, we present a motivating example showing the need for discovering transactional behavior to improve workflow reliability.

1.1   Motivating Example

Consider a simplified example of a bank loan request processing. The workflow for the loan request processing is represented in graphical form in figure 1. First, the customer specifies the amount and loan terms through the $EnLR$ activity. Then the workflow instance retrieves client information and assesses the credit worthiness through the CCW activity. After that, the bank makes its decision choosing exclusively between these three options : (i) It evaluates the risk to the bank (REv) and updates the value of the banks total involvement (RUp) or (ii) The client may be important and have a convincing argument why the loan should be granted, without risk evaluation or despite the evaluation failing. Under these circumstances an executive officer of the loan department would have to approve the risk (REx) or (iii) rejects the loan (ReL). Then the EnD activity enters the decision to either grant or reject the loan request and records the relevant information on the agreed terms of the loan in bank database. The delivery of decision to the customer (loan contract or reject notification) ICl cannot be done without supervisor (bank direction) approval. Indeed, a human agent activity SuD is present along the decision process to supervise these activities. Thus, the loan can only be granted if the supervisor agent agrees. This agent may reject the loan request even if we have a positive decision in EnD. The supervisor can give freely his decision at any time during the loan process and thus achieve the SuD activity.

To deal with workflow failures and ensure a reliable execution, designers specify additional transactional workflow interactions (dotted arrow). In our example, it was specified that if EnD fails then $RUpC$ compensates the effects RUp and the loan decision process should be restarted. Also, in case of REv failure, the workflow continues the execution by an exceptional loan grant (REx) or a loan reject decision (ReL). Besides, ICl has the capability to be (re)executed until success in case of failure. As for CCW failure, the workflow terminates its execution by performing ICl and aborting the execution of SuD (the loan request is rejected). Finally, workflow designers did not provide failure handling mechanisms for the other activities and suppose that these activities are reliable (i.e. never fail).

Let suppose now that in reality (by observation of sufficient execution cases) CCW never fails but Sud can fail. This means there is no need to specify a recovery mech-

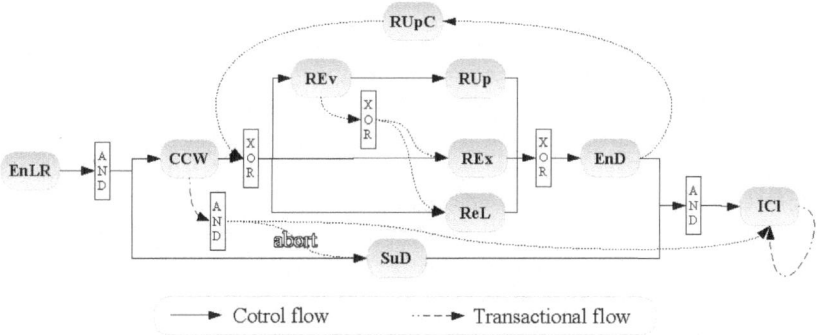

Fig. 1. Example of workflow

anism for CCW and we should abort concurrent activities of decision process and resume workflow execution when SuD fails. Starting from workflow logs (sufficient workflow execution cases), we propose workflow mining techniques detecting these transactional design gaps and providing help to correct them.

The remainder of this paper is structured as follows. First, we describe the structure of workflow event logs. After that, we detail our approach for mining workflow recovery techniques. We mainly proceed in three steps. First, we discover workflow patterns (see section 3) using logs statistical analysis. After that, we extract, in section 4, workflow transactional dependencies after failure. Then based in these mined results, we use a set of rules to improve workflow failures handling and recovery, and finally the application reliability. We discuss the related work, in section 5, and we conclude the paper by summarizing the main results and describing our future work. We illustrate the applicability of each one of these mining points through the previous motivating example.

2   Workflow Event Logs

The workflow specification might not be concerned with the details of the activities however it would have to at least deal with the externally visible completion events of activities (such as aborted, failed and completed). Currently, most of WfMSs log every events occurring during process execution. We expect the activities to be traceable, meaning that the system should in one way or another keep track of ongoing and past executions. As shown in the UML class diagram in figure 2, WorkflowLog is composed of a set of EventStreams (definition 1). Each EventStream traces the execution of one case (instance). It consists of a set of events (Event) that captures the activities life cycle performed in a particular workflow instance. An Event is described by the activity identifier that it concerns, the current activity state (aborted, failed and completed) and the time when it occurs (TimeStamp). A Window defines a set of Events over an EventStream. Finally, A Partition builds a set of partially overlapping Windows partition over an EventStream.

Definition 1. *(EventStream)*
An EventStream *represents the history of a worflow instance events as a tuple* stream=
*(*begin, end, sequenceLog, SOccurrence*) where:*
   ✓begin : TimeStamp *is the moment of log beginning of the workflow instance ;*
   ✓end : TimeStamp *is the moment of log end of the workflow instance;*
   ✓sequenceLog : Event* *is an ordered* Event *set belonging to a workflow instance;*
   ✓SOccurrence : int *is the instance number.*
A WorkflowLog *is a set of* EventStreams. WorkflowLog=(workflowID, {EventStream$_i$,
$0 \leq i \leq$ number of workflow instances}) where EventStream$_i$ is the event stream of the
i^{th} workflow instance.*

An example of 5th EventStream extracted from the workflow example of figure 1
in its 5th instantiation:

L = EventStream((13/5,5:42:12),(14/5, 14:01:54), [**Event**("EnLR", completed, (13/5,
   5:42:12)), **Event**("CCW", completed, (13/5,11:11:12)), **Event**("REv", completed,
       (13/5,14:01:54)), **Event**("SuD", completed, (14/5, 00:01:54)), **Event**("RUp",
completed, (14/5,5:45:54)), **Event**("EnD", completed, (14/5,10:32:55)), **Event**("ICl",
                   completed,(14/5,14:01:54))],5)

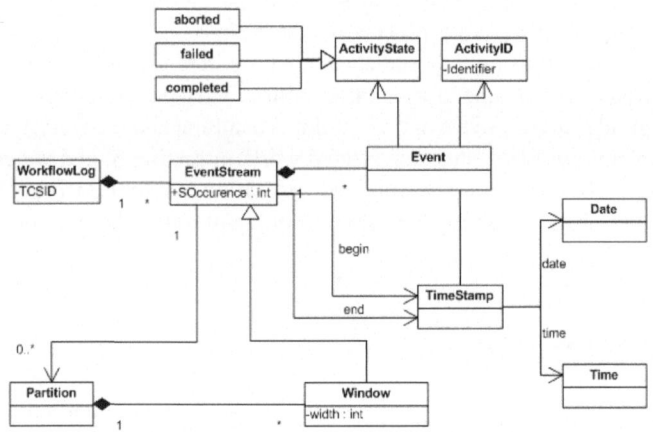

Fig. 2. Structure of a workflow event Logs

3   Control Flow Mining

The workflow mining process should be able to discover and analyze efficiently events
dependencies (see definition 1). Based on event states, there is several kinds of events
dependencies. However, we are interested, in this section, in discovering "elementary"
routing workflow patterns: Sequence, AND-split, OR-split, XOR-split, AND-join, OR-
join, and M-out-of-N Join patterns inspired from workflow patterns [7]. These patterns
describe control flow interactions for activities executed without "exceptions" (*i.e.* they

reached successfully their **completed** state). Thus, there is no need to use the events dependencies relating to failure executions (failed or aborted) states which concern only workflow transactional behavior (see section 4). For these reasons, we need to filter workflow logs and take only EventStreams of instances executed without activities failure or abortion. We denote by $workflowLog_{completed}$ this workflow logs projection. Thus, the minimal condition to discover workflow patterns is to have workflow logs containing at least the completed event state. This feature allows us to mine control flow from "poor" logs which contain only completed event state. Any information system using transactional systems such as ERP, CRM, or workflow management systems offer this information in some form [8].

Definition 1 *Events Dependency*
Let e_i, e_j be two events in WorkflowLog, *e_j depends on e_i iff there is an* EventStream *where the occurrence of e_i provokes* **directly** *the occurrence of e_j.*

Our control flow mining approach proceeds in two steps : Step (i) the construction of statistical dependency table SDT, and Step (ii) the mining of workflow patterns through a set of rules using these statistical calculus.

3.1 Construction of the Statistical Dependency Table SDT

As we state before, we build through statistical techniques the statistical dependency table SDT that expresses control flow dependencies. Since discovered workflow patterns specify interactions for activities executed without "exceptions", SDT captures only events dependencies with completed state in successful executions. In consequence we confuse, in this section, between activities dependencies and events dependencies.

For each activity A, we extract From $workflowLog_{completed}$ the following information : (i) The overall frequency of this activity (denoted $\#A$) and (ii) The causal dependencies to previous activities B_i (denoted $P(A/B_i)$). The size of SDT is N*N, where N is the number of workflow activities. The (m,n) table entry (notation P(m/n)) is the frequency of the nth activity **immediately preceding** the mth activity. Table 5 represents a fraction of the initial SDT of our motivating example workflow. For instance, in this table P(RUp/REv)=0.67 expresses that if RUp occurs then we have 67% of chance that REv occurs directly before RUp in the workflow logs.

As it is computed, the initial SDT presents some problems to express correctly activities dependencies especially relating to concurrent or parallel behavior. In the following, we detail these issues and propose solutions to correct them.

Erroneous dependencies: If we assume that each EventStream from WorkflowLog comes from a sequential (i.e no concurrent behaviour) workflow, a zero entry in SDT represents a causal independence and a non-zero entry means a causal dependency (*i.e.* sequential or conditional dependency). But, in case of concurrent behavior EventStreams may contain interleaved events sequences from concurrent threads. As consequence, some entries, in initial SDT, can indicate non-zero entries that do not correspond to dependencies. For example, the EventStream given in section 2 "suggests" erroneous causal dependencies between REv and SuD in one side and SuD and RUp in another side. Indeed, REv comes just before SuD and SuD comes immediately before RUp in this EventStream. These erroneous entries are reported by *P(SuD/REv)* and $P(RUp/SuD)$ which are different to zero. These entries are erroneous because there is no causal dependencies between these activities. Underlined values in SDT report this behavior for other similar cases.

Formally, two activities A and B are in concurrence *iff* $P(A/B)$ and $P(B/A)$ entries in SDT are different from zero. Based on this definition, we propose an algorithm to discover activities parallelism and then mark the erroneous entries in SDT. Through this marking, we can eliminate the confusion caused by the concurrent behaviour producing these erroneous non-zero entries. The algorithm (A) in figure 3 scans the initial SDT and marks concurrent activities dependencies by changing their values to (-1).

Table 1. Fraction of Statistical Dependencies Table SDT (**P(x,y)**) and activities Frequencies (#)

	Initial SDT								Final SDT							
P(x,y)	EnLR	CCW	REv	RUp	REx	ReL	EnD	SuD	EnLR	CCW	REv	RUp	REx	ReL	EnD	SuD
EnLR	0	0	0	0	0	0	0	0	0	0	0	0	0	0	0	0
CCW	**0.94**	0	0	0	0	0	0	0.06	1	0	0	0	0	0	0	-1
REv	0	**0.71**	0	0	0	0	0	0.29	0	1	0	0	0	0	0	-1
RUp	0	0	**0.67**	0	0	0	0	0.33	0	0	1	0	0	0	0	-1
REx	0	**0.87**	0	0	0	0	0	0.13	0	1	0	0	0	0	0	-1
ReL	0	**0.92**	0	0	0	0	0	0.08	0	1	0	0	0	0	0	-1
EnD	0	0	0	0.48	0.22	0.18	0	0.12	0	0	0	0.58	0.23	0.19	0	-1
SuD	**0.06**	0.1	0.14	0.2	0.04	0.06	0.4	0	1	-1	-1	-1	-1	-1	-1	0

$$EnLR = CCW = EnD = SuD = 100$$
$$REv = RUp = 69 REx = 21 ReL = 10$$

Undetectable dependencies: For concurrency reasons, an activity might not depend on its immediate predecessor in the EventStream, but it might depend on another "indirectly" preceding activity. As an example of this behavior, SuD is logged between REv and RUp in the EventStream given in section 2. As consequence, REv does not occur always immediately before RUp in the workflow logs. Thus we have only $P(RUp/REv) = 0.67$ that is an under evaluated dependency frequency. In fact, the right value is 1 because the execution of RUp depends exclusively on REv. Similarly, values in bold in SDT report this behavior for other cases. To discover these indirect dependencies, we introduce the notion of activity concurrent window (definition 2). An activity concurrent window (ACW) is related to the activity of its last event and covers its directly and indirectly preceding activities. Initially, the width of ACW of an activity (*i.e.* the number of activities within) is equal to 2. Every time this activity is in concurrence with an other activity we add 1 to this width. If this activity is not in concurrence with other activities and has preceding concurrent activities, then we add their number to ACW width. For example the activity RUp is in concurrence with SuD the width of its ACW is equal to 3. Based on this the algorithm (B) in figure 3 calculates the ACW width for each activity and regroups them in the ACW table. This algorithm scans the "marked" **SDT** and updates the ACW table.

```
Input: SDT : Statistic Dependecy Table
Output: MSDT : Marked Statistic Dependecy Table
Var:
  SDT_size : int;
  Dependency : int
begin
  MSDT=SDT;
  MSDT_size = Size_tab(MSDT); /* the function
Size_tab returns the size of  SDT */
  For int i=0; i<MSDT_size; i++;
    For int j=0; j<i; j++;
      If MSDT[i][j] > 0 and MSDT[j][i] > 0 then
        MSDT[i][j] =-1;
        MSDT[j][i] =-1;
      endIf
    endFor
  endFor
end
```

A- Marking Concurrent Behavior in SDT Algorithm

```
Input: SDT : Statistic Dependecy Table
Output: ACWT /*Activity Concurrent Width table*/
Var:
  SDT_size : int;
begin
  SDT_size = Size_tab(SDT);
  For int i=0; i<SDT_size; i++;
    ACW[i]=2;
  endFor;
  For int i=0; i<SDT_size; i++;
    For int j=0; j<i; j++;
      If MSDT[i][j] =-1 then
        ACW[i]++;
        ACW[j]++;
        For int k=0; k<SDT_size; i++;
          If MSDT[k][i] >0 then ACW[k]++; endIf
        endFor
      endIf
    endFor
  endFor
end
```

B- Activity Concurrent Width Algorithm

```
Input: Wlog : TCSLog_terminated (TCSLog)
       MSDT : Marked Statistic Dependecy Table
Output: FSDT : Final Statistic Dependecy Table
Var:
  t_reference: int;
  t_preceded : int;
  fWin : window;
  depFreq :int[][];
  freq :int
begin
  For all win:window in partition(Wlog)
    t_reference = last_activity(win)  /* the function
    last_activity(win) returns the activityId of the
    last event in win.wLog */
    win = preceded_Events(win); /* the function
    preceded_Events(win) returns win without
    the last event*/
    for all e:event in (win.wLog;
      t_preceded= e.activityId;
      If ( MSDT[t_reference][t_preceded] >0) then
        depFreq[t_reference][t_preceded]++;
      endIf
    endFor
  endFor

/*Final step: construction of statistical dependency  table*/
  FSDT=MSDT;
  For all freq=depFreq[t_reference][t_preceded] in depFreq
    FSDT[t_reference t_preceded]=freq #t_reference;
  endFor
end
```

C- Final SDT Algorithm

Fig. 3. SDT Algorithms

Definition 2. Window

*A log window defines a log slide over an events stream S : **stream** (bStream, eStream, sLog, workflowocc). Formally, we define a log window as a triplet **window**(wLog, bWin, eWin) :*

✓ *(bWin : TimeStamp) and (eWin : TimeStamp) are the moment of the window beginning and end (with bStream ≤ bWin and eWin ≤ eStream)*

✓ *wLog ⊂ sLog and ∀ e: **event** ∈ S.sLog where bWin ≤ e.TimeStamp ≤ eWin ⇒ e ∈ wLog.*

After that, we proceed through an EventStream partition (definition 3) that builds a set of partially overlapping Windows over the EventStream using the ACW table. Finally, the algorithm (C) of figure 3 computes the final SDT. For each ACW, It computes for its last activity the frequencies of its preceded activities. The final SDT will be found by dividing each row entry by the frequency of its activity. Note that, our approach adjust **dynamically**, through the width of ACW, the process calculating activities dependencies. Indeed, this width is sensible to concurrent behavior : it increases in case of concurrence and is "neutral" in case on concurrent behavior absence. Now by applying these algorithms, we can compute the final SDT (table 5) which will be used to discover workflow patterns.

Definition 3. Partition
*A **partition** builds a set of partially overlapping* Windows *partition over an events stream.*
Partition : WorkflowLog → (Window)*
S : EventStream(*bStr, eStr, sLog: (Evt$_i$ 1≤i≤n), wocc)* → {w_i :Window; *1≤i≤n*}
where : Evt_i= *the last event in* w_i ∧ width(w_i)= *ACWT[Evt$_i$.ActivityID].*

3.2   Workflow Patterns Mining

The last step is the identification of workflow patterns through a set of rules. In fact, each pattern has its own statistical features which abstract statistically its causal dependencies, and represent its unique identifier. These rules allow, if workflow logs is completed, the discovery of the whole workflow patterns included in the mined workflow. Our control flow mining rules are characterized by a "local" workflow patterns discovery. Indeed, these rules proceed through a **local log analyzing** that allows us to **recover partial results** of mining workflow patterns. In fact, to discover a particular workflow pattern we need only events relating to pattern's elements. Thus, even using only fractions of workflow logs, we can discover correctly corresponding workflow patterns (which their events belong to these fractions). We divided the workflow patterns in three categories : sequence, fork and join patterns. In the following we present rules to discover the most interesting workflow patterns belonging to these three categories.

Table 2. Rules of sequence workflow pattern

Rules	workflow patterns
$(\#B = \#A)$	Sequence pattern
$(P(B/A) = 1)$	

Sequence pattern: In this category, we find only the sequence pattern (table 2) where the enactment of the activity B depends only on the completion of activity A.

Fork patterns: The three patterns of this category (table 3) have a "fork" point where a single thread of control splits into multiple threads of control which can be, according to the used pattern, executed or not. The causality between the activities A and B_i before and after "fork" point is shared by the three patterns of this category. This causality is ensured by the statistical property $(\forall 0 \leq i \leq n; P(B_i/A) = 1)$. The non-parallelism between B_i, in the xor-split pattern are ensured by $(\forall 0 \leq i, j \leq n; P(B_i/B_j) = 0)$. The or-split and and-split patterns differentiate themselves by the frequencies relation between the activity A and the activities B_i. Effectively, only a part of activities are executed in the or-split pattern after "fork" point, while all the B_i activities are executed in and-split pattern.

Join patterns: The three patterns of this category (table 4) has a "join" point where multiple threads of control merge in a single thread of control. The number of necessary

Table 3. Rules of fork workflow patterns

Rules	workflow patterns
$(\Sigma_{i=0}^{n}\,(\#B_i){=}\#A)$	xor-split pattern
$(\forall 0 \leq i \leq n; P(B_i/A) = 1) \wedge$ $(\forall 0 \leq i,j \leq n;\, P(B_i/B_j) = 0)$	
$(\forall 0 \leq i \leq n;\, \#B_i{=}\#A)$	and-split pattern
$(\forall 0 \leq i \leq n;\, P(B_i/A) = 1) \wedge$ $(\forall 0 \leq i,j \leq n\; P(B_i/B_j) = -1)$	
$(\#A \leq \Sigma_{i=0}^{n}\,(\#B_i)) \wedge$ $(\forall 0 \leq i \leq n;\, \#B_i \leq \#A)$	or-split pattern
$(\forall 0 \leq i \leq n;\, P(B_i/A) = 1) \wedge$ $(\exists 0 \leq i,j \leq n; P(B_i/B_j) = -1)$	

branches for the activation of the activity B after the "join" point depends on the used pattern. The enactment of activity B after the "join" point in the and-join requires the execution of all the A_i activities $(\forall 0 \leq i \leq n; P(B/A_i) = 1)$. In contrary of xor-join and M-out-of-N-Join patterns where a "partial" parallelism between A_i activities can be only seen in the M-out-of-N-Join pattern $(\exists 0 \leq i,j \leq n; P(A_i/A_j) = -1)$.

4 Workflow Recovery Mining

4.1 Workflow Transactional Behavior

The integration of transactions into workflows was motivated by research efforts concerning database transaction models for advanced applications, as for example summarized in [9]. The term "transactional workflows" has been introduced in [10] to clearly recognize the relevance of transactions in the context of workflows. Transactional workflows involve coordinated execution and suggest selective use of transactional properties for individual activities or entire workflows. Basically, they use advanced transaction models as a core concept to specify workflow correctness, data consistency and reliability [11,12,13]. The motivation behind modelling workflow transactional behavior is to add the capability in the workflow to handle exceptional circumstances that would otherwise leave the workflow in an unacceptable state. Within transactional workflow behavior, we distinguish between *activity transactional properties* and *transactional flow (interactions)*.

Table 4. Rules of join workflow patterns

Rules	workflow patterns
$(\Sigma_{i=0}^{n}\,(\#A_i)=\#B)$	xor-join pattern
$(\Sigma_{i=0}^{n}\,P(B/A_i)=1)\,\wedge$ $(\forall 0 \leq i,j \leq n;\,P(A_i/A_j)=0)$	A_1, A_2, ... A_n → XOR → B
$(\forall 0 \leq i \leq n;\,\#A_i=\#B)$	and-join pattern
$(\forall 0 \leq i \leq n;\,P(B/A_i)=1)\wedge$ $(\forall 0 \leq i,j \leq n\,\,P(A_i/A_j)=-1)$	A_1, A_2, ... A_n → AND → B
$(m * \#B \leq \Sigma_{i=0}^{n}\,(\#A_i))$ $\wedge\,(\forall 0 \leq i \leq n;\,\#A_i \leq \#B)$ $(m \leq \Sigma_{i=0}^{n}\,P(B/A_i) \leq n)$ $\wedge\,(\exists 0 \leq i,j \leq n;\,P(A_i/A_j)=-1)$	M-out-of-N-Join pattern A_1, A_2, ... A_n → OR → B

Activities transactional properties: During the workflow execution, an activity can pass through several stages defined as activity states (aborted, failed and completed). The transactional properties of an activity depend on the set of its internal states transitions. The main transactional properties that we are considering are *retriable* and *pivot* [14]. An activity a is said to be retriable (a^r) *iff* it is sure to complete even if it fails. a is said to be pivot (a^p) *iff* if once it successfully completes, its effects remain for ever and cannot be semantically undone or reactivated.

Transactional flow: After failure, transactional external transitions are fired by external entities (scheduler, human intervention, etc.) and allow to the failed activity to interact with the outside to recover a consistent state. The goal is to bring the failed process back to some semantically acceptable state. Thus, failure inconsistent state can then be fixed and the execution resumed with the hope that it will then complete successfully. For instance, in our example, in case of CCW failure, it was specified that the workflow terminates its execution by performing ICl and aborting the execution of SuD.

4.2 Mining Transactional Dependencies

As we have done to discover workflow patterns, we build statistical transactional dependencies tables STrD that report only events dependencies captured after activities failures. These dependencies provide a convenient way to specify and reason about workflow transactional behavior expressed in terms of activities transactional properties and transactional flow. To calculate these dependencies we use the same definition

Table 5. Fractions of Statistical Transactional Dependencies tables of End Activity

ITR_{End} table

ITR_{EnD}	EnD,f	RUpC,c	REv,c	REx,c	ReL,c	RUp,c
EnD,f	0	0	0	0.56	0.34	0.1
RUpC,c	1	0	0	0	0	0
REv,c	0	1				
REx,c	0	1				
ReL,c	0	1				
RUp,c	0	0				

ATR_{End} tables

ATR_{End}^{REv}	c	f	a
c	1	0	0
f	0	0	0
a	0	0	0

ATR_{End}^{REx}	c	f	a
c	1	0	0
f	0	0	0
a	0	0	0

ATR_{End}^{RUp}	c	f	a
c	1	0	0
f	0	0	0
a	0	0	0

ATR_{End}^{ReL}	c	f	a
c	1	0	0
f	0	0	0
a	0	0	0

c=completed, f=failed, a=aborted

(used in section 3.1), except that we capture only event dependencies after activities failures. In practical terms, each STrD is related to an activity "act" and captures statistically workflow behavior after "act"'s failures. Within we distinguished between inter activities transactional dependencies and intra activity transactional dependencies (see definition 2).

Definition 2 *Activity Transactional Dependencies*
We denote by ITR_{act} the inter activities transactional dependencies table that reports event dependencies after the "act"'s failures. Each $ITR_{act}(e_1, e_2)$ entry is an event dependency where :
  ✓ ($e_i.state=$ failed and $e_i.activity =$ "act"; i=1,2) **OR**
  ✓ ($e_i.activity$; is a "new" activity (i.e not appear in control flow mining); i=1,2)
We denote by ATR_{act}^A the intra activities transactional dependencies table that reports the states transitions of the "A" activity after "act"'s failures. These transitions are extracted from a workflow logs projection that take only events of "A".

The table 5 represents a fraction of the STrD tables of our motivating example workflow after End's failures. For instance, in the ITR_{EnD} there is a "new" $RUpC$ activity that is executed after the End's failures. ATR_{EnD}^{REv} indicates that REv activity is re-executed after End's failures.

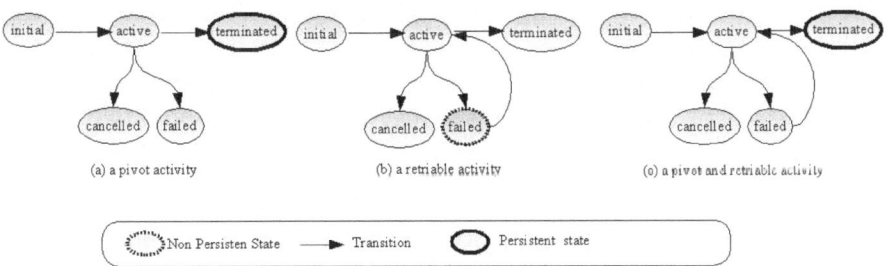

(a) a pivot activity   (b) a retriable activity   (o) a pivot and retriable activity

Non Persisten State   ➤ Transition   ◯ Persistent state

Fig. 4. Activity transactional properties

Discovering activity transactional properties: Every activity can be associated to a life cycle statechart that models the possible states through which the executions of this activity can go, and the possible transitions between these states. This structure, *i.e.* life cycle statechart, has an initial state and, on the start transition, moves into the executing state. There could be one or more transitions after this. Thus, the relationships between the significant events of an activity can be represented by a state transition diagram, which serves as an abstraction for the actual activity by hiding irrelevant details of its internal computations. The significant events transitions of an activity depend on the characteristics of its transactional proprieties.

Based on this, *retriable* and *pivot* are identified by an unique statechart life cycle. Figure 4.a illustrates the states/transitions diagram of a pivot activity. When a pivot activity is activated, the instance can normally continue its execution or it can be *cancelled* during its execution. In the first case, it can achieve its objective and successfully *completes* or it can *fail*. Once it successfully completes, its effects remains for ever and cannot be semantically undone or reactivated. Thus, if it completes successfully it keeps forever the completed state (i.e the completed state is a persistent state that can not be compensated by the workflow). The states/transition diagram of a retriable activity (figure 4.b) has in addition a transition that specifies a retry operation after observing failure. Indeed, the failure state in retriable activity can not be a persistent state, the activity should be re-executed after each failure until success. We can specify statistically these properties :

✓ An activity a is said to be retriable (a^r) iff $ATR_a^a(completed, failed)=1 \wedge ITR_a((a, completed), (a, failed)) = 1$.

✓ An activity a is said to be pivot (a^p) iff $\nexists\ act$: activity where $act \neq a \wedge ATR_{act}^a("x", c) \neq 0$; $"x"=$ completed or aborted or failed.

Now using these statistical specifications we can discover activities transactional properties from StrD tables. For instance, we can deduce from table 5 that the REv, RUp, REx and ReL activities are not pivot.

Discovering transactional flow: Basically, the workflow has to decide, after an activity failure, whether an inconsistent state is reached. Depending on this decision either a complex **recovery procedure** has to be started or the process execution can continues. The main challenge thereby is to identify and reach a consistent state from where the workflow can be continued. A consistent state point is an execution step of the workflow (equivalent to a save point in database transactions) which represents an acceptable intermediate execution state that is determined to be acceptable from a business perspective and hopefully also a decision point where certain actions can be taken to either fix the problem that caused the failure or choose an alternative execution path to avoid this problem [15].

The recovery procedure is initialized by an alternative dependency (see definition 3). Depending on the localization of the consistent point, we have identified two kinds of alternative dependencies : backward alternative and forward alternative. Bringing the workflow back to a semantically acceptable state can also entail compensating the already completed activities until the acceptable state is reached through "new" compensation activities which semantically undo the failed activity [16]. Furthermore, an activity failure can cause a non regular, abnormal termination (abort) of one or more active activities. If such a situation happens, the failure of an activity induces the abortion of other activities. This behavior is described through the abortion dependency (see definition 4). For instance, table 5 indicates that the bank loan process in figure 1 performs a backward recovery after the failure of EnD and there is an alternative dependency between EnD and $RUpC$. $RUpC$ is a compensation activity that it is executed for compensating the failure of EnD.

Definition 3 *Alternative Dependency: There is an alternative dependency from A_1 to A_2 if the failure of A_1 can fire the activation of A_2. Statistically expressed we have ITR_{A_1} ((A_2,completed), (A_1,failed)) $\neq 0$. We distinguished two kind of alternative dependencies:*
   ✓ *a backward alternative : if the consistent state is located before the failed activity*
   ✓ *a forward alternative : if the consistent state is located after the failed activity*

Definition 4 *Abortion Dependency: There is an abortion dependency from A_1 to A_2 if the failure of A_1 can fire the abortion of A_2. Using statistical transactional dependencies we have: $ITR_{A_1}(($A_2$,aborted), ($A_1$,failed)) \neq 1$*

Summarizing up, after an activity failure, we have the following possibilities:

- The activity is no vital. The workflow can continue without any specific recovery mechanism.
- The activity is vital and a recovery mechanism is activated. Then, the activity becomes recoverable (see definition 5). This investigation distinguishes between three different recovery mechanisms:
  - The activity is *retriable* : It is rolled back automatically until success. Such mechanism is generally used if the failed activity is idempotent. An idempotent activity can be executed one or more times without changing the result which is a very comfortable feature within workflow execution.
  - Backward recovery : If the failed activity has a vital relationship then a complex recovery procedure is necessary in order to reach a **previous** consistent state again. It may be necessary to start a compensation activity which removes inconsistent side effects and semantically "undoes" the effect of the corresponding failed activity. This backward recovery is initialized through a backward alternative dependency.
  - Forward recovery : If the failed activity has no vital relationship then a positive and consistent termination of the corresponding activity can be achieved by making forward progress. Instead, it may be necessary to start another activity which tries to terminate correctly the workflow execution in an alternative way. Thus the workflow will enforce regular process execution, probably along another execution path. This forward recovery is initialized through a forward alternative dependency.

Definition 5: *a is said to be recoverable (a^{rc}) iff if this activity fails then there is a recovery mechanism to resume the workflow execution.*

4.3 Improving Workflow Recovery

The goal of recovery techniques is to minimize the amount of effort resuming the workflow execution. However, the applicability of these techniques depends on the semantics of the process. In this section, we propose techniques giving help for that and correcting potential design errors detected after workflow mining. In fact, we use the transactional behavior mining as a feed back loop to correct wrong transactional behaviors. By wrong transactional behaviors we mean activity transactional properties and transactional flow initially specified and which do not coincide with the reality. These wrong transactional behaviors can be simply costly but also source of error. We proceed through a set of rules that allow us to :

  ✓ correct or remove the wrong transactional behavior,
  ✓ add relevant transactional behavior for a better failure handling and recovery.

These rules depend on both discovered workflow patterns and discovered transactional behavior. Indeed, workflow transaction behavior specification must respect some semantic "regulations" partially depending on the discovered control flow. Concretely, workflow recovery is tightly related to the control flow (*i.e* workflow patterns) as following rules:

1. **R1** : For all workflow patterns : **if a vital activity can fail then it must be recoverable**;
2. **R2** : For all workflow patterns except some parallel threads of M-out-of-N Join and OR-Join patterns : **all the activities are essential and therefore vital**;
3. **R3** : For XOR-split, XOR-join, sequence patterns : **There is no cancellation dependencies between the activities.** *A cancellation dependency can exist only in case of failure and between parallel activities*;
4. **R4** : For AND-split, AND-join patterns combination : **If an activity fails then the consistent point must be out of the parallel threads between the "join" and "fork" point**;
5. **R5** :For all workflow patterns : **If we have a backward recovery then the consistent point cannot be before an executed pivot activity**.

To illustrate the applicability of our rules we go back to our motivating example. We have discovered that CCW never fails and SuD can fail. Then we can suggest that:

  ✓By applying **R1** there is no need for CCW to be recoverable.

  ✓By applying **R3** we remove the abortion dependency between CCW and SuD.

  ✓By applying **R1** and **R2** we should specify that SuD is recoverable.

  ✓By applying **R4** the consistent point after SuD failure is localized just before ICl or just after $EnLR$

  ✓By applying **R5** we can suggest only a forward recovery, after SuD failure, if EnD or CCW are pivot activities.

  ✓By applying **R3** we can suggest to specify abortion dependencies between SuD in one side and (CCW, REv, RUp, REx and ReL) activities in another side.

5 Related Work

Prior art in process mining field focus on control flow mining perspectives. Van der Aalst et al. propose in [6] an exhaustive survey. Compared to these previous work

[17,18,19,8,20] our control flow mining (see section 3) approach deal **dynamically** with concurrent behavior through *concurrent* windows that gives necessary additional calculus only where we need that. Furthermore, we give an original control flow mining approach through the discovery of workflow patterns witch are well-formed structures giving an abstract description of recurrent class of control flow interactions. Besides, we propose a set of control flow mining rules that are characterized by a "local" work-flow patterns discovery. These rules proceed through a **local log analyzing** that allow us, if we have only fractions of workflow log, to recover correctly **partial results**.

Mining transactional workflows is still in its early phase. So far previous works in workflow mining seem to emphasize only flow control within applications. To the best of our knowledge, there are practically no approaches to transactional workflow mining and correction based on event based logs that discuss the correctness of transactional interactions or address the issue of failures handling and recovery, except [21] which proposed techniques for discovering workflow transactional behaviour.

Concerning workflow recovery there are only a few research activities to name. A first discussion was presented in [22] and the necessity of workflow recovery concepts is slightly addressed in [1]. Especially, the concept of business transactions, introduced in [23], describes some basic workflow recovery ideas in detail (above all partial backward recovery). More recent work in this area is presented in [23,24,5,25,26]. Nevertheless, there exists no broad discussion about the mining of workflow recovery and this paper may be seen as a first deep step in this important area.

6   Conclusion

In this paper we presented an original approach for ensuring reliable workflow exe-cutions. Different from previous works, our approach starts from effective executions, while previous works use only specification properties (which remain assumptions). In-deed, our approach starts from workflow executions log and uses a set of mining tech-niques to discover the workflow control flow and the workflow transactional behavior. Then, based on this mining step, we use a set of rules to improve workflow recovery.

However, the work described in this paper represents an initial investigation. In our future works, we hope to enhance workflow recovery mining techniques by enriching workflow logs and extracting data flow dependencies. We are also interested in the mod-elling and the discovery of more complex transactional characteristics of cooperative workflows.

Acknowledments. The authors would like to thank Sami Bhiri for the fruitful discus-sions we had during the writing of this paper.

References

1. Dimitrios Georgakopoulos, Mark Hornick, and Amit Sheth. An overview of workflow man-agement: from process modeling to workflow automation infrastructure. *Distrib. Parallel Databases*, 3(2):119–153, 1995.
2. A. H. M. ter Hofstede, M. E. Orlowska, and J. Rajapakse. Verification problems in conceptual workflow specifications. *Data Knowl. Eng.*, 24(3):239–256, 1998.

3. W.M.P. van der Aalst. The Application of Petri Nets to Workflow Management. *The Journal of Circuits, Systems and Computers*, 8(1):21–66, 1998.
4. Nabil R. Adam, Vijayalakshmi Atluri, and Wei-Kuang Huang. Modeling and analysis of workflows using petri nets. *J. Intell. Inf. Syst.*, 10(2):131–158, 1998.
5. Johann Eder and Walter Liebhart. Workflow recovery. In *Conference on Cooperative Information Systems*, pages 124–134, 1996.
6. Wil M. P. van der Aalst and B. F. van Dongen. Workflow mining: A survey of issues and approaches. In *Data and Knowledge Engineering*, 2003.
7. W. M. P. Van Der Aalst, A. H. M. Ter Hofstede, B. Kiepuszewski, and A. P. Barros. Workflow patterns. *Distrib. Parallel Databases*, 14(1):5–51, 2003.
8. W.M.P. van der Aalst and L. Maruster. Workflow mining: Discovering process models from event logs. In *QUT Technical report, FIT-TR-2003-03, Queensland University of Technology*, Brisbane, 2003.
9. Ahmed K. Elmagarmid. *Database transaction models for advanced applications*. Morgan Kaufmann Publishers Inc., 1992.
10. Sheth A and Rusinkiewicz M. On transactional workflows. *Special Issue on Workflow and Extended Transaction Systems IEEE Computer Society , Washington DC*, 1993.
11. D. Agrawal and A. El. Abbadi. Transaction Management in Database Systems. In A. K. Elmagarmid, editor, *Database transaction models for advanced applications*. Morgan Kauffman, 1990.
12. Moss J. Nested transactions and reliable distributed computing. In *Proceedings Of The 2nd Symposium on Reliability in Distributed Software and database Systems*. IEEE CS Press, 1982.
13. Helmut Wachter and Andreas Reuter. The contract model. pages 219–263, 1992.
14. A. Elmagarmid, Y. Leu, W. Litwin, and Marek Rusinkiewicz. A multidatabase transaction model for interbase. In *Proceedings of the sixteenth international conference on Very large databases*, pages 507–518. Morgan Kaufmann Publishers Inc., 1990.
15. Weimin Du, Jim Davis, and Ming-Chien Shan. Flexible specification of workflow compensation scopes. In *Proceedings of the international ACM SIGGROUP conference on Supporting group work : the integration challenge*, pages 309–316. ACM Press, 1997.
16. Bartek Kiepuszewski, Ralf Muhlberger, and Maria E. Orlowska. Flowback: providing backward recovery for workflow management systems. In *Proceedings of the 1998 ACM SIGMOD international conference on Management of data*, pages 555–557. ACM Press, 1998.
17. Rakesh Agrawal, Dimitrios Gunopulos, and Frank Leymann. Mining process models from workflow logs. *Lecture Notes in Computer Science*, 1377:469–498, 1998.
18. Jonathan E. Cook and Alexander L. Wolf. Discovering models of software processes from event-based data. *ACM Transactions on Software Engineering and Methodology (TOSEM)*, 7(3):215–249, 1998.
19. Joachim Herbst. A machine learning approach to workflow management. In *Machine Learning: ECML 2000, 11th European Conference on Machine Learning, Barcelona, Catalonia, Spain*, volume 1810, pages 183–194. Springer, Berlin, May 2000.
20. Guido Schimm. Mining exact models of concurrent workflows. *Comput. Ind.*, 53(3):265–281, 2004.
21. W. Gaaloul, S. Bhiri, and C. Godart. Discovering workflow transactional behaviour event-based log. In *12th International Conference on Cooperative Information Systems (CoopIS'04)*, LNCS, Larnaca, Cyprus, October 25-29, 2004. Springer-Verlag.
22. W. Woody Jin, Marek Rusinkiewicz, Linda Ness, and Amit Sheth. Concurrency control and recovery of multidatabase work flows in telecommunication applications. In *Proceedings of the 1993 ACM SIGMOD international conference on Management of data*, pages 456–459. ACM Press, 1993.

23. F. Leymann. Supporting business transactions via partial backward recovery in workflow management systems. In *Proceedings of BTW95*, pages 51–70. Springer, 1995.
24. G. Alonso, M. Kamath, D. Agrawal, A. E. Abbadi, R. Gunthor, and C. Mohan. Failure handling in large scale workflow management systems. Technical report, IBM Almaden Research Center, 95.
25. Qiming Chen and Umeshwar Dayal. Failure handling for transaction hierarchies. In *ICDE '97: Proceedings of the Thirteenth International Conference on Data Engineering*, pages 245–254. IEEE Computer Society, 1997.
26. Jian Tang and San-Yih Hwang. A scheme to specify and implement ad-hoc recovery in workflow systems. In *EDBT '98: Proceedings of the 6th International Conference on Extending Database Technology*, pages 484–498. Springer-Verlag, 1998.

Behavior Based Integration of Composite Business Processes*

Georg Grossmann, Yikai Ren, Michael Schrefl, and Markus Stumptner

University of South Australia, Advanced Computing Research Centre,
Mawson Lakes, SA 5095, Adelaide, Australia
{cisgg, reny, cismis, mst}@cs.unisa.edu.au

Abstract. Integration of autonomous object-oriented systems requires the integration of object structure and object behavior. Past research in the integration of autonomous object-oriented systems has so far mainly addressed integration of object structure. During our research we have identified business process correspondences and have given proper integration operators. So far these integration operators are suited for creating generalized models but not for creating or dealing with the composition of business processes. In this paper we propose integration operators which are able to create, deal, and finalize composition between them. For modeling purposes we use the Unified Modeling Language (UML), especially activity diagrams.

Keywords: business process modeling, behavior based integration, federated information systems, business process integration, web services.

1 Introduction

"Integration" is one of the driving themes in current database and applied computing research in general. A special issue of the *Communications of the ACM* [2] and several articles in subsequent issues dealt with integration topics. Whether at the level of classical database applications, web services, workflows, integration of applications is a matter of significant concern. However, past research in this area has concentrated almost exclusively on structural aspects, e.g.[1,4,9,11,14,20,21]). Integration of object behaviour has received some attention, but only at the level of single operations (or "activities" at the conceptual level) [5,27].

In [25], we described a generic approach and resulting architecture for the behaviour oriented integration of business processes. It is based on a meta-class architecture that uses inheritance and instantiation relationships to describe high-level integration operators that can adapt and produce individualized integration plans (i.e., groups of operations) for the integration of processes from a particular domain.

In [10], we have described an integration process which consists of the identification of business process correspondences and associated integration operators. The correspondences are specified via relationships between equivalent and non equivalent business processes and their activities respectively. For each identified relationship we

---

* This research was partially supported by the Australian Research Council under Discovery Grant DP0210654.

W.M.P. van der Aalst et al. (Eds.): BPM 2005, LNCS 3649, pp. 186–204, 2005.

proposed proper integration options which build a new integrating model. This resulting integrated model is a generalization of the input models. Our approach resulted in the first coherent categorisation of integration options [10], building on a history of detailed examination of individual options using generalisation [16,15,7].

However, while generalisation (or inheritance) is a crucial and (due to its unique characteristics) extensively studied structural building block of current day information system models, it is not the only one, rather it stands besides the two other major kinds of relationships solidly embedded in conceptual models for more than a decade: *composition* (part-of relationships) and *association* (the generic category of all other, usually domain-dependent relationships). The goal of this paper is to extend the rigorous categorisation approach employed in [10], to adapt it to composition, and to examine its effects on a process-oriented view of the integration task.

The concept of composition is the creation of a part-of hierarchy of business processes. A main business process consists of subprocesses which may again have subprocesses of their own. The integration task is to build an integrated hierarchy of such processes and coordinate the control and data flow between them. As preliminary work in this area, [17,18] dealt with the composition of internal and external services and verification of consistency criteria. The recent special issue on "Coordination" in *Data & Knowledge Engineering* [26] pointed out that actual coordination of processes is still a missing aspect in current integration research, and suggested solutions such as that of [23] who propose event-based coordination and [22], who use state- and control flow dependencies between public and private workflows.

In this paper we use a similar integration approach that we used for generalization [10] but for the composite business processes. This approach consists of the following steps:

– **Observation**: At the beginning we examine possible relationships between two business processes.
– **Integration options**: The next step offers several integration options. These integration options contain basic integration operators which can be combined to composite operators. The integration options are independent from the relationships identified in the observation step.
– **Integration option mapping**: In a following step the results of the observation and the integration options are combined together. We receive integration choices which are the requirement for the last step in the integration process, the model transformation. For each relationship we suggest preferred and alternative integration options.
– **Model transformation**: For each identified relationship the model is transformed by using the integration operators. The proper operators are selected given by the integration choices.

We now describe the different steps in more detail.

2   Observation

The observation is the first step in the integration approach and deals with the analysis of semantic relationships between business processes, their activities, and their states.

In [10] we examined pairs of objects which represent the same real world object or have common behavior. In the composition and association of business processes, the semantic relationship between different objects are observed.

We assume that the composite object and its business process are known which means that the structure of the composite object, i.e. its relationship to the types of components, and its behavior are given. The business process of each component is also given but the connection between the composite and the component business processes are unknown. This is the part where our approach sets in. It supports the business process designer in identifying the places where synchronization between the composite and the component business processes is needed to achieve a proper communication. This support is also applicable to the case of associations where only loosely connections between business processes are created.

In the next section we give for each proposed synchronization point a list of synchronization options which can be achieved. Our approach is a first step towards an automatic integration of composite and component business processes and leads to a faster and less error-prone integration.

First we list relationships between activities and states within a composition and second we discuss relationships between business processes in an association.

2.1 Observation of Composite Business Processes

We mentioned in the previous section that the concept of composition is the creation of a part-of hierarchy between objects, the assumption being that there exists one composite object which normally is related to multiple component objects. Depending on how many different types of components are included in a composite object, we distinguish between *heterogeneous* and *homogeneous* relationships. A homogeneous relationship means that an object consists of only one object type of components, e.g., a complete set of car wheels consists of 4 wheels. If a composite object is made of several different types of components, e.g., a computer is made of a hard disk, a cdrom drive, etc., then we talk about a heterogeneous part-of relationship. Note that, as shown above, this is independent of the cardinality of the relationship at instance level, i.e., the homogeneous car-wheel relationship still is a 1:n relation ship. Although the cardinality of most part-of relationships will be 1:n, 1:1 and m:n relationships could be considered; although the latter will usually be limited to organisational contexts. In such a m:n relationship the components are shared by different composite objects, e.g., two car rental companies share the same pool of cars.

In this section we are dealing with integration of business processes which belong to composite and component objects. Because one business process may deal with several types of objects, we suggest that in a projection phase all activities which are dealing with the same type of objects are identified and build a sub business processes [10]. So in the following we are talking about business processes which deal with only one type of object. In this paper a business process that belongs to a composite object is called "composite business process" and a business process which belongs to a component is called "component business process". For space reasons we only deal with composite business processes that have a 1:1 or a 1:n relationship to component business processes, but the results generalize directly.

We explain our approach based on the example of a computer retailer BP_C. BP_C sells cheap computers by assembling computer parts from cheap vendors together to a complete system. The process of identifying which is the cheapest vendor and which parts of which vendor fit together are given. However the prices of the computer parts change rapidly and so the vendors who deliver the parts may change often. It is important for BP_C to have a model which tells them when a communication to potential new vendors should be established and how this communication can be coordinated. Furthermore BP_C offers a second hand market for computer parts and a service center for warranty purposes.

The life cycle of a business process composition consists of three phases, (a) the **construction**, (b) the **coordination** of composite and component business processes, and (c) the **destruction**. Each phase is partitioned into subparts and for each subpart we define a semantic relationship between activities and business processes:

1. **Construction**: At the beginning the composite object is constructed. This means in the context of business processes that a first connection between the composite and the component business processes is established. The construction is separated into two parts:
   - *Creation*: A create operation is needed if the component objects do not exist yet.
   - *Assembly*: If the component objects have existed before the composition was formed, an assembly function fits the components to a composite object together.
2. **Coordination**: After the composite object is created, the composite and component business processes must be coordinated so the coherence and the communication between the components and their composite object can be administrated and is observable. The coordination occurs along two dimensions:
   - *Complex activity decomposition*: This construct deals with the synchronization between composite and component activities.
   - *State synchronization*: deals with the coordination of composite or component activity execution with state conditions.
3. **Destruction**: The last phase of a composite object life cycle is the destruction of the composite object. Depending on the future existence of the components there are two possibilities:
   - *Disassembly*: If the components continue to exist after the decomposition they will be separated from the composite object.
   - *Destruction*: If the components are not going to persist they will be destroyed.

In the following we explain the three phases in more detail and give examples.

2.2 Construction

To create a composite object, all parts must be obtained first and are assembled together. In our computer retailer example we assume that BP_C is aware of the components, how they fit together, and of the order in which to assemble them. An overview of a possible sequence of create and assembly activities between BP_C and three different vendors BP_1, BP_2, and BP_3 is shown in Figure 1. The construction consists of two steps, the creation and the assembly step:

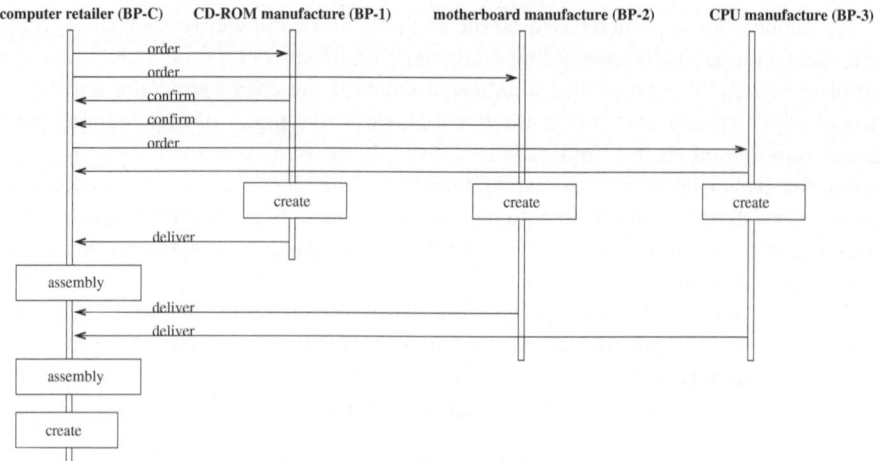

Fig. 1. Example for the sequence of order, create, and assembly operations

Creation: In the computer retailer example, creation consists of obtaining the components from different vendors who create and deliver them to BP_C. Ordering the composite object consists of ordering each component. Figure 2 illustrates a section of the order process for PC components. In this example, BP_C orders some motherboards, CPUs, and graphic cards. Between the order activities we identified the relationship **Component_commit (*cn_comm*)**. A CPU depends on the type of motherboard because it does not fit on all sockets. We assume for the BP_C example that the retailer does not want to order components on stock. So if no motherboard is available on which the chosen CPU fits, the CPU should not be ordered. A *cn_comm* relationship holds between two activities if both activities must be executed successfully or none of them. If an activity A1 succeeds, an activity A2 fails, and there exists a *cn_comm* relationship between them, then A1 must be canceled. This relationship holds between the activities A1 and A4 in Figure 2.

Assembly: The assembly activities for each component are placed after receiving the component from the vendors as shown in Figure 2. Depending on the composite object, the components must be assembled in a specific order, e.g., before installing interface cards on the motherboard, it must be installed in the case first. For specifying this order, we have identified the relationship **component_history (*cn_hist*)** between the activities A3 and A6, as well as between A8 and A3 shown in Figure 2. The arrow indicates the direction of the dependency. Two activities A1 and A2 in different component business processes hold a *cn_hist* relationship if A1 is only allowed to be executed after A2 has finished execution successfully. A similar approach is described in [24,23] where sequence constraints define the possible orders of activity execution. An object life cycle consists of a set of states which are connected through activities. Certain activities can be applied on an object depending on its state and either changes the state of the object or not. The sequence constraints in [24,23] are modeled by state machines which assure that only certain activities can be executed in each state and cannot be executed in another state again. These constraints hold in *cn_hist* as well where

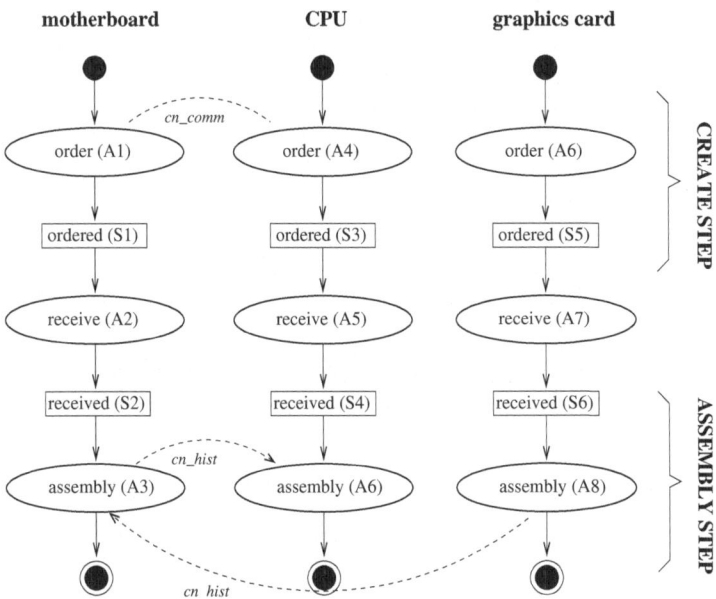

Fig. 2. Construction phase: Examples for *cn_hist* and *cn_comm*

activities are executed in a specific order but can only be carried out once. So *cn_hist* relationships do not allow modelling loops which are possible in [24,23] as long as an object stays in the same state.

At the end of the assembly phase, the composite object is created. The next phase in the life cycle of a composite object deals with relationships between the composite objects and its components. Therefore the activities of the composite and the component business processes need to be coordinated.

2.3 Coordination of Composite and Component Business Processes

For modeling purposes we use the activity diagrams of the UML 2.0 specification [13] and describe the coordination on level of activities. The base concept is that there is an activity A_c in the composite business process BP_c which consists of one or several other activities $A_1, \ldots, A_n$ located in component business processes $BP_1, \ldots, BP_n$.

We extend this hierarchical concept by introducing states to the activity diagrams. By adding the stereotype «state» to the UML metamodel as explained in Section 5.1 we can build conditions between activities and states of two business processes, e.g., while the composite business process remains in a certain state, component activities can be executed.

We define the two expressions "situation" and "Situation Invariant" which are used later in the paper:

- **Situation**: A situation is defined as the set of nodes which are occupied by an object at a specific point of time in a business process.

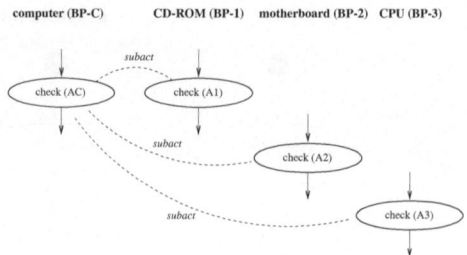

Fig. 3. Example for subactivity condition

- **Situation Invariant**: Describes a state which must be present in any situation during the synchronization of a given set of composite and component business processes.

In the following we describe these two coordination aspects in more detail:

1. **Complex activity decomposition (*subact*)**: The first part deals with the coordination of activities in composite and in component business processes. A composite activity, which is an activity executed in a composite business process, consists of subactivities which may again consist of subactivities, a hierarchical structure that frequently occurs, e.g., in supply chains [12]. We refer to this as a *subact* hierarchy. Figure 3 shows an example of the service centre of the computer retailer BP_C. A customer brought his computer to the service because it does not boot anymore. At the service centre an activity "check" starts which consists of the subactivities "check the CD-ROM" in the component process BP_1, "check the motherboard" in BP_2, and "check the CPU" in BP_3. BP_1, BP_2, and BP_3 are performed at the manufacturer of each component. We define the *subact* hierarchy as follows: Two activities A_C and A_1 in the business process BP_C and BP_1 hold a *subact* hierarchy where BP_C represents the composite and BP_1 a component business process if A_1 is started by A_C and has to be finished before A_C can finish. If the object in BP_C is only present in A_C, then we call A_C the "Situation Invariant".
The order of subactivity execution plays an important role in complex activity decomposition. It is decided during the business process design and restricted by constraints [12]. In the example shown in Figure 3, the motherboard might be checked first, because in most cases of a non-bootable computer, the motherboard is damaged.

2. **State condition (*stat_cond*)**: If a business process must stay in a specific state so that another business process can start an activity or enter a state, we talk of a state condition. Because the condition is determined by another business process we also call it an **interprocess dependency (*IPD*)**. There exist three different kinds of IPDs.
Composite state IPDs (*cs_ipd*): A *cs_ipd* is given if the execution of a component activity depends on the state of the composite object, e.g., to call the support of a software, the computer's OEM licence must be valid. The example shown in Figure 4 describes the business process of computer retailer BP_C at the instant of checking the software of a customer's computer. For further help the process needs to call the software support of the installed software.

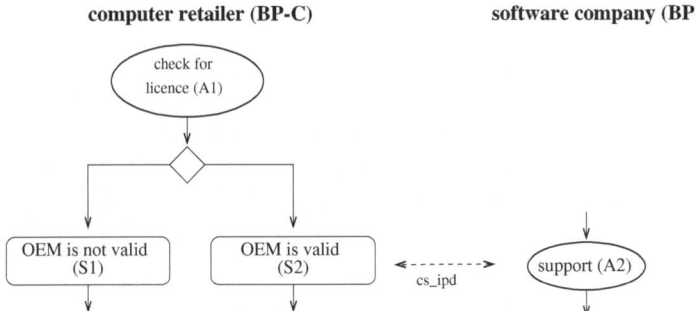

Fig. 4. Example for *cs_ipd* condition

Component state IPDs (*cn_ipd*): A *cn_ipd* between a composite activity and component state holds if the composite activity can only be executed if the component is in a certain state, e.g., a computer can only sell computers with a OEM licence if he has bought the licence at the software company first, as shown in Figure 5.

State component state IPDs (*st_cn_ipd*): The *st_cn_ipd* condition between a composite state SC and a component state $S1$ holds, if as a result of the component entering $S1$, the composite object must enter SC, e.g., if a CD-ROM is detected as damaged, then computer is damaged as well, as shown in Figure 6.

2.4   Destruction

The last phase of a composite object's lifecycle is the destruction phase which covers the disassembly and the demolition of the components. As a result, the composite object is destroyed and the components might be destroyed as well or kept for other purposes. In the BP_C example, the retailer offers PC hardware on a second hand market. He buys old computers, dismantles them, sorts out which hardware parts can still be used, and resells them. Similar to the example of the construction phase, Figure 7 illustrates part of the destruction phase of a PC.

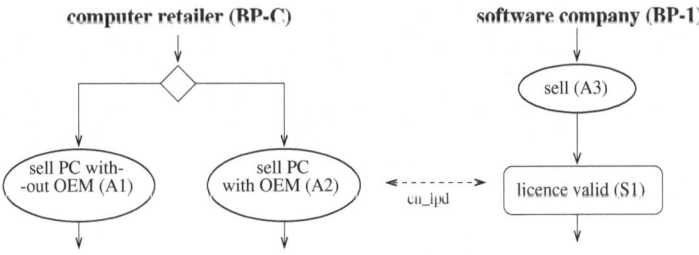

Fig. 5. Example for *cn_ipd* condition

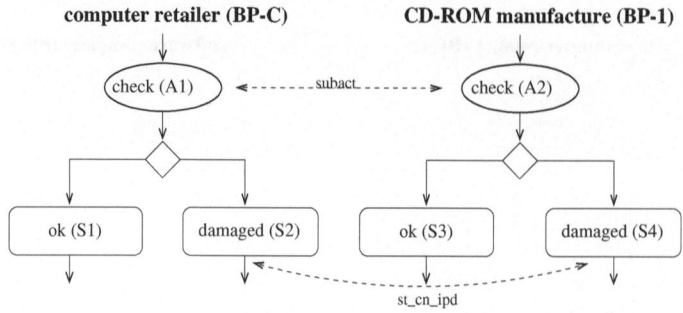

Fig. 6. Example for *st_cs_ipd* condition

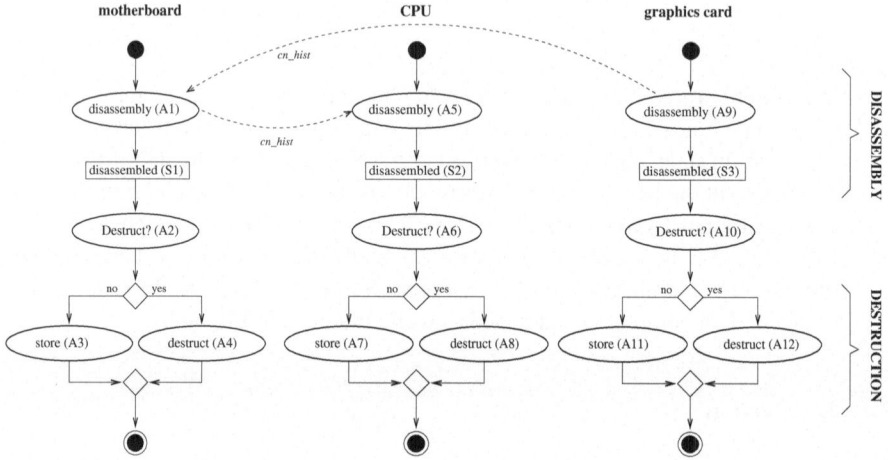

Fig. 7. Destruction phase: Examples for *cn_hist* relationship

1. Disassembly: The Disassembly step destroys the composite object by removing all components from it. Figure 7 shows the disassembly activities of each component according to the example which we demonstrated in the construction phase. Compared to this phase, the assembly activities have to be executed in reverse order. We have identified the semantic relationship *cn_hist* which is described in Section 2.2 between the pairs of activities (A1, A5) and (A9, A1).

2. Demolition: After the components are separated from the composite object, they can be destroyed. In case of the BP_C example the different parts are either stored and filed for the second hand market, or they are recycled as shown in Figure 7.

3   Integration Options

In the previous observation step we have examined different dependencies between business processes and described them as semantic relationships. In the next step we list integration options which are later mapped to the semantic relationships. They transform the models to an integrated business process and create the actual composition.

The integration options depict three different situations where synchronization is needed, (1) synchronization of activities, (2) synchronization of activities and states, and (3) synchronization of states. We explain the situations on three business processes BP_1, BP_2, and BP_3 where (w.l.o.g.) BP_1 is always holding the "Situation Invariant".

3.1   Synchronization of Activities (*synch_act*)

The first case of integration options deals with the synchronization of activities *synch_act*. Depending on the behavior of the object in the BP_1 ("Situation Invariant") we distinguish between blocking, non-blocking, and future synchronization integration options. The fourth synchronization option deals with ordering invocations of activities in different business processes. Finally, the last two more synchronization options ensure atomicity: either all activities are executed successfully or none of them are.

1. **Blocking (*block*)**: A blocking synchronization holds the object in activity A_1 in BP_1 as long as several other activities $A_2, \ldots, A_n$ in BP_2 and BP_3 are executed. After $A_2, \ldots, A_n$ have finished execution, the object is able to continue the flow in BP_1. We use the UML construct *link* for realizing this type of synchronization. A link is a tuple of object references that is an instance of an association or of a connector [19]. Semantically, a link is an individual connection among two or more objects and may be used for navigation and sending messages. We adopted this construct for Activity Diagrams by extending the UML meta model with stereotypes as explained in Section 5.1. A link defined in our Activity Diagram extension represents a connection between two activities of two business processes over which messages can be sent. For realizing *block* we define two types of links:

   - An *invoke link* $L_{invoke} = (N_1, N_2)$ between two business processes BP_1 and BP_2 specifies an activity or a state N_1 in BP_1 which triggers an activity or a state N_2 in BP_2. The link is directed from N_1 to N_2 where N_1 represents the "Situation Invariant". If an object reaches N_1, a message is sent over the link and executes N_2.
   - A *finished link* $L_{finished} = (A_3, N_1)$ between two business processes BP_1 and BP_2 consists of an activity A_3 in BP_2 which was directly or indirectly triggered by an activity or state N_1 in BP_1. After A_3 has finished execution a message is sent over $L_{finished}$ to N_1. The object waiting in N_1 can only continue the flow in BP_1 if a message was received before.

   The *block* synchronization is defined as $S_{block}(L_{invoke}(A_1, A_2), L_{finished}(A_3, A_1))$ and consists of an *invoke link* L_{invoke} and an *finished link* $L_{finished}$. A_1 is an activity in BP_1 and A_2 and A_3 are activities in BP_2 such that $A_2 = A_3$ or $A_2 \neq A_3$. If $A_2 \neq A_3$ then all objects which have entered A_2 must reach A_3.

2. **Non blocking (*nblock*)**: A non blocking synchronization between two activities A_1 and A_2 where A_1 belongs to BP_1 and A_2 to BP_2 ensures that starts of A_1 and A_2 execution is synchronized. There exist three different cases of *nblock* depending on the execution state of A_1:

- A_1 has not started: *nblock_s*: The executions of A_1 and A_2 are synchronized by a join and fork combination JF. JF synchronizes the edges leading to A_1 and A_2 and ensures the synchronous execution of both activities according to the UML specification.
- A_1 has started *nblock_sed*: In this case a message is sent from A_1 to A_2 over an *invoke link* $L_{invoke}(A_1, A_2)$.
- A_1 has finished *nblock_f*: This synchronization option is used for enabling the execution of A_2 after A_1 has finished its execution. *nblock_f* is realized by join and fork combination JF where JF synchronizes the edges E_1 and E_2 where E_1 leaves A_1 and E_2 leads to A_2. From JF two edges E_3 and E_4 leaves where E_3 leads to the state following A_1 and E_4 leads to A_2.

3. **Future synchronization (*future*)**: The future synchronization offers the opportunity to finish the synchronization at a later point of time. The involved business processes can continue their execution between the start and end of synchronization. A similar example is shown in [8] but in that case the execution of the business processes is finished after the synchronization has ended.

   The *future* synchronization is defined as $S_{future}(L_{invoke}(A_1, A_3), L_{finished}$ $(A_4, A_2))$ where A_1 and A_2 are activities in BP_1, and A_3 and A_4 are activities in BP_2.

   Four constraints are associated with S_{future}: (a) $N_1 \neq N_2$, (b) $N_3 \neq N_4$, (c) All objects which have reached N_3 must reach N_4, and (d) All objects which have reached N_1 must reach N_2.

4. **Ordering (*order_act*)**: For ordering the invocation of activities we use the *order_act* synchronization. It is defined as $S_{order}(C, I, F)$ and consists of a set of activity calls $C \neq 0$, set of numbers $I \neq 0$, and an allocation of numbers to activity calls $F \subseteq (I \times C)$ where $\forall c \in C: (\exists o \in O: (c, o) \in F$. A number $n \in O$ represents the order in which an activity call should be executed. An activity call $c \in C$ is an activity which invokes another activity e in a different business process. c is synchronized with e either by a *block*, *nblock*, or *future* synchronization option.

   order_act represents a separate business process where two activity calls $c \in C$ and $d \in C$ are executed synchronously if $(c, o) \in F \land (d, o) \in F$. The activity c and d are set in sequence if $(c, o) \in F \land (d, p) \in F \land o < p$.

5. **Execute if available (*avail*)**: This option is used for synchronizing activities where either both must finish execution successfully or none of them. We use a business process template which first checks the availability of a successful execution of an activity A and then executes A as shown in Figure 8(a). In this example two activities $A1$ and $A2$ from two business processes $BP1$ and $BP2$ are checked for availability first and then executed synchronously similar to a two phase commit protocol. If a further activity $A3$ need to be synchronized following steps must be conducted:

   - Insert an activity $A3_{avail}$ that checks the availability of $A3$ and leads to $S1$.
   - Insert new edge $E1$ leading from $F2$ to a new inserted merge $M3$ that again leads to $A3_{avail}$.
   - Insert a new edge $E2$ leading from $F1$ to activity $A3$.

6. **Cancel if unsuccessful (*cancel*)**: An alternative solution to *avail* is proposed by *cancel*. Here as well the synchronization of two activities $A1$ and $A2$ ensures that

either both activities finish successfully or none of them. The difference to *avail* is that instead of checking the availability the activities are executed first and if one of them finishes with an error, all other activities are canceled, similar to a rollback. An example is shown in Figure 8(b). If a third activity $A3$ need to be synchronized following changes to Figure 8(b) must be carried out:

- Insert a new edge $E1$ leading from $A3$ to $J1$.
- Insert a new edge $E2$ leading from $D1$ to a new inserted activity $A3_{cancel}$ that cancels $A3$.
- Insert a new edge $E3$ leading from $A3_{cancel}$ to $M3$.
- Insert a new edge $E4$ leading from $M3$ to a new inserted merge $M4$ that again leads to $A3$.

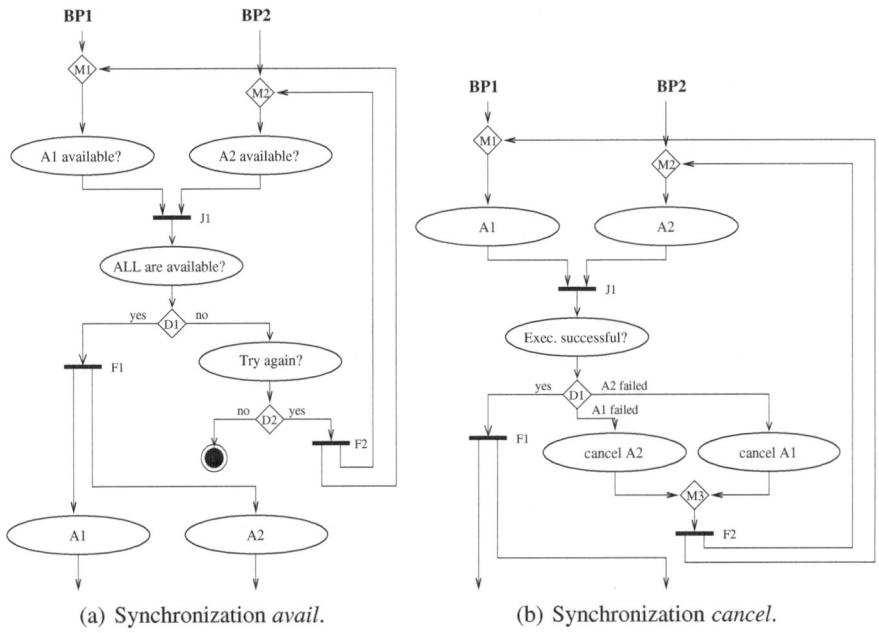

(a) Synchronization *avail*.   (b) Synchronization *cancel*.

Fig. 8. Synchronization of activity execution

3.2 Synchronization of Activities and States

The synchronization of activities and states deals with the synchronization of one activity and several states or one state and several activities. In this section we explain the examples with three different business processes BP_1, BP_2, and BP_3. Again, BP_1 is the "Situation Invariant", but it could be represented by a state or an activity in the diagram.

1. **synch_state_act** stands for the synchronization of a state S in BP_1 and one or several activities $A_1, \ldots, A_n$ in BP_2 and BP_3. If an object is staying in S, $A_1, \ldots, A_n$ are invoked or enabled by sending a message over a link. We define two different

synchronization possibilities within *synch_state_act* depending on the active or passive invocation of $A_1, \ldots, A_n$:

The synchronization with an active invocation **active** is defined as $S_{active}(L_{invoke}(S, A_1), L_{finished}(A_2, S))$. S_{active} consists of an *invoke link* L_{invoke} and an *hasFinised link* $L_{finished}$. S represents the state in BP_1. A_1 and A_2 represents an activity in BP_2 where $A_1 = A_2$ or $A_1 \neq A_2$. If $A_1 \neq A_2$, it must be ensured that all objects reaching A_1 must reach A_2.

The synchronization including a passive invocation **enable** consists of three different links, (a) $L_{enable}(N, A)$, (b) $L_{started}(N_1, N_2)$, and (c) $L_{finished}$. L_{enable} is defined by a state or activity N in BP_1 and an activity A in BP_2. As long as an object is staying in N, the link L_{enable} to A is active and enables the execution of A. $L_{started}$ consists of two activities or states N_1 and N_2 in two different business processes. N_1 sends a message over $L_{started}$ to N_2 when an object has entered N_1, i.e., entered a state or started the execution of an activity. $L_{finished}$ is defined in Section 3.1. We define the synchronization as $S_{enable}(L_{enable}(S, A_1), L_{started}(A_1, S), L_{finished}(A_2, S))$ where S is a state in BP_1 and A_1 and A_2 are two activities in BP_2. $A_1 = A_2$ is possible. If $A_1 \neq A_2$ then all objects which has entered A_1 must reach A_2. A_1 can only be executed if L_{enable} is active. An object can only leave S if either there was no message received from $L_{started}$ or there was a message received from $L_{started}$ and from $L_{finished}$.

2. The synchronization **synch_act_states** deals with synchronization of an activity A in BP_1 and a state S in BP_2. If an object is entering A, the object in BP_2 is set to the state S. *synch_act_states* is realized with a link $L_{invoke}(A, S)$ where A and S are the corresponding activity and state.

3.3   Synchronization of States

The last case of integration options deals with the synchronization of states *s_states* in different business processes. Like in previous sections we use three different business processes BP_1, BP_2, and BP_3 as an example where BP_1 contains a state S_1, BP_2 contains S_2, and BP_3 contains S_3. S_1 signifies the "Situation Invariant".

We define *s_states* as a set of *invoke links* $S_L \neq 0$. If the function $getSource(L_{invoke})$ returns the source node of the *invoke link* L_{invoke} then $\forall L_{invoke} \in S_L$: $getSource$ $(L_{invoke}) = S_1$. In the example of the three business processes, the *s_states* consists of $L_{invoke}(S_1, S_2)$ and $L_{invoke}(S_1, S_3)$.

4   Integration Option Mapping

In this step we choose the proper integration option explained in Section 3 on the basis of the semantic relationships described in Section 2. For each relationship we propose one or more synchronization solutions which integrate the business processes.

For composite business processes as described in Section 2.1, states and activities must be synchronized. An overview of the integration decision for particular relationships is shown in Table 1. The semantic relationships are identified by the line header and the integration decision by the column header. P stands for "preferred integration option" and A for "alternative integration option".

Table 1. Integration decision for particular semantic relationship

	block	nblock_s	nblock_sed	nblock_f	future	order_act	avail	cancel	active	passive	s_states
cn_comm							P	P			
cn_hist						P					
cs_ipd									A	P	
cn_ipd									A	P	
st_cn_ipd											P
subact	P	A	A	A	P	P					

5   Model Transformation

The last step deals with the integration decision for each identified relationship and the application of the chosen synchronization option on the model. According to Table 1 we have chosen a preferred integration options for each semantic relationship that we explained in Section 2:

- **cn_comm**: For the *cn_comm* relationship that we identified in the example shown Figure 2 we have chosen the *avail* synchronization. Because of limited space we cannot provide a figure of the solution but refer to Figure 8(a) which shows a similar example. The activities $A1$ and $A2$ in Figure 8(a) correspond to shortcuts of the order activities in Figure 2.
- **cn_hist**: The *cn_hist* identified in Figure 2 is integrated by *order_act*. S_{order} (C, I, F) consists of $C = C_{A3}, C_{A6}, C_{A8}$ where C_{A3} stands for invoking $A3$, $I = 1, 2, 3$, and $F = 1-> A6, 2-> A3, 3-> A8$. The activity calls are synchronized with the corresponding activity by *block* as shown in Figure 9.
- **subact**: In Figure 3 we identified the *subact* relationship. The activities in this example are synchronized by the integration option *block* as shown in Figure 10.
- **cs_ipd**: For the *cs_ipd* relationship that we have identified in the example shown Figure 4 we have chosen the *passive* synchronization. The result is illustrated in Figure 11.
- **cs_ipd**: The *cs_ipd* identified in Figure 4 is integrated by *passive* as well. The same links are used as in the integrated model shown in Figure 11.
- **st_cn_ipd**: In Figure 6 we identified the *st_cn_ipd* relationship. The activities in this example are synchronized by the integration option *s_states* as shown in Figure 12.
- **cn_hist**: The *cn_hist* identified in Figure 7 is integrated by *order_act*. The destruction is arranged in the opposite order to the assembly step of the composite object. $S_{order}(C, I, F)$ consists of $C = C_{A1}, C_{A5}, C_{A9}$, $I = 1, 2, 3$, and $F = 1-> A9, 2-> A1, 3-> A5$. The activity calls are synchronized with the corresponding activity by *block*.

5.1   A Note on Notation

The representation we use is based on UML 2.0 Activity Diagrams, one of the standard representations for software system behavior. However, we make certain changes. First, of the nodes defined as part of the UML 2.0 AD standard, we use a simplified subset that is sufficient to express business process semantics and preserves or even enhances the underlying Petri-net-resembling semantics of activity diagrams.

On the other hand, we have to make a significant extension. Diagram notations used for conceptual behavior modeling are based on two complementary notational primitives, states and activities. UML behavior representation notations either emphasise states (in the case of Statecharts) or activities (in the case of activity diagrams) and attempt to minimise the use of the respective opposite primitive. While these representations are useful for particular facets of the software development process they typically cannot be used in pure form without overly restricting the modeler, resulting in the introduction of different kinds of "pseudostates" (really particular types of transitions) in Statechart modeling and different types of "locations" (really, intermediate states) in Activity Diagrams. Instead, we use a symmetric notation that permits both states and activities to occur explicitly in the same diagram. As we have found, this significantly facilitates the clear separation between the different integration options compared to "state-biased" or "activity-biased" diagrams. It also helps in several other aspects.

First, states represent a situation of an object between two activities. The activity before the state S has finished and the activity following S has not started yet. In contrast to Petri net based models activities need time in UML-AD. However an activity edge just emit a token from one activity to another without taking time. If synchronization between the execution of two activities is needed we need a state element that represents this situation. One possibility is to insert a join node but in this case at least a second token is needed which is may not available, e.g., send a message to business process P when activity A has finished but activity B has not started yet. Second, Activities are

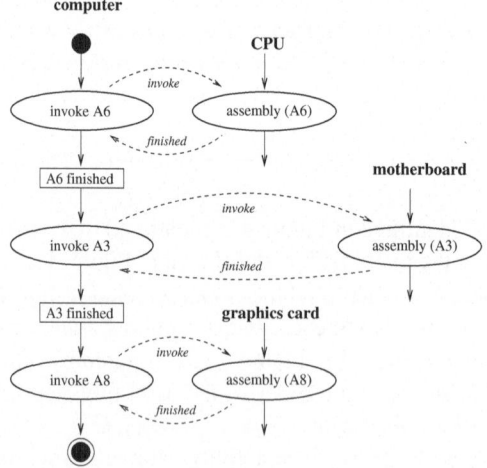

Fig. 9. Synchronization *order_act* of *cn_hist* relationship

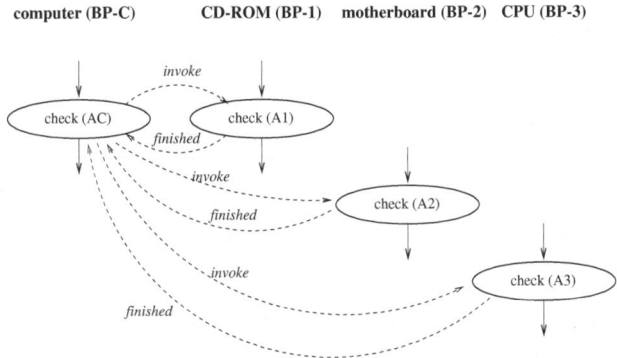

Fig. 10. Synchronization *block* of *subact* relationship

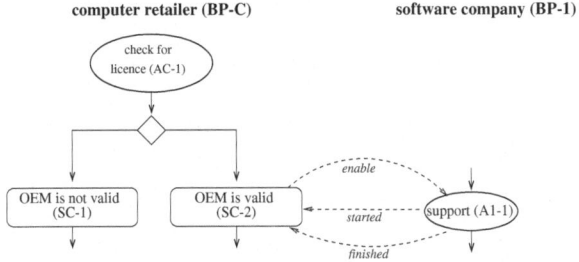

Fig. 11. Synchronization *passive* of *cs_ipd* relationship

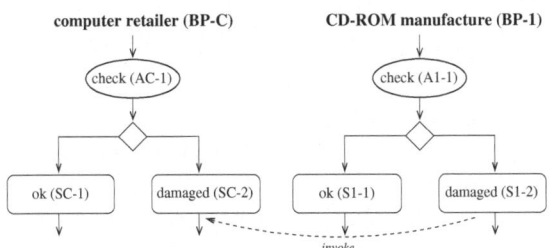

Fig. 12. Synchronization *s_states* of *st_cn_ipd* relationship

executed by events. So we do not know when an activity will be executed. A state can hold a token which is waiting for an event. In UML-AD is not defined where to find a token after an activity has finished but the event for the next activity has not been triggered yet. Lastly,, States offer the possibility of interrupting a business process before an activity can start. By synchronization techniques that we explained in this paper a business process can be forced to stay in a state even if an event triggers an activity. Note that in [6] states were represented by "wait activities". This is a possible solution for introducing states to activity diagrams but the expression "activity" is misleading in this context and "wait activities" also do not posses their own shape in the model. So it is difficult to distinguish them from normal activities.

The actual extension was handled by adding an appropriate stereotype to the activity diagram metamodel. For space reasons, this topic is dealt with in more detail in the technical report version of this paper.

6   Related Work

Related work can be found in [22,23,3]. The integration approaches in [22,3] deals with the coordination of public and internal processes by inter-organizational workflows. Compared to our research the integration is not based on semantic relationships and does not support the reuse of integration options on the level of activities. Furthermore the integration of composite business processes where one activity in a business process consist of several subactivities located in different business processes is not mentioned there.

An interesting approach is explained in [23] where object events between components are coordinated. However because of verification issues there is no graphical notation used for integration which might be difficult to understand for the process designer.

7   Conclusion and Further Research

In this paper, we have described an approach to categorize integration situations between business processes that are a related by a part-of or composition relationship, in a manner similar to our work on integrating business processes that are involved in generalisation relationships [10]. We have given a structured sequence of integration steps that analyzes the relationships between these process. Based on the behavior of the processes involved, this results in limiting the choice from a set of integration options, thus giving direct guidance to the integration designer. Choice of an integration option then leads to the use of a specific high-level integration operator that applies the required modifications and guarantees synchronisation conditions. Together with [10], this work forms part of our metaclass architecture for semantic, i.e., behavior-based, integration [25]. Part of our ongoing work is the implementation of our framework in a meta modeling tool and the application of our results to Web services.

References

1. Omran A. Bukhres and Ahmed Elmagarmid. *Object-Oriented Multidatabase Systems: A Solution for Advanced Applications*. Prentice Hall, 1996.
2. CACM. Special Issue on Enterprise Application Integration. *CACM*, 45(10), October 2002.
3. Issam Chebbi, Schahram Dustdar, and Samir Tata. The view-based approach to dynamic inter-organizational workflow cooperation. *Data and Knowledge Engineering*, 2005. Article in press. Available online at http://www.sciencedirect.com.

4. Stefan Conrad. *Föderierte Datenbanksysteme. Konzepte der Datenintegration.* Springer Verlag, 1997. Only available in German.
5. Stefan Conrad, Barry Eaglestone, Wilhelm Hasselbring, Mark Roantree, Felix Saltor, Martin Schonhoff, Markus Strassler, and Mark W. W. Vermeer. Research issues in federated database systems: Report of EFDBS '97 workshop. *SIGMOD Record,* 26(4):54–56, 1997.
6. H. Eshuis. *Semantics and Verification of UML Activity Diagrams for Workflow Modelling.* PhD thesis, University of Twente, Enschede, The Netherlands, 2002.
7. Heinz Frank and Johann Eder. Towards an Automatic Integration of Statecharts. In *Proc. ER'99,* LNCS 1728, pages 430–444, Paris, 1999. Springer-Verlag.
8. G. Kappel G. Engels, L. Groenewegen. *Coordinated Collobaration of Objects,* chapter 1, pages 307–332. Advances in Object-Oriented Modeling. MIT Press, 2000.
9. Manuel García-Solaco, Fèlix Saltor, and Malú Castellanos. A structure based schema integration methodology. In Philip S. Yu and Arbee L. P. Chen, editors, *Proceedings of the Eleventh International Conference on Data Engineering, March 6-10, 1995, Taipei, Taiwan,* pages 505–512. IEEE Computer Society, 1995.
10. Georg Grossmann, Michael Schrefl, and Markus Stumptner. Classification of business process correspondences and associated integration operators. In *Proc. Int'l Workshop on Conceptual Modeling Approaches for e-Business (eCOMO),* LNCS 3289, pages 653–666, November 2004.
11. W. Klas and M. Schrefl. *Metaclasses and their Applications: Data Model Tailoring and Database Integration.* LNCS 943. Springer-Verlag, Berlin, Heidelberg, 1995.
12. Takashi Kobayashi, Masato Tamaki, and Norihisa Komoda. Business Process Integration as a Solution to the Implementation of Supply Chain Management Systems. *Information Management,* 40(8):769–780, 2003.
13. Object Management Group (OMG). UML 2 Superstructure Final Adopted specification, August 2003. http://www.omg.org/uml, 2003-08-02.
14. Christine Parent and Stefano Spaccapietra. Issues and approaches of database integration. *Communications of the ACM,* 41(5es):166–178, 1998.
15. G. Preuner, S. Conrad, and M. Schrefl. View Integration of Behavior in Object-Oriented Databases. *Data and Knowledge Engineering,* 36(2):153–183, 2001.
16. G. Preuner and M. Schrefl. Observation consistent integration of views of object life-cycles. In *Proceedings of the 16th British National Conferenc on Databases,* pages 32–48. Springer-Verlag, 1998.
17. G. Preuner and M. Schrefl. Behavior-consistent composition of business processes from internal and external services. In *Proc. Int'l Workshop on Conceptual Modeling Approaches for e-Business (eCOMO),* Lecture Notes in Computer Science. Springer-Verlag, October 2002.
18. M. Preuner and M. Schrefl. Requester-centered Composition of Business Processes from Internal and External Services. *Data and Knowledge Engineering,* 2004.
19. James Rumbaugh, Ivar Jacobson, and Grady Booch. *The Unified Modeling Language Reference Manual, 2nd edition.* Object Technology Series. Addison-Wesley Publishing Company, 2004.
20. I. Schmitt. *Schema Integration for the Design of Federated Databases.* Dissertationen zu Datenbanken und Informationssystemen, Vol. 43. infix-Verlag, Sankt Augustin, 1998.
21. Michael Schrefl and Erich J. Neuhold. Object class definition by generalization using upward inheritance. In *Proceedings IEEE ICDE,* pages 4–13. IEEE Computer Society, 1988.
22. Karsten A. Schulz and Maria E. Orlowska. Facilitating cross-organisational Workflows with a Workflow View Approach. *Data and Knowledge Engineering,* 51(1):109–147, October 2004.
23. M. Snoeck, W. Lemahieu, F. Goethals, G. Dedene, and J. Vandenbulcke. Events as atomic contracts for component integration. *Data and Knowledge Engineering,* 51(1):81–107, October 2004.

24. Monique Snoeck. Sequence constraints in business modelling and business process modelling. pages 194–201, 2003.
25. Markus Stumptner, Michael Schrefl, and Georg Grossmann. On the road to behavior-based integration. In *Proceedings 1st Asia-Pacific Conference on Conceptual Modelling*, pages 15–22, 2004.
26. Willem-Jan van den Heuvel and Hans Weigand. Contract-driven coordination and collaboration in the internet context. *Data and Knowledge Engineering*, 51(1):1–3, October 2004.
27. Mark W. W. Vermeer and Peter M. G. Apers. Behaviour specification in database interoperation. In *Conference on Advanced Information Systems Engineering*, pages 61–74, 1997.

Visualization Support for Managing Large Business Process Specifications

Alexander Streit, Binh Pham, and Ross Brown

Faculty of Information Technology,
Queensland University of Technology,
2 George St, Brisbane Australia
{a.streit, b.pham, r.brown}@qut.edu.au

Abstract. This paper proposes a visualization technique to support the modelling and management of large business process specifications. The technique uses a set of criteria to produce views of the specification that exclude less relevant features. The proposed approach consists of three steps: assessing the relevance of nodes, reducing the specification, and presenting the results. Algorithms and methods are presented for these steps along with examples.

1 Introduction

There are multiple graphical business process modelling techniques such as EPC (Event-driven Process Chain) and YAWL (Yet Another Workflow Language). Graphical business process modelling languages are elegant solutions because the user can visually interpret the process. For a more detailed discussion of graphical modelling languages see [1] pp.3. However, as the process grows in size the graph becomes difficult to deal with. This problem is well known to fields that use graphical languages [2]. While zooming initially solves the issue of gaining an overview perspective, there is a finite limit to the amount of zooming that can be performed before information becomes obscured. Screen real estate is limited and the specification given in Figure 7, for example, does not fit clearly on a standard computer display.

Features requiring controlled visual processing, such as interpretation of text, are dominant in business process modelling languages. The ability to interpret controlled visual processing is particularly affected as more information is added. Automatic processing features, such as colour, find limited use in specification languages such as EPC and YAWL. In the case of EPC, where colour is used, colour does not contribute to the overall structural interpretation of the graph.

The traditional solution to this has been to allow decomposition of tasks to sub-specifications. This approach requires that the user construct a deliberate hierarchical structure to support what is in essence a multi-resolution model. Another approach is the conversion of the information into another format, but this loses the benefits of user familiarity, requiring users to learn a new representation.

W.M.P. van der Aalst et al. (Eds.): BPM 2005, LNCS 3649, pp. 205–219, 2005.

For large models to be understood it is necessary that the level of controlled processing required is reduced. The approach explored in this paper is to provide views of the specification that exclude less relevant information. This filtering of information produces a model with lower complexity, but introduces a degree of uncertainty. This uncertainty reflects the lower resolution model's potential for representing variations of the original model. This use of uncertainty mimics human reasoning [3], where decisions are made on relevant information instead of relying upon a detailed and precise model.

The discipline of 3D computer graphics has conducted extensive research into level of detail algorithms [4]. These algorithms construct simplified representations of a full scale model. The purpose of simplification is to maintain a representation of the model that is recognisable while reducing the processing and data requirements of the system (see Figure 1). Typically, lower level detail versions of a model are substituted for the object when it is further away from the observer, where the change is indiscernible.

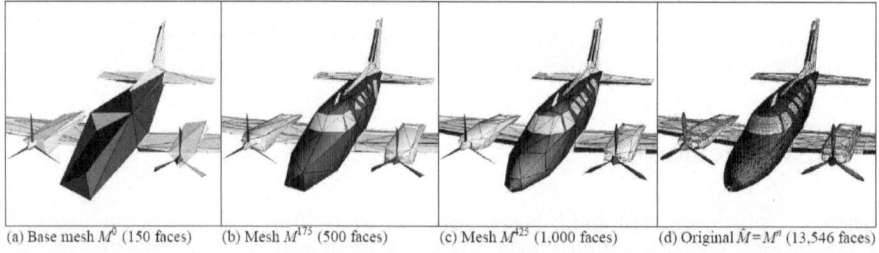

(a) Base mesh M^0 (150 faces)   (b) Mesh M^{175} (500 faces)   (c) Mesh M^{425} (1,000 faces)   (d) Original $\hat{M}=M^n$ (13,546 faces)

Fig. 1. The structure of the 3D model of a plane is evident, even at four different levels of detail. (from [5])

The approach in this paper is motivated by the success of level of detail methods in the 3D graphics field. The proposal is a simplification approach for business process specifications by constructing a *reduced graph* that captures the most relevant information of the original graph. By using this approach the user avoids learning a new notation, because it uses the same graphical notation as the original graph. However, the reduced graph must also preserve the semantics of the original graph to avoid being misleading.

This reduction process presents an opportunity to not only preserve the overview of structure, but to actually provide different views of the same graph according to different interests of the user. Reduction should therefore be directed by criteria that represent the interest of the user, which is governed by the task of the user. For example, the user may wish to see only those processes that are involved in a possible dead-lock situation, or alternatively the user may wish to see nodes that are relevant to a text search term. A graphical search engine can be constructed by creating *reduced views* of business process models according to search terms. This effectively allows the user to browse the business process similar to using a web search engine.

To expand on the example of the search engine, consider the prototype shown in Figure 2. The user is able to enter a search term that is used to direct the criterion function. The resulting display is a *reduced view* of the specification that includes only the most relevant nodes and their relationships to one another. Should the user enter a different term, the process is repeated, starting from the original specification every time. No changes are made to the original specification, instead a temporary reduced view of the specification is constructed for display to the user. Such a tool might be incorporated into the modelling package, to aid the user's understanding and construction of large or complicated specifications.

The mechanics of the reduction algorithm is based on first determining a relevancy factor for each node, followed by analysing the paths through the process model and removing the least relevant nodes. Once the graph has been reduced it must be prepared for display, which requires an aesthetically pleasing and intuitive layout for the graph.

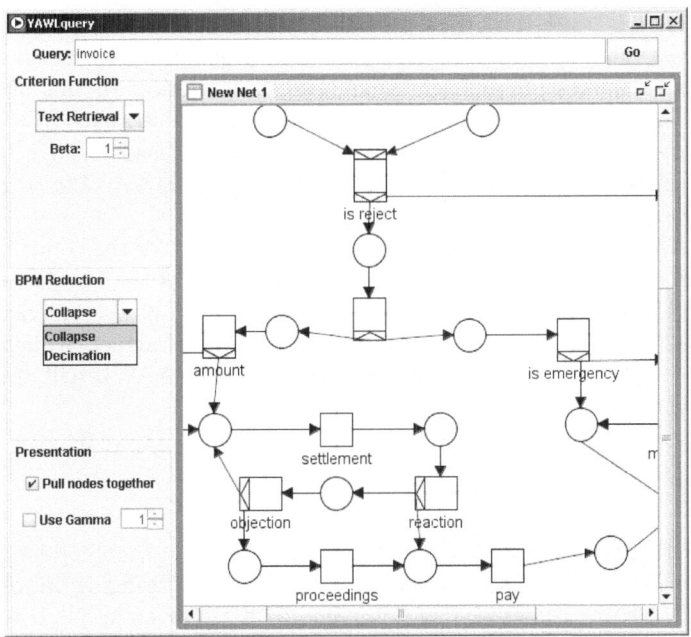

Fig. 2. Prototype for a system that allows users to query specifications in a similar manner to a web search engine

Section 2 provides background material, section 3 details the techniques and approach, while section 4 provides a summary of the work and points to future work.

2   Background

2.1   Workflow Specifications

Business process management (BPM) is about the management of business processes. BPM is receiving increased attention due to improvements in information systems [6]. Workflow management systems (WFMS) are computerised tools to support BPM and workflow specifications drive the WFMS.

Workflow specifications can be observed from different perspectives: control-flow, data, resource, and operational. The *control-flow perspective* describes the order of execution of tasks. Tasks can either be atomic or decompose to sub-specifications, which creates a hierarchical view of the process. The *data perspective* deals with the flow of objects such as documents and can overlay the control flow perspective. The *resource perspective* links tasks to the resources required to perform them. The *operational perspective* details the practical execution of tasks, such as the underlying software services involved.

There are a number of workflow specification languages, both commercial and academic (see [1]). WF-nets were proposed [6] as a specification language based on Petri-nets. The advantage of using Petri-nets is that they provide a formal basis, which enforces precise definition. The disadvantage to this approach is that some patterns do not map well onto high-level Petri-nets [7]. YAWL [7] is a progression from WF-nets that overcomes these disadvantages by adding mechanisms to support the workflow patterns in [8]. The YAWL environment is freely available[1].

The YAWL environment currently provides support for the control-flow perspective, data perspective, and the operational perspective. The formal underpinnings and expressiveness of YAWL make it an ideal choice for visualization research. The former allows for a formal analysis of techniques, while the latter implies that successful development of techniques for YAWL will translate to other workflow specification languages.

The constructs of the YAWL language are given in Figure 3.

2.2   Visualization

A visualization program is analogous to a looking glass through which the user inspects an underlying system. In other words, it is the "bringing out of meaning in data" [9]. Examples of visualization techniques are given in [9]. Traditionally, visualization research has produced visualization techniques that were classified according to data type [10,11,12,13]. However, recent opinion has criticised this approach as producing "showy" images that are insufficiently useful to the user [14].

Suggestions for overcoming this include working more closely with the application domain [14] and creating task-oriented visualization systems [15]. Task-oriented visualizations are driven by the task of the user rather than the composition of the underlying data. The user-centric approach of task-oriented vi-

---

[1] The YAWL environment is available through http://www.yawl-system.com

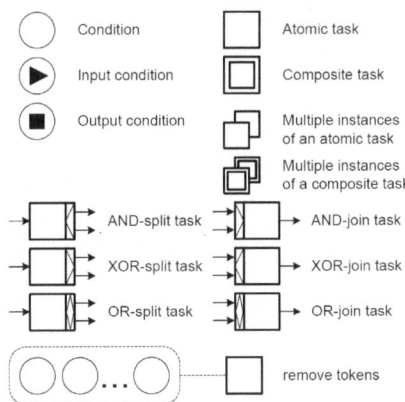

Fig. 3. Constructs of the YAWL language [7]

sualization requires an understanding of the user's requirements. This in turn requires closer cooperation with the application domain.

Visual elements can be classified into two categories [12]: automatic visual processing elements are easily interpreted and include colour, shape, and width, whereas controlled visual processing requires additional user interpretation and include features such a text, icons, and arrows.

2.3   Mesh Simplification Algorithms

Real-time computer graphics applications use 3D mesh structures to model 3D objects. The mesh structure consists of a collection of convex surfaces defined by their vertices. The visible surfaces of the mesh are rasterized, to produced a raster image, which is subsequently shown to the user. To maintain interactive frame rates, this process must be performed for every visible 3D object, in under 83 milliseconds. The sheer mesh complexity required for acceptably accurate models creates processing challenges and has lead to the creation of novel techniques to reduce complexity.

Simplification algorithms reduce the mesh complexity while maintaining the important characteristics of the model. These techniques are used to reduce computation requirements for uses such as fluid flow simulation, shadow volume extrusion, and particularly preserving visual appearance. Several techniques exist (see [4]), which can be placed into two broad categories: *decimation* and *collapse*. Decimation techniques remove numerous elements and reconstruct surfaces over the holes this creates, whereas collapse methods incrementally reduce the mesh through atomic operations.

The progressive mesh [5] is a collapse technique designed for progressive transmission of mesh data. Partially received progressive meshes can be displayed to give the user a low resolution model and the model is refined as more data is received. The edge collapse technique used has an inverse operation, called a vertex split. Given vertex split information, in the correct order, the mesh can be reconstructed to the desired level of detail.

All simplification algorithms make use of an error metric to choose the appropriate reductions. The error metric varies depending upon the intended application of the simplified mesh. For example, the error metric used to generate the appearance preserving meshes given in Figure 1 uses an energy function that measures the squared distance of the proposed vertices to the original mesh, tempered by a spring function to distribute collapses across the mesh [5]. For further reading on this topic see [4].

2.4   Other Related Work

Researchers have previously identified comprehension issues with large conceptual schemas. Their solution builds abstractions for conceptual schemas through recursive derivation of simplified representations [16]. Each derived representation is termed an abstraction level. The abstraction mechanism introduces an importance rating for roles. Objects are weighted according to the sum of their anchored role weights. Object weights that exceed the current abstraction level threshold are identified as important and are included in the abstraction level, whereas a series of production rules are used to remove the remaining objects and their associated roles.

3   Approach

This section details the approach including the underlying algorithms. Section 3.1 describes methods for assessing the relevance of nodes, section 3.2 provides algorithms for reducing graphs based on the relevance of nodes, and section 3.3 discusses methods for presenting the results to the user.

The aim is to construct a reduced representation for a given input specification. Two methods are proposed to achieve this construction, both of which are guided by a criterion function that reflects the requirements of the user. The reduced graph is then presented to the user, who may alter their requirements or request a different level of detail in response. The input specification is hereafter referred to as the original graph. This process is called the visualization process and is illustrated in Figure 4.

This paper proposes the following visualization process:

1: Calculate the relevance of each node according to the criteria
2: Reduce the graph by either the collapse or decimation methods
3: Display the graph to the user for inspection

The original graph is a graph $G(V, E)$, where V is the set of vertices and E is the set of edges. Each node $v \in V$ is either a condition or a task. There is always one *start* condition, s, and one *end* condition, t. For the purposes of this paper, a *completed* graph has at least one task and every node v can be reached on a directed path from s to t. In other words $\forall v \in V : p(s, v) \neq$

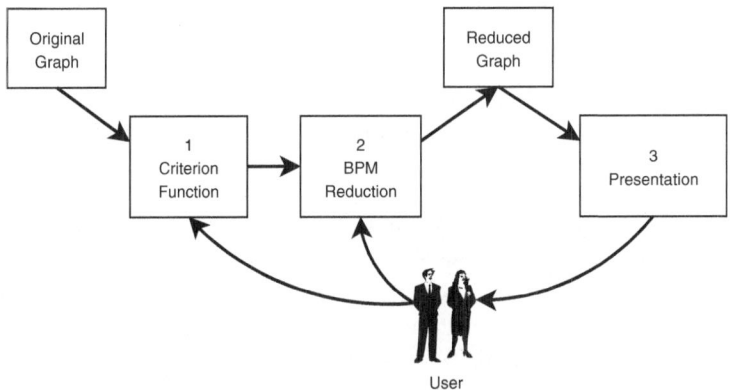

Fig. 4. Visualization support using the reduced graph approach

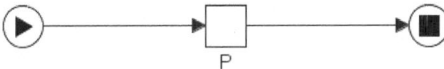

Fig. 5. Any valid process model can be abstracted to the level of a single task P, which stands for "execute the process"

$\emptyset \wedge p(v, t) \neq \emptyset$, where $p(v_i, v_j) \subset E$ returns the edges on a directed path from v_i to v_j. Partial graphs are those graphs where not all nodes can be reached from the start condition, or the end condition is not reachable from any node, or both. During the modelling process, where the user is still building the process specification, the graph may not necessarily be completed. Both of the graph reduction strategies proposed here support these partial graphs, however, the criterion function requires additional care to ensure that it also supports partial graphs. In practice the end condition is typically unreachable in a partial graph.

Any valid workflow specification, which must be a completed graph, can be abstracted to the level $start \rightarrow P \rightarrow end$ (see Figure 5), which is the minimum valid specification possible.

The aim is to build a *reduced* graph $G_R(V_R, E_R)$, where the R subscript denotes reduction. The reduced graph is built such that it contains a subset of the nodes of the original graph. In other words, a reduced graph $G_R(V_R, E_R)$ is built from an original graph $G(V, E)$, such that $V_R \subset V$. A relevance factor, ϵ, is calculated by $\epsilon_i = C(v_i)$ for each node $v_i \in V$, where C is the criterion function $C : V \rightarrow \mathbb{R}$ and $\mathbb{R}$ is the set of real numbers. C orders the nodes according to their relevance to the task of the user.

3.1   The Criterion Function

The criterion function is formulated according to the task of the user. For example, if the task of the user is to identify deadlocks, then neighbourhood nodes that contribute to the deadlock state are of greater interest to the user than the overall graph structure. Contrast this with a user that wishes to see only those nodes that contain a particular search term and closely related nodes. Consequently we assign each task a different criterion function, whose effects dictate the degree to which the preservation of structure overrides the relevance of neighbouring nodes.

Structural Importance. Preservation of the overall structure of the graph is achieved through identifying important control flow nodes. The control perspective defines the flow of control through the graph.

A promising structural importance heuristic is based on the connectedness, $\chi : V \to \mathbb{Z}$, of the node and its estimated position in the routing hierarchy, $\phi : V \to \mathbb{Z}$. $\mathbb{Z}$ is the set of integers and ϕ is calculated by counting the number of splits and subtracting the number of joins on the *shortest* path from s to the node, excluding this node. χ is simply the sum of all connected nodes to this node. ϵ is calculated as follows:

$$\epsilon_i = \frac{\chi(v_i)}{\min(\phi(v_i), 1)}$$

An example application of the heuristic structural importance criteria is shown in Figure 8.

Text Retrieval. Text retrieval algorithms perform best when there are a number of words in a document. Business process models rarely include much text for each node, limiting the applicability of traditional text retrieval ranking methods. However, the context for a node can be viewed as the neighbouring nodes.

One approach to take advantage of this neighbourhood is to introduce a notion of *relevance flow*, which increases the relevance of nearby nodes. The amount of the contribution drops off with distance travelled including loops. The amount of the drop off is arbitrary and a constant rate, β, gives adequate results. This algorithm to assign relevance factors based on a text search term for graph $G(V, E)$ is as follows:

1: Find S_T, the set of all nodes that contain the search term.
2: For each $v \in S_T$,
3:     Initialise the contribution value, $c \leftarrow 1$.
4:     Initialise the neighbourhood node set, $S_N \leftarrow \{v\}$.
5:     While $c > 0$ and $S_N \neq \emptyset$,
6:         update ϵ for all neighbours: $\epsilon'(n) \leftarrow \epsilon(n) + c$ for all $n \in S_N$.
7:         reduce future contributions: $c' \leftarrow c - \beta$.
8:         update neighbour list: $S_N \leftarrow \{n \in S_N : w \in V, \{nw\} \in E\}$.
9:     End while.
10: End for.

An example using the text retrieval criteria is shown in Figure 9.

Graphical Considerations. The business process model is a graphical representation, meaning that the modeller has assigned the positions of the nodes. These positions hold meaning, for example, invoicing related tasks will commonly be grouped together spatially. This meaning can be included in the criterion function by measuring the relative change in the position of a node.

3.2   Business Process Model Reduction

This section describes model transformation techniques that produce reduced models based on the criterion function.

The reduced graph must preserve the semantics of the original graph to avoid being misleading. Semantics are preserved if all *possible orders of execution* of the remaining nodes are unchanged from the original graph. In other words, the dependencies between nodes cannot change.

Two methods are described: the *collapse* method, which incrementally reduces the graph until a threshold value for ϵ is reached, and the *decimation* method, which removes all nodes below a threshold value and reconstructs the paths between remaining nodes.

The threshold value is assigned by the user and is called the alpha-cut value, denoted α.

Collapse. The principle behind the collapse technique is to incrementally reduce the graph. Each incremental change in the graph is selected on the basis of removing the least relevant (minimum ϵ) node from the current model $G_R{}^n$ to produce next $G_R{}^{n+1}$, according to conditions described next.

A non-join node is selected for removal at each increment. A split node is only selected if its predecessor is a task. The removal is performed by merging the node with its predecessor. Figure 6 illustrates how this is done under various circumstances. Split and join decorators are removed from a node when a single inflow or outflow, respectively, results from the collapse, yielding a sequence operation. Given the selection pattern under the heading 'Original YAWL' in Figure 6, the first selected node is y, which is merged with a to produce the version shown under the heading 'Reduced (introduce ϵ)'. Subsequently, x is chosen and merged with a to produce the sequence pattern of $a \to b$.

One advantage of the collapse technique is that the order of collapses can be stored. The inverse operation of a collapse, called a node-split, can then be performed to restore $G_R{}^{n+1}$ to $G_R{}^n$. Another advantage is that since collapses relate one level of detail to another, the presentation can animate changes to increase interpretability of the technique. The calculation of collapses can be performed in a pre-processing step and since the actual collapse operation requires minimal processing, the visualization system can allow interactive navigation between various levels of detail.

An example application of the collapse algorithm is shown in Figure 8.

Decimation. The decimation approach selects a number of nodes that will be included in G_R. All other nodes are *removed*. The original graph is then analysed

Pattern	Original YAWL	Reduced (introduce ε)

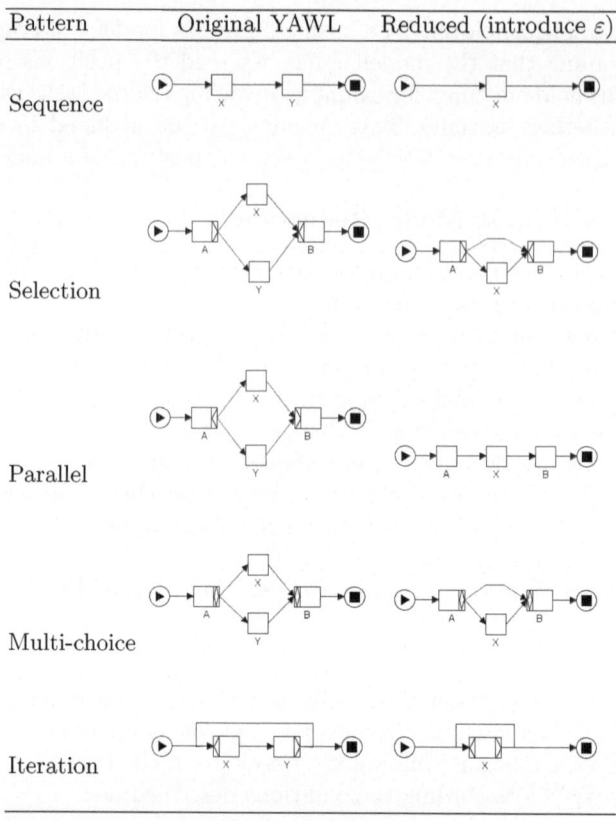

Sequence

Selection

Parallel

Multi-choice

Iteration

Fig. 6. Selected reduction patterns for the collapse technique

to reconstruct the paths between the remaining nodes. Nodes are selected for inclusion if their relevance is α or higher. A *concurrent* path is defined as any path from one node to another where a split exists on the path that was *not* synchronised before reaching the destination node. A *direct* path from $x \in S$ to $y \in S$ is a path from x to y without going through any other element of S.

The decimation-construction algorithm is given as follows:

1: Initialise the set of included nodes, $S_I \leftarrow \{s, t\}$
2: add all v_i where $C(v_i) > \alpha$ to S_I.
3: Initialise output edges, $E_R \leftarrow \emptyset$
4: For $x \in S_I, y \in S_I, x \neq y,$
5:    $V_R' \leftarrow V_R \cup \{y\}$
6:    if there is a direct path from y to y, $E_R' \leftarrow E_R \cup \{yy\}$
7:    if there is a direct path from x to y, $E_R' \leftarrow E_R \cup \{xy\}$
8:    if a concurrent path $\{x..y\}$ includes any $z \in S_I$ ($z \neq x \neq y$),
          add the offending split node(s) before x and y to S_I,
          add the matching join node(s) after x and y to S_I.
9: End for

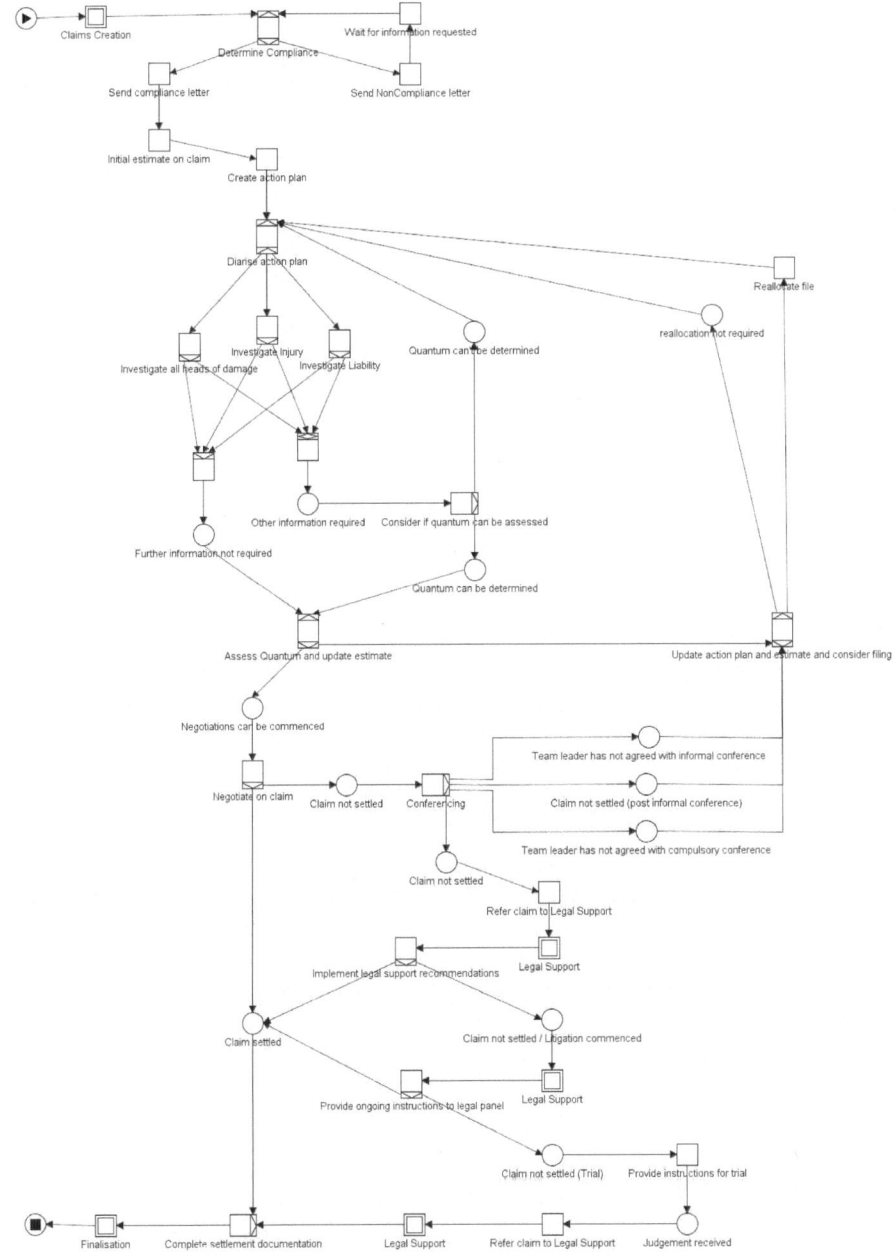

Fig. 7. An example workflow specification that is too large to fit on a computer screen

In practice, the detection of concurrent paths in step 8 is implemented by performing a search for elements of $S_I - \{x\}$ in the original graph G, starting at x and following the directions of edges in E. Split nodes will spawn multiple paths

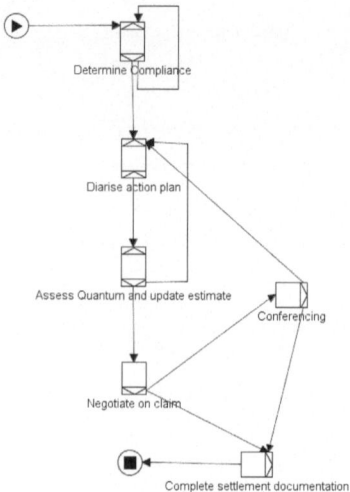

Fig. 8. A reduced version of specification graph in Figure 7 showing structure built using the collapse approach ($\alpha = 2.5$)

leading away from them, whereas join nodes reunite paths back together. Since the graph is cyclic, the search must keep a set of traversed edges to ensure they are not pursued again. Any path is terminated if it reaches y, or there are no untraversed outflows to follow. If any path finds an element of $S_I - \{x, y\}$, the path is recorded then terminated. The search is terminated when all paths terminate. The offending split nodes are found by backtracking recorded paths until a forward path to y is found. The matching join nodes are found using a similar approach that starts a path at every element of $S_I - \{x\}$ that was previously found, including y. The algorithm continues until all paths terminate or there is a single path. It records join nodes that combine the paths along the way.

An example application of the decimation algorithm is shown in Figure 9 where the task 'Negotiate on claim' is an offending split node and 'Complete settlement documentation' is the matching join node.

3.3   Presentation Techniques

This section considers issues of presenting the reduced graph to the user. The reduction techniques described previously produce a reduced graph, but do not alter the position of the nodes. The role of the presentation algorithm is to produce a visually pleasing layout of the reduced graph while preserving the intuitiveness of the result.

Producing an appropriate layout is the subject of ongoing work and preliminary results are described here. The method used in figures 8 and 9 adjusts the length of the edges in E_R to seek an even length among all edges. Nodes also carry a localised repellent force that push nodes apart and keeps them from overlapping. The system is then allowed to stabilise, which will occur when all

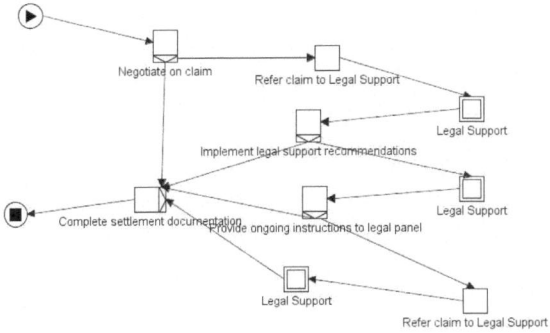

Fig. 9. A reduced version of the specification in Figure 7 for the text query 'legal' built using the decimation approach ($\beta = 0.5$, $\alpha = 1$)

edges have expanded or contracted until the repellent force of each node equals the force on its edges.

The aim of presentation is to ensure that the user is able to interpret the results, which depends largely on the user easily being able to relate the graph back to the original. The original position of the nodes was chosen by the modeller and holds associated meaning. Whenever automated changes to the graph are performed, it is necessary to preserve the user's mental map [17]. The relative position of nodes to one another should not alter considerably, since this would also reduce the ability of users to relate the reduced graph to the original graph.

To improve fidelity to the original graph, another force can be modelled as an attracting force between the node and its relative position in the original graph, called the scaled original position. The scaled original position $P_s(i)$ of node i is the position in the original graph $P_o(i)$ scaled by the ratio of the extents of the graphs γ:

$$P_s(i) = \gamma P_o(i)$$

where γ is given by the ratio of the extents of the current graph E_c to the extents of the original graph E_o:

$$\gamma = \frac{E_c}{E_o}$$

4   Conclusions and Future Work

This paper proposed the use of a visualization process to support the understanding of large business process specifications. The process was divided into three steps that together provide the user with a simplified specification that is relevant to the task of the user.

The visualization process can be used to allow users to browse the specification in a similar manner to the way in which the web is explored through web search engines. The techniques presented in section 3 work with partial graphs, allowing the process to be used during modelling of the business process specification.

The notion of a criterion function allows for rich and flexible expression of relevance. Section 3.1 provided two criterion functions: one favouring the overall structure of the graph and another for locating nodes related to a text search term. Section 3.2 described two methods for actually building the reduced graphs: the *collapse* technique, which is incremental, and the *decimation* technique. Section 3.3 covered aesthetic and interpretability issues involved with presenting the reduced graph to the user. Two examples were given in figures 8 and 9 for the original graph in figure 7.

Future work would explore additional criteria functions and presentation algorithms. User interaction with the system can also be extended, such as to allow users to 'brush' over nodes to reveal their neighbourhood. The applicability of this visualization process is not limited to modelling large processes. It can be extended to work with run-time data or process mining results.

Acknowledgements. The authors wish to acknowledge Arthur ter Hofstede for his feedback and Michael Roseman for providing access to real world large business process specifications. Thanks also go to Alexander Campbell, Rune Rasmussen, and Frederic Maire for their suggestions on graph theory.

References

1. W. van der Aalst, "Business process management demystified: A tutorial on models, systems and standards for workflow management," in *Lectures on Concurrency and Petri Nets*, vol. 3098, pp. 1–65, Springer Verlag, Berlin, 2004.
2. M. M. Burnett, M. J. Baker, C. Bohus, P. Carlson, S. Yang, and P. van Zee, "Scaling up visual programming languages," *Computer*, vol. 28, no. 3, pp. 45–54, 1995.
3. J. M. Mendel, *Uncertain Rule-Based Fuzzy Logic Systems*. Prentice Hall PTR, 2001.
4. D. Luebke, M. Reddy, J. D. Cohen, A. Varshney, B. Watson, and R. Huebner, *Level of Detail for 3D Graphics*. Morgan Kaufman Publishers, 2003.
5. H. Hoppe, "Progressive meshes," in *SIGGRAPH '96: Proceedings of the 23rd annual conference on Computer graphics and interactive techniques*, (New York, NY, USA), pp. 99–108, ACM Press, 1996.
6. W. van der Aalst and K. van Hee, *Workflow Management: Models, Methods, and Systems*. MIT Press, 2002.
7. W. van der Aalst and A. ter Hofstede, "Yawl: Yet another workflow language," in *Information Systems*, vol. 30, June 2005.
8. W. van der Aalst, A. ter Hofstede, B. Kiepuszewski, and A. P. Barros, "Workflow patterns," *Distrib. Parallel Databases*, vol. 14, no. 1, pp. 5–51, 2003.
9. P. Keller and M. Keller, *Visual Cues*. IEEE Press, 1992.
10. S. Henderson, "Vised: Visaulization techniques." Retrieved 25 June 2004 from http://www.siggraph.org/education/materials/HyperVis/vised/VisTech/vtmain.html, 1996.
11. M. Reed and D. Heller, "Olive: Online library of information visualization environments." Retrieved 15 May 2004 from http://www.otal.umd.edu/Olive/, 1997.

12. S. Card and J. Mackinlay, "The structure of the information visualization design space," in *IEEE Symposium on Information Visualization*, pp. 92–99, IEEE Press, Oct 1997.

13. E. Chi, "A taxonomy of visualization techniques using the data state reference model," in *IEEE Symposium on Information Visualization*, pp. 69–75, IEEE Press, Oct 2000.

14. K.-L. Ma, "Visualization - a quickly emerging field," *ACM Computer Graphics*, vol. February, pp. 4–7, 2004.

15. R. Brown and B. Pham, "Visualisation of fuzzy decision support information: A case study," in *IEEE International Conference on Fuzzy Systems*, 2003.

16. L. J. Campbell, T. A. Halpin, and H. A. Proper, "Conceptual schemas with abstractions: Making flat conceptual schemas more comprehensible.," *Data Knowl. Eng.*, vol. 20, no. 1, pp. 39–85, 1996.

17. K. Misue, P. Eades, W. Lai, and K. Sugiyama, "Layout adjustment and the mental map," *Visual Languages and Computing*, vol. 6, no. 2, pp. 183–210, 1995.

Transforming BPEL to Petri Nets

Sebastian Hinz, Karsten Schmidt, and Christian Stahl

Humboldt–Universität zu Berlin,
Institut für Informatik, D–10099 Berlin
{hinz, kschmidt, stahl}@informatik.hu-berlin.de

Abstract. We present a Petri net semantics for the Business Process Execution Language for Web Services (BPEL). Our semantics covers the standard behaviour of BPEL as well as the exceptional behaviour (e.g. faults, events, compensation). The semantics is implemented as a parser that translates BPEL specifications into the input language of the Petri net model checking tool LoLA. We demonstrate that the semantics is well suited for computer aided verification purposes.

keywords: Business process modeling and analysis, Formal models in business process management, Process verification and validation, BPEL, Petri nets.

1 Introduction

The *Business Process Execution Language for Web Services* (BPEL) is part of ongoing activities to standardize a family of technologies for web services. A textual specification [1] appeared in 2003 and is subject to further revisions. The language contains features from previous languages, for instance IBM's WSFL [2] and Microsoft's XLANG [3]. The textual specification is, of course, not suitable for formal methods such as computer aided verification. With computer aided verification, in particular model checking, it would be possible to decide crucial properties such as composability of processes, soundness, and controllability (the possibility to communicate with the process such that the process terminates in a desired end state). For a formal treatment, it is necessary to resolve the ambiguities and inconsistencies of the language which occurred particularly due to the unification of rather different concepts in WSFL and XLANG.

Several groups have proposed formal semantics for BPEL. Among the existing attempts, there are some based on finite state machines [4,5], process algebras [6], and abstract state machines [7,8]. Though all of them are successful in unravelling weaknesses in the informal specification, they are of different significance for formal verification. The semantics based on abstract state machines are feature-complete. However, Petri nets provide a much broader basis for computer aided verification than abstract state machines. Most of the other approaches typically do not support some of BPEL's most interesting features such as fault, compensation, and event handling.

In this paper, we consider a *Petri net semantics* for BPEL. The semantics is *complete* (i.e., covers all the standard and exceptional behaviour of BPEL),

W.M.P. van der Aalst et al. (Eds.): BPM 2005, LNCS 3649, pp. 220–235, 2005.

and *formal* (i.e., feasible for model checking). With Petri nets, several elegant technologies such as the theory of workflow nets [9], a theory of controllability [10,11], a long list of verification techniques [12] and tools [13,14,12] become directly applicable. The Petri net semantics provides patterns for each BPEL activity. Compound activities contain slots for the patterns of their subactivities. This way, it is possible to translate BPEL processes automatically into Petri nets. Using high-level Petri nets, data aspects can be fully incorporated while these aspects can as well be ignored by switching to low-level Petri nets.

We first explain the general concepts of BPEL. Afterwards we introduce the principles of our Petri net semantics and explain the Petri net patterns for a few typical BPEL activities. Then we report first experiences with an automated translation of BPEL into Petri nets, and subsequent model checking. Finally, we discuss some ideas for an extension of our technology that aims at models which are better suitable for model checking.

2   Introduction to BPEL

BPEL is a language for describing the behaviour of business processes based on web services. Such a business process can be described in two different ways: either as *executable business process* or as *business protocol*. An executable business process which is the focus of this paper models the behaviour and the interface of a *partner* (a participant), in a business interaction. A business protocol, in contrast, only models the interface and the message exchange of a partner. The rest of its internal behaviour is hidden. Throughout this paper, we will use the term *BPEL process* instead of "executable business process specified in BPEL". Executing a BPEL process means to create an *instance* of this process which is executed.

For the specification of the internal behaviour of a business process, BPEL provides two kinds of *activities*. An activity is either an *elementary activity* or a *structured activity*. The set of elementary activities includes: empty [1] (do nothing), wait (wait for some time), assign (copy a value from one place to another), receive (wait for a message from a partner), invoke (invoke a partner), reply (reply a message to a partner), throw (signal a fault) and terminate (terminate the entire process instance).

A structured activity defines a causal order on the elementary activities. It can be nested with other structured activities. The set of structured activities includes: sequence (nested activities are ordered sequentially), flow (nested activities occur concurrently to each other), while (while loop), switch (selects one control path depending on data) and pick (selects one control path depending either on timeouts or external messages). The most important structured activity is a scope. It links an activity to a transaction management. It provides a fault handler, a compensation handler, an event handler, correlation sets and data variables. A process is a special scope. More precisely, it is the outmost scope of the business process.

---

[1] We use this type-writer font for BPEL constructs.

A `fault handler` is a component that provides methods to handle faults which may occur during the execution of its enclosing `scope`. In contrast, a `compensation handler` is used to reverse some effects which happened during the execution of activities. With the help of an `event handler`, external message events and specified timeouts can be handled. A `correlation set` is used for identifying the instance of a BPEL `process` only by the content of a message. Thus, a `correlation set` is an identifier – more precisely, it is a collection of properties – and all messages of an instance must contain it. It is either initialized by the first incoming or outgoing message.

Another important concept in BPEL are `links`. A `link` can be used to define an order between two concurrent activities in a `flow`. It has a *source* activity and a *target* activity. The source may specify a boolean expression, the status of the `link`. The target may also specify a boolean expression (the `join` condition) which evaluates the status of all incoming `links`. The target activity is only executed when it evaluates its join condition to true. BPEL provides *dead-path-elimination* [15], i.e. the status of all outgoing `links` of a source activity that is not executed anymore is set to negative. Consider, for instance, an activity within a branch that is not taken in a `switch` activity.

3   Petri Net Semantics for BPEL

Our goal is to translate every BPEL process into a Petri net. The translation is guided by the syntax of BPEL. In BPEL, a process is built by plugging instances of language constructs together. Accordingly, we translate each construct of the language separately into a Petri net. Such a net forms a *pattern* of the respective BPEL construct. Each pattern has an *interface* for joining it with other patterns as is done with BPEL constructs. Some of the patterns are used with a parameter, e.g. there are some constructs that have inner constructs. The respective pattern must be able to carry any number of inner constructs as its equivalent in BPEL can do. We aim at keeping all properties of the constructs in the patterns. The collection of patterns forms our *Petri net semantics* for BPEL.

In the following subsections, we give a glimpse on our semantics, using a basic activity (receive), a structured activity (flow) and the stop pattern as examples. The complete version of the Petri net semantics is reported in [16,17].

3.1   Example of a Basic Activity

Let us have a more detailed look at the general design of a pattern. Figure 1 depicts the pattern for the BPEL's `receive` activity. `receive` is responsible for receiving a partner's request. To identify whether the request is sent to this receive pattern and not to another instance of the process, BPEL's `receive` specifies at least one `correlation set`. The pattern in Fig. 1 presents a `receive` with one `correlation set` which is already initialized[2].

---

[2] The pattern of BPEL's `receive` where a `correlation set` is initialized by the incoming message is very similar to Fig. 1 and can be found in [17].

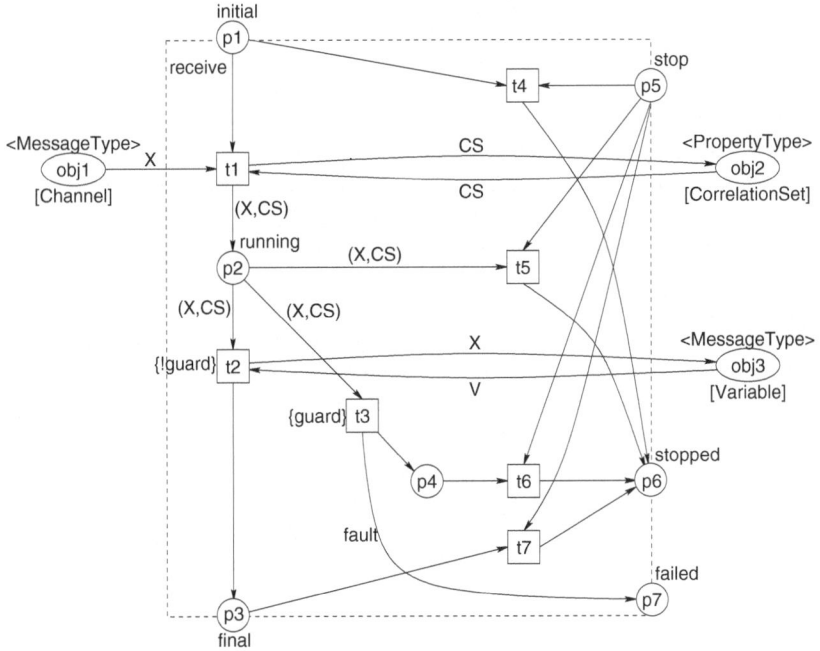

Fig. 1. Pattern for BPEL's `receive`. When the pattern is activated, it is executed in two steps. First, the message is taken from the channel (`obj1`) and the `correlation set` (`obj2`) is read (`t1`). Both values are saved in variables X and CS, respectively. In the second step, this information is analyzed. Either the message is saved in the `variable` (`t2`) or a fault occurs (`t3`). With it variable V holds the old value of `obj3` and `fault` holds the fault information. In both cases, the pattern is finished.

Before we discuss details of the receive pattern, we give some general comments on the notion of patterns. Firstly, we use the common graphical notations for Petri nets. Places and transitions are labelled with an identifier, e.g. `p1` [3] or `t1` which are depicted (contrary to common notation) inside the respective Petri net node. In addition, some nodes have a second label depicted outside the node, e.g. `initial`. This label is used to show the purpose of the node in the net. Secondly, a variable with small letter in arc inscriptions, e.g. `fault`, symbolizes a single variable and a variable with a capital letter, e.g. X, symbolizes a tupel of variables. Thirdly, there are transitions, e.g. `t2` which have a transition guard. Such a transition can only fire when its guard, a boolean expression, is evaluated to true. A guard is depicted (in braces) next to the transition it belongs to, e.g. `{!guard}`.

In general, a pattern is framed by a dashed box. Inside the frame, the structure of the corresponding BPEL construct is modelled. The interface is established by the nodes depicted directly on the frame. Positive control flows from top

---

[3] We use this serif-free font for labels in a Figure.

to bottom while communication between processes flows horizontally. In Fig. 1 the positive control flow starts with a token on initial and it ends either with a token on finish or failed. Outside the frame, there are external objects, e.g. obj1. An object is either a place of a scope pattern (variable, correlation set) or of the process pattern (channel). An activity's pattern as the receive pattern in Fig. 1 relates to those places. The label on the top of an object defines its sort whereas the role is defined at the bottom of the object. A sort is the domain of the tokens lying on and arriving at this place. The object's role is independent of its sort.

The pattern shown in Fig. 1 takes a message from the channel, reads the correlation set and either updates its variable by saving this message or a fault is thrown because of a mismatch between the values of the receive's correlation set and the correlation set in the message or some other error.

The meaning of place stop, stopped and failed in Fig. 1 needs to be explained. In BPEL, a process is forced to stop its positive control flow, e.g., when a fault occurs or activity terminate is activated. However, the BPEL specification [1] tells only informally the requirements how to stop a scope. For instance, activity receive "is interrupted and terminated prematurely" [1, p. 79]. The specification does not describe how to realize those requirements. Thus, we had to make some modelling decisions in our model: The pattern of BPEL's scope is extended by a stop pattern (see Sect. 3.3 for more details), which has no equivalent construct in BPEL. If a scope needs to be stopped, the stop pattern controls this procedure. Our idea is to remove all tokens from the patterns, embedded in the scope pattern; thus the patterns of BPEL's activities and event handler contain a subnet – a so called *stop component*. In contrast, the patterns of BPEL's compensation handler and fault handler do not contain a stop component, because they both need not to be stopped. In [16] we proved that every process can be stopped using stop components. In the case of Fig. 1, the stop component is established by transitions t4 – t7 using the interface stop and stopped. Throughout this paper, we will call this the *negative control flow* of an activity.

In order to explain how a stop component works, consider a scope that contains just a receive and the latter throws a fault. This leads to place failed being marked – the token is an object that consists of the fault's name. This place is joined with a place in the stop pattern; thus this pattern gets the control of the scope. First of all it stops the inner activity of the scope and consequently a token is produced on the receive's stop place. Transition t6 fires and stopped is marked. This place is also joined with a place in the stop pattern. In contrast, transitions t4, t5, t7 consume the token on stop by stopping the receive pattern wherever the control flow is in this pattern. As a result, a token is produced on stopped, too. One might assume that t4 obtains priority before t1 and t5 before t2. Indeed, this would destroy the model's asynchronous behaviour without changing the possible set of runs. We use this asynchronous behaviour in our patterns to model the aspect that sending the stop signal needs time, too. Consider, for instance, two receive patterns executed sequentially. It is possible that the first receive is finished (and so the second receive is activated) exactly in the moment signal stop is sent. In our patterns, however, this possibility is taken into account. Alternatively, a different modelling approach is possible: A transition of the receive pattern's positive control flow is only enabled when no fault has been occurred in the surrounding scope pattern. This fact could be modelled by a place marked when no fault has been occurred. But this, of course, would destroy the asynchronous character of any BPEL process.

3.2   Example of a Structured Activity

Next we show the general pattern of BPEL's flow. flow is used to execute subtasks concurrently. The subtasks can be further synchronized by so-called links.

The pattern in Fig. 2 can carry n inner activities which are executed concurrently. An embedded activity can be any BPEL construct; thus only the interface is visualized and all other information of the pattern is hidden. Therefore only the frame and places initial, final, stop, stopped and if needed negLink are visible (see, for instance innerActivity1 in Fig. 2). The interface of each embedded pattern is joined with the surrounding flow pattern.

negLink is an abbreviation of negative link. It is an optional place that is only part of a pattern's interface when it embeds at least one activity that is source of a link. With the help of negLink the status of all outgoing links of

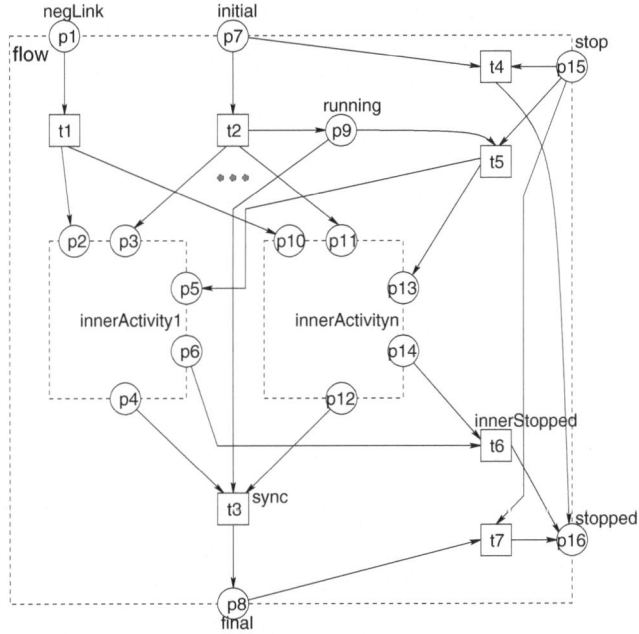

Fig. 2. Pattern for BPEL's flow embeds n inner activities. There are two possible scenarios: Either all inner activities are executed concurrently (t2) and afterwards they are synchronized (t3) or the status of all source links embedded in the flow is set to negative (t1)

an inner activity (i.e. all links for that the inner activity is source) that is not executed anymore are set to negative. Consider an activity within a branch that is not taken in a switch activity. In other words, negLink is a place for modelling dead-path-elimination. In Fig. 2 we assume that innerActivity1 and innerActivityn contain at least one activity that is source of a link.

In our semantics, we model a link by a place of sort Boolean. If the link is set, the place is marked. The value of the token is the status of the link that depends on how the transition condition is evaluated. The join condition determines whether a target activity is executed or not. It is modelled by a transition guard. For modelling dead-path-elimination, we build a link pattern that embeds an activity.

If there is a token on stop, the flow and its embedded activities are stopped. After t5 has fired, the token on running is consumed; thus t3 cannot be activated. Furthermore the stop place of each inner activity is marked. So innerActivity1, ..., innerActivityn can be stopped concurrently. Firing t6 synchronizes them.

3.3   The Stop Pattern

After an activity has thrown a fault, the fault handler of the enclosing scope has firstly to finish the positive control flow inside the scope and secondly it has to handle the fault. We preserve this division and extend every scope by a so-called stop pattern which has no equivalent construct in BPEL. When the stop pattern receives the fault, it finishes its enclosing scope and afterwards it signals the fault to the scope's fault handler. Furthermore the stop pattern is used to realize BPEL's terminate activity, i.e. to stop the entire process.

Figure 3 depicts the pattern of the stop pattern. It is quite complex, because the scope can be in different states when the fault signal occurs. For example, a fault can occur in the positive control flow or in a fault handler. For each scenario the stop pattern behaves differently. In order to explain how this pattern works it is useful to make the following commitment: The pattern we have a look at is embedded in a scope B. B itself embeds a scope C called the *child scope* of B. Furthermore B is child scope of A or in other words: A is the *parent scope* of B.

First of all we have a look at the interface of Fig. 3 which differs from the former patterns. On top there are four important places: ft_in (marked if A wants B to be stopped), fault_in (a fault is occurred in an enclosing activity of B, i.e. either a token on a failed place or C's fault handler rethrows a fault it cannot handle), terminate_up (a terminate activity embedded in A is activated) and terminate (a terminate activity either embedded in C or in B is activated). The place fault_in results from joining the failed places of all activities enclosed by B. All other interface places on top are state places of B. For the most part the state places take inspiration from the *business agreement protocol* (BAP) [18]. The BAP specifies a set of signals serving for communication between a scope and its parent scope. The places on the right are used to remove all tokens in B's compensation handler (cleanCH, ch_cleaned) and to stop the positive control flow of B (stop, stopped). On the bottom there are places to activate other

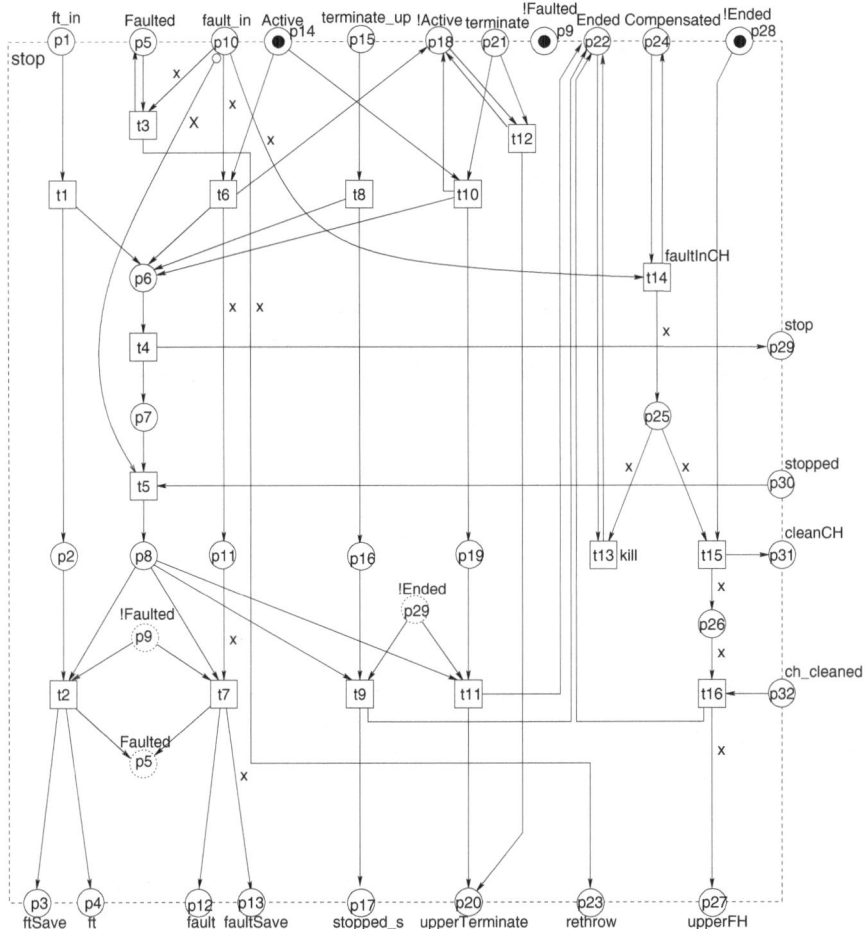

Fig. 3. Stop pattern embedded in a scope

patterns. ft and ft_fault (signalling that A wants to stop B), fault and faultSave (signalling the occurrence of a fault) and rethrow (signalling the occurrence of a fault during the execution of B's fault handler) activate the fault handler of B. In contrast, upperTerminate (signals scope A that it has to be terminated) and upperFH (rethrows a fault to A's fault handler that could not be handled by B's fault handler) activate the parent scope and the parent scope's fault handler, respectively. stopped_s is the stopped place of B.

The arc connecting p10 and t5 differs from the other arcs in its notation (a little circle at its source) and also in its semantics. It consumes all tokens of p10 making no difference if there are 0, 1 or more tokens on this place. In other words, p10 is emptied. This arc is a so-called *reset arc* [19].

Altogether 8 possible scenarios are modelled in this pattern: Either a fault is thrown, an activity **terminate** is activated or A wants to stop B. In the case of an activated `terminate` activity we distinguish if this activity is embedded in an enclosing `scope` (here A) or not and if B's `fault handler` is activated or not. In the case of a thrown fault we distinguish a fault in the positive control flow, in the `compensation handler`, and in the `fault handler`. In this paper, we restrict ourselves to explain how a `scope` can be stopped if a fault in the positive control flow occurs. For details of the remaining scenarios, the interested reader is referred to [17].

Let us continue the scenario described in Sect. 3.1: Let B be the `scope` that encloses the `receive`. If the `receive` throws a fault, its failed place is marked – the token is an object that consists of the fault's name. As already mentioned, the failed place and the place fault_in in Fig. 3 are identical. It is the first fault occurred; thus B is in state Active, i.e. Active is marked. t6 can fire and variable x holds the fault information. Firing t4 produces a token on stop which leads to removing all tokens inside the receive pattern and to produce a token on place stopped. Place stopped in the receive pattern and stopped in Fig. 3 are identical, too. So t5 can fire and the positive control flow of B is finished. By firing t7 the stop pattern invokes the `fault handler` by signalling the fault information.

4   BPEL2PN

In [20], we translated a small BPEL process – it was a modification of the Purchase Order Process presented in the BPEL specification [1, pp. 14] – into a Petri net. This BPEL process consists of 17 activities. The resulting Petri net consists of 158 places and 249 transitions and it was generated manually. In fact this transformation was very laborious and took hours. Therefore tool support was necessary to transform a BPEL process automatically into a Petri net.

We built a parser, *BPEL2PN* [21], that can automatically transform a given BPEL process into a Petri net. The way BPEL2PN works is shown in Fig. 4: It takes a BPEL process process.bpel as an input. Then this process is transformed into a Petri net according to the Petri net semantics. In more detail, for each activity of process.bpel an instance of the corresponding pattern is generated and all these patterns are stuck together as done in the BPEL process. The resulting Petri net, process.lola, is the output of BPEL2PN where .lola is the data format of our model checker LoLA [12]. LoLA offers the user the opportunity to write out the net into the standard interchange format for Petri nets, the *Petri Net Markup Language* (PNML) [22].

As explained in Sect. 3.3, in the stop pattern a reset arc is used to remove all tokens from place fault_in. In the following we draft the idea how such an arc can be modelled as a high-level construct which can be, in turn, unfolded into a low-level construct: It is possible to safely over-approximate the maximal number k of tokens, i.e. the number of faults that can be produced on place fault_in. This is the number of activities of the enclosing scope that can throw a fault. Every scope encloses only a finite number of activities. Consequently k is bounded. So place fault_in is a high-level place that is k-*bounded*, i.e. the number of tokens on fault_in is never greater than k. Then, unfolding the reset arc means to replace

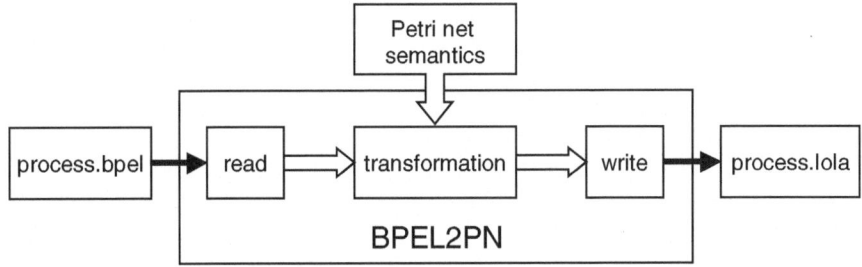

Fig. 4. Mode of operation of BPEL2PN

fault_in by $k+1$ places (0 tokens are possible, too). Furthermore every transition of the pre-set or post-set of fault_in has to be replaced by $k+1$ transitions. It can be easily seen that a reset arc causes an increasing of the net size. The value of k can be narrowed, for instance, in the case of a sequence. Unaffected by the number of its inner activities only one fault can be thrown, because after this fault is thrown the control flow within the sequence is blocked. Calculating the best possible k of place fault_in is ongoing research. In order to avoid an increasing net size due to unfolding we could build an abstract stop pattern. In this pattern we could restrict the number of faults (and therefore k) to 1. Those ideas are explained in more detail in Sect. 6.

The current version of BPEL2PN has the following limitations: Firstly, as already mentioned in [20] we decided to *abstract from data*, i.e. messages and data are modelled as black tokens, because we directed our attention to the control flow. Consequently, all other high-level constructs like transition guards and variables were left out, too. So selecting one of two control pathes in the Petri net semantics, solved by the evaluation of data, is modelled by a nondeterministic choice, e.g. t2 or t3 in Fig. 1. Therefore the resulting Petri net is low-level[4]. Data aspects can be integrated later in our tool or analyzed by methods of static analysis. Secondly, every activity is limited to one `correlation set` (except the synchronous `invoke` that is limited to two `correlation sets`). And last, attribute *enableInstanceCompensation* is ignored. Therefore it is not possible to compensate a process instance, i.e. the entire BPEL process. This is, however, no real limitation: You only need to redefine the process as a `scope` and embed this `scope` in a `process`. Then, the old process can be compensated.

In fact, these are no serious limitations, because the control flow of the BPEL process is preserved. In the next section, we want to give the reader an impression what complex processes can be translated by BPEL2PN and analyzed by our model checker LoLA.

---

[4] Due to the high-level construct of the reset arc the net generated by BPEL2PN is high-level, but it is unfolded to a low-level Petri net by LoLA. Generating a low-level net by BPEL2PN would be possible, too. As a consequence, the complexity of the parser would be increased.

5   Case Study: Online Shop

In this section we present a case study. It shows how a given, realistic BPEL process can be analyzed by the use of our semantics. We generated a business process and verified several relevant properties of this process. We use the Petri net based model checker LoLA that features powerful state space reduction techniques like symmetries [23] partial order reduction using stubborn sets [24] and the sweep-line method [25].

In Fig. 5 our example process is depicted – a modification of the Online Shop Process presented in [10]. A box frames an activity. In the case of a scope or the process itself we use a bold frame. Sequential flow is depicted by dashed arcs, whereas concurrent activities are grouped in parallel. Arcs with solid lines symbolize links. The two nested scopes of the Online Shop Process are depicted in Figures 6(a) and 6(b).

This is a medium-sized example. It consists of 53 activities, yet most of BPEL's activities including fault handler, event handler, nested scopes, and links occur.

The Petri net of the example process consists of 410 places and 1069 transitions. It was generated by our tool BPEL2PN. LoLA takes this Petri net as an input and generates the state space, i.e. it calculates the *reachability graph* of the Petri net. The whole state space consists of 6,261,684 states and is calculated in ca. 96 minutes. By using LoLA's state space reduction techniques (partial order reduction and sweep line method in combination) a reduced state space consisting of 443,218 states could be generated in 50 minutes. More detailed, these reduction techniques do not work on the Petri net patterns, but on the reachability graph of the Petri net. We also generated a variant of the Online Shop Process where every place fault_in was 1-bounded, i.e. safe. That means, in every scope only one fault can occur. As a consequence, the net consists of only 382 places and 495 transitions. The state space reduced to 6,246,601 states (full state space) and 412,731 states (reduced state space), respectively.

If the state space can be fully explored by our tool, it is possible to analyze Petri net specific properties like dead places and dead transitions as well as any temporal property of the underlying process that can be expressed by a formula of the temporal logic CTL.

LoLA calculated dead places and dead transitions. These resulting places and transitions show which aspects of the patterns have been unused. Furthermore this result was used to prove whether there are activities inside the process that can never be activated. In fact, this is possible due to incorrect use of links. As an example consider the switch in Fig. 6(b). If the two assigns were ordered by a link, the target activity would never be activated: On the one hand the branch of the source activity is chosen and so the target activity is not executed. On the other hand the branch of the target activity is chosen, but due to dead-path-elimination the link is set to false. Thus, this activity is never activated, but the process will deadlock neither. In our example all activities can be activated.

We further verified relevant properties of the Online Shop Process like termination and "the customer will always get an answer". Of course, the formula

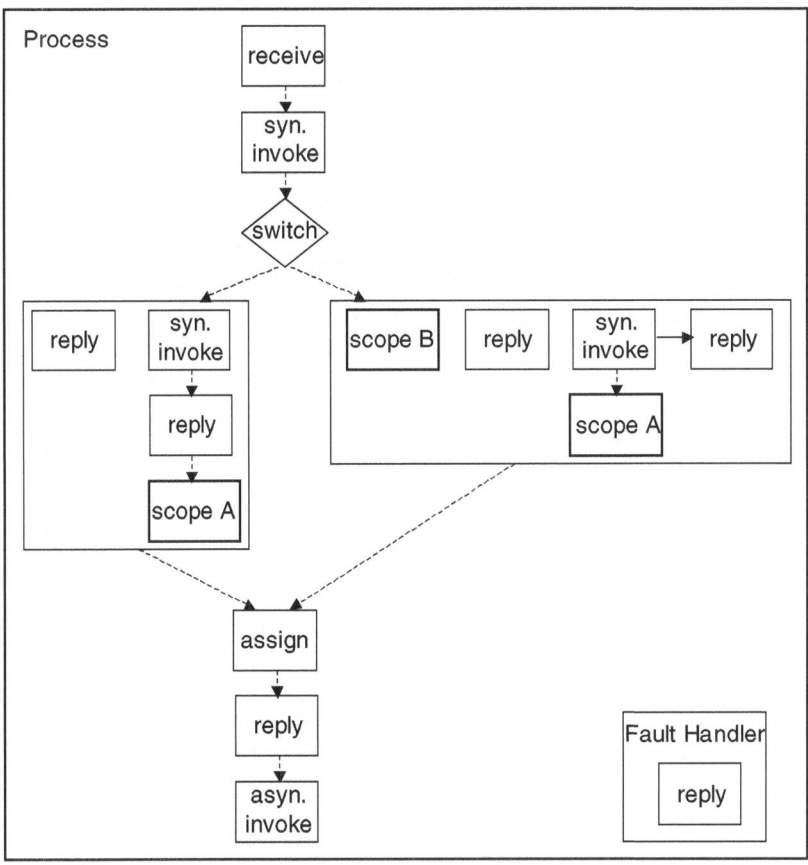

Fig. 5. When the Online Shop Process receives an order from a customer, it retrieves the customer's data. These data are analyzed, because the business strategy of the shop distinguishes new and already known customers. If it is a known customer (left switch branch) the shop initiates two tasks concurrently: The marketing department sends a special offer (on the left) and the customer department (right sequence) firstly takes the order and secondly send its discount level. Afterwards the shop invites offers from the suppliers (scope A). In the case of a new customer (right switch branch), the shop initiates four tasks concurrently: It collects the customer's bank data (scope B), the marketing department sends the customer a special offer (second task on the left). Furthermore the shop takes the order and then it invites offers from the suppliers (scope A). In addition, the terms of trade are sent to the customer (right task). After the completion of the flow the tasks of both, new and known customer are joined. The price information are saved and then the shop sends the supply information to the customer. The process finishes after the shop has invoked the shipper. There is a dependency between two tasks in the case of a new customer, realized by a link: The terms of trade are only sent after the shop has received the customer's order.

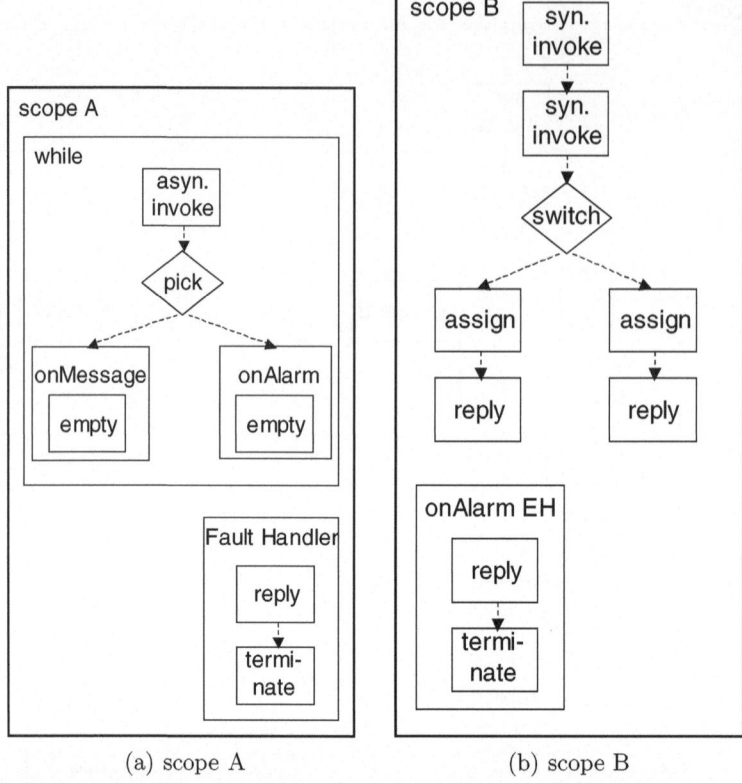

(a) scope A   (b) scope B

Fig. 6. (a) The shop invokes one supplier after the other to invite offers for the product. If the supplier does not answer in time, the next supplier is invoked. Additionally, if a fault occurs during the execution of the process, the customer is informed and the process instance is terminated. (b) The customer's bank data are analyzed whether he is credit worthy. The result is sent to the customer. If he is credit worthy, the process goes on. Otherwise, the process stops. If the bank does not reply in time, the process is terminated after the customer gets a message. One further dependency is modelled by a link (not depicted in the figure): The shop starts invoking the suppliers (scope A) only when the customer is credit worthy.

of the respective temporal properties were generated manually by ourself. The Online Shop behaves as expected: it always comes to an end and the costumer will always get an answer. By abstracting from data aspects as we did, a single process must always terminate, because a deadlock is not possible. Termination plays a more important role if we compose several BPEL processes. Then, it is possible that the composed processes run into a deadlock.

We also tried to analyze a business process that consists of 132 activities. Due to the huge net size we were not able to calculate the full state space of this process. In order to check such an extremely huge process it is necessary to get a smaller model. The next section presents some ideas how to do so.

6   Advanced Translation

The models generated by the present version of our parser can be seen as brute force models. The generated models are significantly larger than typical manually generated models. This is due to the fact that the Petri net patterns are complete, i.e. applicable in every context. For a particular process, many of the modelled features are unused. For instance, if a basic activity cannot throw any error, many of the error handling mechanisms in the surrounding compound activity can be spared. Furthermore, to decide if a specific property holds, it is often sufficient to restrict the patterns to specific aspects. To prove the correct inter-operation of two BPEL processes, for instance, it is sufficient to restrict the attention to communication aspects of the patterns, while internal actions can be abstracted away.

In ongoing projects, we aim at an improved translation where several Petri net patterns with different degree of abstraction are available for each BPEL activity. Using static analysis on the BPEL code, we want to select the most abstract pattern applicable in a given context. We believe that model sizes can be drastically reduced this way thus alleviating the state explosion problem inherent to model checking.

Data flow equations, the basis of static analysis, are already available for many features of BPEL [26]. It is, however, still necessary to select suitable abstraction techniques in order to make static analysis run.

7   Conclusion

We presented first experimental results for generating Petri net models of BPEL processes. The translation of BPEL to Petri nets follows a feature-complete Petri net semantics of BPEL. The translation is implemented, and we were able to present first results. The results show that it is necessary to complement the technology with an improved model generation. We have proposed the use of static analysis as a tool for providing process-specific information that can be exploited in a flexible model generator.

Our goal is a technology chain that, starting at a BPEL process, performs static analysis. Based on the analyzed information, the translator selects the most abstract pattern for each activity that is feasible in the analyzed context and synthesizes a Petri net model. On the Petri net model, a model checker evaluates relevant properties. The analysis results (e.g., counter example paths) are translated back to the BPEL source code.

References

1. Curbera, Goland, Klein, Leymann, Roller, Thatte, Weerawarana: Business Process Execution Language for Web Services, Version 1.1. Technical report, BEA Systems, International Business Machines Corporation, Microsoft Corporation (2003)

2. Leymann, F.: WSFL – Web Services Flow Language. IBM Software Group, Whitepaper. (2001) http://ibm.com/webservices/pdf/WSFL.pdf.
3. Thatte, S.: XLANG – Web Services for Business Process Design. Microsoft Corporation, Initial Public Draft. (2001) http://www.gotdotnet.com/team/xml_wsspecs/xlang-c.
4. Fisteus, J.A., Fernández, L.S., Kloos, C.D.: Formal Verification of BPEL4WS Business Collaborations. In: Proceedings of the 5th International Conference on Electronic Commerce and Web Technologies (EC-Web '04). LNCS, Springer (2004)
5. Fu, X., Bultan, T., Su, J.: Analysis of interacting BPEL web services. In: WWW '04: Proceedings of the 13th international conference on World Wide Web, ACM Press (2004) 621–630
6. Ferrara, A.: Web services: a process algebra approach. In: ICSOC, ACM (2004) 242–251
7. Fahland, D., Reisig, W.: ASM-based semantics for BPEL: The negative Control Flow. In D. Beauquier, E.B., Slissenko, A., eds.: Proc. 12th International Workshop on Abstract State Machines, Paris, March 2005. Lecture Notes in Computer Science, Springer-Verlag (to appear, 2005)
8. Farahbod, R., Glässer, U., Vajihollahi, M.: Specification and Validation of the Business Process Execution Language for Web Services. In: Abstract State Machines. Volume 3052 of Lecture Notes in Computer Science., Springer (2004) 78–94
9. van der Aalst, W.M.P.: The Application of Petri Nets to Workflow Management. Journal of Circuits, Systems and Computers 8 (1998) 21–66
10. Martens, A.: Verteilte Geschäftsprozesse – Modellierung und Verifikation mit Hilfe von Web Services. Dissertation, WiKu-Verlag Stuttgart (2004)
11. Schmidt, K.: Controlability of Business Processes. Technical Report 180, Humboldt-Universität zu Berlin (2004)
12. Schmidt, K.: LoLA – A Low Level Analyser. In Nielsen, M., Simpson, D., eds.: International Conference on Application and Theory of Petri Nets. LNCS 1825, Springer-Verlag (2000) 465 ff.
13. Ratzer, A.V., Wells, L., Lassen, H.M., Laursen, M., Qvortrup, J.F., Stissing, M.S., Westergaard, M., Christensen, S., Jensen, K.: CPN Tools for Editing, Simulating, and Analysing Coloured Petri Nets. In: Proceedings of the 24th International Conference on Applications and Theory of Petri Nets (ICATPN 2003), Eindhoven, The Netherlands, June 23-27, 2003 — Volume 2679 of Lecture Notes in Computer Science / Wil M. P. van der Aalst and Eike Best (Eds.), Springer-Verlag (2003) 450–462
14. Starke, P.H., Roch, S.: Ina et al. In Mortensen, K.H., ed.: Tool Demonstrations 21st International Conference on Application and Theory of Petri Nets, Department of Computer Science, University of Aarhus (2000) 51–56
15. Leymann, F., Roller, D.: Production Workflow – Concepts and Techniques. Prentice Hall (1999)
16. Stahl, C.: Transformation von BPEL4WS in Petrinetze. Diplomarbeit, Humboldt-Universität zu Berlin (2004)
17. Stahl, C.: A Petri Net Semantics for BPEL. Technical report, Humboldt-Universität zu Berlin (to appear June, 2005)
18. Cabrera, Copeland, Cox, Freund, Klein, Storey, Thatte: Web Services Transaction. Vorschlag zur Standardisierung, Version 1.0. (2002) http://ibm.com/developerworks/webservices/library/ws-transpec/.

19. Dufourd, C., Finkel, A., Schnoebelen, P.: Reset nets between decidability and undecidability. In Spies, K., Schätz, B., eds.: Proc. 25th Int. Coll. Automata, Languages, and Programming (ICALP'98), Aalborg, Denmark, July 1998. Lecture Notes in Computer Science 1443, Springer (1998) 103–115

20. Schmidt, K., Stahl, C.: A Petri net semantic for BPEL4WS - validation and application. In Kindler, E., ed.: Proceedings of the 11th Workshop on Algorithms and Tools for Petri Nets (AWPN'04), Universität Paderborn (2004) 1–6

21. Hinz, S.: Implementation einer Petrinetz-Semantik für BPEL4WS. Diplomarbeit, Humboldt-Universität zu Berlin (2005)

22. Billington et al., J.: The Petri Net Markup Language: Concepts, Technology, and Tools (2003)

23. Schmidt, K.: How to calculate symmetries of petri nets. Acta Informatica (2000) 545–590

24. Schmidt, K.: Stubborn set for standard properties. In: Proc. 20th Int. Conf. Application and Theory of Petri nets. Volume 1639 of LNCS., Springer-Verlag (1999) 46–65

25. Schmidt, K.: Automated Generation of a Progress Measure for the Sweep-Line Method. In: Proc. 10th Conf. Tools and Algorithms for the Construction and Analysis of Systems (TACAS). Volume 2988 of LNCS., Springer-Verlag (2004) 192–204

26. Heidinger, T.: Statische Analyse von BPEL4WS-Prozessmodellen. Studienarbeit, Humboldt-Universität zu Berlin (2003)

Event-Based Coordination of Process-Oriented Composite Applications

Marlon Dumas[1], Tore Fjellheim[1], Stephen Milliner[1], and Julien Vayssière[2]

[1] Queensland University of Technology, Australia
(t.fjellheim, s.milliner, m.dumas)@qut.edu.au
[2] SAP Research Centre, Brisbane, Australia
julien.vayssiere@sap.com

Abstract. A process-oriented composite application aggregates functionality from a number of other applications and coordinates these applications according to a process model. Traditional approaches to develop process-oriented composite application rely on statically defined process models that are deployed into a process management engine. This approach has the advantage that application designers and users can comprehend the dependencies between the applications involved in the composition by referring to the process model. A major disadvantage however is that once deployed the behaviour of every execution of the composite application is expected to abide by its process model until this model is changed and re-deployed. This makes it difficult to enrich the application with even minor features, to plug-in new applications into the composition, or to hot-fix the composite application to meet special circumstances or demands (e.g. to personalise the application). This paper describes a technique for translating a process-oriented application into an event-based application which is more amenable to such runtime adaptation. The process-based and event-based views of the application can then co-exist and be synchronised offline if the changes become permanent and it is found desirable to reflect them in the process model.
Keywords: flexible process execution, activity diagram, event-based coordination, coordination middleware, object space.

1 Introduction

Process-oriented composite applications aggregate functionality from a number of other applications by specifying interconnections between these applications through a process model. This model determines how the underlying applications should be orchestrated, most notably their dependencies in terms of flow of control and data. Mainstream infrastructures for developing and executing process-oriented composite application include workflow management systems and process management modules embedded within Enterprise Application Integration (EAI) solutions. Predefined process models can be deployed into the runtime environments associated to these infrastructures for execution.

A major advantage of using a process-oriented approach for composite application development is that it provides an easy-to-comprehend and global view

W.M.P. van der Aalst et al. (Eds.): BPM 2005, LNCS 3649, pp. 236–251, 2005.

of the dependencies between the underlying applications. However, in existing process-oriented systems these dependencies have to be completely specified before deployment [1]. In certain environments, such as mobile computing, changes occur frequently and exceptions are numerous. A just-in-case approach where the designer specifies all possible paths in the process model is impractical, leading to models that are large and unintelligible. Applications operating in such environments may be better served by a just-in-time approach, where adaptation and personalization may be done after the process has been deployed and without requiring all executions to perfectly align with the process model.

Existing methods and techniques in the area of adaptive, dynamic, and flexible workflow systems have addressed issues such as specifying exception handling mechanisms within process models [6,13] or migrating running processes when replacing a previously deployed process model with a new one [1,14]. However, these prior proposals do not provide mechanisms to alter the behaviour of process-oriented composite applications after deployment *without changing the process model*, that is, without requiring alignment between each execution of the composite application and its process model (whether the originally deployed model or a modified version of it). Such ad hoc flexibility mechanisms are instrumental for a number of purposes including: (i) personalising applications to suit the requirements or preferences of specific users; (ii) adapting the behaviour of composite applications based on the users' context (e.g. location, device or network connection) without overloading the process model with such details; and (iii) hot-fixing the composite application to address unforeseen errors, as opposed to predicted exceptions, or to add new features (e.g. to plug-in new applications or to re-route tasks and data).

To overcome the above limitations of existing systems, we propose to adopt an event-based coordination approach to execute process-oriented composite applications. Due to its finer-grained nature, event-based coordination approaches has several advantages over process-based ones when it comes to runtime adaptation and re-configuration [12]. By translating process models of composite applications into event-based models and using the latter in the runtime environment, it becomes possible by adding and removing event-based rules (e.g. event subscriptions related to a specific task) to overlay behaviour on top of already deployed composite applications in response to special requirements or unforeseen situations. In this way, users, administrators and/or developers can re-route data and control in an already deployed composite application in order to steer it into executions paths not foreseen in the process model, thereby facilitating the personalization and adaptation of these applications.

The main contribution of this paper is a technique for translating a process model described in a mainstream process modelling notation (UML Activity Diagrams) into an event-based model described through coordination rules made up of composite event specifications, predicates, and a small number of publishing/sharing primitives. We also discuss how the resulting event-based model can be executed on top of a shared object space infrastructure and how adaptation and personalization is achieved by adding rules (encoded as active objects) into the shared space. We illustrate the proposed technique and its supporting infrastructure through a use case scenario drawn from the area of mobile computing, where the need for adaptation and personalization is often prominent.

The paper is structured as follows. First, we outline a use case scenario (Section 2) and describe the coordination primitives and infrastructure upon which our proposal relies (Section 3). In Section 4 we introduce a technique for translating a process model captured as a UML activity diagram into an event-based coordination model. We then discuss how adaptation can be achieved by adding, enabling and disabling rules in the event-based model (Section 5). Finally, we discuss related work (Section 6) and conclude (Section 7).

2   Use Case Scenario

This section presents a use case that will be used as a motivating and working example in the rest of the paper. The scenario is described as a UML activity diagram[1] in Figure 1. We chose UML activity diagrams as a process modelling notation because of its status as a *de jure* standard and because its constructs are representative of those found in other process modelling and process execution languages (e.g. sequence, fork, join, decision and merge nodes). Thus the proposed techniques can be adapted to other languages that rely on these constructs. Moreover, a recent study shows that UML activity diagrams (version 2.0) provide direct support for many common workflow patterns [17].

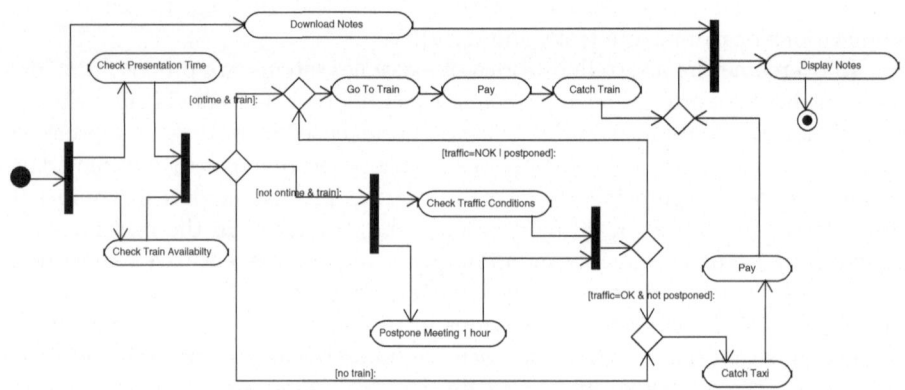

Fig. 1. UML Activity Diagram describing the working example

The scenario is an example of a *personal workflow* [11], i.e. a process aimed at assisting a user in the achievement of a goal that requires the execution of a number of tasks. Most of the tasks composing the process (but not necessarily the process itself) are intended to be executed in a mobile device. Thus the scenario is also an example of a *mobile workflow*. We have chosen this scenario because,

---

[1] http://www.uml.org. Note that in this paper, we refer to UML version 2.0.

putting aside their futuristic nature, mobile and personal workflows constitute a class of process-oriented composite applications in which personalization and runtime adaptation are prominent requirements. Such requirements can also be found to varying degrees in more traditional applications (e.g. order handling) and the proposed techniques are also applicable in these settings.

In this scenario, a user is on a trip to attend a meeting. Before the meeting commences he runs this process-oriented application so that it assists him in the lead-up to the meeting. The process starts with three activities in parallel: 1) checking the presentation time, 2) checking the availability of trains to the destination, and 3) downloading meeting notes to the user's device (which may take some time due to low bandwidth).

After the presentation time and the train availability of the train have been checked three options are available: 1) If the user is "on time" AND "there is a train" that would take the user near the meeting's location, the user is directed to the train station; 2) If there is "no train", a taxi is automatically ordered; 3) If the user is "late" AND "there is a train", two new activities are started to determine if a taxi or a train is the best option for the user. At this point, the process checks the traffic conditions and tries to postpone the meeting by one hour (both actions in parallel). If the traffic is adverse, there is no point in catching a taxi, and the application will advise the user to catch the train. The same applies if the meeting is postponed. If however, there is favorable traffic and the meeting can not be postponed, the user will catch a taxi to get there sooner. Each transportation requires a payment. Payment is automatically arranged by the application and the details of the payment are sent to the finance department to arrange for a refund (both of these steps are modelled as a single task "pay"). Finally, once all the user is on his/her way to the meeting and the meeting notes have been downloaded, the application displays the notes.

3   Infrastructure for Event-Based Model Execution

This section presents the coordination infrastructure upon which the proposed technique relies and the framework to describe event-based coordination models.

3.1   The Active Object Space

To be able to execute the event-based coordination models that will be derived from process models, we require an execution infrastructure with support for : (i) event publishing, data transfer/sharing, and complex event subscription; (ii) association of reactions to event occurrences; and (iii) runtime re-configuration so that new event subscriptions and reaction rules can be added anytime. For reasons outlined below, we have chosen the *Active Object Space* (AOS) [7,8] as our target infrastructure. The AOS is an exemplar of a family of communication infrastructure known as *coordination middleware* which has its roots in the tuple space model underlying the Linda system [10]. Other exemplars of coordination middleware include Sun's JavaSpaces[2] and IBM's TSpaces[3].

At the centre of the AOS is a shared memory (the *space*). Coordination between applications occurs through objects being written and taken from the space. Some of these objects may correspond to data that needs to flow from one application to another, while others may serve as signposting, indicating that a given step of work has been completed or that a given step of work is enabled but has not yet started. The AOS supports undirected decoupled communication based on four elementary operations, namely *read, write, take* and *notify*. A read operation copies an object from the space matching a given object template; a take operation moves an object matching a given object template out of the space; a write operation puts an object on the space; and a notify operation registers a subscription for a composite event expressed as a set of object templates: Whenever there is a combination of objects present in the space that matches these object templates, an event occurrence will be raised and a notification will be sent to the subscriber. An *object template* is an expression composed of a class name and a set of equality constraints on the properties of that class. An object matches a template if its class is equal to or is a sub-class of the class designated by the template and it fulfills the template's constraints.

An originality of the AOS with respect to other object-oriented coordination middleware lies in its support for *active objects*, that is, objects with their own thread of control that run on the space. Active objects can be deployed, suspended, resumed, and destroyed by applications running outside the space at any time. Active objects can read and write passive objects to/from the space, subscribe to events, and receive notifications from the space. At the implementation level, the difference between active objects and "passive" objects is that an active object has a special *execute* method that is invoked on a dedicated thread of control when the object is written into the space.

As illustrated in the rest of the paper, the deployment of active objects operating on a shared memory and writing and taking objects to/from this space, constitutes a powerful paradigm not only for executing event-based coordination models, but also for re-configuring these models after their deployment. Reconfiguration is facilitated by two features of the AOS infrastructure: (i) the use of undirected (also known as "generative") communication primitives which allows data and events to be produced and consumed without a priori determined recipients (and thus allows data and control to be "re-routed"); and (ii) the ability to add, remove, suspend and resume individual active objects and thus alter the behaviour of an application.

This having been said, we recognize that other coordination middleware or publish/subscribe middleware supporting composite events (e.g. Elvin[4]) constitute suitable alternatives to the AOS. To adapt our proposal to such alternative infrastructures, active objects would have to be replaced by dedicated applica-

---

[2] http://java.sun.com/developer/products/jini/index.jsp
[3] http://www.almaden.ibm.com/cs/TSpaces
[4] http://elvin.dstc.edu.au

tions operating outside the space (or operating on top of the messaging bus in the case of a pub/sub middleware).

3.2   Coordinators

Having introduced the basic concepts and functionality of the AOS, we now define a higher-level concepts that we use to explicate the execution of event-driven coordination models.

A *coordinator* is an active object that is deployed in the space to coordinate work (e.g. to perform synchronization or data transfer) and operates in an infinite loop until suspended or destroyed, with each iteration comprising three phases:

1. Waiting for an event, which could be either the addition to the space of an object matching a given template or an interaction initiated by an external application;
2. Performing internal processing and/or interacting with external applications;
3. Writing one or several objects to the space.

For methodological reasons, it is useful to distinguish two types of coordinators, namely *connectors* and *routers*. This way, internal coordination steps within the space (which is the responsibility of the routers) are separated from communication with external applications (which is the responsibility of the connectors). The following paragraphs explain these types of coordinators in turn.

Connectors. A *connector* is a type of coordinator dedicated to enabling a connection between the space and one or several external applications. Connectors are necessary because external applications will generally not be programmed to interact with the space but will instead they rely on other communication protocols and interfaces. Thus, a way of wrapping external applications so that they can be coordinated through the space is necessary and this is what connectors achieve. For example, a connector could be placed on the space for the purpose of relaying context data between a sensor and the space. This connector would: (i) receive or poll data from the sensor; (ii) encode these data as a passive object; and (iii) write this object in the space, possibly overriding the object containing the previous known state of the context data. Another example of a connector is an active object that calls an external web service when an object of a certain type is written to the space, like for example an object that indicates that a certain task has completed. This latter example shows that connectors can be used as a mechanism to detect that a given task is enabled and thus that a given application has to be invoked to perform this task.

Control Routers. Control routers (or routers for short) react to the arrival of an object or a combination of objects to the space and perform some processing before producing a set of new objects and writing them onto the space. The processing that a router performs is generally translation of data using a specified operation. This can be a simple operation such as an arithmetic operation, or more complex operations such as checking that a purchase order is valid, but in any case, this operation should not involve interaction with external applications, since interactions with external applications are handled by the connectors.

A router is described by the following elements:

- Input set: A set made up of a combination of object templates and boolean conditions.
- Output: A set of expressions, each of whichs evaluates into an object.
- Stop set: A set containing a combination of object templates and boolean conditions.
- Replace set: A set of coordinators.

The way these elements are used is as follows. Upon creation, the router will place a subscription with the space for the set of object templates contained in its input set (i.e. the set obtained after removing the boolean conditions from the input set). Subsequently, the router will be notified whenever a set of objects matching these templates are available on the space. At this point, the router evaluates the set of conditions in its input set. If all these conditions are true, the router proceeds to "take" the set of objects in question and if it succeeds to take them, it will evaluate the transformation functions (i.e. the expressions in the "Output") taking these objects as input. The objects resulting from the transformation are then written back to the space. The "input set" thus captures the events and conditions that lead to the activation of a router (where an event corresponds to the arrival of an object to the space). The "Output" on the other hand encodes the events that the router will produce upon activation, i.e. the objects to be placed in the space for consumption by other coordinators. Finally, if a set of objects matching the object templates in the stop set is found on the space, the router will terminate its execution and replace itself by the set of routers specified in the replace set.

A set of routers can be deployed and interconnected with existing applications (through connectors) in order to coordinate the execution of the instances of a process. During the execution of a process instance, routers read and take from the space, objects denoting the completion of tasks (i.e. *task completion objects*) and write into the space objects denoting the enabling of tasks (i.e. *task enabling objects*). Connectors on the other hand read and take task enabling objects, execute the corresponding task by interacting with external applications, and eventually write back task completion objects, which are then read by routers. To make sure that routers only correlate task completion events relating to the same instance of a process, every object template in the input set of the router will contain a constraint stating that all the matched task completion objects must have the same value for the attribute corresponding to the process instance identifier (*piid*). In addition, when a router (connector) writes a task enabling (task completion) object to the space, it includes the corresponding piid. A process instance is created when a "process instantiation" object with the corresponding process and process instance identifier is placed on the space by a connector. It is the responsibility of the connectors which place such objects to ensure that process instance identifiers are unique.

4   From Process-Based to Event-Based Models

This section focuses on the issue of generating coordinators for process orchestration from UML activity diagrams. We first describe the technique for generating coordinators from UML activity diagram restricted to control-flow constructs. We then show how data-flow aspects are incorporated.

4.1   Translating Control-Flow Constructs into Input Sets

For each action[5] in an activity diagram, a connector will be generated to handle its execution, which in the case of process-oriented composite applications will involve an interaction with an external application. Connectors thus encapsulate the execution of actions in the process.

On the other hand, a number of routers are generated for each action. The input sets for these routers are generated according to the algorithm sketched using a functional programming notation in Figure 2 and explained below. The main function defined by this algorithm (namely AllInputSets) takes as input an activity diagram represented as a set of nodes (action, decision, merge, fork, join, initial, and final nodes) inter-linked through transitions. From there, it generates a set of input sets (see definition of input set in Section 3.2). The input sets produced by this algorithm can then be used to create a collection of routers (one router per input set) that collectively are able to coordinate the execution of instances of the process in question. Intuitively, each input set encodes one possible way of arriving to a given node in the process.

Given the set of connectors and routers deployed for a process-oriented composite application, execution occurs as follows. A router corresponding to an action node will wait until the object templates in its input set are all matched, at which point if all the boolean expressions in the input set evaluate to true, it will place an object on the space to indicate that the action is enabled and thus that the corresponding external application invocation may be performed by a connector. Once the connector has completed its interaction with the external application, it will put an object in the space to signify this completion. Such *completion objects* will then match the object templates of the input set of another router, eventually causing the activation of this other router. In this way the execution of the process moves from a router corresponding to a given action, to another. The initial and final states are mapped trivially to two routers that respectively detect the commencement of the process instance and perform clean-up (i.e. delete all remaining objects related to the completed instance).

Algorithm for Input Sets Generation. The algorithm focuses on a core subset of activity diagrams covering only initial and final nodes, action nodes, and control nodes (i.e. decision, merge, fork, and join nodes) connected by transitions. In particular, the algorithm does not take into account object flow (which is discussed later) nor swimlanes (which are irrelevant for the purposes of this paper). Without loss of generality, the algorithm assumes that all conditional guards in

---

[5] Action is the term used in UML activity diagrams to refer to a "task".

the activity diagram are specified in disjunctive normal form. Also without loss of generality, the algorithm assumes that there are no "implicit" forks and joins in the diagram. An implicit fork (join) occurs when several transitions leave from (arrive to) an action node. In this case, the semantics of this fragment of the diagram is the same as that of a diagram in which this action node only has one outgoing (incoming) transition leading to (originating from) a fork node (a join node). Thus implicit forks and joins should be eliminated from a diagram and replaced by explicit fork and join nodes prior to applying this algorithm.

```
AllInputSets(p: Process) :
        let {x₁, ..., xₙ} = ActionNodes(p) in
            InputSets(x₁) ∪ ...∪ InputSets(xₙ)
InputSets(x : Node) :
      let {t₁, ...tₙ} = IncomingTrans(x) in
          return InputSetTrans(t₁) ∪ ... InputSetTrans(tₙ)

InputSetsTrans(t : Transition) :
      let x = Source(t)
          if NodeType(x) = "action"
            return CompletionObject(x)
          else if NodeType(x) = "initial"
                  return ProcessInstantiationObject(Process(x))
          else if NodeType(x) ∈ { "decision", "fork" }
              let {c₁, ..., cₙ} = Disjuncts(Guard(t)),
                  {i₁, ..., iₙ} = InputSets(Source(t)) in
                  return {{c₁} ∪ i₁, ..., {c₁} ∪ iₙ},
                                ...
                        {cₙ} ∪ i₁, ..., {cₙ} ∪ iₙ}
          else if NodeType(x) = "merge"
              let {t₁, ..., tₙ} = IncomingTrans(x) in
                  return InputSetsTrans(t₁) ∪ ...∪ InputSetsTrans(tₙ)
          else if NodeType(x) = "join"
              let {t₁, ..., tₙ} = IncomingTrans(x),
                  {⟨ i₁,₁, ..., i₁,ₙ⟩,
                    ...
                  ⟨ iₘ,₁, ..., iₘ,ₙ⟩} =
                  InputSetsTrans(t₁) × ...× InputSetsTrans(tₙ) in
                  return {i₁,₁ ∪ ...∪ i₁,ₙ,
                                ...
                        iₘ,₁ ∪ ...∪ iₘ,ₙ}
```

Fig. 2. Algorithm for deriving input sets from an activity diagram.

Figure 2 defines three functions: the first one, namely AllInputSets generates all the input sets for a process by relying on a second function, namely InputSets, which generates a set of input sets for a given node of the diagram. This latter function relies on a third (auxiliary) function named InputSetsTrans, which produces the same type of output as InputSets but takes as parameter a transition rather than a set. This definition of InputSetsTrans operates based on the node type of the source of the transition, which may be an action node, an initial node, or one of the four types of control nodes. If the transition's source is an action node, a single input set is returned containing a completion object (see Section 3.2) for that action. Intuitively, this means that the transition in question may be taken when a completion object corresponding to that action is placed on the space. Similarly, if the source of the transition is the initial node of the activity diagram, a single input set with a "process instantiation" object is created, indicating that the transition in question will be taken when an object is placed on the space signalling that a new instance of the process must be started. If the transition's source is a control node, the algorithm keeps working backwards through the diagram, traversing other control nodes, until reaching action nodes. In the case of a transition originating from a decision or a fork node, which is generally labelled by a guard (or an implicit "true" guard if no guard is explicitly given), the transition's guard is decomposed into its disjuncts, and an input set is created for each of these guards. This is done because the elements of an input set are linked by an "and" (not an "or") and thus an input set can only capture a conjunction of elementary conditions and completion/instantiation objects (i.e. a disjunct). Finally, in the case of a transition originating from a "merge" (resp. a "join"), the function is recursively called for each of the transitions leading to this merge node (join node), and the resulting sets of input sets are combined to capture the fact that when any (all) of these transitions is (are) taken, the corresponding merge node (join node) may fire.

The following notations are used in the algorithm:

- ActionNodes(p) is the set of action nodes contained in process p (described as an activity diagram).
- Source(t) is the source state of transition t
- Guard(t) is the guard on transition t
- Disjuncts(c) is the set of disjuncts composing condition c
- IncomingTrans(x) is the set of transitions whose target is node x
- NodeType(x) is the type of node x (e.g. "action", "decision", "merge", etc.)
- Process(x) is the process to which node x belongs.

Example. Figure 3 describes the router for the "CheckTraffic" using a concrete XML syntax. This action node will only have one router associated to it because there is only one path leading to the execution of this action. Indeed, to execute this action, it is necessary that both the "check presentation time" and the "check train availability" actions have completed, and in addition that the condition "not ontime and train" evaluates to true, and this condition does not contain any disjunction. When all these conditions are satisfied, the router will produce an enabling object that will eventually be picked up by the connector associated to action "check traffic".

It can be noted in this example that the process instance identifier (piid) attribute of the completion object templates are associated with a variable. In the concrete XML syntax, an XML namespace (aliased "var") is reserved to refer to variables. The AOS is capable of interpreting collections of object templates where some of the atttributes are associated with such variables and to match these templates in a way that if the same variable is associated with attributes of two different templates, then the objects matching these templates should contain the same values for these attributes.

```
<Router name = ''CheckTrafficEnabler''>
  <Input>
   <Template>
   <CompletionObject actionName=''CheckPresentationTime'' piid=''var:X''/>
   </Template>
   <Template>
   <CompletionObject actionName=''CheckTrainAvailability'' piid=''var:X''/>
   </Template>
   <Condition>
    <Equality variable=''ontime'' value=''false''/>
   </Condition>
   <Condition>
    <Equality variable=''train'' value=''true''/>
   </Condition>
  </Input>
  <Output>
   <EnablingObject action=''CheckTraffic'' piid=''var:X''/>
  </Output>
</Router>
```

Fig. 3. Sample router

4.2   Incorporating Data-Flow

Data flow (or more precisely *object flow*) in activity diagrams is represented by object nodes, represented as rectangles as illustrated in Figure 4. Object nodes are directly linked to a "producing" action preceding the object node. They are also linked, either directly or through the intermediary of a number of control nodes, to one or several "consuming" action node(s) following the object node. In the example of Figure 4, the user pays using his mobile device and this produces a receipt object which is then forwarded to the finance department so that the user may obtain a refund.

In terms of the proposed technique, object flows are treated as follows. The production of objects for a given object node is the responsibility of the connector corresponding to the action node directly preceding this object node (i.e. the producing action). In other words, the corresponding object would appear as one of the elements in the "output" of this coordinator (see Section 3.2). In the example at hand, the production of objects of type "Receipt" is done by the connectors of the action nodes labelled "Pay". On the other hand, the consumption of objects corresponding to an object node is carried out by the connectors of action nodes that follow this object node, either directly or through the intermediary of a number of control nodes (i.e. the consuming actions). In the example at hand, this means that the connector of the action node labelled "Request Refund" will take an object of type "Receipt" from the space when this action is enabled.

Since object flow is handled exclusively by connectors, the algorithm presented above does not have to deal with object nodes. Accordingly, object nodes should be removed from the activity diagram before applying the algorithm for deriving input sets. Removing object nodes from an activity diagram is trivial since they always have only one incoming and one outgoing transition.

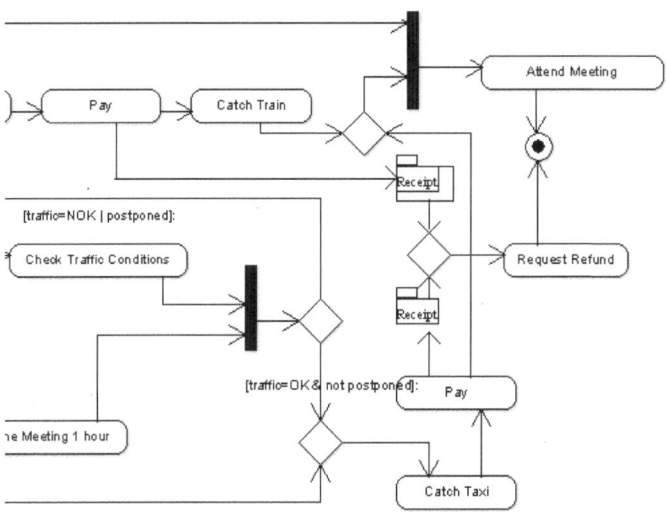

Fig. 4. Working example with object nodes

5   Achieving Adaptation

In certain situations, some functionality may or should be made unavailable. A context change may mean that some processing can not be performed, or a user moving outside a firewall may prevent him/her from executing certain applications. In our working example, it may happen that the system takes too much time to contact the other meeting participants to check if the meeting can be postponed (i.e. the execution of the "postpone meeting" may take more time than the user is willing to wait for). In this case, a user may indicate that (s)he does not wish to be delayed by this action, but instead, if the "Check Traffic" action is completed and if the traffic conditions are OK, (s)he would immediately take a taxi. This adaptation can be achieved by activating the router specified in a concrete XML syntax in Figure 5. In this XML fragment, we assume that the piid of the process instance for which this modification is to be done is 1. The element StopSet indicates that this router is disabled if the "Postpone Meeting" action is completed. Thus this router will only place an enabling object to trigger action "Catch Taxi" if the action "Check Traffic" completes before "Postpone Meeting" and the corresponding boolean expression evaluates to true.

The above adaptation could arguably be achieved by modifying the process model.[6] However, in this case, significant tool support would be required and model versioning may become an issue. In contrast, enabling an event-based rule (encoded as a router) provides a more lightweight adaptation mechanism.

More radical changes may also be made. For example, consider a user that prefers taxis over trains in any case and so would always catch taxis regardless of traffic conditions and amount of time before the meeting. In this case, a router may be introduced that enables the action "CatchTaxi" immediately upon process instantiation when the process instance is started by the user in question. At the same time, all other routers for that process instance would be disabled, except the ones for download notes and display notes.

```
<Coordinator name = 'with participants''>
  <Input>
   <Template>
    <CompletionObject action=''CheckTraffic'' piid=''1''/>
   </Template>
   <Condition>
    <Equality variable=''traffic'' value=''OK''/>
   </Condition>
  </Input>
  <Output>
     <EnablingObject action=''CatchTaxi'' piid=''1''/>
  </Output>
  <StopSet>
   <CompletionObject action=''PostponeMeeting'' piid=''1''/>
  </StopSet>
</Coordinator>
```

Fig. 5. Sample router for process adaptation

How the user actually specifies "dynamic" changes to composite applications is a user interface issue outside the scope of this article. This may be achieved, for example, by means of personalization applications running as active objects and disabling or enabling routers or placing completion or enabling objects according to an adaptation logic previously coded by a developer. Another option is to provide users with options for adapting/personalising applications. When a user manually selects one of these options, a number of coordinators are enabled and/or completion and enabling objects are written to or taken off the space. Of course, this mechanism may be abused and lead to undesirable effects such as deadlocked executions. However, as shown above, adaptation may be scoped

---

[6] Note that expressing this type of discriminator (or 1-out-of-2) join in UML activity diagrams requires the use of advanced constructs (namely signals) not covered by our algorithm [17]. However, this is not the point that we try to make here.

to specific process instances to avoid affecting a wider user base. In addition, as certain adaptations become permanent, they may be propagated back to the process model resulting in a new process model being deployed.

6   Related Work

Process-oriented application development has been the subject of significant attention in the last decade, prompting the emergence of a large number of process modelling and execution languages, some of which have been the subject of standardisation initiatives such as the Business Process Execution Language for Web Services (BPEL4WS).[7] However, the platforms supporting these languages adopt an approach to process-oriented application development that is not suitable in scenarios where personalization and adaptation are prominent requirements. Indeed, these platforms typically rely on the static definition of process models and allow little change to occur without a significant redeployment effort.

As discussed in the Introduction, proposals in the area of adaptive and flexible workflow [14] generally focus either on a priori adaptation (e.g. attaching exception handling policies to a process model) or on dealing with changes in the process model. In contrast, we advocate that adaptation should not be handled at the level of the process model. Our proposal shows that if an event-based coordination model is used at the execution layer, it is possible to make fine-grained changes to specific parts of the process and to confine these changes to specific process instances, without altering the process model. In other words, the process model can be used as a reference to deal with the majority of cases but deviations can occur for specific cases based on the activation or de-activation of the rules composing the event model. Parallels can be drawn between our approach and the one followed in case handling systems [3] where human workers route cases (i.e. process instances) manually based on information associated to each case and contextual information such as workload and resource availability. However, case handling is targeted at processes composed mostly of manual tasks. In contrast, our proposal is targeted at processes in which tasks are delegated to software applications so that it is not possible to count on human intervention at each step of the process.

There exist a large body of proposals in the area of coordination architectures, and in particular space-based ones. Some of these architectures (e.g. Mars [5] and Limone [9]) support the definition of reaction rules to coordinate application components, similar to the way coordinators operate in our framework. However, despite their potential synergies, proposals in the areas of coordination architectures on the one hand, and process-oriented application development on the other, have so far evolved independently – a notable exception being the work by Tolksdorf [16] who describes a space-based architecture for routing XML documents through processing steps encoded in XSL. A major novelty of our proposal is that it seamlessly combines techniques from coordination-based and from process-oriented software architectures.

---

[7] http://www.oasis-open.org/committees/tc_home.php?wg_abbrev=wsbpel

This paper partly builds upon previous work on decentralised orchestration of process-oriented composite services specified as UML statecharts [4]. In this prior work, an algorithm was proposed that bears some similarities with the one presented in Figure 2. In addition to technical differences between the algorithms, stemming in part from the use of activity diagrams (version 2.0) rather than statecharts, the proposal of this paper differs from the previous one in the use that it makes of the output of the algorithm: Instead of using this output for decentralised orchestration, it uses it for event-based centralised orchestration based on coordination middleware. The proposal in this paper can also be seen as a refinement of the architecture presented in [15], where agents and tuple spaces are combined in an architecture for service composition. In the present paper, we have presented a concrete approach to encode and execute event-based models and we have detailed a method for generating event-based models from process-based ones. We have also shown that by encoding event-based models as active objects it is possible to achieve various forms of adaptation.

7   Conclusion and Future Work

This paper has shown how a process model specified using UML activity diagrams can be translated into an event-based model that can be executed on top of a coordination middleware. Specifically, a process model is encoded as a collection of active objects that interact with each other through a shared object space. We have argue and illustrated that this approach is suitable for undertaking post-deployment adaptation of process-oriented composite applications. In particular, new control dependencies can be encoded by dropping new (or enabling existing) active objects into the space and/or disabling existing ones.

A possible direction for future work is to extend the proposed algorithm for input sets generation to cover a larger set of process modelling constructs such as signals in UML activity diagrams or advanced control-flow constructs such as those found in YAWL [2]. Another direction for future work is to design a mapping from event-based models to process models. The idea would be to automatically derive a process model from a collection of routers. This "reverse" mapping would assist developers in propagating changes in the event-based model to the process model, when it is decided that these changes should be made permanent. Techniques such as those developed in the setting of process mining, where process models are derived from causal relations extracted from execution traces, could provide insights for designing this reverse mapping.

Acknowledgments. The first author is funded by a Queensland Government Smart State Fellowship co-sponsored by SAP. The second author is funded by an SAP-sponsored scholarship.

References

1. W. M.P. van der Aalst. How to handle dynamic change and capture management information: An approach based on generic workflow models. *Computer Systems Science and Engineering*, 15(5):295–318, 2001.
2. W.M.P. van der Aalst and A.H.M. ter Hofstede. YAWL: Yet Another Workflow Language. *Information Systems*, 30(4):245–275, 2004.
3. W.M.P. van der Aalst, M. Weske, and D. Grünbauer. Case handling: A new paradigm for business process support. *Data and Knowledge Engineering*, 53(2):129–162, 2005.
4. B. Benatallah, M. Dumas, and Q.Z. Sheng. Facilitating the rapid development and scalable orchestration of composite web services. *Distributed and Parallel Databases*, 15(1):5–37, January 2005.
5. G. Cabri, L. Leonardi, and F. Zambonelli. Reactive tuple spaces for mobile agent coordination. In *Proceedings of the Second International Workshop on Mobile Agents (1998)*, pages 237–248, Stuttgart, Germany, 1999. Springer Verlag.
6. F. Casati, S. Ceri, S. Paraboschi, and G. Pozzi. Specification and implementation of exceptions in workflow management systems. *ACM Transactions on Database Systems*, 24(3):405–451, 1999.
7. K. Elms, S. Milliner, and J. Vayssiere. Object spaces with active objects. U.S. Patent Application # 2004P00851US, filed 29 December 2004.
8. T. Fjellheim, S. Milliner, M. Dumas, and K. Elms. The 3DMA middleware for mobile applications. In *Proceedings of the 2004 International Conference on Embedded and Ubiquitous Computing*, Aizu, Japan, August 2004. Springer Verlag.
9. C-L. Fok, G-C. Roman, and G. Hackmann. A lightweight coordination middleware for mobile computing. In *Proceedings of the 6th International Conference on Coordination Models and Languages*, pages 135–151, Pisa, Italy, February 2004. Springer Verlag.
10. D. Gelernter. Generative communication in Linda. *ACM Transactions on Programming*, 2(1):80–112, January 1985.
11. S-Y. Hwang and Y-F. Chen. Personal workflows: Modeling and management. In *Proceedings of the 4th International Conference on Mobile Data Management*, 2003.
12. D. Luckham. *The Power of Events: An Introduction to Complex Event Processing in Distributed Enterprise Systems*. Addison-Wesley Professional, 2002.
13. R. Muller, U. Greiner, and E. Rahm. AgentWork: a workflow system supporting rule-based workflow adaptation. *Data and Knowledge Engineering*, 51(2):223–256, November 2004.
14. S. Rinderle, M. Reichert, and P. Dadam. Correctness criteria for dynamic changes in workflow systems - a survey. *Data and Knowledge Engineering*, 50(1):9–34, 2004.
15. Q.Z. Sheng, B. Benatallah, Z. Maamar, M. Dumas, and A.H.H. Ngu. Enabling personalized composition and adaptive provisioning of web services. In *Proceedings of the International Conference on Advanced Information Systems Engineering*, pages 322–337, Riga, Latvia, June 2004. Springer Verlag.
16. R. Tolksdorf. Coordination technology for workflows on the web: Workspaces. In *Proceedings of the 4th International Conference on Coordination Models and Languages*, pages 36–50, Limassol, Cyprus, September 2000. Springer Verlag.
17. P. Wohed, W. M.P. van der Aalst, M. Dumas, A. H.M. ter Hofstede, and N. Russell. Pattern-based Analysis of the Control-flow Perspective of UML Activity Diagrams. In *Proceedings of the International Conference on Conceptual Modelling (ER)*, Klagenfurt, Austria, October 2004. Springer Verlag.

Integrating Process Learning and Process Evolution – A Semantics Based Approach

Stefanie Rinderle[1], Barbara Weber[2], Manfred Reichert[3], and Werner Wild[4]

[1] Dept. Databases and Information Systems, University of Ulm, Germany
rinderle@informatik.uni--ulm.de
[2] Quality Engineering Research Group, University of Innsbruck
Barbara.Weber@uibk.ac.at
[3] Information Systems Group, University of Twente, The Netherlands
m.u.reichert@cs.utwente.nl
[4] Evolution Consulting, Innsbruck, Austria
werner.wild@evolution.at

Abstract. Companies are developing a growing interest in aligning their information systems in a process-oriented way. However, current process-aware information systems (PAIS) fail to meet process flexibility requirements, which reduces the applicability of such systems. To overcome this limitation PAIS should capture the whole process life cycle and all kinds of changes in an integrated way. In this paper we present such a holistic approach providing full process life cycle support by combining the ADEPT framework for dynamic process changes with the concepts and methods provided by case-based reasoning (CBR) technology. This allows expressing the semantics of process changes, their memorization and their reuse to perform similar changes in the future. If the same or similar process instance changes occur frequently, potential process type changes are suggested to the process engineer. The process engineer can then perform a schema evolution and migrate running instances to the new schema version by using the ADEPT framework. Finally, the case–base related to the old schema version is migrated as well.

1 Introduction

Adaptive process management technology offers a promising approach for realizing highly flexible, process–oriented information systems [1,2,3,4]. In particular, it enables dynamic process changes during runtime to handle exceptional situations and changing needs. Basically, such process changes can be made at two levels – the *process type* and *process instance* level [5].

In our experience an adaptive process management system (PMS) must support both kinds of changes in an integrated way [6]. In the ADEPT project we have elaborated a conceptual framework which enables changes at the process instance and at the process type level. For the latter we support the subsequent migration of both *unbiased* and *biased* process instances to the changed process type schema. This is especially important for long–running process instances [7].

W.M.P. van der Aalst et al. (Eds.): BPM 2005, LNCS 3649, pp. 252–267, 2005.

We denote process instances as unbiased if they are running according to the original process schema they were derived from [8], whereas process instances are denoted as biased when they have been individually modified by an ad–hoc change [6].

So far, our work on adaptive processes (e.g., [7,8,9]) has not incorporated application semantics, i.e., it has not considered the reasons for and the context of a change. Instead, very similar to database technology, all checks and procedures necessary to perform a dynamic change have been applied solely at the syntactical level, which, nevertheless, is an important prerequisite for any adaptive process management technology. To provide more intelligent support to its users and to reuse knowledge about previously applied process changes *semantical* aspects must be considered as well. This paper shows how adding semantics contributes to the seamless integration of process changes into the process life cycle (cf. Fig. 1).

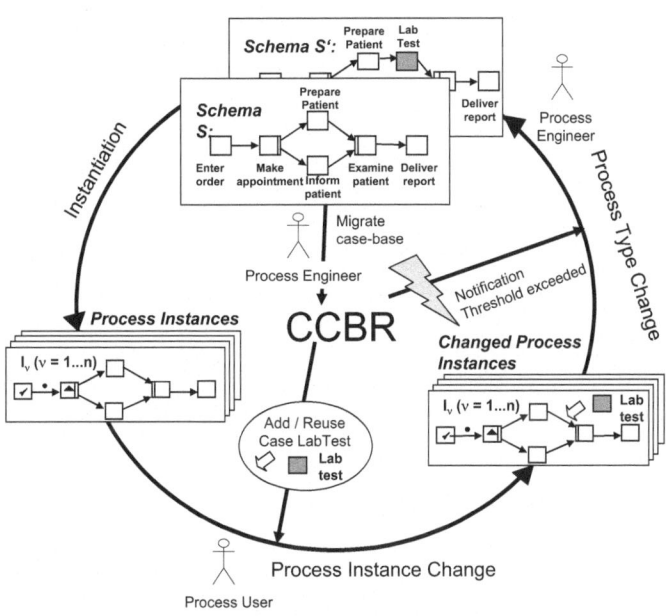

Fig. 1. Process Life Cycle

In this paper we combine the ADEPT framework for process changes with the concepts and methods provided by CBRFlow [10], a process change approach using case-based reasoning (CBR) technology. This allows us to express the semantics of process changes and to provide users with information about the context of and the reasons for process changes, ensuring the traceability of instance changes. The latter is a crucial requirement in many domains (e.g., hospitals have strict guidelines regarding the documentation of deviations from the predefined treatment process). Respective change information is stored as cases in a process schema specific *case–base*. This information can be used to support process actors in reusing information about similar ad-hoc changes which have been applied to previously executed process instances. Change definition requires significant user experience, especially when further adaptations become necessary (e.g., when deleting a particular activity, data-dependent activities may have to be deleted as well [9]). Therefore, reuse of existing knowledge about previous ad-hoc changes is highly desirable.

Furthermore, case–bases are continuously monitored in order to automatically derive suggestions for process type changes from previously applied process instance changes. In practice, necessary type changes are frequently indicated by process instances whose execution deviates from the original process type schema over and over again. In such a situation a process type change is desirable to move the respective optimizations up to the process type level.

In the ADEPT framework a process type change is performed by deriving a new version of the process type schema and – if possible and desired – by automatically migrating the already running process instances to this new schema version [7,8]. This may include the migration of both unbiased and biased process instances. Interestingly, not only processes but also case-bases evolve over time. When a case–base is "migrated" to a new process schema version, it should only keep information which is still relevant for instances of the new process schema (and for changes to them). In our approach a process schema evolution is always accompanied by the evolution of the related case-base. This poses several challenges which will be discussed later in this paper.

In Section 2 we provide background information. Section 3 presents fundamentals of CBR and their application to process instance changes. In Section 4 we show how to learn from instance changes by using CBR techniques and we provide an overview of our process evolution approach. The evolution of case–bases is described in Section 5. We discuss related work in Section 6 and close with a summary and an outlook in Section 7.

2   Background Information

In this section we give basic definitions for process type schemes and process instances as needed for our further considerations. To improve readability we restrict the discussion to Activity Nets [11]; however, our approach is applicable to more complex process meta models as well (e.g., WSM Nets [7]).

For each business process to be supported a process type T is defined. It is represented by a *process type schema* which may exist in different versions.

Definition 1 (Process Type Schema). *A tuple S with S = (N, D, CtrlEdges, DataEdges, EC) is called a process schema, if the following holds:*

- *N is a set of activities and D a set of data elements*
- *CtrlEdges ⊂ N × N is a precedence relation*
  (notation: $n_{src} \rightarrow n_{dst} \equiv (n_{src}, n_{dst}) \in CtrlEdges$)

- *DataEdges $\subseteq N \times D \times \{read, write\}$ is a set of read/write data links between activities and data elements*
- *EC: CtrlEdges $\mapsto$ Conds(D) where Conds(D) denotes the set of all valid transition conditions on data elements from D.*

For a process type schema several correctness constraints exist, e.g., (N, CtrlEdges) must be an acyclic graph to ensure the absence of deadlocks.

At runtime new process instances are created from a process schema S and are then executed. Each instance I is associated with an instance-specific schema $S_I := S + \Delta_I{}^1$. $S = S(T,V)$ denotes the original process schema S from which instance I was derived, whereby T is the process type of I and V the version of the process type schema. $\Delta_I = \{op_1, ..., op_n\}$ comprises the change operations op_i (i = 1, ..., n) applied to I so far (cf. Fig. 4). In this context a change operation $op_i = (opType, s, paramList)$ (i = 1, ..., n) is specified by an operation type (e.g., insertion of an activity), a subject (e.g., the newly inserted activity), and a list of parameters (e.g., position of the newly inserted activity). Selected change operations are shown in Table 1.

The execution state of I is captured by marking function $M^{S_I} = (NS^{S_I}, ES^{S_I})$. It assigns to each activity n its current status $NS(n)$ and to each control edge e its marking $ES(e)$. Markings are determined according to well defined rules, markings of already passed regions and skipped branches are preserved.

Definition 2 (Process Instance). *A process instance I is defined by a tuple (T, V, Δ_I, M^{S_I}, Val^{S_I}) where*

- *T denotes the process type of I and V the version of the process schema S := S(T,V) = (N, D, CtrlEdges, ...) instance I was derived from. We call S the <u>original schema</u> of I.*
- *Δ_I comprises the instance-specific changes $op_1, ..., op_m$ that have been applied to I so far[2]. We call Δ_I the <u>bias</u> of I. Schema $S_I := S + \Delta_I$ (with $S_I = (N_I, D_I, ...)$) which results from the application of Δ_I to S, is called the <u>instance–specific schema</u> of I.*
- *$M^{S_I} = (NS^{S_I}, ES^{S_I})$ describes node and edge markings of I:*
  NS^{S_I}: $N_I \mapsto$ {NotActivated, Activated, Running, Completed, Skipped}
  ES^{S_I}: $(CtrlEdges_I) \mapsto$ {NotSignaled, TrueSignaled, FalseSignaled}
- *Val^{S_I} is a function on D_I. It reflects for each data element $d \in D_I$ either its current value or the value UNDEFINED (if d has not been written yet).*

Table 1 presents a selection of *high-level change operations* on process schemes which can be used to create or modify schemes at the type as well as at the instance level. These change operations also include formal pre- and

---

[1] For unbiased instances $\Delta_I(S) - \emptyset$ and consequently $S_I = S$ holds.
[2] Thereby an operation $op_j := (opType_j, s_j, paramList_j)$ (j = 1, ..., m) consists of an operation type $opType_j$, the subject s_j of the change, and a parameter list $paramList_j$ (cf. Tab. 1).

post-conditions. They automatically perform the necessary schema transformations while ensuring schema correctness. An example for such a change operation is the insertion of an activity and its embedding into the process context.

Table 1. *A Selection of High-Level Change Operations on Process Schemes*

Change Operation op Applied to S	opType	subject	paramList	Effects on S
	Additive Change Operations			
sInsert(S, X, A, B)	Insert	X	S, A, B	inserts activity X between two directly succeeding activities A and B
cInsert(S, X, A, B, sc)	Insert	X	S, A, B, sc	inserts activity X between two directly succeeding activities A and B as a new branch with selection code sc; edge (A, B) gets selection code "default"
	Subtractive Change Operations			
delAct(S, X)	Delete	X	S	deletes activity X from schema S
	Order-Changing Operations			
sMove(S, X, A, B)	Move	X	S, A, B	moves activity X from its current position to position between directly succeeding activities A and B

3   Providing Change Semantics Through CCBR

In this section we describe how CBR can be used to capture the semantics of process changes, how these changes can be memorized, and how they can be retrieved and reused when similar changes become necessary in the future.

3.1   Introduction to Case-Based Reasoning

Case-based reasoning (CBR) is a contemporary approach to problem solving and learning [12]. New problems are dealt with by drawing on past experiences – described in cases and stored in case–bases – and by adapting their solutions to the new problem situation. Reasoning by using past experiences is a powerful and frequently applied way to solve problems by humans [13]. A physician, for example, remembers previous cases to determine the disease of a new patient. A banker working on a difficult loan decision uses her experiences about previous cases in order to decide whether to grant a loan or not.

A case is a contextualized piece of knowledge representing an experience [12], which typically consists of a problem description and the corresponding solution. As opposed to most other approaches in Artificial Intelligence, CBR uses specific knowledge of past experiences to solve new problems. CBR also contributes to incremental and sustained learning: every time a new problem is solved, information about it and its solution is retained and therefore immediately made available for solving future problems [13].

Conversational CBR (CCBR) is an extension of the CBR paradigm, which actively involves users in the inference process [14]. A CCBR system can be characterized as an interactive system that, via a mixed-initiative dialogue, guides users through a question-answering sequence in a case retrieval context (cf.

Title			Select Operation Type	Insert	▼
Description			Select Activity/Edge	Lab Test	▼

Question-Answer Pairs ... **Please Answer the Questions**

Question	Answer
Patient has diabetes?	Yes
What is the patient's age?	> 40

Question	Answer
Patient has diabetes?	Yes
What is the patient's age?	> 40

Actions ... **Display List of Cases**

Operation Type	Subject	Parameters
sInsert	LabTest	S, PreparePatient, Examine Patient

Case ID	Title	Similarity	Reputation Score
125	Lab Test required	100%	25

Fig. 2. CCBR User Dialogs - Adding a New Case and Retrieving Similar Cases

Fig 2). Unlike traditional CBR, CCBR neither requires the user to provide a complete a priori problem specification for case retrieval nor requires him to provide knowledge about the relevance of each feature for problem solving. Instead, the system assists the user in finding relevant cases by presenting a set of questions to assess the given situation. It guides users who can supply already known information on their initiative. Therefore, CCBR is especially suitable for handling exceptional or unanticipated situations that cannot be dealt with in a fully automated way.

3.2 Conversational Case-Based Reasoning and Adaptive Workflows

In our approach CCBR is used to provide semantical information about changes to process instances. This information can be reused when either similar ad-hoc modifications become necessary or to trigger process type changes.

Case Representation. In our context, a case c represents a concrete ad-hoc modification to one or more process instances providing the context of and the reasons for the deviation (cf. Fig. 2). It consists of a textual problem description pd which briefly describes the exceptional or unanticipated situation which made the ad-hoc modification necessary. The reasons for the change are described as question-answer pairs $\{q_1 an_1, \ldots, q_n an_n\}$. Each question-answer pair denotes a condition under which the modification has become necessary; they are used to retrieve similar cases when a problem arises in the future. The solution part sol (i.e., the list of actions) contains the change operations (and related context information) to be executed to deal with the exceptional or unanticipated situation. Finally, the reputation score $rScore$ of a case indicates how successfully it has been reused in the past, i.e., the trust users can have into the semantical correctness of this case. The reputation score is calculated as the sum of the feedback scores (see below).

Definition 3 (Case). *A case c is a tuple (pd, $\{q_1 an_1, \ldots, q_n an_n\}$, sol, rScore) where*

- *pd is a textual problem description*
- *$\{q_1 an_1, \ldots, q_n an_n\}$ denotes a set of question-answer pairs*

- $sol = \{\, op_j \mid op_j = (opType_j, s_j, paramList_j),\ j = 1,\ \ldots,\ k\}$ is the solution part of the case denoting a list of change operations (i.e., the changes that have been applied to one or more process instances; cf. Def. 2)
- $rScore$ indicates how successfully case c has been applied in the past. It is calculated as the sum of the feedback scores.

All information on process instance changes related to a process schema version S are stored as cases in the associated case-base of S.

Definition 4 (Case–Base). A case–base $cb_{T,V}$ is a tuple $(T,\ V,\ \{c_1, \ldots, c_m\},\ freq_{T,V}))$ where

- $S := S(T, V)$ denotes the schema version $cb_{T,V}$ is currently associated with
- $\{c_1, \ldots, c_m\}$ denotes a set of cases
- $freq_{T,V}(c_i)$ denotes the frequency c_i was reused in connection with schema version $S(T, V)$ in the past, formally:
  $freq_{T,V}: \{c_1, \ldots, c_m\} \mapsto \mathbb{N}$

Case Retrieval. When deviations from the predefined process schema become necessary, the user initiates a case retrieval dialog in the CCBR component. The system then assists her in finding already stored similar cases (i.e., change scenarios in our context) by presenting a set of questions. The user may apply a filter to the case-base and/or answer any of the questions in arbitrary order. Filtering is done by specifiying an operation type $opType$ and a subject s on which the operation is supposed to operate on. Cases not matching the filter criteria are removed from the displayed list of cases. The system then searches for similar cases by calculating the similarity for each case in the filtered case-base and displays the top n ranked cases (ordered by decreasing similarity) as well as their reputation scores. This is repeated for any other question answered by the user. Case similarity is calculated by dividing the number of correctly answered questions minus the number of incorrectly answered questions by the total number of questions in the case [15].

Case Reuse. When a user decides to reuse an existing case, the actions specified in the solution part of the case are forwarded to and performed by the ADEPT change engine. The reuse counter is incremented and a work item is created for this user for evaluating the ad-hoc change later on to maintain the quality of the case-base.

Adding a New Case. If no similar cases can be found when performing a process instance change, the user adds a new case $c = (pd, \{q_1an_1, \ldots,\}, sol, 1)$ to case-base $cb_{T,V}$. She enters this case by briefly describing the current problem pd and by entering a set of question-answer pairs to describe the reasons for the ad-hoc deviation. Question-answer pairs can be entered either by selecting the question from a list of previously defined questions (i.e., reusing questions from existing cases) or, if there is no suitable question in the system, by defining a new one and by giving the appropriate answer. The user then specifies the actions to perform by selecting the desired operation types $opType_1, \ldots, opType_p$. She further defines the subjects $s_1, \ldots, s_p$ they operate on (e.g., activities and control edges), and provides the parameters for each selected operation. Moreover, the reuse counter of the case is initialized to 1. Finally, the case is stored in case-base $cb_{T,V}$ of process schema $S = S(T, V)$ and thus immediately made available for future reuse.

Ensuring Semantical Correctness through Evaluation Mechanisms. In our approach we use the concept of reputation to indicate how successfully an ad-hoc modification represented by a case has been applied in the past. Whenever a user adds or reuses a case she is encouraged to provide feedback on the performed process instance change. For this, a work item representing the feedback task is generated and inserted in the worklist of this particular user. She then can later rate the performance of the respective ad-hoc modification either with 2 (highly positive), 1 (positive), 0 (neutral) , -1 (negative) or -2 (highly negative), and may optionally specify additional comments. The reputation score of a case is calculated as the sum of feedback scores regarding the ad-hoc modification specified in this case. While a high reputation score of a case is an indicator for its semantical correctness, negative feedback probably results from problems after performing a process instance change. As ADEPT ensures the syntactical correctness of changes, a negative feedback thus indicates semantical problems. Negative feedback therefore results in an immediate notification of the process engineer, who can then deactivate the case to prevent its further reuse. The case itself, however, remains in the system to allow for learning from failures as well as to maintain traceability.

Example 1. As depicted in Fig. 3 the examination of a patient usually takes place after a preparation step. Before an examination the physician recognizes that the patient suffers from diabetes and he detects several other important risk factors. The physician decides to request an additional lab test for this patient to be performed after activity Prepare patient *and before activity* Examine Patient. *As the system contains no similar cases, the physician enters a new case describing the situation and the action to be taken (cf. Fig. 2). ADEPT then checks whether inserting activity* Lab Test *is possible for the respective process instance, and, if so, applies the specified insert operation to that instance. The latter includes updating the instance markings and all user worklists. If, for example,* Prepare patient *is completed and* Examine Patient *is activated, this activation will be undone (i.e., respective work items are removed from all user worklists) and the newly inserted activity* Lab test *becomes immediately activated. In any case, the newly inserted activity is treated like any other process step, i.e., the same scheduling and monitoring facilities exist.*

When talking with another diabetic patient later on, the physician vaguely remembers that there has been a similar situation before and initiates the CCBR sub-system to retrieve similar cases. For example, as he still remembers that he had performed an additional lab test he selects the Insert *operation type as well as the* Lab Test *activity to optionally filter the case-base. He then answers the questions presented by the system, finds the previously added case, and reuses it (cf. Fig. 2).*

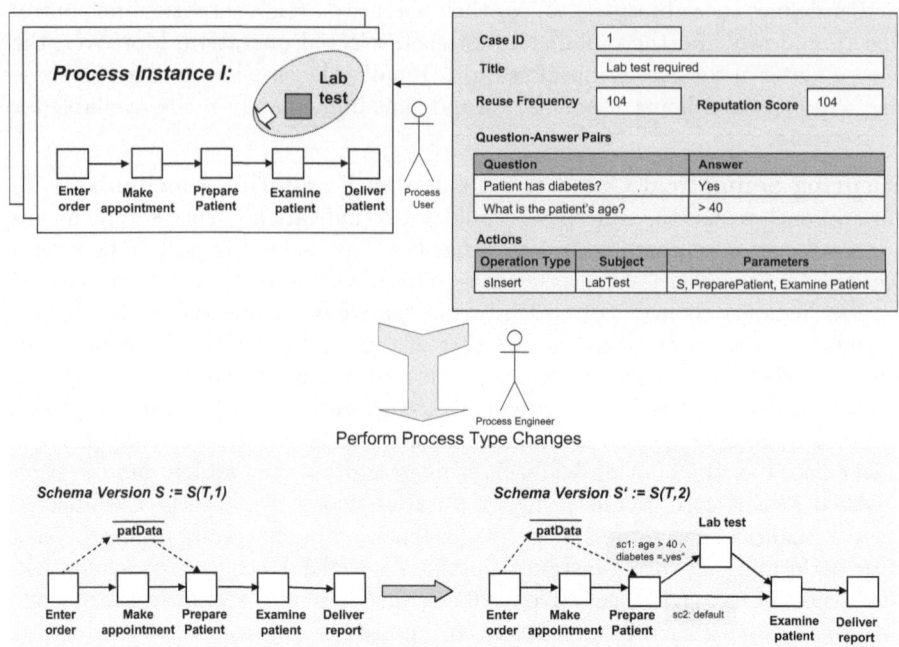

Fig. 3. Perform Process Type Change

4 Process Learning and Seamless Process Evolution

When the same or similar changes occur frequently, the process engineer is notified about the potential need for a process type change (Sect. 4.1). The process engineer can then perform a change of the process type schema and migrate running instances to the new schema version by using the ADEPT framework (Sect. 4.2).

4.1 On Suggesting Process Optimizations

To derive suggestions for process type changes from a collection of previous instance changes we need the following information:

- Case–base $cb_{T,V} = (T, V, \{c_1, ..., c_m\}, freq_{T,V})$
- $rI_{T,V}$ denoting the number of process instances created from schema version $S := S(T, V)$
- thr denoting a configurable threshold ($0 \leq thr \leq 1$)

If there is a case $c_j \in cb_{T,V}$ with

$$\frac{freq_{T,V}(c_j)}{rI_{T,V}} \geq thr \tag{1}$$

the process engineer is notified that a process type change should be considered. In this notification the system suggests the solution part sol_j of the respective case c_j as the process type change to be applied, but allows the process engineer to customize it if desired.

Example 2. Let $rI_{T,V} = 10397$, $thr = 0.01$, $cb_{T,V} = \{c_1 = (..., sol_1 = \{\texttt{insert(S,}$ $\texttt{X, A, B)}\}, ...)\}$, and $freq_{T,V}(c_1) = 104$. As $\frac{freq_{T,V}(c_j)}{rI_{T,V}} = \frac{104}{10397}$ exceeds threshold 0.01, *the system suggests to pull change operation sInsert(S, X, A, B) up to the process type level.*

Generally, the situation is more complex, as a certain change operation may have been applied to several instances for different reasons. Note that in our approach this is reflected by sets of different question-answer pairs in separate cases. As a consequence the respective case–base contains distinct cases, i.e., cases with the identical solution parts but different question-answer pairs. Then equation (1) is no longer adequate and must be adapted for a set of cases.

Example 3. Let $rI_{T,V} = 10397$ and $thr = 0.01$; $cb_{T,V}$ now becomes:
$$cb_{T,V} = \{c_1 = (..., sol_1 = \{\texttt{sInsert(S, X, A, B)}\}, ...),$$
$$c_2 = (..., sol_2 = \{\texttt{sInsert(S, X, A, B)}\}, ...),$$
$$c_3 = (..., sol_3 = \{\texttt{sInsert(S, X, A, B)}\}, ...), ... \} \; with$$
$$freq_{T,V}(c_1) = 48, \; freq_{T,V}(c_2) = 23, \; and \; freq_{T,V}(c_3) = 33$$

Exceeding the threshold *thr* can be determined by summing over the frequencies of all cases which have the same solution part, i.e., if there is a set of cases $cb_{T,V}(sol) := \{c^{sol} = (..., sol, ...) \in cb_{T,V})\}$ with

$$\frac{\sum_{c_j \in cb_{T,V}(sol)} freq_{T,V}(c_j)}{rI_{T,V}} \geq thr \tag{2}$$

Thus, in the example above a process type change is indicated and the system suggests *sol* as process type change to the process engineer (e.g., $\texttt{sInsert(S,}$ $\texttt{X, A, B)}$).

Equation (2) still does not reflect the most general scenario as the solution parts of the cases indicating a process type change may not be identical but may have one or more overlapping changes.

Example 4. Assume the following example ($rI_{T,V} = 10397$ and $thr = 0.01$):
$$cb_{T,V} = \{c_1 = (..., sol_1 = \{\texttt{sInsert(S, X, A, B), delAct(S, C)}\}, ...),$$
$$c_2 = (..., sol_2 = \{\texttt{sInsert(S, X, A, B)}\}, ...),$$
$$c_3 = (..., sol_3 = \{\texttt{sInsert(S, X, A, B)}\}, ...), ... \} \; with$$
$$freq_{T,V}(c_1) = 48, \; freq_{T,V}(c_2) = 23, \; and \; freq_{T,V}(c_3) = 33$$

All cases contain the same change operation $\texttt{sInsert(S, X, A, B)}$ but case c_1 contains an additional change operation $\texttt{delAct(S, C)}$. This is not relevant for the evaluation of the case-base and the suggested process type change. (Note that the migration of the respective process instances is handled by the process schema evolution framework [7].) This can be taken into account by determining the following set

$$cb_{T,V}(\text{sub\_sol}) := \{c = (..., sol, ...) \in cb_{T,V}) \mid \text{sub\_sol} \subseteq sol \}$$

and by applying equation (3):

$$\frac{\sum_{c_j \in cb_{T,V}(sub\_sol)} freq_{T,V}(c_j)}{rI_{T,V}} \geq thr \tag{3}$$

In this case sub_sol is suggested to the process engineer as the process type change to perform (sInsert(S, X, A, B) in our example).

The process engineer has to examine the cases that exceeded the threshold as well as the cases with overlapping solution parts in order to decide how to perform the process type change. When a change operation is relevant for all process instances it can be pulled up to the process type level. In most situations the change operations cannot directly be applied to the process schema as the changes have been performed in a particular context. Examining the question-answer pairs allows the process engineer to gain valuable insights into the context of a change operation as each question-answer pair represents a (semantic) condition under which the case was applied. Assume, for example, that a particular change operation has been primarily performed for patients older than 40 years who suffer from diabetes. Therefore, the solution part of the case is not directly pulled up to the process type level, but the process engineer inserts the necessary XOr nodes and transitions (cf. Fig. 3). However, it must be ensured that the necessary data, e.g., patient's age and diabetes (yes/no), is provided to the running process instances after applying the respective process type change, i.e., to guarantee that all necessary data is available within the system.

In the ADEPT approach change operations have formal pre- and postconditions which ensure their correct application, in particular a correct data flow after applying the changes. Therefore, if we can insert a new XOrSplit at the process type level it is guaranteed that all necessary data is available at runtime, as ADEPT allows the application of correct changes only. Data availability can be achieved if either the activities preceding the XOr-Split set the respective data elements (e.g., activity admit patient writes patient age and diseases) or parameter provisioning services are inserted directly before the XOrSplit. At runtime these services ask the user for the missing information.

4.2   Process Schema Evolution and Process Instance Migration

Assume that the process engineer decides to apply change operation cInsert(S, Lab test, Prepare Patient, sc1) as depicted in Fig. 3 and 4. The challenge is then to migrate the already running process instances to the new schema version S'. As we can see from Fig. 4 we are confronted with different kinds of running process instances: instances still running according to their original schema (unbiased instances, e.g., I_1) and instances which have already been individually modified (biased instances, e.g., $I_2 - I_4$). We further have to distinguish between biased instances for which their instance–specific change (bias) overlaps the process type change (e.g., I_3, I_4) and biased instances with a disjoint bias

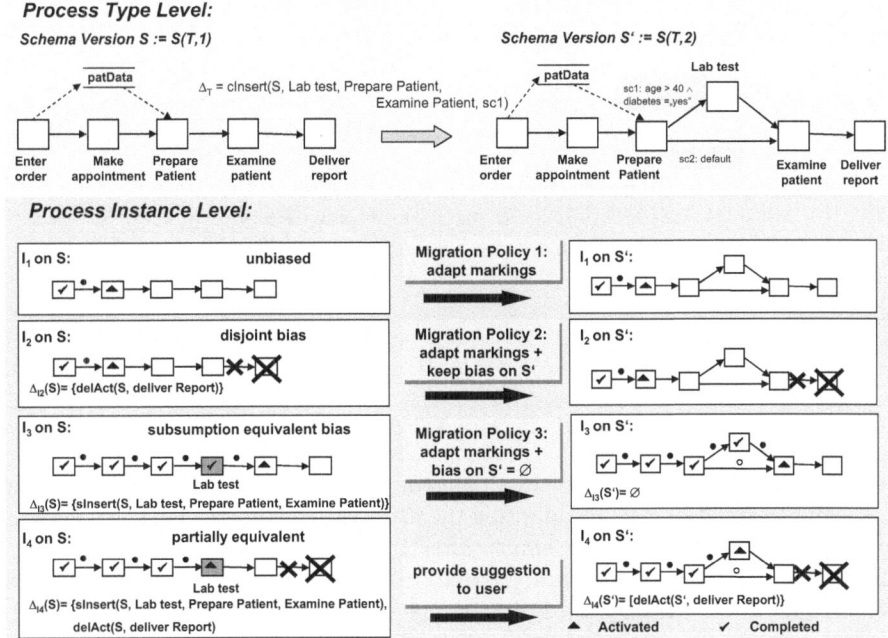

Fig. 4. Process Instance Migration

(e.g., I_2). Process instances with overlapping bias have already anticipated the process type change (cf. Sect. 4.1) and require a different migration policy than the process instances with disjoint bias (for details see [16]).

For unbiased process instances state–related compliance with the new schema version has to be checked [8]. Compliant instances are then migrated to the new schema version by applying marking adaptations (as, for example, depicted in Fig. 4 for instance I_1). Process instances with disjoint bias can be migrated to the new schema version S' if they are compliant regarding their state and their structure [6]. In Fig. 4 instance I_2 has a disjoint bias and is compliant with S'. Therefore I_2 is migrated to the new schema version by adapting its marking and keeping its instance–specific bias in the new schema version (cf. Tab. 2).

The most interesting question is how to deal with the process instances which have totally or partially anticipated the process type change (resulting in an overlap of process type and instance–specific changes). The migration policy to be applied to such instances depends on the particular *degree of overlap* between process type and instance–specific changes, which can be determined precisely by a *hybrid approach* (for details see [7,16]). Table 2 shows the different degrees of overlap and the related migration policies; default migration policies for partially equivalent changes can only be provided in certain situations.

In Fig. 4, for example, instance I_3 would be classified as having a subsumption equivalent bias related to type change Δ_T. Δ_{I_3} and Δ_T both insert activity lab

Table 2. Degrees of Overlap Between Changes and Related Migration Policies

Degree of Overlap between Δ_T and Δ_I	Migration Policy
Δ_T and Δ_I disjoint, i.e., $\Delta_T \cap \Delta_I = \emptyset$	• apply Δ_T on $S_I := S + \Delta_I$ • migrate I to S' • $\Delta_I(S') = \Delta_I(S)$
Δ_T and Δ_I equivalent, i.e., $\Delta_T \equiv \Delta_I$	• migrate I to S' • $\Delta_I(S') = \emptyset$
Δ_T subsumes Δ_I, i.e., $\Delta_T \prec \Delta_I$	• migrate I to S' • calculate $\Delta_I(S')$
Δ_I subsumes Δ_T, i.e., $\Delta_I \prec \Delta_T$	• migrate I to S' • $\Delta_I(S') = \emptyset$
Δ_T and Δ_I partially equivalent, i.e., $\Delta_T \between \Delta_I$	default policies not always possible $\to$ provide suggestion to user

test, but Δ_T additionally creates an alternative branching. According to Table 2, I_3 can be migrated to S' by adapting the instance markings of I_3. The instance-specific bias $\Delta_{I_3}(S')$ becomes empty after the migration. Comparing Δ_T with Δ_{I_4} we see that both changes are partially equivalent, thus we cannot provide a default migration strategy [7], but only make a suggestion to the user (cf. Fig. 4). Note that there are optimizations regarding the determination of the precise degree of overlap [7].

In total, we provide a complete framework for migrating process instances to a new schema version even if they have anticipated the type change. This closes the process life cycle depicted in Fig. 1.

5   Case–Base Evolution

Process type changes are accompanied by migrating compliant process instances to the new schema version S' as well as migrating the associated case-base cb to cb'. The challenge is to decide which cases of case-base cb should be transfered to cb' and which ones are already covered by the new schema version S' and can therefore be dropped.

If a case or a group of cases exceeds the predefined threshold the resulting process type change can either be relevant for all process instances or only for a particular subset (cf. Section 4.1). In the former scenario the solution parts of the cases that triggered the change are directly reflected in the new process schema S'. Therefore, cases whose solution part is a subset of Δ_T are not transferred to the new case-base version cb'. Cases whose solution parts are a true superset of Δ_T are presented to the process engineer who then decides whether to transfer these cases or not. Cases without overlapping solution parts are automatically transferred to cb' as they are not covered by S'. In the latter scenario the migration from cb to cb' is more complicated. It involves finding the regions of the process graph that are affected by the process type change Δ_T. In our example the change region corresponds to the subgraph induced by the newly

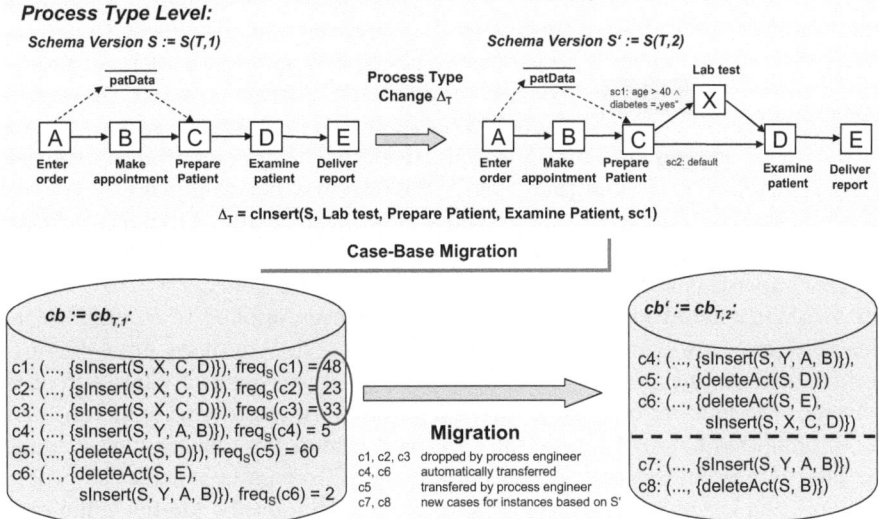

Fig. 5. Case–Base Evolution

inserted activity X and the insertion context (i.e., activities C and D). It can be determined by applying the hybrid approach described in [7]. The cases which contain change operations referring to activities or edges within these regions as subjects or parameters are presented to the process engineer who then can manually transfer relevant cases. All other case are automatically transferred to case-base cb'.

Example 5. As illustrated in Fig. 5 the process engineer has been notified to perform a schema evolution as cases $c_1, ..., c_3$ exceed the predefined threshold value. After migrating schema S to S' the process engineer has to migrate case-base cb to cb' as well. Cases c_4 and c_6 are automatically transferred to cb' as they do not use activities or edges within the affected process graph region. All other cases are presented to the process engineer, who then decides to drop cases $c_1, ..., c_3$ which are already covered by the process type change and to transfer case c_5. Of course, new cases may be added to cb' due to ongoing ad-hoc changes of instances based on S'. Later on, migrating cb' will become necessary when another process schema evolution takes place.

6   Related Work

Process Mining [17] and Delta Analysis [18,19] are techniques to improve the quality of business processes. Though these approaches are very inspiring, they do not answer how they feed the improved process type schemes into the system. To our best knowledge this is accomplished by establishing the mined process schemes as new process type schema versions. Already running process instances are then completed according to the "old" (suboptimal) process type schema and new instances are started according to the improved one. This leads to a "gap" within the process life cycle which can be closed by applying the derived process optimization to the current process type schema. Already running process instances are then smoothly migrated to the improved process type schema [15,20].

Related work also includes approaches dealing with process schema evolution [1,2,4,21]. However, none of them covers the interplay between process type and process instance changes, i.e., there is no approach which allows to migrate biased process instances to a changed process type schema.

This paper is based on the idea of integrating ADEPT [9] and CCBR [10] (see also [15]). In related work traditional CBR has been applied to configure complex core processes by using process components [22]. Workflows are configured during their instantiation by combining predefined process components in order to reduce the number of possible process variants. As each process instance has to be configured before its start, this approach is more suitable for long-running, complex core processes with a limited number of process instances; the process configuration is similar to a project planning task. Similarily, Madhusudan et al. [23] use CBR to provide workflow modeling support by facilitating the reuse of existing models and their components. In contrast to our approach, CBR techniques are applied to support the modeling of business processes and not their execution.

7   Summary and Outlook

The integration of ADEPT and CBRFlow offers promising perspectives, as process instance changes are enriched with semantic information. This, on the one hand ensures the traceability of instance changes and on the other hand supports users in reusing information about previous instance changes. Furthermore, our approach provides techniques to automatically derive suggestions for process type changes from previously applied instance changes. If the process engineer decides to pull up an instance change to the process type level, already running process instances can be smoothly migrated to the new process schema version. Finally an evolution of the associated case–base is done.

Currently we implement a prototype integrating the concepts of ADEPT and CBRFlow and plan to evaluate the resulting prototype in different application scenarios. Future work will focus on the semantic compliance of process type and process instance changes when they are concurrently applied to the same process schema. In this context the representation and the evaluation of semantic information stored for process changes are challenging research topics.

References

1. v.d. Aalst, W., Basten, T.: Inheritance of workflows: An approach to tackling problems related to change. Theoret. Comp. Science **270** (2002) 125–203
2. Ellis, C., Keddara, K., Rozenberg, G.: Dynamic change within workflow systems. In: Proc. Int'l COOCS'95, Milpitas, CA (1995) 10–21

3. Rinderle, S., Reichert, M., Dadam, P.: Correctness criteria for dynamic changes in workflow systems – a survey. DKE **50** (2004) 9–34

4. Weske, M.: Formal foundation and conceptual design of dynamic adaptations in a workflow management system. In: Proc. HICSS-34. (2001)

5. Reichert, M., Rinderle, S., Dadam, P.: On the common support of workflow type and instance changes under correctness constraints. In: Proc. Int'l CoopIS'03. LNCS 2888, Catania, Italy (2003) 407–425

6. Rinderle, S., Reichert, M., Dadam, P.: On dealing with structural conflicts between process type and instance changes. In: Proc. BPM'04, Potsdam (2004) 274–289

7. Rinderle, S.: Schema Evolution in Process Management Systems. PhD thesis, University of Ulm (2004)

8. Rinderle, S., Reichert, M., Dadam, P.: Flexible support of team processes by adaptive workflow systems. DPD **16** (2004) 91–116

9. Reichert, M., Dadam, P.: ADEPT$_{flex}$ - supporting dynamic changes of workflows without losing control. JIIS **10** (1998) 93–129

10. Weber, B., Wild, W., Breu, R.: CBRFlow: Enabling adaptive workflow management through conversational case-based reasoning. In: Proc. ECCBR'04, Madrid (2004) 434–448

11. Leymann, F., Altenhuber, W.: Managing business processes as an information ressource. IBM Systems Journal **33** (1994) 326–348

12. Kolodner, J.L.: Case-Based Reasoning. Morgan Kaufmann (1993)

13. A. Aamodt, E.P.: Case-based reasoning: Foundational issues, methodological variations and system approaches. AI Communications **7** (1994) 39–59

14. Aha, D.W., Muñoz-Avila, H.: Introduction: Interactive case-based reasoning. Applied Intelligence **14** (2001) 7–8

15. Weber, B., Rinderle, S., Wild, W., Reichert, M.: CCBR–driven business process evolution. In: Proc. Int. Conf. on Cased based Reasoning (ICCBR'05), Chicago (2005)

16. Rinderle, S., Reichert, M., Dadam, P.: Disjoint and overlapping process changes: Challenges, solutions, applications. In: Proc. CoopIS'04, Cyprus (2004) 101–120

17. v.d. Aalst, W., van Dongen, B., Herbst, J., Maruster, L., Schimm, G., Weijters, A.: Workflow mining: A survey of issues and approaches. DKE **27** (2003) 237–267

18. v.d. Aalst, W.: Inheritance of business processes: A journey visiting four notorious problems. In: Proc. Petri Net Technology for Communication Based Systems. LNCS 2472 (2003) 383–408

19. Guth, V., Oberweis, A.: Delta analysis of petri net based models for business processes. In: Proc. Applied Informatics. (1997) 23–32

20. Weber, B., Reichert, M., Rinderle, S., Wild, W.: Towards a framework for the agile mining of business processes. In: Proc. of BPM 05 BPI workshop. (2005)

21. Casati, F., Ceri, S., Pernici, B., Pozzi, G.: Workflow evolution. DKE **24** (1998) 211–238

22. Wargitsch, C., Wewers, T., Theisinger, F.: An organizational memory-based approach for an evolutionary workflow management system. In: Proc. HICCS-31. (1998) 174–183

23. Madhusudan, T., Zhao, J.: A case-based framework for workflow model management. In: Proc. Int'l Conf. BPM'03, Eindhoven (2003) 354–369

An Analysis and Taxonomy of Unstructured Workflows

Rong Liu and Akhil Kumar

Smeal College of Business, Penn State University, University Park, PA 16802, USA
{rongliu, akhilkumar}@psu.edu

Abstract. Most workflow tools support structured workflows despite the fact that unstructured workflows can be more expressive. The reason for this is that unstructured workflows are more prone to errors. In this paper, we describe a taxonomy that serves as a framework for analyzing unstructured workflows. The taxonomy organizes unstructured workflows in terms of two considerations: improper nesting and mismatched split-join pairs. Based on this taxonomy we characterize situations that are well-behaved and others that are not. We also discuss well-behaved unstructured workflows that have equivalent structured mappings. Finally, we also introduce a relaxed notion of correctness called quasi-equivalence that is based on one-directional bisimulation. The results of our research will be useful for researchers investigating expressiveness and correctness issues in unstructured workflows.

1 Introduction

Workflow technology has emerged as an important tool for businesses to integrate and automate business processes, not only within the company, but also across the entire supply chain, giving rise to complex inter-organizational processes. Process modeling involves methodologies for designing business process models [5]. In doing so, business processes must be properly modeled before they are implemented as workflows. It is essential that process models not only precisely capture business requirements but also ensure successful workflow execution. If a process is put into production before being properly checked and verified, it could fail to execute properly and cause considerable loss to a business. A correct process model is one without structural flaws, such as deadlocks, dead-end paths, incomplete terminations, etc [8]. Therefore, it is very important that the correctness of workflows be verified systematically before the process models are implemented.

Workflows allow coordination of various activities in a process through control elements such as AND-splits, AND-joins, OR-splits, OR-joins, etc. One accepted notion of correctness is structuredness. A *structured workflow* is one in which each split control element (e.g., AND, OR) is matched with a join control element of the same type, and such split-join pairs are also properly nested. However, not all workflows are structured; some unstructured workflows give more expressive power than structured ones, and are also well behaved. Thus, the requirement of structuredness is restrictive. Following the pioneering work of Kiepuszewski, Hofstede, and Bussler [6], our objective is to study what kinds of unstructured workflows are well-behaved; which ones have equivalent structured mappings; and,

W.M.P. van der Aalst et al. (Eds.): BPM 2005, LNCS 3649, pp. 268–284, 2005.

which ones have what we call *quasi equivalent* structured mappings. Quasi-equivalence is a relaxed notion of equivalence based on uni-directional bisimulation [6], and it allows multiple instances of an activity to exist concurrently. We will show later that in certain situations, multiple instances do not cause correctness problems. We also introduce notions of *improper nesting* and *mismatched pairs* as a means to organize our taxonomy of unstructured workflows, and to understand them in a systematic way. The taxonomy serves as a means of analyzing the main building blocks that constitute a given workflow model. Moreover, this approach helps us in identifying causes of various flaws in workflow models, and also determining cases in which unstructured workflows have equivalent structured mappings.

Related work in this area is still limited, some notable studies being [6, 7, 9]. A graph reduction technique is proposed in [9]. Although this technique can detect structural conflicts through a reduction process, it gives no details about the causes of these conflicts, and, therefore, provides no help for further improvement. [6] defines a restrictive group of workflows as structured workflows, which never lead to structural flaws. It also addresses the possibility that an unstructured workflow can be mapped to a structured one through equivalence preserving transformations. Aalst et al [3] have compared 15 main workflow management systems in terms of a set of selected workflow patterns, and showed none of these systems supports all these patterns, but all of them can support structured workflows. Since most workflow products impose different structural constraints, while structured workflows are widely supported, the mapping may bridge the gap between process modeling and workflow implementation. However, in [6] the possibility of the transformation is discussed mainly through examples, and lacks generality. Our goal in this paper is to mainly extend previous efforts and develop a more general framework to describe and understand the situations considered there. Structural consistency and correctness is also addressed in the context of the ADEPTflex model [7]. Logic-based approaches for workflow verification are discussed by Bi and Zhao [4]. The approach of Aalst and Verbeek is based on converting a workflow into a Petri net and then checking correctness using a tool like Woflan [2, 10]. The drawback with this approach is that after the workflow is converted into a Petri-net, it loses its natural structure.

This paper is organized as follows. Section 2 describes our notations and the assumptions made in this paper. In addition, the concept of structured workflows and workflow correctness are reviewed. In Section 3, we introduce our taxonomy for unstructured workflows and give some results. Section 4 considers situations where loops are present. Section 5 discusses possible future work and concludes the paper.

2 Workflows and Correctness Issues

2.1 Workflow Definition and Semantics

Definition 1 (Workflows). A workflow is a directed graph consisting of *activities, arcs,* and *control elements.* The *control elements* are of the following types: *start, finish, split-parallel, join-parallel, split-choice,* and *join-choice* where:

(1) Arcs are used to connect activities and control elements.
(2) *In-degree $d^-(n)$* and *out-degree $d^+(n)$* indicate the number of arcs entering and leaving a workflow node n respectively.
(3) Each workflow has only *one* start node and *one* end node. For a start node s, $d^-(s)=0$ and $d^+(s)=1$; For an end node e, $d^-(e)=1$ and $d^+(e)=0$.
(4) For any activity a, $d^-(a)=1$ and $d^+(a)=1$.
(5) For any join element j, we require $d^-(j)=2$ and $d^+(j)=1$. Similarly, for any split element s, $d^-(s)=1$ and $d^+(s)=2$. The in-degree (d^-) and out-degree (d^+) of an element can determine whether it is a split or a join.
(6) To avoid triviality, in any path from a split element to a join element, there exists at least *one* activity.
(7) Every activity or control element is in at least one path from the start node to the end node.   ∎

Fig. 1 shows the graphical notation for workflow nodes. For simplicity, and without loss of generality, we assume that each control element has only two incoming (or outgoing, as the case may be) branches. A join with $d^-(j)>2$ can be represented by a combination of multiple join elements. Similarly, a split with $d^+(j)>2$ can be achieved by a combination of multiple split elements. In addition, Definition 1 also implies that, in a workflow, there is at least one path between any two nodes. Based on this definition, we will develop a systematic approach to verifying workflows, detecting causes of structural flaws (if any), and showing *equivalent* structured mappings. This rigorous workflow definition by no means limits the application of this approach. Other relaxed workflow models [4, 9] can also be verified by this approach after simple conversions in accordance with Definition 1.

(a) Start (b) End (c) Activity (d) Split-parallel (e) Join-parallel (f) Split-choice (g) Join-choice

Fig. 1. Graphical representations of workflow control elements (or nodes)

2.2 Semantics of Control Elements

Next, we discuss the semantics of the workflow control elements. Semantically, after a *split-parallel* element, both of its branches can be executed concurrently. Moreover, a *join-parallel* element can be executed only after both of its incoming branches have been executed. We also assume a split-choice element has the semantics of *exclusive choice*, i.e., both branches of a split-choice are exclusive and only one branch can be executed at one time. An *inclusive choice* can be achieved by a combination of split-parallel and exclusive choice [3]. The *semantics of a join-choice element* can be one of the following:

(1) *Single execution*: The join-choice element is executed only once after whichever branch is done first. The other branches are discarded when they finish.
(2) *Multiple executions*: Whenever any of its incoming branches is done, the join-choice element is executed. This may create multiple instances.

If both the incoming branches of a join-choice can be active at the same time, the "multiple executions" semantics may cause correctness problems, as we will discuss.

2.3 Structured Workflows

Structured workflows impose certain restrictions on the relationships between control elements. In particular, in a structured workflow, each split-parallel element must have a corresponding join-parallel element, and each split-choice element has a corresponding join-choice element. There are four *basic types of structured workflows* [6]: *sequence, decision structure, parallel structure*, and *structured loop*, as shown in Fig. 2. A structured workflow can be composed inductively based on these four types or patterns. In addition, any structured workflow (or sub-workflow) can be treated as a single composite activity [9]. Therefore, we will also use the symbol of activity (see Fig. 1(c)), to denote any structured workflow, or sub-workflow.

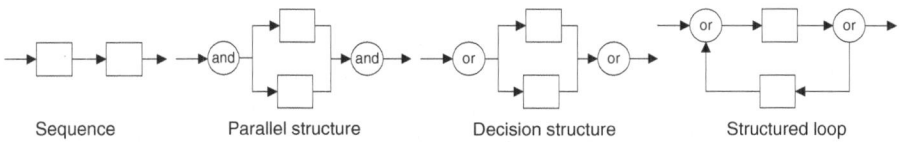

| Sequence | Parallel structure | Decision structure | Structured loop |

Fig. 2. Basic types (or patterns) of structured workflows

2.4 Notions of Correctness

In general, in an unstructured workflow, there may not be a strict *one-to-one correspondence* between a split control element and a join control element. Hence, it is called unstructured, and this lack of structuredness may cause structural flaws that may lead to problems in execution. There are two typical structural flaws in workflows: *deadlocks* and *multiple active instances of the same activity* [6, 9]. A workflow is considered to be *well behaved* if it can be shown that it does not produce deadlocks and also does not allow multiple instances of the same activity [6]. A *deadlock* means that the workflow will never end. Multiple instances can lead to some undesirable results, such as redundant activities, competition for resources, and dangling activities (e.g. one instance is synchronized with an activity, and then the other instances are left dangling). A structured workflow, on the other hand, is well behaved and does not suffer from these problems. Therefore, unstructured workflows must be carefully verified. In addition, although some unstructured workflows do not have any structural flaws, yet workflow products may not support execution of such workflows. However, by transforming them into their *equivalent* structured mappings, they can be implemented by such products.

Typically, a deadlock results from a split choice element being followed by a join-parallel element (see Workflow *wf1* in Fig. 3), where a deadlock may occur at the join-parallel element. This simple example shows how an unstructured workflow might fail to execute. On the other hand, a join-choice element following a split-parallel element (see Workflow *wf2* in Fig. 3) could result in multiple instances of the same activity. However, this structure should be analyzed in detail based on the *semantics of the join-choice element* as follows:

Case 1: If the join-choice has the "*single execution*" semantics, only one instance of *B* is created, after either *A1* or *A2* is done. Therefore, there are no structural flaws.

Case 2: If the join-choice element has the semantics of "*multiple executions*" (see Section 2.2), *A1* and *A2* will produce two instances of activity *B*.

The first case may sometimes be very useful in business processes because it can make workflows more expressive. For example, in a process, suppose a job can be handled by two alternative approaches, say, *A1* and *A2*. To save time, these two approaches can be tried in parallel. Whenever one approach is successful, the other approach is ignored, and the process proceeds with its subsequent tasks.

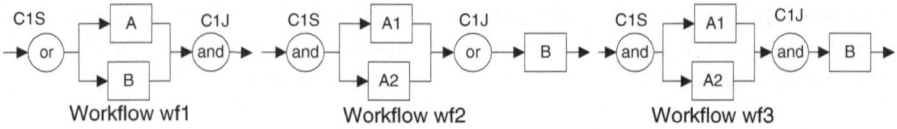

Fig. 3. Workflows with correctness issues present

Since the semantics of the join-choice in the first case is somewhat similar to a join-parallel, in that it proceeds only once, although both its incoming branches can be executed, to help our correctness analysis, we can temporarily map it to a join-parallel element. This mapping is called a *quasi-equivalent mapping*. In Fig. 3, Workflow *wf3* is the quasi-equivalent mapping of the Workflow *wf2*.

Definition 2 (Quasi-equivalent or q-equivalent mapping). A mapping from workflow *A* to workflow *B* is quasi-equivalent if *A* can simulate *B*, but *B* cannot simulate *A*.   ■

Based on the concept of *bisimulation games*, workflow *A* can simulate workflow *B* if *A* can imitate any movement (e.g., starting or finishing the workflow, or completing an activity,) of *B*. If *A* can simulate *B*, and *B* can also simulate *A*, then we say *A* and *B* are equivalent, or *A* is the *equivalent mapping* of *B* [6]. As an example, consider the two workflow patterns in Fig. 3. Here, based on bisimulation games, Workflow *wf2* can simulate Workflow *wf3*, but *wf3* cannot simulate *wf2*. In *wf2*, the possible completed execution paths (or interleavings) are *A1A2B*, *A2A1B*, *A1BA2*, and *A2BA1*, but only *A1A2B* and *A2A1B* are the possible paths of *wf3*. In other words, *wf2* is more expressive than *wf3*. Thus, this mapping is not *completely equivalence preserving*. However, such a *quasi-equivalent mapping* can help in the verification of complicated workflows involving join-choice elements as we will see later.

2.5 Structural Flaws – Notions of Corresponding Control Elements

Next, we study causes of structural flaws with the help of some definitions.

Definition 3 (Corresponding control elements). A split element *s* corresponds to a join element *j*, if two *minimal* paths, starting along two different outgoing arcs of *s*, first join at *j*. This corresponding pair *s* and *j* is denoted by (*s*, *j*).   ■

By *minimal* we mean that any node on a subpath of these paths does not have the correspondence property with *s*. For example, in Fig. 4(a), Workflow *wf1* has (*C1S*,

C1J), (*C2S, C2J*) and (*C3S, C3J*), but not (*C1S, C2J*), since a path from *C1* to *C2J* (*C1S→A2→C3S→A5→C1J→A6→C2J*) is not minimal as it already contains *C1J*, a corresponding element of *C1S*. Similarly, we do not have a correspondence (*C1S, C3J*). Next, we state some simple lemmas without proof.

Lemma 1: Every split control element must have at least one corresponding join control element.

Lemma 2: In general (unstructured) workflows, the split and join control elements need not be of the same type

Lemma 3: A split control element may have multiple unique corresponding join control elements.

Lemma 4: A join control element may correspond to more than one split control elements.

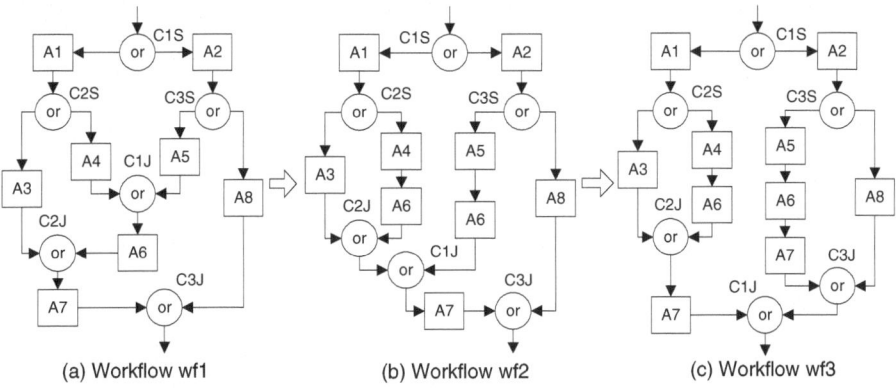

(a) Workflow wf1   (b) Workflow wf2   (c) Workflow wf3

Fig. 4. Steps in mapping an unstructured workflow *wf1* into a structured *wf3*

Definition 4 (Mismatched pair). A pair of corresponding control elements (s, j) is mismatched if s is a split-choice and j is a join-parallel element, or s is a split-parallel and j is a join-choice element. (s, j) is called a *mismatched pair*. ∎

Definition 5 (Improper nesting). A pair of matching control elements (s, j) is *improperly nested* with another pair of matching control elements (u, v), if s (or j) is in a path from u to v, but j (or s) is not in this path. This is denoted as $(u, v)\, {}^{[}_{]}\,(s, j)$. ∎

Definition 6 (Order of Improper nesting). $(u, v)\, {}^{[}_{]}\,(s, j)$ is called *first-order improper nesting*, if there is no (x, y), such that $(u, v)\, {}^{[}_{]}\,(x, y)$, where $(x, y) \neq (s, j)$. $(u, v)\, {}^{[}_{]}\,(s, j)$ is called *nth-order improper nesting* if there exist $(u, v)\, {}^{[}_{]}\,(x_i, y_i)$, where $(x_i, y_i) \neq (s, j)$ and $i=1, 2, ..., n-1$. We use $(u, v)\, {}^{[}_{]}\, \{(x_1, y_1), (x_2, y_2),..., (x_n, y_n)\}$ to denote n pairs of corresponding elements nested into (u, v). ∎

Fig. 5 shows two examples of improper nesting. In both examples, a pair of *corresponding* control elements (*C2S, C2J*) is nested with another pair of *corresponding* control elements (*C1S, C1J*), constructing a first-order improper nesting (*C1S, C1J*)$\,{}^{[}_{]}\,$(*C2S, C2J*). In this figure, *A, B, C, D* represent any activities or tasks, while the dotted lines represent other activities or control elements. In addition, mismatched pairs can be combined with improper nesting to create other combinations of pairs as we will see in the next section.

3 Scenarios for Unstructured Workflows

3.1 General Forms of Unstructured Workflows

As discussed above, split and join control elements are always paired, i.e. every split control element has a corresponding join control element. If the control elements are properly nested, and the corresponding elements are of the same type, then it is a structured workflow; however, if they are of different types, then it is an unstructured workflow. An (And, Or) pair produces a *q-equivalent* mapping. However, an (Or, And) pair will lead to a deadlock. Therefore, if control elements are properly nested, then it is possible to determine whether a workflow is *q-equivalent* or whether it will deadlock by looking at all pairs that are present.

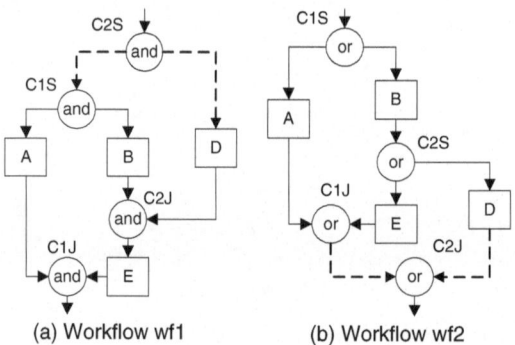

(a) Workflow wf1   (b) Workflow wf2

Fig. 5. Two patterns of improper nesting

Algorithm 1: If a workflow is properly nested, then compare each split and join pair:

(1) If only (And, Or) pairs exist, it has a q-equivalent mapping.
(2) If any (Or, And) pair exists, then the workflow will deadlock and is incorrect.

On the other hand, if the workflow is improperly nested, then we need to consider the nesting patterns based on *split-join combinations* in each pair. In the context of Fig. 5, each control element can either be an AND or an OR. Thus, each control node can take two values, and there are 16 combinations. Next, we will discuss each combination in detail. Our purpose is to investigate the possible correctness issues for each combination, determine which combinations can be mapped to structured ones, and develop *equivalent* or *quasi-equivalent mappings* for them. We also want to determine whether an unstructured workflow is well behaved, and if it is well behaved, whether a structured mapping exists for it. We will show later how this can be done in some cases.

3.2 Enumerating All Combinations

Table 1 shows the 16 different scenarios for improperly nested structures, both matched and mismatched. Here we consider a $(C1S,C1J)$ $\rceil_\rceil$ $(C2S,C2J)$ control structure. We consider all possible combinations of the four control elements involved. We also assume that there is a *one-to-one correspondence* between the split and the join control elements. This means that $C1J(C2J)$ is the only join element corresponding to $C1S(C2S)$, and vice-versa.

Table 1. First-order improper nesting and mismatched pairs and their behaviors

Type	(C1S	C1J)	(C2S	C2J)	Correctness issues	Structured transformation
1	OR	OR	OR	OR	well-behaved	Yes
2	OR	OR	OR	AND	deadlock	No
3	OR	OR	AND	OR	multiple instances	q-equivalent mapping
4	OR	OR	AND	AND	deadlock	No
5	AND	AND	OR	OR	deadlock	No
6	AND	AND	OR	AND	deadlock	No
7	AND	AND	AND	OR	multiple instances	q-equivalent mapping
8	AND	AND	AND	AND	well-behaved	No
9	OR	AND	OR	OR	deadlock	No
10	OR	AND	OR	AND	deadlock	No
11	OR	AND	AND	OR	deadlock	No
12	OR	AND	AND	AND	deadlock	No
13	AND	OR	OR	OR	multiple instances	q-equivalent mapping
14	AND	OR	OR	AND	deadlock	No
15	AND	OR	AND	OR	multiple instances	q-equivalent mapping
16	AND	OR	AND	AND	multiple instances	q-equivalent mapping

As the table shows, all combinations where there a split-choice node is followed by a join-parallel node lead to deadlock. This explains the behavior in 9 of the 16 cases. Among the remaining 7 cases, two are well-behaved, and 5 involve multiple instances. The well-behaved ones are those where all four control elements are identical (either all ANDs or all ORs). Next we look at some examples to illustrate some cases from Table 1.

3.3 Approach and Examples

Here we first give some examples to illustrate our approach.

Example 1: Fig. 6 shows a workflow example of Type 1 (from Table 1) and its equivalent structured mapping. In this mapping, the activity that lies between the two join-choice nodes (activity E) is duplicated and pushed up. Thus the two join-choice nodes are together, and can be interchanged. In this way, improper nesting $(C1S, C1J)$ $\rceil_\rceil$ $(C2S, C2J)$ is corrected.

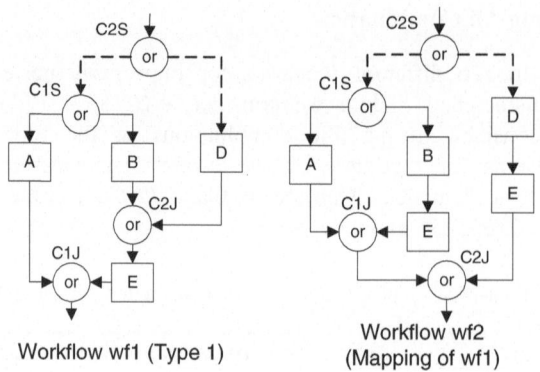

Workflow wf1 (Type 1)

Workflow wf2
(Mapping of wf1)

Fig. 6. Example workflow to illustrate case 1 from Table 1

The general approach applicable to Type 1 workflows is to push the activities that lie between two join-choice nodes up, and thus create a structured mapping.

Example 2: Fig. 4(a) shows a workflow with 3 split and 3 join control elements. The correspondence between the three pairs of control elements is denoted as: $(C1S, C1J)$, $(C2S, C2J)$, $(C3S, C3J)$. The improper nesting relationships are as follows:

$(C1S, C1J) \ {}^{[}_{]} \ \{(C2S, C2J), (C3S, C3J)\}$,
$(C2S, C2J) \ {}^{[}_{]} \ (C1S, C1J)$, and
$(C3S, C3J) \ {}^{[}_{]} \ \{(C1S, C1J), (C2S, C2J)\}$.

Only $(C2S, C2J) \ {}^{[}_{]} \ (C1S, C1J)$ belongs to first-order improper nesting, and it is of Type 1 in Table 1. Therefore, we can remove this improper nesting using the technique above, and get workflow *wf2* of Fig. 4(b). In *wf2*, only one improper nesting $(C3S, C3J) \ {}^{[}_{]} \ (C1S, C1J)$ (or say $(C1S, C1J) \ {}^{[}_{]} \ (C3S, C3J)$) remains. Again, this improper nesting is of Type 1. We continue this procedure and get the final transformation shown as Workflow *wf3* in Fig. 4(c).

Therefore, this approach for detecting structural flaws and developing a structured mapping is as follows:

(1) Find all pairs of corresponding elements and determine improper nesting and mismatching pair relationships.
(2) For any first-order improper nesting or mismatched pairs, look up Table 1 for its corresponding type and correct it, if possible.
(3) Repeat this process until all improper nestings have been fixed.

However, some unstructured workflows may only have second-order and higher improper nestings. These cases are discussed next.

3.4 Results and Handling Second- and Higher-Order Improper Nesting

Before discussing our approach for handling higher-order improper nesting, we give the definition of *adjacent join nodes*.

Definition 7 (Adjacent join nodes): Two *join nodes* are *adjacent*, if there are only activity nodes, but no other control nodes between them.   ■

For example, in Fig. 4(a), (*C1J, C2J*) and (*C2J, C3J*) are pairs of adjacent join nodes. An intuitive observation is that, for example, if *C1J* can be pushed down below *C2J*, at least we can remove one improper nesting (*C2S, C2J*) $[_]$ (*C1S, C1J*). By this approach, we can reduce a higher order of improper nesting eventually to a first order, and then determine the correctness of this workflow by using Table 1. However, we need certain rules to switch locations of two adjacent join nodes in order to preserve the equivalence. In a workflow with improper nesting, two adjacent join nodes always exist, and we need to determine if they can be switched.

For two adjacent join nodes, there are four possible cases. Fig. 7 shows these four cases. For simplicity, this figure only shows the join nodes and omits all the corresponding split nodes. These cases will be illustrated with examples. Note that Case 4 requires two upstream join-choice nodes adjacent to a join-parallel node downstream. Later on, we will show that when parallel structures are nested into decision structures, only improper nesting of Case 4 possibly has transformations. Next, we handle each case one by one and we discuss some of our results.

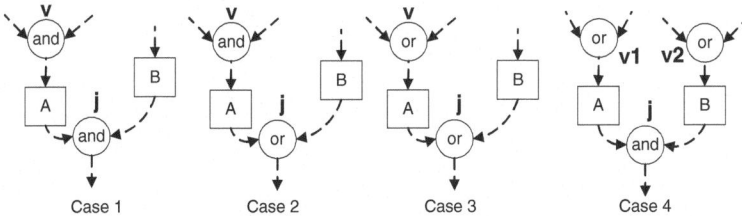

Fig. 7. Different cases of *adjacent join nodes*

3.4.1 Cases 1 and 2
Theorem 1: In a workflow that contains a (u, v) $[_]$ (s, j) nesting, where u is a split-parallel, v is a join-parallel, and s is in a path from u to v, if there is at least one activity between u and s, and at least one activity between v and j, this workflow does not have an equivalent structured mapping.

Proof sketch: The proof is based on construction and uses Fig. 8. Two scenarios are constructed for creating a structured mapping by: removing activity D and pushing v down (Fig. 8(b)), or removing activity B and pushing s up (Fig. 8 (c)). Then, we argue that in either case the removed activity cannot be reinserted without disturbing the order of activities in Fig. 8 (a). Therefore, a structured mapping is not possible.   ■

Theorem 1 can be used to check whether an unstructured workflow has an equivalent structured mapping, and shows that Cases 1 and 2 (of Fig. 7) do not have structured mappings. However, Theorem 1 says nothing about situations in Fig. 8(a) where either B or D does not exist. In such cases, structured mappings are possible as shown in Fig. 9. This is possible because there is no activity between the two split-parallel nodes u and s.

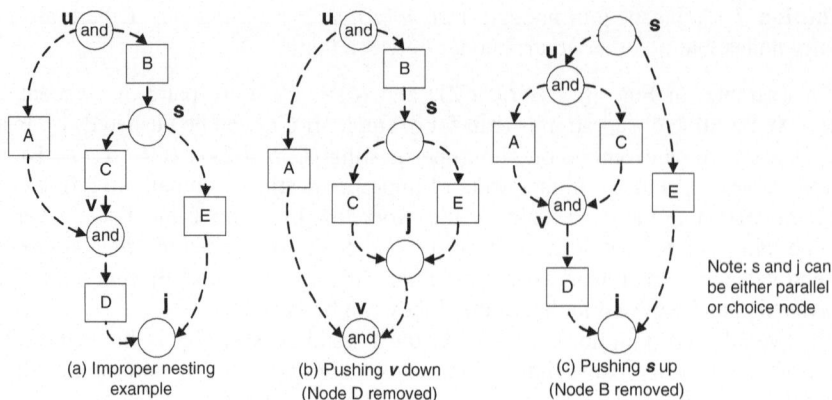

(a) Improper nesting
example

(b) Pushing **v** down
(Node D removed)

(c) Pushing **s** up
(Node B removed)

Note: s and j can
be either parallel
or choice node

Fig. 8. A workflow with improper nesting inside AND elements, and its unsuccessful structured mappings

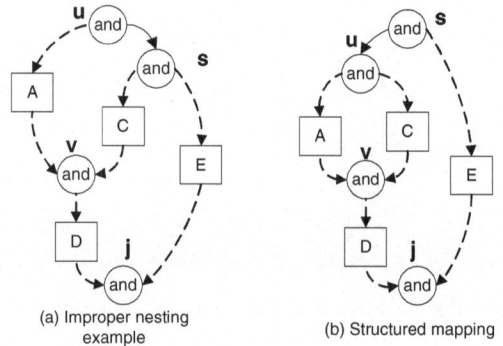

(a) Improper nesting
example

(b) Structured mapping

Fig. 9. Workflow with improper nesting and its equivalent structured mapping

Note that while we just showed Cases 1 and 2 do not have structured mapping, yet they may still be well behaved. We further note that problems of deadlocks and multiple instances arise only when choice nodes are present. For example, in Figure 8(a), if s and j are OR nodes, this workflow leads to deadlocks at node v. However, a workflow without choice nodes is free of these structural flaws. Therefore, we state Lemma 5 (without proof) as follows.

Lemma 5: If a workflow contains only AND nodes, it must be well behaved. However, it does not have a structured mapping if it contains improper nesting (u, v) ⌈⌉ (s, j) *and* there is at least one activity between u and s and at least one activity between v and j.

3.4.2 Case 3
Case 3 is a generalization of an example already discussed in Fig. 4, and it does have an equivalent structured mapping. As discussed there, the general strategy for this is as follows: *the activities that lie between the two adjacent join-choice nodes are*

duplicated and pushed up above the upper join-choice, and the two join-choice nodes are interchanged. This produces an equivalent structured mapping.

3.4.3 Case 4

This is a case of second order nesting such as $(s, j)\, ^{[}_{]} \{(u, v), (x, y)\}$ where (s, j) are AND nodes and the other two pairs are OR nodes (Fig. 7). We first state some definitions.

Definition 8 (parallel paths and exclusive paths): Two paths p and q are *parallel* if, when p is taken, then q must be taken simultaneously; p and q are *exclusive* if, when p is taken, it implies that q *cannot* be taken.   ∎

In this situation, a structured mapping is possible if the adjacent join-choice and join-parallel nodes can be interchanged. To check if this is possible, we consider all combinations of path pairs from a split-parallel to the join-parallel node. The combinations that consist of *parallel* paths between the split-parallel and join-parallel nodes are feasible. Each path combination can thus form a structured AND sub-workflow, which can be inserted inside the original split-choice and join-choice nodes, to give an equivalent structured mapping. If even one such combination exists, then a structured mapping is possible. If no such combination exists, it means the workflow will deadlock and is incorrect. More details of the algorithm are omitted for space reasons, but an example will illustrate our idea.

Fig. 10(a) is an example of a $(s, j)\, ^{[}_{]} \{(u, v), (x, y)\}$ workflow from [6]. We apply our method to analyze this workflow by finding all paths between a split-parallel (*C2S* or *C2S′*) and the join-parallel (*C2J*) nodes. Next, we find combinations where all paths within a combination are in parallel with each other, and exclude the other combinations of paths. Obviously, *DI* and *GJ*, and *FI* and *EJ* are two pairs of *exclusive* paths, and can be eliminated. However, there are other two combinations with two *parallel* paths in each. These are shown along with the final structured mapping in Fig. 10(b).

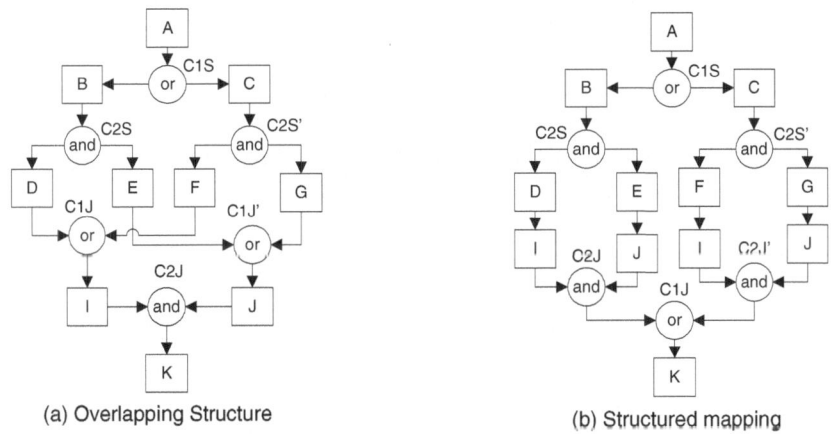

(a) Overlapping Structure   (b) Structured mapping

Fig. 10. An overlapping structure and its mapping

4 Introducing Loops

So far we only considered acyclic workflows, i.e. in these workflows, there were no paths that created cycles. Next, we turn to consider workflows where cycles or loops exist. Loops are normally created between a join-choice element and a split-choice element as shown in Fig. 11. These two elements do not have corresponding elements to them in the sense of Definition 3. However, they are said to correspond to each other (for loops), and every loop will have such a pair of choice elements. In Fig. 11, *C1S* and *C1J* are such distinguished nodes and they are said to correspond to each other. In a structured loop (see Fig. 2), there are no other exits from or entrances into the loop path. However, in a general loop additional entrances and exits may exist. We call these as situations of improper nesting into the loop. In this section, we are primarily interested in loops with at least one join-choice element that serves as an entrance, and one split-choice that serves as an exit from the loop. Then, we consider scenarios involving additional entrances and exits.

4.1 Scenarios and Taxonomy

Fig. 11 shows the corresponding scenarios of interest. In both these figures there is a correspondence between *C2S* and *C2J* nodes. In Fig. 11(a), *C2J* lies on the loop and *C2S* is outside the loop, while in Fig. 11(b), it is the other way around. There are four combinations of values for the *C2S* and *C2J* pairs, and these are considered in Tables 2 and 3, which correspond to Fig. 11(a) and Fig. 11(b) respectively. The tables show that in both scenarios, 2 out of 4 cases behave similarly and are acceptable. When the split-join combination is OR-OR, the workflow is well-behaved and also has a corresponding structured representation. The AND-OR combination leads to multiple instances, and in an entering structure, it has a q-equivalent mapping. A third combination AND-AND, causes a deadlock for an entering structure, but works well in the exit structure. The semantics in this case is as follows: if there are multiple passes through the loop, then activity *D* will get invoked repeatedly. However, when the loop is exited, then the AND control element *C2J* will be activated. From a

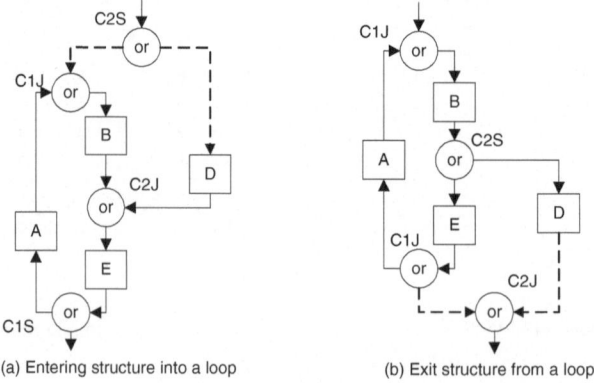

(a) Entering structure into a loop   (b) Exit structure from a loop

Fig. 11. Structures entering and leaving loops

semantic perspective, the results from the most recent execution of *D* should be regarded, while the earlier ones can be ignored. The last structure in the table is OR-AND that leads to deadlock. Clearly the behavior in the case of an entrance versus an exit from the loop is not symmetric. Next, we see how equivalent mappings of structures with loops can be created.

4.2 Mappings

Next, to appreciate how a q-equivalent structured mapping can be created for a Type 1N workflow from Table 2, consider Fig. 12(a). In the mapping shown here, the loop $E{\rightarrow}A{\rightarrow}B{\rightarrow}E$ is duplicated and *C2J*, a join-choice node is mapped to a join-parallel node. In this mapping, multiple instances can arise of activities *E*, *A* and *B*, which lie inside the loop.

Similarly, a type 3N structure from Table 2 has an equivalent structured mapping as shown in Fig. 13. A Type 3X workflow, as shown in Fig. 14, is well behaved, but we will show that it has no structured mapping without using auxiliary variables.

Table 2. Behavior of structures entering a loop

Type	(C2S	C2J)	Correctness issues	Structured Transformation
1N	AND	OR	multiple instances	q-equivalent mapping
2N	OR	AND	deadlock	No
3N	OR	OR	well-behaved	Yes
4N	AND	AND	deadlock	No

Table 3. Behavior of structures exiting a loop

Type	(C2S	C2J)	Correctness issues	Structured Transformation
1X	AND	OR	multiple instances	No
2X	OR	AND	deadlock	No
3X	OR	OR	well-behaved	Yes
4X	AND	AND	well-behaved	No

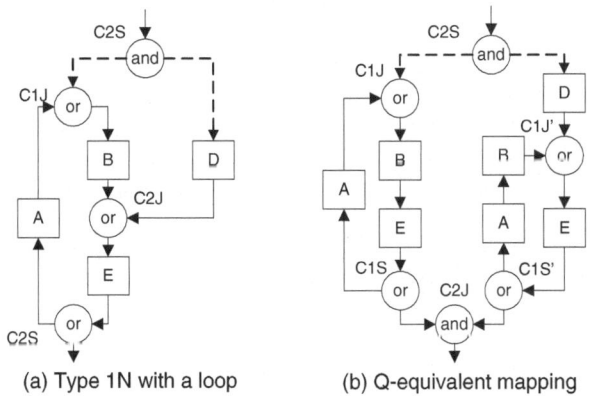

(a) Type 1N with a loop   (b) Q-equivalent mapping

Fig. 12. Type 1N with a loop and its quasi-equivalent mapping

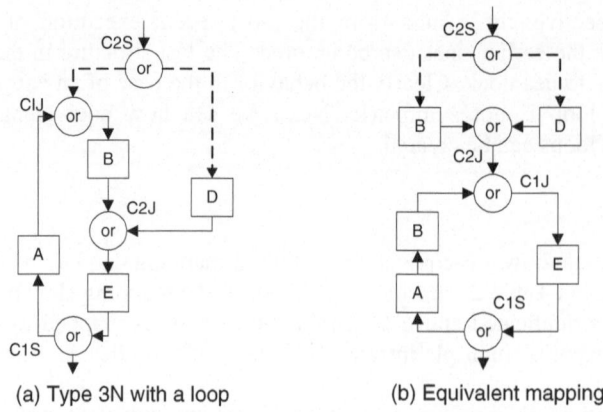

(a) Type 3N with a loop   (b) Equivalent mapping

Fig. 13. Type 3N with a loop and its equivalent structured mapping

4.3 Results

Finally, we give two main results related to loops.

Lemma 6: A workflow pattern of type 3X with a loop cannot be mapped to structured workflows (without using auxiliary variables).

Proof sketch: (by contradiction) The proof is based on arguing that this loop has two exit nodes, and a different activity follows after each of these exit nodes. A structured mapping of the loop will have only one exit node, and an auxiliary variable would be required to determine which of the two exits was taken in order to make sure that the correct activity follows the exit.   □

Fig. 14(a) gives an example of how a mapping for this situation can be produced with auxiliary variables. It is also observed in [6] that certain forms of unstructured workflows cannot be transformed without the use of auxiliary variables.

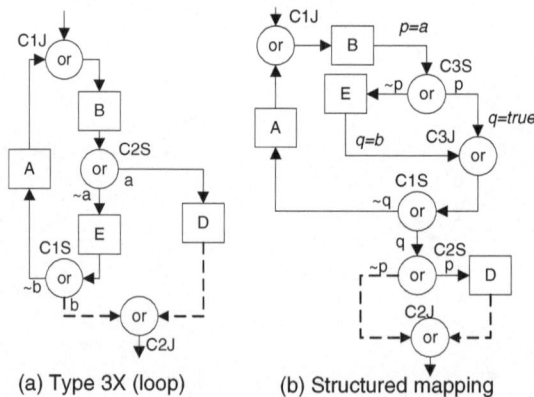

(a) Type 3X (loop)   (b) Structured mapping

Fig. 14. Type 3X (loop) and its structured mapping using auxiliary variables

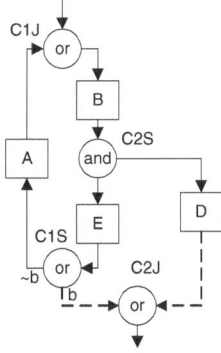

Fig. 15. Type 1X (loop)

Lemma 7: A Type 1X (or 4X) workflow cannot have a q-equivalent (or equivalent) structured mapping.

Proof sketch: (by contradiction) This result is proved by arguing that for a workflow of Type 1X (see Fig. 15), any structured mapping must contain a structured loop and a parallel structure. The parallel structure would contain E and D in parallel. If such a structure were inside the loop, then both E and D would be part of the loop (but D is not); while, if it were outside the loop, then both would be outside (but E is in the loop)! □

5 Discussion and Conclusions

In this paper, we have tried to create formal taxonomy of unstructured workflows based on a notion of improper nestings and mismatched pairs. The notion of nestings is developed in terms of first- and higher-order nestings. We have shown how this taxonomy can help in analyzing unstructured workflows and determining whether they are well-behaved, and if so, whether they can be transformed into equivalent structured mappings. Such an equivalent structured mapping may involve some redundancies, but it allows us to verify that a workflow is correct. Moreover, as mentioned earlier most tools do not support unstructured workflows; however, if certain kinds of unstructured workflows can be mapped into structured ones (even at the cost of duplicating some activites), this can provide an easy way to increase the expressive power of the workflows that can be supported by these tools. We have extended previous research by putting it into a framework, and also developed some new results.

We have also given results for analyzing situations with higher-order nestings. However, there are still some open issues. One conjecture we are still trying to verify is that: if the structures involved in a higher-order nesting have one-to-one correspondence between them, then each pair of nesting can be analyzed separately as a first-order nesting, using Table 1, to determine whether the workflow is correct. Some examples bear this out but more work is required to prove it formally.

Acknowledgment. The authors thank Henry H. Bi for discussions and comments.

References

1. Aalst, W.M.P. van der, "The application of Petri nets to workflow management," *The journal of Circuits, Systems and Computes*, 7(1):21-66, 1997.
2. Aalst, W.M.P. van der, and B., Hofstede. "Verification of Workflow Task Structures: A Petri-net- based approach," *Information Systems*, 25(1):43-69, 2000.
3. Aalst, W.M.P. van der, Hofstede, A.H.M. ter, Kiepuszewski, B., and Barro, A.P. "Workflow patterns," *Distributed and Parallel Databases,* 14(3):5-51, July 2003.
4. Bi, H. and Zhao, L. "Process logic for verifying the correctness of business process models," *Proceedings of International Conference on Information Systems (ICIS 2004)*, Washington, D.C. , December 12-15, 2004.
5. Georgakopoulos, D. and Hornick, Mark "An Overview of Workflow Management From Process Modeling to Workflow Automation Infrastructure," *Distributed and Parallel Database*, 3:119-153, 1995
6. Kiepuszewski, B., Hofstede, A.H.M, and Bussler, C. "On Structured Workflow Modeling" *In Proceedings CAiSE'2000,* LNCS Vol. 1797, Springer Verlag.
7. M. Reichert and P. Dadam, "ADEPTflex---Supporting dynamic changes of workflows without losing control," *Journal of Intelligent Information Systems---Special Issue on Workflow Managament*, 10(2):93-129, 1998.
8. Sadiq, W. and Orlowska, M. E. "On correctness issues in conceptual modeling of workflows," In *Proceedings of the 5th European Conference on Information Systems (ECIS `97)*, Cork, Ireland, June 19-21, 1997, pp. 943-964.
9. Sadiq, W, and Orlowska, M. E. "Analyzing process models using graph reduction techniques," *Information Systems*, 25(2):117-134, 2000.
10. Verbeek, H.M.W., Basten, T. and Aalst, W.M.P. van der. "Diagnosing Workflow Processes using Woflan," *The Computer Journal*, 44(4):246-279. British Computer Society, 2001.

A Framework for Document-Driven Workflow Systems

Jianrui Wang and Akhil Kumar

Smeal College of Business,
Pennsylvania State University, University Park, PA 16802, U.S.A.
{JerryWang, AkhilKumar}@psu.edu

Abstract. We propose and demonstrate the feasibility of a framework for document-driven workflow systems that requires no explicit control flow and the execution of the process is driven by input documents. The framework can assist workflow designers to discover the data dependencies between tasks in a process and achieve more efficient control flow design. The framework also provides an architecture to separate the workflow system from application data and facilitate inter-organizational processes. Document-driven workflow systems are more flexible than traditional control flow processes, easier to verify and work better for ad hoc workflows. We also implemented a prototype workflow system using the framework entirely in a RDBMS using Transact-SQL in Microsoft SQL Server 2000. A detailed comparison with control driven workflows has also been done.

1   Introduction

Academic interest in workflow systems has increased considerably in the past decade, especially with the boom in e-business and supply chain management. Workflow is built into most commercial e-business and supply chain management software, and functions as a foundation module to support business process performance and coordination.

ARIS (Architecture of Integrated Information Systems) [17] developed a pioneering approach to model business processes, and also served as a foundation of SAP/R3. ARIS takes five views of business processes: *functional*, *organizational*, *data*, *output*, and *control*. The Workflow Management Coalition views workflows as interactions of process, information and resource [9]. Depending on the dimension used for modeling, workflow systems can be viewed from one of the following perspectives:

1. Process based perspective. This perspective tends to emphasize process as the dominant dimension; processes consume, produce or transform information under a set of business rules.
2. Information based architectures. This perspective emphasizes the information dimension, viewing processes as operations that are triggered as a result of information changes.

W.M.P. van der Aalst et al. (Eds.): BPM 2005, LNCS 3649, pp. 285–301, 2005.

3. Organization perspective. This perspective views workflow as a mapping of organization structures and focuses on the utilization of organization resource.

Unfortunately, although it is well accepted that workflow systems are an integration of data, control, and resource, most workflow modeling languages such as WSBPEL [16] (formerly BPEL4WS) and XPDL [19] focus on control flow, and give less attention to other dimensions. One popular control flow study is the one on workflow patterns by Aalst [1]. There are only a few studies on data flow modelling [4,6,7,11,12]. However, for the most part, data and resource flow research has received little attention compared with control flow [15].

In this paper, we take the information based perspective, and extend the ideas in the WIDE approach [8]. As noted there, workflow systems must be able to respond to data events, temporal events and external events. One logical development of this idea is to consider the possibility of implementing a complete workflow system inside a database using events as the main mechanism to drive the workflow. In our study, we propose a framework and implementation of document-driven workflow systems. This framework is more flexible than control flow oriented workflow systems and works much better for ad hoc workflows. The rest of the paper is organized as follows. In Section 2 we provide a motivation for our approach with a clear example. Then in Section 3, we give a framework and meta-models for document-driven workflow systems. An implementation of this framework is described in Section 4. Here we discuss our SQL-based implementation for a document-driven workflow system. Finally, in section 5 we discuss the advantages and disadvantages of document-driven workflow systems compared with control flow based systems. The paper is concluded in Section 6.

2   Motivation

In this section, we motivate our approach with a detailed example that compares a control flow based workflow with the corresponding data flow based approach. Fig. 1(a) shows an order process using control flow design. In this process, an order is received, and then the customer's credit rating is checked. Based on the result of the credit check, either the order is cancelled or the steps of warehouse pickup, shipping, invoicing and close order are performed. (To simplify the case, we ignore the exception handling issues.)

The control flow design puts emphasis on the process, that is, the execution sequence of the tasks. It does not explicitly explain why a task should be performed before another. For example, it is not clear why the *Warehouse Pickup* task is done before *Ship* (in Fig. 1(a)), or *Invoice* is done after *Ship*. In general, control flow diagrams assume that the process designer has the business knowledge to layout the task sequence. Tasks have various kinds of dependencies between them. Zlotkin [20] summarizes three basic types of dependencies: *Fit*, *Flow*, and *Sharing*, as shown in Fig. 2. Using Zlotkin's dependency theory, we can find that the tasks Warehouse Pickup and Ship have a *flow dependency* between them, i.e. the output of task *Warehouse Pickup* is one of the required inputs of task *Ship*. A *sharing dependency* arises when several tasks compete for the same resource. *Fit dependencies* arise when multiple activities collectively produce a single resource, and they do not occur very often in workflow situations.

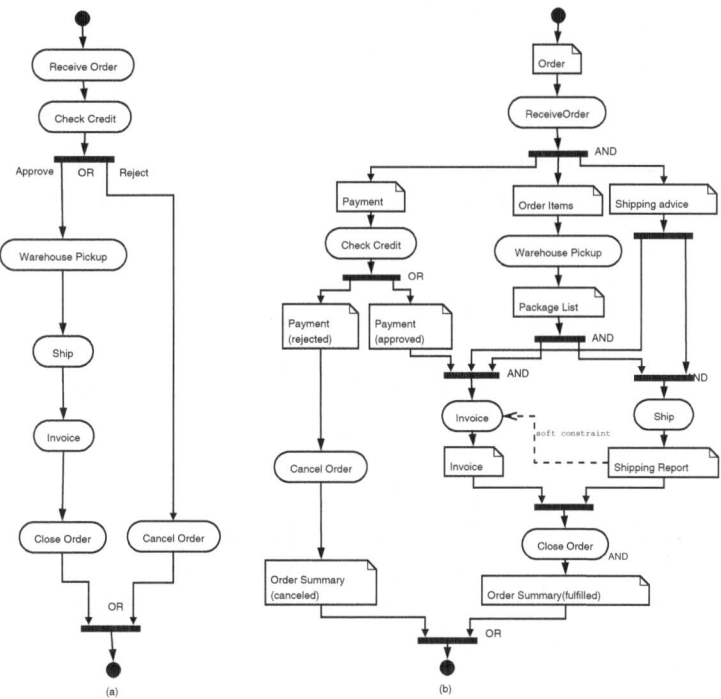

Fig. 1. Order processing workflow with the control and document flow approaches

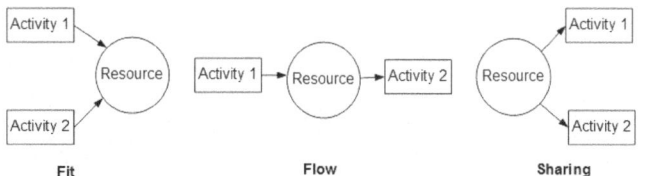

Fig. 2. Three basic types of dependencies among activities (Zlotkin [20])

If we take the dependency analysis approach one step further, and focus on data dependencies, then we can develop a data flow chart as shown in Table 1 for the order process of Fig. 1(a). The data flow analysis provides the input data for a task to be executed, and its output data. Then we can draw a new process diagram using data flow analysis. This is shown in Fig. 1(b). As can be seen from Fig. 1(b), the task *Invoice* does not have to be performed after task *Ship* because there is no data dependency between them. However, a seller may have a policy that invoicing can only be done after shipment. Thus, we have two types of constraints which determine the sequence of tasks: *data dependency constraints* and *business policy constraints*. We call data dependency as a *hard constraint* and business policy as a *soft constraint* because the former applies to all organizations, while the latter may vary from one organization to another.

Table 1. Data flow analysis for tasks in an order process

Task	Input Data	Output Data

Receive order	Order Information:
	  − Payment information(i.e. Customer ID, credit card.)
	  − Order items(i.e. SKUs, unit price, quantity.)
	  − Shipping Advice(i.e. UPS ground.)
		The order information in the input document is split into three documents:
		  − Payment information
		  − Order items
		  − Shipping Advice

Check credit	Payment	Approved or rejected
Warehouse pickup	Order items	Pickup List
Invoice	Payment, Package List, and Shipping Advice	Invoice
Ship	Pickup List and Shipping Advice	Proof of Shipment

The process in Fig. 1(b) also raises two important questions about information flow. The first question is: Why did the task *Receive Order* split the original order data into three documents (payment, order items and shipping advice), instead of handling it as one document? There are two advantages of doing so. First, it is more efficient. If we simply send the whole order to task *Warehouse Pickup*, then the whole order is locked when the task is executing, which prevents others from making changes to any part of the order. However, such a lock is unnecessary because change of shipping advice has nothing to do with Warehouse Pickup. Second, it is more secure. The payment information is sensitive and should be only released to relevant staff, i.e. the *Credit Check* staff. The second question is: Can the two tasks, *Invoice* and *Ship*, be performed concurrently? Since both tasks require Shipping Advice and Package List information, the question actually is, can Shipping Advice and Package List information be accessed at the same time? The answer in this case is yes, because both tasks only need read access to the data in the documents. In the next section, we will introduce a document meta-model to go deeper into these issues.

The above data flow analysis has two advantages. First, it provides a partial ordering for the tasks. Second, it imposes restrictions on the way in which the process can be reconfigured because of soft constraints.

3   A Framework for Document-Driven Workflow Systems

We propose a four-layer architecture for modeling document-driven workflow systems as shown in Fig. 3. The four layers are *schema*, *runtime*, *scheduling*, and *application layer*. The *schema layer* defines workflow processes, which consist of tasks, documents and resources. The *runtime layer* specifies how processes and tasks are started and ended. The *scheduling layer* contains algorithms to assign documents and resources to a task so they can be executed. The *application layer* provides links between the workflow system and the applications. It defines how application data can be linked to the corresponding documents. Since there is a clear separation between workflow data and application data, the details of the application data are not important in the context of the workflow architecture and are not discussed in detail here. The significant differences between our

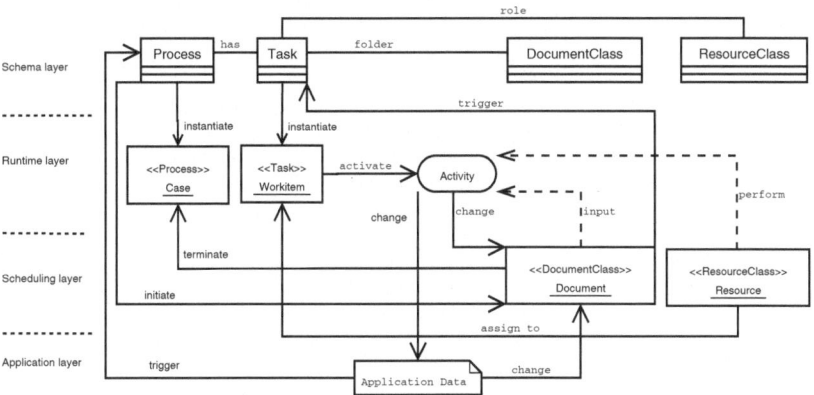

Fig. 3. Document-driven workflow framework

document-driven workflow systems and conventional *control flow based workflow systems* lie in the runtime and the application layers. In document-driven workflow systems, a process is instantiated into a case when certain external events arrive (say, along with a message or a document). The process also creates a set of initial documents of the *process instance* (or *case*). A *task* is instantiated into a *workitem* when its input documents exist. The input documents required by one task are usually the output documents from a previous task, except the initial documents for the first task, which are generated by the process repository when the process is instantiated. After a workitem gets its input documents and associated resources (at the scheduling layer), it becomes an *activity*, which can be executed. An activity changes input documents or produces new documents, which drives the next task. A process ends when its desired documents are produced and its exit constraints are satisfied.

It is important to realize that the input documents may not be available for the workitem when it is instantiated (because someone else may be using them). Therefore, multiple workitems can be created concurrently if their input documents exist, and they will compete for both resources and documents to become executable activities.

The application layer serves as a bridge between the workflow system and the applications. It should be noted that users cannot change application documents directly, rather it is done under the control of the workflow system. When an attempt is made to change a document (by a user or another application), each document has its own event adapter that will capture the changes and check the associated constraints, and then update the document if the constraints are satisfied. Although the architecture encompasses resource and scheduling, the main focus is on documents in this paper. However, it is important to incorporate resource and scheduling in the future research.

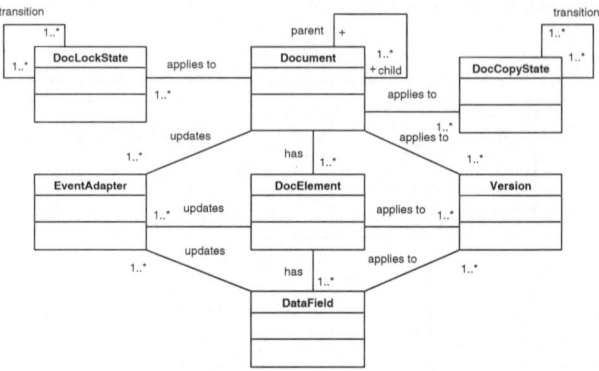

Fig. 4. Document meta-model

To support this framework, we develop a meta-model for documents as shown in Fig. 4. A *Document* is a set of information pieces which are composed together to serve a well defined business purpose. A Document consists of several *DocElements*. A DocElement is a group of *DataFields* which have certain business meaning. For example, address can be a DocElement which may consist of street number, street, city, state, and zip code. A DataField is a piece of information which is treated atomically. Moreover, changing a zip code from 16801 to 16802 is an example of an atomic change. A document may receive events from its DocElements and DataFields, or from other documents. Not all the received events will produce changes in the document. For example, an order form will not be changed if the order has already been shipped. This is managed by the use of constraints (to be discussed shortly). A document generates update *events* if its content is changed. *EventAdapters* along with theirs constraints are used to determine which events documents should respond to.

Fig. 4 also shows that a document has *Versions*. A *version* is used to model the traceable history of workflow data. Since workflow transactions have long transaction time, it is very common that some of the original data have been changed before the transaction is completed. Therefore, keeping the data traceable is necessary and helpful. An order may also be *split* into two sub orders, which result in two new documents with their own versions. A version may apply to a DataField, DocElement, or Document, but it is most relevant for a Document. In general, a series of related data field changes will lead to a new document version. Documents can also have a link or parent-child relationship between them. It is the responsibility of the application to keep the historical data; the workflow system only provides a link to the application.

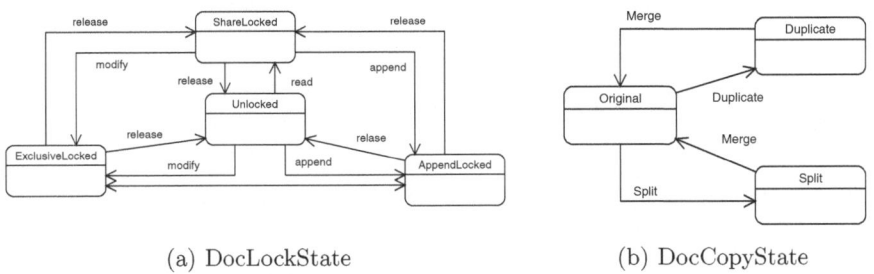

(a) DocLockState   (b) DocCopyState

Fig. 5. State diagram of DocLockState and DocCopyState

We also introduce *DocCopyState* and *DocLockState* to model concurrent access to a Document. The different values of DocLockState (*ShareLocked*, *ExclusiveLocked* and *AppendLocked*) and the permissible transitions between them can be found in Fig. 5(a). The DocCopyStates are *Duplicate* (which means an identical copy of the original) and *Split* which divides the DocFields into separate documents. ShareLocked means that the lock mode is shared between multiple tasks, while ExclusiveLock mode can be held by only one task. Finally AppendLock mode means that the task can attach (or append) data into the document, but not change any existing data. Multiple access is allowed in this mode. DocCopyState can be used to trace and monitor documents when several copies are distributed in the workflow system. The two types of states are related; however, in a manual workflow system or in an inter-organizational system, the DocCopyState is needed in addition to the DocLockState to keep track of how many copies of a physical document are in circulation.

Fig. 5(b) shows the state diagram of DocCopyState. If a document has to be used by more than one task (say, in a manual system), copies must be made, thus the document enters the Duplicate state. Once a task is done, its duplicate copy must be destroyed to avoid inconsistency. A document can also be split, for example, if an order is partially fulfilled, then the order can be divided into two parts: the fulfilled part and the back order part. The former can be shipped immediately and the latter will still remain in process. In this case, the split documents require no merge. However, there are other situations in which merge may be necessary. For example, if the customer asks for all items to be sent in one shipment, then split documents (corresponding to individual item orders) should be merged.

DocLockState and DocCopyState together play a key role in determining the control flow. A parallel split is only feasible when the document supports certain state combinations given in Table 2. For example, row 1 of this table shows that if the DocCopyState is Duplicate and the DocLockState is ShareLocked, then

Table 2. State combinations that support parallel split

DocCopyState	DocLockState
Duplicate	SharedLocked
Duplicate	AppendLocked
Split	SharedLocked
Split	AppendLocked
Split	ExclusiveLocked

it is possible to access the document simultaneously in parallel. However, if the DocCopyState is Duplicate and the DocLockState is Exclusive then sharing is not possible. Hence, there is no entry for this combination. On the other hand, when the DocCopyState is Split, then all three lock states are permissible.

The impact of application data changes on documents can be very complex and has not been fully studied. From a systems perspective, the application data is dynamic and subject to change over time. Any time some application data changes, all associated tasks may be triggered. For example, a customer may change his shipping address after he submits the order. Then, the order form may be changed depending on the order status and the seller's business policy (i.e. the order form will not be changed if the order has already been shipped; otherwise, it can be changed).

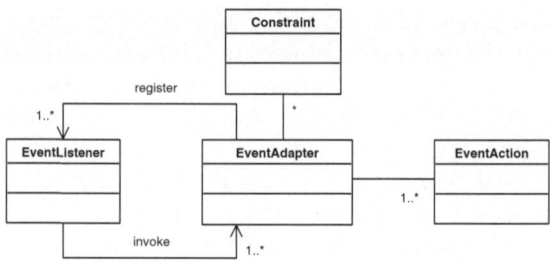

Fig. 6. EventAdapter and Constraint meta-model

Next, we turn to the *EventAdapter* and *Constraint* meta-model shown in Fig. 6. All the workflow entities (e.g. process, task, resource, document) in a document-driven workflow system communicate with each other through events. Fig. 6 shows that the EventAdaptor registers itself with EventListener and receives specified events. If an event arrives and all the constraints are satisfied, then an EventAdaptor performs one or more EventActions. An EventAction is a set of SQL statements. A constraint is an SQL statement that returns a Boolean value. For example, we may have a constraint that says that an order can only be changed when it is open. The constraint for order #99 can be written in a Transact-SQL [14] statement as:

```
Exists(Select * from Orders Where ID= 99 and State='open')
```

If this constraint returns FALSE, the EventAdaptor will ignore the order change event, and the corresponding action will not be executed. In general, this constraint language is powerful because any kind of constraint that can be expressed in SQL can be handled by this system. In the next section, we turn to implement this framework.

4 Implementation

We implemented the document-driven workflow system using Transact-SQL on Microsoft SQL Server 2000. We use triggers to enact the workflow system. The framework presented in Fig. 3 is mapped into a RDBMS using the architecture described in Fig. 7. It shows that when a database table is changed (through an insert, update, or delete operation), a corresponding trigger is fired. This trigger generates appropriate events and puts them into the event queue table. Then the trigger associated with the event queue table sends new event messages to event listeners and the listeners execute all the event adapters that registered for these events. Finally, the event adapters update the associated tables and start the next iteration. The architecture shown in Fig. 7 consists of two loops: *workflow layer loop* and *application layer loop*. The workflow layer loop updates the workflow tables (through the workflow event adapter) and the application layer loop updates the application table (through the application event adapter). There are two types of triggers shown in Fig. 7, the system triggers (i.e. workflow and event triggers) and application triggers. Both use the same underlying technology. However, it is the user's responsibility to supply application triggers, event adapters and tables for the application layer.

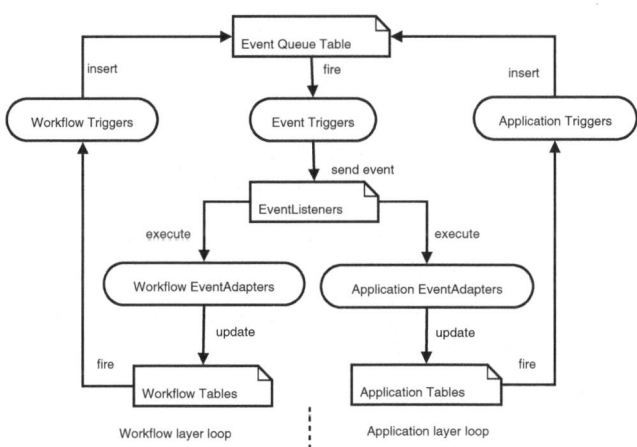

Fig. 7. Workflow system execution architecture in RDBMS

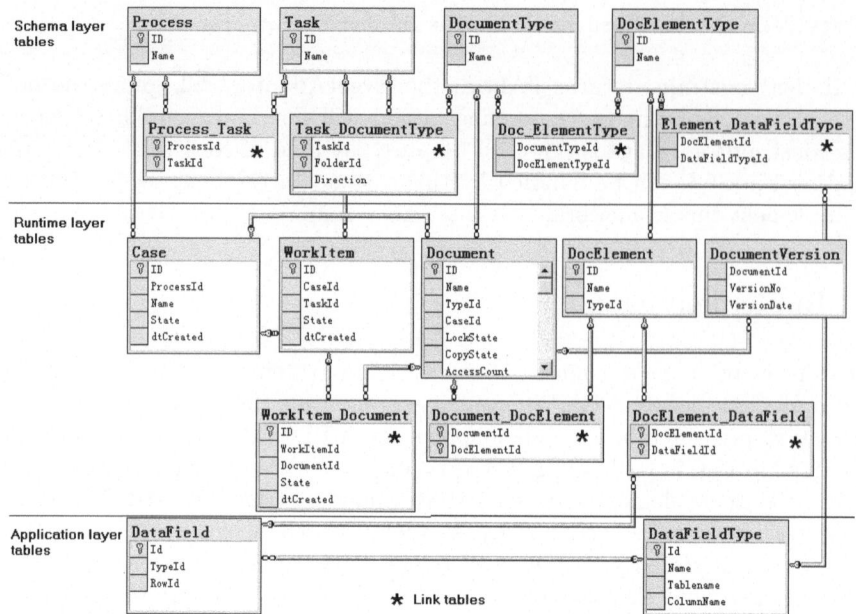

Fig. 8. Workflow system entity schema

Fig. 8 shows the tables for entities presented in the document meta-model arranged by the layers of the framework (the scheduling layer is not shown because it was not implemented). The entities in the schema layer are *Process, Tasks, DocumentType* and *associated tables*. The entities in the runtime layer are: *Case, Workitem, Document* and *associated tables*. There are no entities at the scheduling level. Finally, the entities at the application layer are: *DataField* and *DataFieldType*. DataField also serves as a link between the workflow system and the application data by mapping a document data field into a cell in the application data tables. This is indicated by the *TableName* and *ColumnName* attributes in the DataFieldType table, and the *RowId* in the DataField table. Therefore, we can link application data back to the workflow system by looking up these two tables. This is also how a clean separation between the workflow system and the application is achieved.

Fig. 9 shows the mechanism that drives the workflow system and belongs to the runtime layer. It shows the implementation of the architecture in Fig. 7. The main entities are: *Event, EventAdapter* and *EventListener*. The EventAction entity in the meta-model is implemented as an attribute (called *CommandText*) in the EventAdapter table.

Fig. 10 shows the trigger used to fire events when new documents arrive. The trigger is fired when a new record is inserted into table *Document* by the workflow system. It first retrieves the new document context into the *@docId* variable using the select query in line 10. Then it retrieves the corresponding

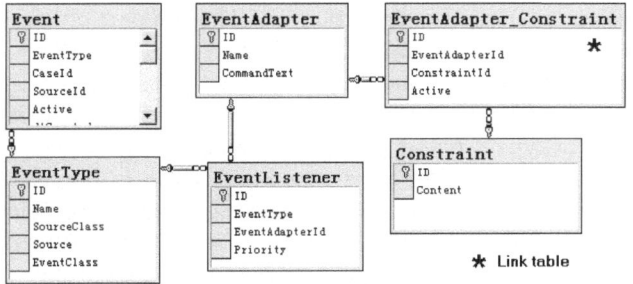

Fig. 9. Workflow system event schema

```
1   CREATE TRIGGER [onNewDocument] ON dbo.Document
2   FOR INSERT
3   AS
4   BEGIN
5       DECLARE @typeId int
6       DECLARE @docId int
7       DECLARE @caseId int
8       DECLARE @eventType int
9       /* retrieve document context */
10      SELECT @docId=[ID],@typeId=TypeId,@caseId=CaseId FROM inserted
11      /* lookup associated event type*/
12      SELECT @eventType=[ID] FROM dbo.EventType
13          WHERE SourceClass='Document' AND Source=@typeId AND EventClass='New'
14      /* generate event */
15      INSERT dbo.Event (EventType,CaseId,SourceId,dtCreated)
16          VALUES (@eventType,@caseId,@docId,getdate())
17  END
```

Fig. 10. System trigger for new document arrival

event into the variable *@eventType* using the select query in lines 12-13. Then it generates a new event and inserts it into the event queue (i.e. the Events table) in lines 15-16.

Fig. 11 demonstrates another system trigger used to activate event adapters when events arrive. A cursor is declared to retrieve all the event adapters registered to listen to this event in lines 11-16. Then a loop is used to execute each entry in the cursor in lines 18-27. Each iteration through the loop will retrieve the event action stored in the variable @commandText, and attach the case Id to the event action as a part of the SQL statement stored in the new string @commandText (line 22). Then this SQL statement is executed in line 24. It should be noted that no application data or tables are directly touched in the above system triggers.

The workflow designer has to supply a *process definition file* to load the process into the workflow system. The process is defined in an XML file that includes three sections: *interfaces*, *documents* and *tasks*. The interfaces section describes the events and event adapters for the process repository. The documents and tasks sections define all the documents and tasks related to the process. Fig. 12

shows parts of an order process definition file. The top part of this file describes the interfaces section. There can be multiple interfaces in this section. Each interface consists of an event element and the associated eventAdapter element. For example, the definition shows that the *order Arrives* event is an external event and the event type is *Order.* The corresponding event adapter has an action called *dbo.wfEAOrderArrives.* This event adapter looks for an external event of type order in listen mode (as opposed to send mode). This interface defines the event that triggers the process repository to instantiate the process into a case. The actual code of instantiating the process is implemented in the adapter action (e.g. dbo.wfEAOrderArrives). The constraints associated with each adapter can also be defined as child elements of the eventAdapter element. Each constraint has a name attribute and a *constraintText* attribute as shown in line 12.

```
1    CREATE  TRIGGER [onNewEvent] ON dbo.Event
2    FOR INSERT
3    AS
4    BEGIN
5        DECLARE @eventType int
6        DECLARE @eventId int
7        DECLARE @commandText varchar(3000)
8        /* retrieve event context */
9        SELECT @eventId=ID, @eventType=eventType FROM inserted
10       /* retrieve all adapters listen to the event */
11       DECLARE cur_adapters CURSOR LOCAL For
12       SELECT dbo.EventAdapter.CommandText
13           FROM dbo.EventListener INNER JOIN
14               dbo.EventAdapter ON dbo.EventListener.EventAdapterId = dbo.EventAdapter.ID
15           WHERE (dbo.EventListener.EventType = @eventType)
16           ORDER BY dbo.EventListener.Priority
17       OPEN cur_adapters
18       FETCH NEXT FROM cur_adapters INTO @commandText
19       WHILE @@fetch_status=0
20       BEGIN
21           /* get SQL statement and attach input parameter */
22           SET @commandText=@commandText+ ' ' + cast(@eventId as varchar(50))
23           /* run SQL (event action)*/
24           EXEC (@commandText)
25           /* next adapter */
26           FETCH NEXT FROM cur_adapters INTO @commandText
27       END
28       CLOSE cur_adapters
29       DEALLOCATE cur_adapters
30   END
```

Fig. 11. System trigger for new event arrival

The next section of the file describes an Order Form document. Two events are associated with this document. When a new instance of this document is created, it will generate the *New Document* event. When an existing instance of this document is updated, it will generate the *Document Updated* event. A document may also have other events such as *Split* and *Duplicate* as described in the document meta-model.

The last section of the file describes a *Receive Order* task. A task has at least three definition subsections: *inputDocument, outputDocument,* and *eventAdapter.* The inputDocument subsection specifies the input documents for a

```
 1  <?xml version="1.0" encoding="utf-8" ?>
 2  <process name="Express Order Process">
 3    <interfaces>
 4      <interface>
 5        <events>
 6          <event name="Order Arrives" sourceClass="External" source="Order"/>
 7        </events>
 8        <eventAdapters>
 9          <eventAdapter sourceClass="External" sourceId="Order" mode="listen"
10            action="dbo.wfEAOrderArrives">
11            <constraints>
12            <constraint name="IsExpressOrder" constraintText="dbo.isExpressOrder"
13            </constraints>
14          </eventAdapter>
15        </eventAdapters>
16      </interface>
17        ...
18    </interfaces>
19    <documents>
20      <document name="Order Form">
21        <events>
22          <event name="Document New" eventClass="New" />
23          <event name="Document Updated" eventClass="Update" />
24        </events>
25        <docElement>
26          <dataField name="Cusomer ID" tableName="Customers" columnName="ID"/>
27            ...
28        </docElement>
29          ...
30      </document>
31        ...
32    </documents>
33    <tasks>
34      <task name="Receive Order">
35        <inputDocument>
36          <item document="Order Form"/>
37        </inputDocument>
38        <outputDocument>
39          <item document="Payment"/>
40          <item document="Pickup List"/>
41          <item document="Shipping Advice"/>
42        </outputDocument>
43        <eventAdapters>
44          <eventAdapter sourceType="Document" sourceId="Order" mode="listen"
45            action="dbo.wfEANewOrderDoc"/>
46        </eventAdapters>
47      </task>
48        ...
49    </tasks>
50  </process>
```

Fig. 12. Sample process description file

task (i.e. the Receive Order task requires an Order Form document as its input document). In general, a task must have at least one input document. The *out putDocument* subsection defines the output of the task. There are three output documents: *Payment*, *Pickup List*, and *Shipping Advice*. The last subsection is eventAdapters, which describes all the eventAdapters for the task. A task has at least one eventAdapter that instantiates the task into a workitem. For example, the eventAdapter waiting for a new document event of type Order (lines 44-45) creates an instance for the task (e.g. workitem) when it receives such an event. A task may have *entryConstraints* and *exitConstraints* which are not shown in Fig. 12. An example of an entryConstraint could be some business rules such as: if the shipping option is UPS ground, then the Receive Order will wait for one day to instantiate.

In this section we have demonstrated that the framework proposed in Section 3 can be implemented entirely inside a database system using SQL. The general methodology consists of the workflow designer creating a process definition file such as the one in Fig. 12 (using a text editor), and loading it into the workflow system. Then the application developer supplies the application triggers and event adapters to operate on the application data. After that the system triggers enable the execution of the process, and react to events from the application.

5   Discussion and Related Work

It is evident that researchers are realizing the importance of integrating data flow (document) into workflow modeling as indicated by some recent work on document-centric workflows [6,7,11,12]. These studies present useful concepts for modeling aspects of documents; however, the role of document (and it's more general concept, resource) and the dependencies between documents are not addressed. Therefore, it is unclear how documents can be integrated fully into workflow systems in the design stage.

Flexibility is one of the most important issues in a workflow management system. Different approaches such as structured processes [10], workflow patterns [1], and Petri-Nets [2,3] offer varying degrees of flexibility. All these approaches are based on control flow, and try to achieve better flexibility by using complex flow structures built upon split, join, loop, and wait-for constructs [3]. However, they cannot predict the upper boundary of the flexibility. In the document-driven architecture, this upper boundary is obviously the dependency between documents. For example, a customer cannot start to eat unless the food is produced, but whether the customer should pay before or after eating may vary from one restaurant to another. Therefore, the food dependency is a hard dependency, and the payment policy is a soft one. The document-driven design can easily discover the hard dependencies and provides the upper boundary of flexibility.

This flexibility makes document-driven workflows especially suitable for *ad hoc workflow* processes as opposed to *production workflows*. Voorhoeve and Aalst [18] define an ad-hoc workflow as an intermediate between well-structured, high volume production workflow and less-structured cooperative groupware systems. They also note that traditional workflow management systems (mostly based on control flow) could be error-prone when the processes require frequent changes. As an example, document-driven workflows may work well for most maintenance processes which have a few tasks and no strict control flow. There are two issues here. On the one hand, verification of document driven workflow is easier than

Table 3. Comparison between document-driven workflow and control flow workflow

Document-Driven Workflow	Control Flow Workflow
Process is driven by the documents.	Process is driven by the control flow.
The process is very flexible and can be changed instantly by changing constraints.	The process is less flexible because of the limitation of flow patterns. It is difficult to change a control flow of an instance because all the instances share the same control flow pattern.
There are no fork/join design issues since there is no control flow.	Fork/join elements are used to describe control flow. The control flow may not be feasible because resource dependencies are ignored.
Application Data is separated from the process.	In most case, the application data are attached to the control flow.
Suited for ad hoc workflows.	Good for production workflows with mature processes and a large number of tasks.
Verification is relatively easy.	Verification could be hard.
Need conflict resolution in complex workflows.	No need for conflict resolution.
Difficult to visualize the process.	Process can be visualized easily.

for a control driven workflow. However, document driven workflows can lead to multiple triggers being enabled simultaneously when a document is created or updated. In this situation, it is necessary to use priorities with triggers in order to choose between enabled triggers. Table 3 shows a detailed comparison between document-driven and control flow methodologies.

A drawback of document-driven workflows is that they lack visualization of processes since there is no explicit control flow. The lack of visualization makes document-driven workflows unsuitable for modeling processes with a large number of tasks. Besides, a control flow diagram is still needed to get the big picture when designing the process.

Another advantage of our approach is that it produces a clean separation of application data from workflow processes. The *DataField* and *DataFieldType* tables with associated events and eventAdpaters act as a middle layer between the workflow system and the application. Any changes on each side only require the middle layer to be changed and will not affect the other side. This strict separation of application data from the process can facilitate the implementation of inter-organizational processes because no control information needs to be exchanged between organizations.

The use of triggers in workflow system has been discussed in the WIDE project [8]. Triggers are used to capture events and handle exceptions in addition to the normal workflow which is designed as a control flow. However, our study takes this approach one step further by the use of database triggers as mechanisms to drive and enact the workflow system, and removing the need for a workflow "engine". As a result, the workflow system can be implemented entirely inside the database. The concept of constraints has been studied in most workflow systems and they are usually represented as ECA (Event-Condition-Action) rules [13]. Finally, a State Entity Activity Model (SEAM) was developed by Bajaj and Ram [5] to model workflows at the conceptual level. While SEAM provides a direct mapping between the SEAM models and RDBMS, it is unclear whether SEAM could support complex workflows such as the ones in [1].

6   Conclusion and Future Research

We proposed a framework for document-driven workflow systems. In addition we implemented this framework to demonstrate that it is feasible to build a workflow system entirely inside a RDBMS. The most important difference between this approach and conventional workflows is that, in document-driven workflow systems there is no predefined control flow. All the tasks are executed based on the availability of their input documents and associated resources. Therefore, the extra work of checking the correctness of control flow can be reduced. In addition, the framework provides a simple way to find deadlock and dangling tasks through the input/output documents analysis.

The prototype workflow system was implemented entirely in a RDBMS using Transact-SQL. The enactment depends on various events fired by database triggers. The RDBMS based workflow system can be embedded into any database application easily.

The document-driven approach may not work well when a large number of data changes occur concurrently. This may lead to concurrency and conflict resolution issues. Moreover, as mentioned above, lack of visualization is also a drawback. While this approach may not be appropriate in all environments, we feel that for the most part it is especially suitable for ad hoc workflows. The concept of a document-driven model can be extended into a more comprehensive resource-based model by viewing documents as a type of resource along with other resources such as people, machines, facilities and equipment. Information about all these resources can be kept in a database, and tasks would be enabled only when all the resources are available. Thus, the same approach presented here can be easily extended to encompass multiple types of resources.

References

1. Aalst, W.M.P. van der: Workflow Patterns. http://is.tm.tue.nl/research/patterns/.
2. Aalst,W.M.P. van der and Hee, K.van: Workflow Management: Models, Methods, and Systems. The MIT Press, January,2002.
3. Aalst, W.M.P. van der, and Kumar, A.: XML-Based Schema Definition for Support of Interorganizational Workflow. Information System Research, Vol. 14, No. 1, March 2003, 23-46.
4. Bae, H., et al.: Document configuration control process captured in a workflow. Computers in Industry, No. 53, 2004, 117-131.
5. Bajaj, A. and Ram, S.: SEAM: A state-entity-activity-model for a well-defined workflow development methodology. Knowledge and Data Engineering, IEEE Transactions on Volume 14, Issue 2, March-April 2002 Page(s):415 - 431.

6. Botha, R.A., Eloff, J.H.P., 2001. Access control in document-centric workflow systems–an agent-based approach. Computers and Security 20 (6), 525-532.

7. Paul Dourish, W. Keith Edwards, Anthony LaMarca, et al., Extending document management systems with user-specific active properties. ACM Trans. Inf. Syst. 18(2): 140-170 (2000).

8. Grefen, P., et al.: Database Support for Workflow management - The WIDE Project. Kluwer Academic Publishers, 1999.

9. Hollingsworth, D.: The Workflow Reference Model 10 Years On. http://www.wfmc.org/standards/model.htm.

10. Kiepuszewski, B., Hofstede, A.H.M, and Bussler, C.: On Structured Workflow Modeling. In Proceedings CAiSE'2000, LNCS Vol. **1797**, Springer Verlag.

11. Krishnan, R., Munaga, L., and Karlapalem, K., 2002, "XDoC-WFMS: A Framework for Document Centric Workflow Management System," Lecture Notes on Computer Science, 2465, pp. 348-362.

12. Mazumdar, S. and AbuSafiya, M., 2004. A Document-Centric Approach to Business Process Management. In Proc. Intl. Conf. on Information and Knowledge Engineering, pages 461-466.

13. McCarthy, D.R. and Dayal, U.: The Architecture of an Active Database System. in Proc. ACM SIGMOD Conf. on Management of Data, Portland, 1989, pp. 215-224.

14. Micorsoft Corporation: SQL Server Books Online:Transact-SQL Reference. 2000.

15. Muehlen, M. zur: Resource Modeling in Workflow Applications. http://www.workflow-research.de/Publications/PDF/MIZU-WF99.PDF.

16. OASIS: Web Services Business Process Execution Language (WSBPEL). http://www.oasis-open.org.,fig:Architecture

17. Scheer, A.W.: ARIS - Business Process Frameworks. 2ed, Springer, 1998.

18. Voorhoeve, M. and van der Aalst, W.: Ad-hoc workflow: Problems and solutions. International Conference on Database and Expert Systems Applications - DEXA, 1997, p 36-40

19. WFMC: XML Processing Description Language (XPDL). http://www.wfmc.org/standards/XPDL.htm.

20. Zlotkin, G.: Organizing Business Knowledge - The MIT Process Hand-book. Edited by Malone, T.W., et al, The MIT Press, 2003, pp. 20.

Service Interaction Patterns

Alistair Barros[1], Marlon Dumas[2], and Arthur H.M. ter Hofstede[2]

[1] SAP Research Centre, Brisbane, Australia
alistair.barros@sap.com
[2] Queensland University of Technology, Australia
{m.dumas, a.terhofstede}@qut.edu.au

Abstract. With increased sophistication and standardization of modeling languages and execution platforms supporting business process management (BPM) across traditional boundaries, has come the need for consolidated insights into their exploitation from a business perspective. Key technology developments in BPM bear this out, with several web services-related initiatives investing significant effort in the collection of compelling use cases to heighten the exploitation of BPM in multi-party collaborative environments. In this setting, we present a collection of patterns of service interactions which allow emerging web services functionality, especially that pertaining to choreography and orchestration, to be benchmarked against abstracted forms of representative scenarios. Beyond bilateral interactions, these patterns cover multilateral, competing, atomic and causally related interactions. Issues related to the implementation of these patterns using established and emerging web services standards, most notably BPEL, are discussed.

1 Introduction

Process modeling languages have emerged as a key instrument for achieving integration of business applications both within and across organizations in a service-oriented architecture (SOA) setting. This trend is reflected in a number of standardization initiatives such as the set of WS-* Specifications [11], OMG's Enterprise Collaboration Architecture[1] and RosettaNet[2], all of which position processes at the highest level of abstraction. Process modeling languages provide an abstract means of specifying complex sequences of execution steps, leaving lower layers to deal with details like software interfacing, quality of messaging and transport protocol binding. From the SOA prism, process steps result in interactions with (web) services that encapsulate the business logic associated to the step. Processes that rely on services to realize process steps can themselves be deployed as services, a practice known as *process-based service composition*.

Through different insights from various initiatives over the last few years, different aspects of process-based service composition have evolved. In partic-

---

[1] http://www.omg.org/technology/documents/formal/edoc.htm
[2] http://www.rosettanet.org

W.M.P. van der Aalst et al. (Eds.): BPM 2005, LNCS 3649, pp. 302–318, 2005.

ular, the developments of the Business Process Execution Language (BPEL)[3] and W3C's Web Services Choreography Definition Language (WS-CDL)[4], have been accompanied by requirements and use cases gathering. However these have largely steered towards technical concepts and implementation concerns, with documented use cases and examples reflecting little more than simple processes involving basic "buyer-supplier-shipper" interactions.

For service composition technology to progress further, more requirements gathering is needed to shed light into the nature of *service interactions* in collaborative business processes. In particular, it must be considered that there is often a large number of parties in such collaborative processes and thus the nature of interactions may be multilateral rather than bilateral. Furthermore, the assumption of strict synchronization of all canvassed responses breaks down due to the independence of the parties. More realistically, responses are accepted as they arrive or a minimum number is required for an interaction to be successful. Another crucial feature is that not all service providers have comparative advantage and collaborate. Not untypically, they compete. Hence, canvassed requests to competing service providers may require exclusivity – e.g the first response is accepted and the rest ignored. Finally, not all interactions follow a requestor-respondent-requestor structure. Instead, a sender may redirect interactions to nominated delegates and services may outsource requests choosing to "stay in the loop" and partially observe follow-ups. More generally, it may only be possibly to determine the order of interactions at runtime given the message contents.

This paper aims at contributing to this requirements gathering activity by proposing a set of *service interaction patterns*. Patterns have proved invaluable in the reuse of requirements, design and programming knowledge. They were traditionally the province of software design, but have recently emerged in the BPM field [1]. The collected service interaction patterns apply primarily to the service composition layer (orchestration, and choreography) but also to lower layers (e.g. message typing and addressing). They have been derived and extrapolated from insights into real-scale B2B transaction processing, use cases gathered by standardization committees (e.g. BPEL and WS-CDL), generic scenarios identified in industry standards (e.g. RosettaNet Partner Interface Protocols), and case studies reported in the literature. It is not claimed that the proposed set of patterns is complete: the aim is rather to consolidate recurrent scenarios and abstract them in a way that provides reusable knowledge. Furthermore, the patterns allow the assessment of emerging web services standards. Specifically, we use the patterns to analyze the scope and capabilities of BPEL and to some extent of related specifications such as WSDL and WS-Addressing (WS-A) [11].

The proposed patterns are classified according to the following dimensions:

- The maximum number of parties involved in an exchange, which may be either two (*bilateral interactions*, covering both *one-way* and *two-way* interactions) or unbounded (*multilateral interactions*).

---

[3] http://www.oasis-open.org/committees/tc_home.php?wg_abbrev=wsbpel. In this paper, we use the acronym BPEL to refer to WS-BPEL version 2.0.

[4] http://www.w3.org/TR/ws-cdl-10

- The maximum number of exchanges between two parties involved in a given
  interaction, which may be either two (in which case we use the term *single-
  transmission interactions*) or unbounded (*multi-transmission interactions*).
- In the case of two-way interactions (or aggregations thereof) whether the receiver of the "response" is necessarily the same as the sender of the "request"
  (*round-trip interactions*) or not (*routed interactions*).

Based on these dimensions, we identify four groups of patterns. The first one
encompasses single-transmission bilateral interaction patterns. These correspond
to elementary interactions where a party sends (receives) a message, and as a
result expects a reply (sends a reply). This group covers one-way and round-trip
bilateral interactions but not routed interactions which are covered in a separate
group. The second group of patterns stays in the scope of single-transmission
non-routed patterns, but deals with multilateral interactions. In this case, a party
may send or receive multiple messages but as part of different interaction threads
dedicated to different parties. The third group is dedicated to multi-transmission
(non-routed) interactions, where a party sends (receives) more than one message
to (from) the same party. The final group is dedicated to routed interactions.

The proposed patterns may be composed through operators expressing flow
dependencies such as sequence, choice, and synchronization. In this paper however, we do not deal with patterns composition. Also, it is not in the scope of
the proposed patterns to capture internal steps performed by a service that do
not directly contribute to nor directly result from interactions. Also, we abstract
from data representation and manipulation issues as these deserve a separate
elaboration. For the same reason, the patterns do not cover security issues.

The structure of the paper follows the groups of patterns outlined above. For
space reasons, we omit the first group which comprises three well-known patterns
(send, receive and send/receive) as detailed in [3]. Thus the next section starts
directly with Pattern 4.

2   Single-Transmission Multilateral Interaction Patterns

Pattern 4: *Racing incoming messages.*

Description. A party expects to receive one among a set of messages. These
messages may be structurally different (i.e. different types) and may come from
different categories of partners. The way a message is processed depends on its
type and/or the category of partner from which it comes.

Example. A manufacturing process involves remote subcontractors and uses
a pull-strategy to streamline its operations. Each step in the manufacturing
process is undertaken by a subcontractor. A subcontractor signals intention to
execute a step when it becomes available through a request. At the same time,
progress is monitored by a quality assurance service. The service randomly issues
quality check requests in addition to the pre-established quality checkpoints in
the process. When a quality check request arrives, it is processed in full before
processing any new quality check request or subcontractor intention. Similarly,

when a subcontractor intention arrives, it is processed in full before processing any other check request or subcontractor intention. Thus, there are points in the process where quality checks and subcontractor intentions compete.

Issues/design choices.

- The incoming messages may be of different types.
- The processing that follows the message consumption (which we term the *continuation*) may be different depending on the consumed message.
- When one of the expected messages is received, the corresponding continuation is triggered. The remaining messages may or may not need to be discarded.
- Depending on the underlying communication infrastructure, several of the expected messages may be simultaneously available for consumption. In this case, two approaches may be adopted: (i) let the system make a non-deterministic choice, or (ii) provide a "ranking" among the competing messages. In any case, only one message is chosen for consumption.

Solution. This pattern is directly captured by the pick activity in BPEL. The pick activity simultaneously enables the consumption of several types of message events and allows at most one message event to be consumed. Specifically, a pick activity is composed of multiple branches, each of which has a corresponding handler which acts as the trigger of the branch. Occurrences of message events are consumed by onMessage handlers. An onMessage handler is associated with a type of message, identified by a partner link and a WSDL operation. When a message of the type associated to an onMessage handler is available for consumption, a message event may occur which is immediately consumed by the handler. The pick enforces that at most one of its associated onMessage handlers will consume an event. It is also possible to associate a timer with a branch of a pick activity through an *onAlarm handler*. The corresponding branch is taken if the timeout event occurs before any of the other branches is taken.

In the current version of BPEL, it is not possible to express a ranking among the competing types of message event handlers under a given pick. Although in the concrete syntax of BPEL the handlers under a pick are ordered, this order is not significant. Hence, should there be several onMessage handlers able to consume message events when the pick activity is executed, the system may choose any of them non-deterministically. What is needed to capture the fourth issue of this pattern is a way of ranking message events so that when several of them enter into a race, the one with highest ranking is chosen.

Related pattern.

- *Deferred choice* [1]. The deferred choice pattern corresponds to a point in a process where one among a set of branches needs to be taken, but the choice is not made by the process execution engine (as in a "normal choice"). Instead, several alternatives are made available to the environment and the environment chooses one of these alternatives. The Racing Messages pattern can be seen as a specialization of the deferred choice where the choice of branch is determined by the receipt of a message.

Pattern 5: *One-to-many send.*

Description. A party sends messages to several parties. The messages all have the same type (although their contents may be different).

Synonyms. Multicast, scatter [10].

Example. A purchasing service sends a call for tender to all known trading parties that provide a given type of product or service.

Issues/design choices.

- The number of parties to whom the message is sent may or may not be known at design time. In the extreme case, it may only be known just before the interaction occurs.
- As for the one-to-one send, reliable delivery may or may not be required. In the case of reliable delivery, the individual send actions may result in faults and thus fault handling routines should be associated to each of the individual send actions. The logic of these fault handlers is application-dependent: some applications may choose to terminate the whole one-to-many send when one of the individual "send actions" fail, while others may simply record the failures that occur and proceed.

Solution. A natural approach to address this pattern is to use the One-to-one Send pattern as a basic building block. Thus, a number of one-to-one send actions are scheduled in parallel or sequentially depending on the capabilities of the underlying language. For example:

- If the number of parties is known at design time, it is possible to capture this pattern in BPEL through a parallel block (i.e. a *flow* activity) such that each thread contains a one-to-one send action with its associated fault handler. Otherwise, the individual send actions would need to be scheduled sequentially (using a *while*) thus contradicting the essence of the pattern.
- In certain proprietary extensions of BPEL, such as Oracle BPEL[5], special constructs are provided to capture the situation where an arbitrary number of executions of an activity need to be performed in parallel, such that this number is only determined when these parallel executions are started (see for example the *FlowN* construct in Oracle BPEL).[6] The pattern can be captured using such a construct.
- In WSCI and BPML[7], a construct known as "spawn" is provided to start an instance of a sub-process asynchronously. By embedding the "spawn" within a "while" loop, it is possible to start a number of "send sub-processes", each of which would be responsible for sending one of the messages and dealing with any possible fault. These sub-processes would execute in parallel and return back to the parent process upon completion through a "signal". These signals can then be gathered by a dedicated activity in the parent process.

---

[5] http://www.oracle.com/technology/products/ias/bpel

[6] A similar construct (namely *parallel foreach*) has been proposed for introduction into the BPEL standard; see Issue 147 in the list of BPEL issues available from http://www.oasis-open.org/committees/tc_home.php?wg_abbrev=wsbpel.

[7] http://www.bpmi.org

This pattern requires a "dynamic binding by reference" mechanism [2] since in some cases the set of potential parties to which messages will be sent is not known at design/build time. Instead, the identity and location of the partners may be given as parameter, or retrieved from a local database, or from a remote service registry. In BPEL, this is achieved by treating *service endpoints references* (described in WS-A) as first-class citizens that can be associated with predefined partner links at runtime.

Related pattern.

– *Multiple instances with a priori runtime knowledge (MIRT)* (van der Aalst et al. 2003). In this pattern, several instances of a task are created and allowed to execute in parallel with synchronization occurring when all instances have completed. The number of task instances to be created is only known at runtime, just before the instantiation starts. The one-to-many send can be expressed by composition of the MIRT pattern and the one-to-one send pattern discussed above. The FlowN construct of Oracle BPEL (see discussion above) is a realization of the MIRT pattern.

Pattern 6: *One-from-many receive.*

Description. A party receives several logically related messages arising from autonomous events occurring at different parties. The arrival of messages must be *timely* so that they can be correlated as a single logical request. The interaction may complete successfully or not depending on the messages gathered.

Synonyms. Event aggregation [8], gather [10].

Example. A group buying service receives requests for buying different types of items. When a request for buying a given type of product is received, and if there are no other pending requests for this type of item, the service waits for other requests for the same type of item. If at least three requests have been received within five days, a "group request" is created and an order handling process is started. If on the other hand less than three requests are received within the five days timeframe, the requests are discarded and a fault notification is sent back to the corresponding requestors.

Issues/design choices:

– Since messages originate from autonomous parties, a mechanism is needed to determine which incoming messages should be grouped together (i.e. correlated). This correlation may be based on the content of the messages (e.g. product identifier).
  Correlation of messages should occur within a given timeframe. The receiver should avoid waiting indefinitely.
– The number of messages to be received may or may not be known at design time or run-time. Instead, after a certain condition is fulfilled, the received messages are processed without waiting for subsequent related messages (i.e. proceed when X amount of orders for a given product have been received).
– In some cases, a timeout occurs before any message is received.

Solution, The first issue implies that the payload of the messages received should contain a piece of information that determines with which other messages it should be grouped (i.e. in which group should it be placed). At an abstract level, this can be captured through a function *Group: Message → GroupID*, which associates a "group identifier" to a message. Messages with the same group identifier are to be correlated. When a message of the expected type is received, its group ID is inspected and one of three options may be taken: (i) a new group is created for the message if no group for that group ID exists; (ii) the message is added to an existing group; (iii) the message may be discarded because the group ID is not valid (e.g. the group existed before but it is no longer accepting new messages). The latter option entails that the recipient should maintain a list of invalid group IDs (or equivalently a set of valid ones).

Because the number of messages to be received is not necessarily known in advance, it is necessary to incorporate a notion of *stop condition*. The stop condition may be expressed as a predicate over the set of messages received. The stop condition is evaluated each time a message is received. As soon as the stop condition evaluates to true, the interaction is considered to be complete. In a tender scenario, to capture that as soon as 5 bids have been received the interaction completes and subsequent bids are ignored, the corresponding stop condition would be $|R| = 5$, where R denotes the set of messages received.

A solution to this pattern should associate timers to message groups. The timer for a group is started when the group is created. A group may be created either explicitly by the service (e.g. when the service enters a given state) or by the receipt of a message which mapps to a group ID for which no corresponding group is open. In the former case, it is possible that a timeout occurs even if no message has been received.

When a timeout occurs, depending on the set of messages gathered at that point, the interaction may be considered to have succeeded or failed. For example, a tender may be considered as successful if there are at least 3 bids and at least one of them is below a given limit price. Thus, a generic solution to the pattern also needs to incorporate a notion of *success condition* which is evaluated when the interaction completes and determines whether the interaction is considered as successful or not. Again, the success condition can be expressed as a predicate over the set of messages received. In the example at hand, the success condition would be: $|R| \geq 3 \wedge \exists r \in R : Price(r) \leq limitPrice$. Note that in theory, it may happen that the stop condition evaluates to true (and thus the interaction stops), while the success condition evaluates to false, so the interaction is considered to have failed. When a group completes successfully, the set of responses gathered for that group constitute the output of the interaction.

In the "group buying" example above, the stop and success conditions are identical ("at least three requests should be received"), the timeframe is five days, groups are created when the first message for the group arrives, and group IDs are never flagged as invalid since it is always possible to process requests for a type of product whether previous groups for this type have been filled or not.
Related pattern.

- *Multiple instances with a priori runtime knowledge* (MIRT). See discussion in the "Related patterns" paragraph of the previous pattern. Note that existing realizations of the MIRT pattern, such as the FlowN construct of Oracle BPEL (see discussion above) do not support arbitrary stop and success conditions as defined above. Instead, these conditions appear as lower and upper bounds on the number of task instances that are required to complete.

Pattern 7: *One-to-many send/receive.*
Description. A party sends a request to several other parties, which may all be identical or logically related. Responses are expected within a given timeframe. However, some responses may not arrive within the timeframe and some parties may even not respond at all. The interaction may complete successfully or not depending on the set of responses gathered.
Synonyms. Scatter-gather [10,6].
Example. An insurance company outsources some aspects of its claims validation to its external search brokers. Brokers are typically small agencies and have variable demands. For efficiency, the insurance company sends search requests to all the brokers, and accepts the first three responses to undertake the search.
Issues/design choices.

- The number of parties to which messages are sent may or may not be known at design time.
- Responses need to be correlated to their corresponding request.
- The sender should avoid waiting indefinitely or "unnecessarily" for responses.
- It is possible that no response is received.
- Reliable delivery may or may not be required during sending. In the case of reliable delivery, the individual send actions may result in faults.

Solution. A solution to this pattern can be obtained by combining patterns one-to-many send and one-from-many receive through parallel composition (e.g. "flow" construct in BPEL). Since outgoing and incoming messages need to be correlated, it is necessary to include correlation data in the outgoing messages and retrieve these data from the incoming messages. BPEL provides a declarative mechanism, namely correlation sets, for correlating communication actions (e.g. correlating an invoke action with a receive action). Unfortunately, this mechanism can not be employed if the actions to be correlated are executed in different loops located in different branches of a flow activity[8], which is the case for this pattern since an *a priori* unknown number of invoke and receive actions need to be executed in an arbitrary order. Thus the correlation between the send and the receive actions implied by this pattern needs to be handled at the application level, i.e. by introducing actions that insert and extract the correlation data into/out of the incoming/outgoing messages.

The "stop condition" and the "success condition" for the one-from-many receive may involve both the set of requests (to be) sent (say RQ) and the set of responses gathered at a certain point (say RS). For example, to capture that

---

[8] Specifically, in BPEL the invoke and the receive actions to be correlated must be enclosed under a common scope activity such that each of these actions is executed at most once per execution of the scope.

as soon as 10 responses have been received the interaction stops and subsequent responses are ignored, the stop condition can be set to: $|RS| = 10$. Meanwhile, to ensure that at least 50% of the parties need to respond the success predicate should be set to: $|RS| = 0.5 \times |RQ|$.

In the absence of a "stop condition" (i.e. if the stop condition is always true) the pattern can be expressed by combining several elementary send and receive actions through parallel composition which may be preempted by a timeout. As discussed in the previous pattern, this would mean that the underlying language provides a mechanism for executing an a priori unknown number of activities in parallel, such as for example the "FlowN" construct in Oracle BPEL or the "spawn" construct in BPML. Such a mechanism is not present in standard BPEL and a workaround solution where the various one-to-one send/receive would be executed sequentially does not properly address the pattern.

In the case of reliable delivery, fault handling routines (BPEL fault handlers) may be attached either to each individual send actions or to the whole set of send actions. A possible fault handling routine is to record that the message in question was not delivered so that this information can be used in the stop and success conditions. This way, it is possible to express conditions such as "stop as soon as half of the parties who actually received a request have responded".

Related pattern.

- *Scatter-gather* [6]. The scatter-gather pattern is a special case of the one-to-many send/receive. The scatter-gather assumes that all parties respond in a timely manner and that all responses must be gathered. Thus it does not address issues related to timeout, stop and success conditions.
- *One-from-many receive/send*. This is the dual of the One-to-many send/receive. Its description, issues, design choices, and solution are analogue to those of the One-to-many send/receive.

3   Multi-transmission Interaction Patterns

Pattern 8: *Multi-responses.*
Description. A party X sends a request to another party Y. Subsequently, X receives any number of responses from Y until no further responses are required. The trigger of no further responses can arise from a temporal condition or message content, and can arise from either X or Y's side. Responses are no longer expected from Y after one or a combination of the following events: (i) X sends a notification to stop; (ii) a relative or absolute deadline indicated by X; (iii) an interval of inactivity during which X does not receive any response from Y; (iv) a message from Y indicating to X that no further responses will follow. From this point on, no further messages from Y will be accepted by X.
Synonyms. Streamed responses, message stream
Example. A goods deliverer provides an urgent transportation service on behalf of suppliers to customers in a city. For optimization of travel, it subscribes to a local traffic reporting service provides its destination nodes (goods dispatch

and customer locations) and obtains regular feeds on traffic bottlenecks, until it indicates that no feeds are required.

Issues/design choices.

- Party X should be capable of receiving multiple messages from party Y including ones that arrive simultaneously. The number of responses accepted will depend on a condition to be evaluated at runtime.
- As with Pattern 4, the messages may be of different types. The way each message is processed depends on its type.
- As with the One-from-many Receive pattern, a stop condition is pertinent. However, unlike the One-from-many Receive, a success condition does not apply since faults messages received by X are treated individually just as "normal" messages. It is assumed that X and Y establish an *a priori* understanding of the stop condition.
- In the case where X determines when the multi-transmission should stop, there is an interval between the moment when X decides to stop and the moment when Y becomes aware of this decision. During this interval, Y may send messages that will then be rejected by X. Hence, a mechanism should be in place for Y to know that its messages have been rejected.

Solution. As for Pattern 4, the core of this pattern can be captured in BPEL through a pick activity with a onMessage handler per type of message (whether a normal message or a fault message). To capture the fact that several messages may be accepted, the pick activity must be embedded within a "while" activity. The encoding of the stop condition depends on its nature:

- If the stop condition is based on data available at the receiver's side and/or messages' content, the stop condition can be encoded as the exit condition of the while loop (like in the One-from-many receive pattern).
- If the stop condition is an absolute or a relative deadline (with respect to the beginning of the interaction), the while activity must itself be embedded in a scope activity containing an onAlarm handler corresponding to the deadline.
- If the stop condition corresponds to a period of inactivity between responses, it can be captured as a branch in the pick activity associated with an on-Alarm handler capturing the maximum duration of inactivity. If this branch is taken, the while loop is interrupted (e.g. by setting an appropriate flag).
- If the stop condition is determined by the Y, a pre-agreed type of message will signal the end of the interaction to X and thus the stop condition will be encoded as an onMessage handler corresponding to this type of message.

In the case where the stop condition is determined by X, or in the case where it is determined by Y but the underlying messaging infrastructure or interaction policies do not guarantee ordered delivery of messages, X should be able to return fault messages to Y for responses that are ignored. In BPEL, this can be done by activating a thread of control after bespoke while/scope activity, which upon receiving any of the expected types of messages from Y, returns a fault message. This additional coding is necessary because in BPEL, while it is possible to state that a process is expecting a type of message from a given party, it is not possible to express that a process expects not to receive a given type of message and that such messages should be discarded and a fault returned to their sender.

Pattern 9: *Contingent requests.*
Description. A party X makes a request to another party Y. If X does not receive a response within a certain timeframe, X sends a request to another party Z, and so on.
Synonyms. Send with failovers.
Example. A travel agency allows contingent reservations of flights in particular situations - urgent requests and busy flight paths. Customers nominate the preference of flight carriers. In order of preference, reservations are sought in short-timeframes. If a reservation is secured, the interaction ends.
Issues/design choices.

- There is a race between receiving a response and a timer.
- After a contingency request has been issued, it may be possible that a response arrives (late) from a previous request. This means that more than one response may arrive; in all, as many responses may potentially arrive as requests have been sent. The question is when to accept a response if more than one request has been made and more than one response arrives.

Solution. The first issue is generally well-understood and in fact BPEL provides direct support for it through the pick construct containing onMessage and onAlarm handlers. For the second issue, several choices are available. One is to accept the first response even if it is late and stop outstanding requests. Another is to accept the first arriving response, trigger the end of outstanding requests, but receive any further responses that arrive (before the "contingent send" process terminates). Yet another possibility is to disallow late arrivals altogether, and receive only the response of the current request. For these choices, the pattern does not pre-dispose which prevails. In some situations accepting late responses is desirable, while in others it may cause problems of integrity in remote parties particularly if requests are non-idempotent (involving database updates and extending interactions even further with other parties).

Pattern 10: *Atomic multicast notification.*
Description. A party sends notifications to several parties such that a certain number of parties are required to accept the notification within a certain timeframe. For example, all parties or just one party are required to accept the notification. In general, the constraint for successful notification applies over a range between a minimum and maximum number.
Synonyms. Transactional notification
Examples.

- *Classical "all-or-none" atomicity.* A business venture service[9] supports the process of business license applications for small business endeavors (e.g.

---

[9] This example reflects the Queensland Government's SmartLicence initiative (http://www.sd.qld.gov.au/dsdweb/htdocs/slol/)

opening a restaurant). After the steps of obtaining and verifying application details, relevant agencies involved in the approval or registration of the application are notified. All of them must receive notification as there are inter-dependent aspects of the application leading to cross-consultation. There may also be competing applications for the same business. Therefore, all agencies should receive the notification in a timely fashion. In this example, the minimum and maximum equal the number of all agencies notified.

- *Exclusive choice*. A legal firm has automated its property conveyance process for various loan types. The process utilizes a number of search brokers who have the same level of service agreements with the firm. Each of the brokers competes for conveyance applications. Therefore, only one of the notified brokers is selected, namely the first to accept the request. The minimum and maximum both are one.

Issues/design choices.

- The set of parties to which the notification will be sent may not be known at design time nor a priori at run-time.
- Specification for the minimum and maximum bounds should be supported.
- The constraint that all parties should have received the notification, means that if any one party received the notification, all the other parties also received it. Thus, some kind of transactional support is required for this aspect of the interaction.
- Following from the above point, two steps in the interaction can be seen, both of which need to be formalized. The first send-receive establishes the intention to accept a request while the second acts of the decision following an examination of received intentions - parties are notified about whether they have been selected or not.
- The maximum number of parties required to accept the notification may be less than the number of parties that notifications were sent to. Thus, more responses than the maximum allowed may be willing to accept the notification and a *preference function* may be needed to prune some of them.

Solution. The central issue of this pattern (third issue above) clearly relates to transactional atomicity. At present, BPEL does not provide support for transactional atomicity. However, it does provide support for a related notion, known as quasi-atomicity [5] through the notion of *compensation handler*. Quasi-atomicity refers to the ability to "undo" certain parts of a process execution. Using this mechanism, the receiving parties, when they receive the initial request, may actually perform the work associated with this request. Later on during the second round, if the sender decides not to proceed with the request to a given party, then that party may compensate for the work that it had previously done. However, in between these two rounds, the effects of the initial request would be visible to other parties, thus violating the principle of atomicity underlying this pattern. Supporting atomic interactions is the aim of a dedicated WS specification known as WS-AtomicTransaction[10], which provides a realization of the

---

[10] http://msdn.microsoft.com/library/default.asp?url=/library/en-us/
dnwebsrv/html/wsacoord.asp

distributed two-phase commit (2PC) protocol. However, this specification has not yet matured into a standardization initiative.

4   Routing Patterns

Pattern 11: *Request with referral.*

Description. Party A sends a request to party B indicating that any follow-up should be sent to a number of other parties (P1, P2, ..., Pn) depending on the evaluation of certain conditions. By default, faults are sent to these parties, but they could alternatively be sent to another nominated party (possibly party A).

Examples.

– *Referral to single party*: As part of a purchase order processing, a supplier sends a shipment request to a transport service. Subsequently, the transport service reports shipment status (e.g. as per RosettaNet's PIP 3B1) directly to the customer who then correlates these with its initial purchase order.
– *Referral to multiple parties*: After processing its inventory re-stocking for a week, a supermarket's warehouse contacts a supplier for order and dispatch of goods, notifying it of the different transport services available (different services specialize in transport of different sorts of goods). The supplier directly interacts with these transport services regarding the scheduled dispatch times (arranged by the supermarket). Faults related to order fulfillment are sent by the supplier to the warehouse, while faults related to delivery are sent by the corresponding transport services to the warehouse.

Issues/design choices.

– Party B may or may not have prior knowledge of the identity of the other parties. The information transferred from A to B must therefore allow B to fully identify and to interact with the other parties.
– The referred parties (P1, ..., Pn) and the party nominated to process faults (if different from A) may receive messages related to interactions that they did not initiate. These messages should then be related to internal processes at these parties. Sometimes, messages received through referral trigger new process instances, while other times, they will be routed to an activity within an already running process instance. The data transferred must allow the referred parties to route the message to the correct internal process.

Solution. At the messaging level, this pattern is partially addressed by WS-A which defined (among others) two fields that can be included in SOAP message headers, namely reply-to and fault-to. Using these fields, it is possible to specify the service endpoint(s) to which replies and faults should be sent. The information allowing the referred service to correlate the incoming message with its internal processes may be transferred in one of two ways depending of the adopted *state representation style* [4]: (i) it may be encoded in the endpoint reference itself (as per the REST architectural style); or (ii) it may be encoded somewhere else in the message (e.g. in the message body). In the supplier-shipper-customer example, the supplier passes to the transport service, a reference to the customer's procurement service endpoint. In the first style above, this endpoint reference would contain a data item (e.g. the original purchase order ID) allowing the customer to correlate the message with its internal activities, while in the second style, this data item would be encoded inside the shipment notification.

At the service composition level (specifically in BPEL), endpoint references can be manipulated as ordinary data. They can be included in the contents of a message and can be dynamically bound with partner links (e.g. the partner link defined between the transport service and the customer). In addition, BPEL offers a notion of correlation set, which corresponds to information sent along a message that is used on the receiver's end to correlate that message with its internal process instance. Correlation sets can thus be used to encode correlation-related information that it not included as part of the endpoint reference.

Related pattern.

– *Channel mobility.* Channel mobility in pi-Calculus [9] refers to the ability for a process X to pass a channel name to another process Y. Passing channel names along with requests provides a means of realizing the Request with Referral pattern. In fact, this is the way the pattern is captured in BPEL, where channels names are coded as endpoint references and correlation data.

Pattern 12: *Relayed request.*
Description. Party A makes a request to party B which delegates the request to other parties (P1, ..., Pn). Parties P1, ..., Pn then continue interactions with party A while party B observes a *view* of the interactions including faults. The interacting parties are aware of this view (as part of the condition to interact).
Example. Some supportive work of managing regulatory provisions outsourced by government agencies to external agencies fits this pattern. Party A is a client seeking some outcome pending regulation, e.g. obtaining particular land tenure. Party B is the government authority concerned with the regulation. e.g. lands department. Parties P1, ..., Pn are outsourced service providers from the government authority's regulation process, e.g. brokers who validate applications and external land management experts who can provide independent audit of applications. The government authority stipulates that interactions between the client and outsourced service providers associated with key points of processing, such as the start and end of activities, and key reports, be sent to it.
Issues/design choices.

– The delegated parties (P1, ..., Pn) may or may not have prior knowledge of the identity of the request originator, party A.
– A mechanism is needed to express party B's view of interactions between party A and the delegated parties. This may include all interactions or specific ones deemed to be of interest as indicated by the content of the messages.
– The view is defined at design time, but may be modified at run-time (party B may adjust what it needs to see depending on progress of activities).

- Party B could apply referrals for redirecting interactions or faults to other parties, however this issue is orthogonal to this pattern and is covered in Pattern 11.

Solution. This pattern, like the request with referral (pattern 11), involves indirection through delegation (party B passes party A's endpoint service reference to delegated parties for further interactions) and can be effected through WS-A or exchanged message data as previously discussed. The correlation strategies similarly apply. The comparative requirement for relayed requests is representing party B's view and enforcing it, including changing it, as interactions execute as identified through the second and third issues above.

Unfortunately, WS-A does not provide direct support for including party B in the interactions due to its lack of a "Cc field". But even if WS-A offered such Cc field, it would not cover a key requirement of the pattern: The messages passed between party A and the delegated parties would be exactly the same as what party B sees. Of course, not all messages have to be "Cc-ed" to party B, but this remains a rather limited solution since whole message, not filtered messages, are transmitted to B. It is furthermore possible that B do the filtering rather than pushing this up to the level where interactions are generated. We argue, however, that view filtering decoupled from interaction generation, is deficient since party A and the delegated parties no longer have an understanding of what they are obliged to reveal to B, as required by the pattern.

This brings us to the core issue of how to specify views such that they could be deployed and utilized as part of the interaction cycle. Simple views could be specified through a querying language like XPATH while more sophisticated ones could be supported through XQuery. Party A and the delegated parties would either have static view definitions prior to run-time or they would be passed at run-time when B establishes delegation.

For dynamically modified views, B would issue new views. These need to be coordinated with A, so that both ends of interactions are subject to the new version of the view. An obvious solution is to accompany a send in an interaction with a second send for party B, conditional upon the view filter applied to the message passed through the first sent. The two sends must be atomic.

Pattern 13: *Dynamic routing.*
Description. A request is required to be routed to several parties based on a routing condition. The routing order is flexible and more than one party can be activated to receive a request. When the parties that were issued the request have completed, the next set of parties are passed the request. Routing can be subject to dynamic conditions based on data contained in the original request or obtained in one of the intermediate steps.
Synonyms. Routing slip [6,7].
Example. After processing an order, the sales department sends a request to the finance department to process the invoicing and payment receipt for the order. This request contains a reference to the customer's procurement service and possibly also to a shipping service nominated by the customer. After arranging invoicing and payment by interacting directly with the customer, the finance

service forwards the order to the warehouse service. If the order is marked "for pick-up", the warehouse eventually sends a notification of availability for pick-up to the customer's procurement service. Otherwise, the warehouse issues a request to a shipping service which may be either the company's default shipping service, or the one originally nominated by the customer. The shipping service eventually sends a shipping notification directly to the customer.

Issues/design choices.

- The set of parties through which the request will circulate may not be known in advance and these parties may not know each other at design time.
- The specification of ordering should support service-to-role late binding, parallelism and interleaved parallel routing [1], synchronization points between parallel steps, and dynamic conditions.
- A way of providing relevant (fragments of) documents to different parties needs to be supported as well as a mechanism for controlling read-only and write access to these document (fragments).
- The update of routing should be subject to role access permissions, e.g. only a project coordinator is allowed to re-route a proposal review through work-package leaders.

Solution. The requirements for dynamic routing are outside the scope of direct support through BPEL. BPEL solutions are possible but would necessarily be ad hoc and require significant amounts of hand-crafted application code. WS-Routing[11] (a proposal not yet under standardization) can serve to implement some aspects of this pattern: Parallel routing, but not interleaved parallel routing, is possible; static, but not dynamic, conditions are supported, although this and the relevant routing role matching becomes supplementary coding for the full solution. Thus, WS-Routing can support simple dynamic orders, like those of the Routing slip pattern [6]. However, the complex dynamic routes required by our examples above, cannot currently be supported.

5   Conclusion

As service composition developments unfold in their objectives of making real-scale B2B transactions a reality and ushering in newer exploitations of service interoperability, it is striking how insufficiently guided these efforts are by well-structured requirements. We sought in this paper to address this gap by establishing a reference for service interactions. We did so by distilling insights from the literature, standardization activities, and use case scenarios, to derive a set of patterns. These patterns allow relevant technologies to be benchmarked. In this paper, we have investigated BPEL's capabilities in terms of the patterns.

BPEL directly supports *single-transmission bilateral patterns*. For *single-transmission multi-lateral patterns*, BPEL restricts the send-receives to be sequential and requires "house-keeping" code for correlation and for capturing stop

---

[11] http://msdn.microsoft.com/library/en-us/dnglobspec/html/ws-routing.asp

and success conditions. We recommend more effective support for these patterns through a construct capturing parallel composition of an a priori unknown number of send-receives. Of the *multi-transmission patterns*, BPEL event handling capabilities provide support for the *multi-responses* and *contingent sends*. However, lack of sufficient transaction support significantly compromises a BPEL solution for atomic multi-cast. For the *routing patterns*, simple *request referrals* are possible by passing endpoint references and implementing indirect interactions through correlation identifiers. This also serves *request relaying*. In addition, WS-A provides some support for request referrals and relaying although this support would be more direct if a Cc field was available. *Dynamic routing* is outside the scope of BPEL but WS-Routing can serve to implement some aspects of it, though not the flexible ordering and dynamic routing conditions.

Future work will extend the patterns by further extrapolations and will consider conversation management, viz. create, cancel, undo, suspend, and resume conversations. We are also drawing on insights from the patterns to design a framework for conceptual modeling of service interactions.

Acknowledgments. The authors wish to thank Phillipa Oaks, Helen Paik and Ivana Trickovic for their input and feedback. The second author is funded by a Queensland Government "Smart State" Fellowship co-sponsored by SAP.

References

1. W. M.P. van der Aalst, A. H.M. ter Hofstede, B. Kiepuszewski, and A. Barros. Workflow Patterns. *Distributed and Parallel Databases*, 14(1):5-51, 2003.
2. G. Alonso, F. Casati, H. Kuno, and V. Machiraju. *Web services: Concepts, architectures and applications*. Springer Verlag, 2003.
3. A. Barros, M. Dumas, and A. H.M. ter Hofstede. Service Interaction Patterns: Towards a Reference Framework for Service-based Business Process Interconnection. Technical Report FIT-TR-2005-02, Faculty of IT, Queensland University of Technology, 2005. See: http://www.serviceinteraction.com.
4. R. Fielding. *Architectural Styles and the Design of Network-based Software Architectures*. PhD thesis, University of California, Irvine, 2000.
5. C. Hagen, and G. Alonso. Exception Handling in Workflow Management Systems. *IEEE Transactions on Software Engineering* 26(10): 943-958, 2000.
6. G. Hohpe and B. Woolf. *Enterprise integration patterns: Designing, building, and deploying messaging solutions*. Addison-Wesley, 2004.
7. A. Kumar and J.L. Zhao. Workflow Support for Electronic Commerce Applications. *Decision Support Systems* 32: 265-278, 2002.
8. D. Luckham. *The Power of Events: An Introduction to Complex Event Processing in Distributed Enterprise Systems*. Addison-Wesley, 2002.
9. R. Milner. *Communicating and Mobile Systems: The Pi-Calculus*. Cambridge University Press, 1999.
10. M. Snir and W. Gropp. *MPI: The Complete Reference*. MIT Press, 2nd edition, 1998.
11. S. Weerawarana, F. Curbera, F. Leymann, T. Storey, D.F. Ferguson (Editors). *Web Services Platform Architecture: SOAP, WSDL, WS-Policy, WS-Addressing, WS-BPEL, WS-Reliable Messaging, and More*. Prentice Hall, 2005.

Modeling and Assessment of Production Printing Workflows Using Petri Nets

Raju N. Gottumukkala [1], and Dr. Tong Sun[2]

[1] Louisiana Tech University, Computer Science Department,
Ruston, LA 71270, USA
nrg003@latech.edu
[2] Principal Scientist, Adaptive & Smart Document System Lab, Xerox Innovation Group,
Xerox Corporation
Webster, NY 14580, USA
Tong.Sun@xeroxlabs.com

Abstract. Production printing workflow is a high-volume and high-speed print-ing process normally consisting of a set of complex and inter-related tasks namely pre-press, press and post-press procedures. Today many production printing vendors are increasingly offering heterogeneous devices and related software products that autonomously interoperate as a production printing workflow in a digital distributed environment. It is highly desirable in such en-vironment that a detailed workflow assessment is performed either prior to the deployment or during real-time operations. A formal workflow model and as-sessment capability would ultimately benefit the customers who directly man-age these production printing workflows to make better-informed decisions, understand the efficiency of to-be-purchased or already-deployed workflows, foresee the performance implications under a variety of business conditions. Therefore in this paper, we have developed formal workflow models (in both abstract and execution) based on the colored Petri nets [4] that incorporate pro-duction printing semantics. Based on these formal representations, we show how a production printing workflow can be assessed both analytically and quantitatively by leveraging existing Petri net tools.

1 Introduction

Production Printing Workflow (PPW) is a high-volume and high-speed printing proc-ess where various printing related tasks namely pre-press, press and post-press coop-erate with each other and perform collectively for producing printed materials like books, catalogs, manuals, financial statements, collaterals, etc.. Pre-press refers to a set of procedures that process documents before printing, such as composition, pre-flight, imposition etc.. Press is the actual procedure of printing documents on a spe-cific media. And post-press is the procedure involves finishing the printed materials, such as cutting, folding, and binding, etc.. Traditional production printing workflows are more of human-controlled manufacturing processes that are encompassed by a complex set of hardware machineries and manual interventions across all tasks.

W.M.P. van der Aalst et al. (Eds.): BPM 2005, LNCS 3649, pp. 319–333, 2005.

Nowadays, the entire printing business is undergoing a major paradigm shift into a marketplace centered with more automated digital workflows, in which heterogeneous devices and increasing number of software applications seamlessly interoperate with each other. The ever-increasing challenges for today's production printing business include: 1) to make well-informed business decisions as to what products or workflow technologies to invest or acquire in order to improve their current production process efficiency; 2) to maximize devices productivity and maintain the cost effectiveness; 3) to foresee the performance implications under a variety of conditions to meet changing business needs. Normally, static information regards each individual product capability and cost matrix is easily accessible, but the ability to effectively model and assess the implications of these interoperated products in the context of end-to-end production printing workflows against measurable indicators still remains a difficult and costly task. Most production printing workflows are represented in ad-hoc models with sort of trail-and-error analysis. Therefore, it is highly desirable to have a formal workflow model that well captures the production printing workflow semantics and upon which both analytical and quantitative workflow assessment can be performed either before deployment or during the real-time operations.

Representing a workflow model formally provides a powerful analytical capability in verifying the correctness of a workflow model syntactically and semantically. Petri nets are a well-established formalism for modeling concurrency, synchronization and non-determinism in distributed systems [12,13]. In this paper, we use colored Petri nets [4] to represent the basic building blocks of production printing workflow and the workflow itself in both abstract and execution perspectives. Based on these Petri net based models, we are able to leverage some existing Petri net tools [1,5,9,and 11] for formal model verification and discrete event simulation to assess the production printing workflows.

In this paper, we describe the applications of Petri nets in modeling, analysis and simulating production printing workflows. The remainder of this paper is organized as follows: First, we introduce the concepts in production printing workflows (PPW) and Petri nets. In Section 3, we formalize the production printing workflow semantics in both abstract and executable Petri net models. Section 4 discusses both the analytical and quantitative workflow assessment based on the proposed Petri net based models. One example of a production printing workflow is also illustrated. Finally, we conclude our experiences and future works in Section 5. Note, we use "workflow" and "process" inter-changeable in this paper.

2   Preliminaries

This section introduces the basic concepts and semantics in PPW and the Petri net notations used in the remainder of this paper.

2.1   Production Printing Workflow Concepts

PPW usually consists of heterogeneous devices and software application that are inter-connected to streamline a **Print Job** in a centralized or distributed production environment. Each constituent devices and software applications provides one or more well-defined capabilities that accomplish one or more tasks within the PPW. Usually, such a well-defined capability is named as **Service** component. For instance, an imposition software application provides an imposition service that performs imposition procedures on a given **document** before it gets printed. Therefore, the service component is a large-grained logical module that provides the basic building block for any workflows. In other words, a workflow can be viewed as a network of interconnected service components, with each performs certain task(s) in the workflow.

There are three major concepts that provide the semantics foundation in the modeling and assessment of the production printing workflows:

1) **Print Job** is a data structure that contains all the information required to complete the intended output in a printed media [15], such as the sequence of intended steps and their associated resource(s). It normally dictates and initiates an end-to-end control sequence of tasks necessary for satisfying the requested intention. Usually, a particular workflow is instantiated and started when a print job is received.

2) **Service** is a large-grained logical module that provides a well-defined capability and performs a certain task(s) in the workflow. Usually, it is embodied in one or more physical devices or software applications. A workflow model can be constructed by recursively connecting service components in a proper sequence.

3) **Document** is a representation of content element that flows through the entire production printing workflows, while print job is the control data representation. A document could be either electronic or hardcopy, could be in any possible formats (PDF, MS WORD, JPEG, etc.), could be stored in any media (repository, CD ROM, web page, etc.), could be in any size, etc.. A document can be categorized as various types (or document types) by the combination of its characteristics in size, format, finishing type, etc..

2.2 Petri nets

Petri nets are a modeling mechanism that consists of two types of nodes: places and transitions, which are connected by arcs. Places hold tokens which indicate the state of the Petri net and transition change the state of the net which is a result of moving tokens between various places. There have been various extensions to the initial classical Petri nets developed by C.A Petri [8] to include color for modeling token types, time to introduce delays in firing transitions, and hierarchies for modular design of net at various levels of abstraction. We are using colored Petri nets [17] in this paper, which is a generalization of these high level Petri nets.

3 Modeling Production Printing Workflows in Petri Nets

In general, a production printing workflow can be modeled by a network of interconnected (either sequentially or in parallel) service components that are embodied in devices or software applications. Print jobs are processed by these service components concurrently and/or asynchronously with respect to time and other dependencies. While a print job transits through various service components in the workflow, it may be split into multiple sub-jobs, or merge with other job(s) as a single one for load balancing purpose. Also the print job, service component and document all have their own lifecycle semantics, corresponding behaviors and characterization parameters [15]. Therefore, the modeling language should support these semantics at various levels of hierarchy. Petri nets are suitable for modeling PPW, as the characteristics of print job, service and document can be represented intuitively interpreting both abstract and executable scenarios. We describe two sets of models: one at "abstract", one at "execution". The abstract models are a concise model that enables an efficient state space evaluation of the PPW and the executable models represent much detail behavior semantics to enable a discrete event simulation for PPW.

3.1 Abstract Service Net (ASN) and Abstract Workflow Net (AWN)

Each service component conforms to common lifecycle semantics while it delivers its defined capability. Generally, a service component has at least three states in its lifetime: "**ready**", "**processing**" and "**completed**". And upon the arrival of a print job, if the receiving service is "ready", it can "start processing" the job and deliver its capability during the "processing" state, and then it cleans up while "finish processing". Finally, service is "completed" processing, and gets back to "ready" for next job state. In this section, we first define an Abstract Service Net (ASN) that represents three service states (i.e. "ready", "processing" and "completed") in Petri net places, while "start processing" and "finish processing" in Petri net transitions (see Figure 1). The reason we called this model as "abstract" is that it only captures a simplified (or subset of) service semantics to enable an efficient state-space based analysis [1,3,9].

Abstract Service Net (ASN) is a Petri net (as shown in Fig. 1) that contains: a set of service states (i.e. "ready", "processing", "completed") as Places; a finite set of service transitions (i.e. "start processing", "finish processing") as Transitions; a finite set of directed *arcs* connecting these service states and service transitions.

An **Abstract Workflow Net (AWN)** is a high level Petri net representation composed by the ASN of its constituent service components. The AWN topology and the selective connections among the ASNs (i.e. AND_SPLIT, AND_JOIN, SEQ). could be obtained from a pre-defined workflow model (in BPMN, BPML or JDF, created/generated by any existing Workflow Modeling tool). In summary, the Places in AWN is a finite union set of all places of its constituent; ASNs, the Transitions in AWN is a finite union set of all transitions of its constituent ASNs, plus the selective routing logics. An example of an AWN is shown in Fig. A.1. in Appendix 1. By leveraging the Petri nets' powerful formalism, the analytical assessment of PPW can be conducted upon AWN. We will discuss this in Section 3.1.

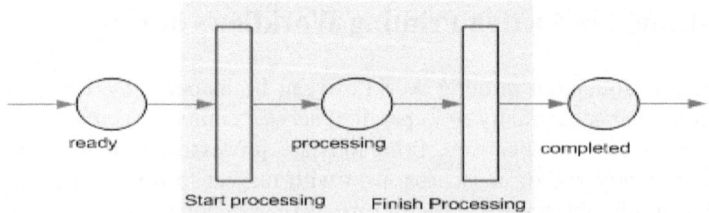

Fig. 1. Abstract Service Net (ASN) : represents abstract service semantics

3.2   Executable Service Net (ESN) and Executable Workflow Net (EWN)

Two executable Petri net representations (Executable Service Net and Executable Workflow Net) defined in this section comprehend the execution semantics of each service component and the composed workflow correspondingly. The motivation behind the development of an execution model is to enable the behavior simulation so that a quantitative workflow assessment can be conducted against measurable indicators (e.g. turn around time, through put, resource utilization, etc.).

The integral components of an executable model include:

1) *The lifecycle semantics* that describe the service's and job's behavior semantics such as the states and transitions during their lifecycle. See Table 1 and 2 for details.
2) *The job parameters* include the parameters describing the characteristics of a particular job (i.e. *job_id* – a unique identity, and its associated document content), and the parameters capturing the dynamics of job execution (i.e. waiting time, turnaround time).
3) *The document parameters* include *doc_id* (a unique identity for specific document type), *no_of_pages* (represents the size of a document in terms of number of pages).

Figure 2 below shows the **Executable Service Net (ESN)**. A job and its associated document are represented by a token in a Petri net as a tuple *[job_id, doc_id, no_of_pages, tatime, wttime]*, which contains the variables job id, doc type, number of pages for a job, current turn around time for the job, and the jobs current waiting time. Some of these variables are updated as the net is simulated. Places represent different states of the service and its corresponding job(s), and additional two places are also added as a parameter container for service's setup time and job's parameters. Transition represents various actions in the service component that change its states and job states as well. The transitions contain operations to be performed on a token, which may be simple arithmetic operations or java method calls to invoke a specific service processing capability. The arc inscriptions indicate the type of token or variable that is being moved as a result of firing a transition.

The execution behaviors that captured in the above ESN include the following five phases:

1) **setup phase**, during which once the job arrives, the service component needs to perform necessary steps to prepare or setup job.
2) **queuing phase**, during which the job is contained in a queue to wait for service to be ready; meanwhile, a queue object needs to be instantiated once service starts up.
3) **processing phase**, during which the job is retrieved from the queue when service is ready, and service processes the job, after processing service needs to cleanup and release the processed the job to the subsequent service component, and make itself ready for next job from the queue.
4) **status feedback phase**, during which the service status such as "Available" is fed back so that the next job can be continuingly retrieved from the queue.
5) **recovery phase**, during which the service recovers from failure condition (such a behavior is modeled as a set of job parameters, such as a failure rate and recovery time for that job).

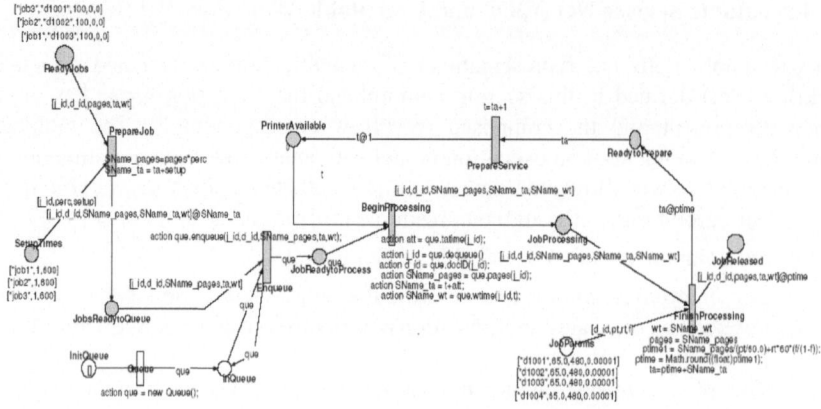

Fig. 2. The Executable Service Net (ESN)

Table 1.   Detailed description of all places in the Executable Service Net (ESN)

Place Name	State Descriptions
JobIn	The initial place when the job arrives to the service component
SetupTimes	The place contains the setup time for a particular Job
JobReadytoQueue	The place contains job(s) that have completed setup and to be queued for processing
JobReadyToProcess	The place contains the queue object with the job(s) ready to be processed
Processing	The place contains the job that has been de-queued and is currently being processed
JobParams	The place contains the job parameters to calculate the average processing time for that job. JobParams are document type, processing time, failure rate and recovery time.
Completed	The place contains the job that has completed processing and is released from service component.
ServiceReadytoPrepare	The place indicates the service state is ready to cleanup.
Ready	The place indicates the service state is available for next job.
QueueInit	The place indicates the initiation of the queue object when service starts up
QueueReady	The place indicates the queue object is created for holding job.

Table 2. Detailed description of all transitions in the Executable Service Net (ESN)

Transition Name	Action Descriptions
PrepareJob	Prepare the job for processing, i.e. it adds the setup times associated with that particular type of job.
EnqueueJob	Queues the prepared job in the order they arrive after the setup.
StartProcessing	De-queue the next job from the queue as long as the job is ready to be processed and service is available. The action inscriptions are used to retrieve the job attribute values. The waiting time is added here.
FinishProcessing	Remove the completed job from the service component and the average processing time is calculated here and added to the total turnaround time of the job.
PrepareService	Prepare the service component for the next job
InitQueue	Initialize the queue object once service starts up

Similarly to the AWN, the EWN can be composed by inter-connecting the ESNs of its constituent services via selective connection logics (i.e. AND_SPLIT, AND_JOIN, and SEQ). An example of an EWN is shown in Fig. A.2 in Appendix 2. In summary, the Places in EWN is a finite union set of all places in its constituent service component's ESN; the Transitions in EWN is a finite union set of all transitions in its constituent service component's ESN, plus the transitions in the selective connection logics.

4 Model Based Assessment for Production Printing Workflows

There are two primary perspectives for workflow assessment: (1) *analytical assessment*, which is to verify the correctness of any given workflow model syntactically and semantically based upon an underlying theoretical model; (2) *quantitative assessment*, which mimics the operation of a real workflow (such as the day-today operation of a print-shop) and gathers/derives quantitative evaluation metrics in a simulation model. In the previous section, we formulate the PPW and its semantics into Petri net representations at both abstract level and execution level, each of which serves a particular purpose in these two perspectives.

4.1 Analytical Workflow Assessment

Both abstract and executable workflow models inherit the powerful Petri net's formalism that enables not only the capability of the syntax model verification, but also the formal analytical techniques to prove certain properties like live-ness, safety, and bounded-ness [3,10,11,14]. However, all these properties are proved by generating an occurrence graph or invariants, which involve generating a state space of the given model. The state space size generated increases exponentially with increase in the number of places. Therefore, analyzing the executable workflow model (EWN) would be very complex and time-consuming for this purpose. However, the Abstract Workflow Net (AWN) is more suitable for analytical workflow assessment purpose.

There are several Petri net tools available [7] for Petri net based analysis. Woflan and Design/CPN are two widely used Petri net based analysis tools. Woflan is used to prove the soundness properties of the workflow. A workflow is sound if the net successfully terminates, there are no tokens left in the net and there are no dead tasks [3,11]. Design/CPN [1,9] generates an occurrence graph to check bounded ness properties like dead locks and live locks.

4.2 Simulation Based Quantitative Workflow Assessment

Opposed to the analytical approach, where the method of analyzing the workflow system is purely theoretical, the simulation approach gives more flexibility and convenience. A simulation is the execution of a workflow model that gives the information about the workflow being investigated [2]. Simulated experimentation accelerates and replaces effectively the "wait and see" anxieties in discovering new insights and explanations of future behavior of the real workflows. Therefore in this paper, the executable workflow Model (EWN) is built for enabling this simulation based quantitative workflow assessment. We also use the Renew tool [5] as our underlying event-driven simulation engine.

4.2.1 Workflow Simulation Architecture

We have built a workflow simulation tool based on the following architecture (see Fig. 3). There are three major parts in this architecture:

1) **Workflow model constructor** - this module could be provided by any existing workflow modeling tool, from which a workflow model can be constructed either manually, or automatically; and the workflow model can be represented in any XML specification (e.g. BPMN, BPEL, JDF, etc.)

2) **Workflow simulation interface** – this module is the core component we built and it contains a set of components to generate WEN from a given workflow model, to collect Simulation Profile Information (see Table 3), to interface the underlying simulation engine, to display the simulation results to the end users.

3) **Event-driven simulation engine** – this module leverages the Renew tool [5], with three added on components: a simulation Trace Generator, a backend SQL database, and a data analysis component.

In order to simulate a variety of customer business conditions, four collections of parameters (or Simulation Profile Information) need to be considered (details see table in Appendix 3):

1. Parameters provided by the end-user directly that describe the variety of business conditions for their PPWs;
2. Parameters derived from the above user input data;
3. Parameters provided by a pre-defined Service Component Capability data source (e.g. product function feature specification, pricing list, etc.)
4. Parameters generated/tracked by simulation engines that describe the desired simulation results or measurement indicators.

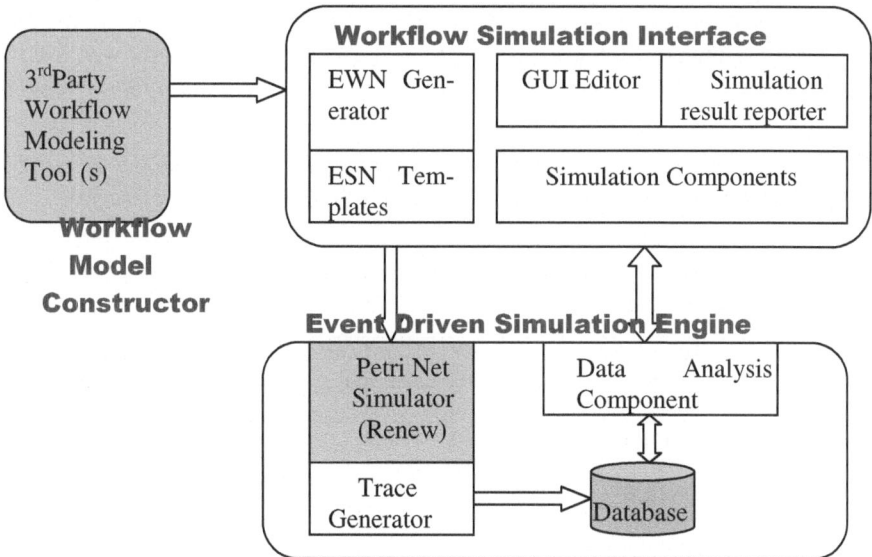

Fig. 3. The Workflow Simulation Architecture

4.2.2 Example Simulation Scenarios

We consider a typical digital production printing workflow – Print On Demand, to illustrate the modeling and simulation of a distributed production workflow scenario. The workflow in Fig 4 consist of a digital library from which documents or images can be retrieved, an imposition software arranges the layout of images, content and page sequence for printing, a production color printer and a monochrome printer to print color pages and black and white pages in parallel, and a finishing device for binding the printed outputs.

For a simple illustration, all simulation parameters used in this example are picked up randomly and simulation instances are small amount. But it doesn't preclude the tool

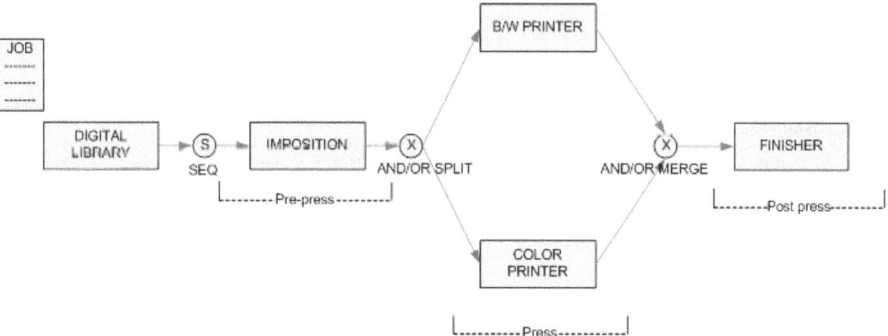

Fig. 4. A typical Print-On-Demand Workflow scenario

to simulate and analyze complex and high print volume PPWs. Assuming the customer want to simulate 2 job instances over the above workflow scenario with following assumed properties: *Types of Jobs, No of pages, Run Length, job split ratio, Document Type,*

- J1 – 100pages -- Run Length 10 -- color 20% BW 80% --doc size 8.0 – 8.5'
- J2 – 80 pages -- Run Length 30 -- color 40% BW 60% -- doc size 9.0 -10.5'

Table 4. Parameters provided from a pre-defined Service Component Capability data source for intended two document types

	Service Component Name	Processing Time (ppm)	Setup Time Per Job (min)	Repair Time (min)	Failure Probability
Doc Type 1	Document Library	500	15	120	0.00014
8.0-8.5	Imposition	500	5	240	0.00014
	BW Print	180	1	60	0.0001
	Color Print	100	1	120	0.0002
	Binder	100	10	240	0.00001
	SaddleStitch	100	10	120	0.00014
Doc Type 2	Document Library	500	15	120	0.00014
9.0-10.25	Imposition	500	5	240	0.00014
	BW Print	154	1	60	0.0001
	Color Print	85	1	120	0.0002
	Binder	100	10	240	0.00001
	SaddleStitch	1000	10	120	0.00014

From simulations of the above two scenarios, Average turn around time, Average waiting time, utilization, Throughput, are obtained by the Workflow Simulation Tool as follows.

Total Turn Around Time=1824.0 seconds:
Max Waiting Time=620.0 seconds
Run Length = 3400
Average Turn Around=45.6 seconds
Page Volume=1240 pages
Throughput=1.47 pages per second
Total Service Time=1340 seconds
Utilization =0.73

5 Conclusions

In this paper, we have demonstrated the formal modeling of production printing workflows by using Petri nets. Two Petri net based PPW models are presented: Abstract Workflow Net (AWN) and Executable Workflow Net (EWN). This separation of the abstract model and the execution model enables two primary perspectives of work-flow assessment seamlessly: analytical assessment and simulation-based quantitative assessment by leverage various existing Petri net tools.   A workflow simulation tool and its architecture are discussed, and a simple workflow simulation scenario is also illustrated. We have successfully demonstrated the flexibility and advantages of mod-eling production printing workflows with colored Petri nets formalism. Concerning the future work, we would like to enrich the formal workflow models to take into account of more complex behaviors such as "pause job", "resume job" semantics, and coordination between service components in a workflow, and also capture other op-erational metrics for production printing workflows, such as operating cost, for further workflow assessment. We also plan to provide a more seamless and automated map-ping from some standard XML-based workflow specifications (e.g. BPMN, BPEL, and JDF) into Petri Net Markup Language (PNML). This could allow a seamless integration between Workflow Simulation Tool with any existing workflow modeling tools that produce these standard workflow specifications.

References

1. K. Jensen, K. Christensen, S., Huber, P and Holla, M., Design/CPN Reference Manual, Department of Computer Science, University of Aarhus, Denmark, http://www.daimi.au.dk/designCPN/, 1995.
2. E. Jim,nez Mac¡as and M. P,rez de la Parte, Simulation and Optimization of Logistic and Production Systems Using Discrete and Continuous Petri Nets SIMULATION, Vol. 80, No. 3, 143-152, 2004.
3. H.M.W. Verbeek, T. Basten, and W.M.P. van der Aalst. Diagnosing Workflow Processes using Woflan. The Computer Journal, 44(4):246–279, 2001.
4. K. Jensen. An Introduction to the Theoretical Aspects of Coloured Petri Nets. In J. W. de Bakker, W.-P. de Roever, and G. Rozenberg, editors, A Decade of Concurrency, volume 803 of Lecture Notes in Computer Science, pages 230{272. Sprinter-Verlag, June 1993.
5. O. Kummer and F. Wienberg. Renew - the Reference Net Workshop. Petri Net Newsletter, 2004. see also:http://www.renew.de/.
6. Miko Mikolajczak Boleslaw, Byrne Debora l, Workflow modeling and diagnosis with Petri nets - a case study of a manufacturing, Proceedings of the second IEEE International Conference on Systems, Man and Cybernetics (SMC'02), Volume: 5 ,6-9 Oct. 2002
7. Petri Nets Tools Database Quick Overview, Petri Net World, 2003 http://www.daimi.au.dk/PetriNets/tools/quick.html.
8. C. A Petri (1962). kommunikation mit Automaten. PhD thesis. Bonn: University of Bonn (In German).
9. Soren Christensen Jens B'k Jorgensen Lars Michael Kristensen, Design/CPN - A Com-puter Tool for Coloured Petri Nets, Proceedings of the Third International Workshop on Tools and Algorithms for Construction and Analysis of Systems, Lecture Notes In Com-puter Science; Vol. 1217, Pages: 209 – 223,1997.
10. W.M.P van der Aalst, and A.H.M. ter Hofstede,   2000. Verification of workflow task structures: A Petri-net-based approach. Information System 25 1, pp. 43-69.
11. W.M.P. van der Aalst. Woflan. A Petri-net-based Workflow Analyzer. Systems Analysis - Modelling - Simulation, 35(3):345–357, 1999.

12. H.M.W. Verbeek, T. Basten, and W.M.P. van der Aalst. Diagnosing Workflow Processes using Woflan. Computing Science Report 99/02, Eindhoven University of Technology, Eindhoven, 1999.
13. W.M.P. van der Aalst. Making Work Flow: On the Application of Petri nets to Business Process Management. In J. Esparza and C. Lakos, editors, Application and Theory of Petri Nets 2002, volume 2360 of Lecture Notes in Computer Science, pages 1–22. Springer-Verlag, Berlin, 2002.
14. Wil M. P. van der Aalst: Workflow Verification: Finding Control-Flow Errors Using Petri-Net-Based Techniques. Business Process Management 2000: 161-183,2000.
15. Job Definition Format Specification, http://www.cip4.org

Appendix 1: Abstract Workflow Net (AWN) Example

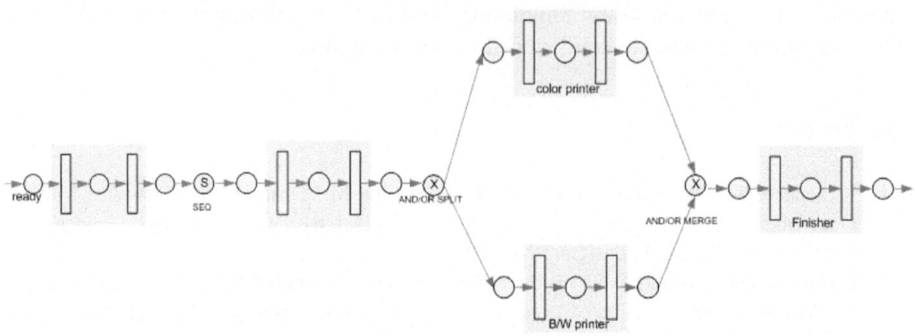

Fig. A-1. An Example of an Abstract Workflow Net (AWN), constructed by connecting 5 ASNs and 3 connection logics (SEQ, AND_SPLIT, AND_JOIN)

Appendix 2: Executable Workflow Net (EWN) Example

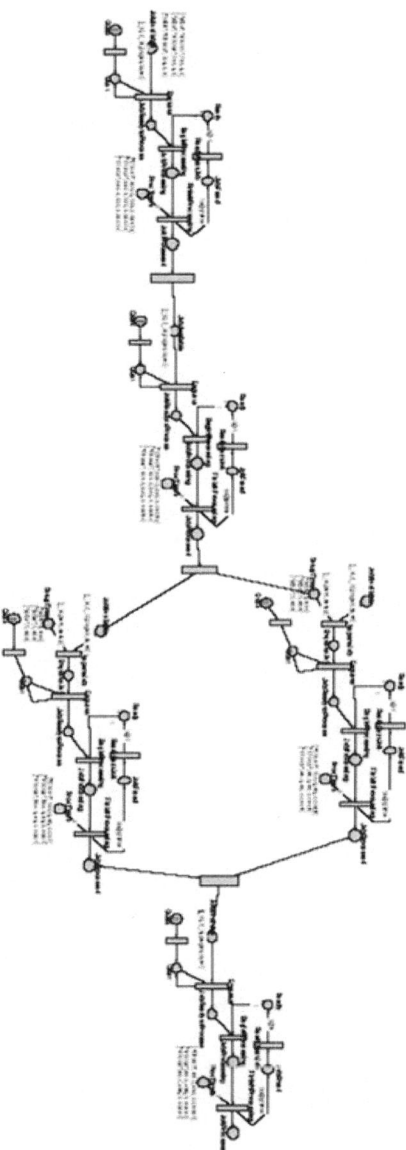

Fig. A-2. An Example of an Executable Workflow Net (EWN), constructed by connecting 5 ESNs and 3 connection logics (SEQ, AND_SPLIT, AND_JOIN

Appendix 3: Simulation Profile Information

Collection Number	Parameter Data Element	Note
Collection-1: Parameters provided by the end-user directly	**Job Arrival Rate (λ) -**	It is the average number of jobs submitted to the workflow per day. The default job arrival rates are assumed to be in Poisson distribution.
	No. of Pages per Job (p)	Number of pages per job
	Run Length per Job (r)	Number of sets of each job
	Document type (d)	This parameter captures the information regards the document output media type, size and finishing option etc.., which have direct impact on the actual processing time in the workflow.
Collection-2: Parameters derived from Collection-1	**Total print volume per day**	$V = \lambda \sum_{i=1}^{n} pi * ri$ where n is the total number of job per day
Collection-3: Parameters provided from a pre-defined Service Component Capability data source	**Processing time (t_p)**	A target processing rate for a specific service component
	Set up Time (t_s)	The time required for preparing the service component to prepare the incoming job, this set uptime for a service may vary with the job type.
	Failure Probability of Service (p_f)	Probability that a service component would fail . Failure probability is given by $p_f = 1 - R$, where R is the reliability of a service component, which is given by $R = e^{-mt}$, where m is the mean of number of failures per year, and t is the time which is one year.
	Repair Time (t_r)	Time elapsed in repairing a failed service component up and making it available for the next job.
Collection-4: Parameters generated and/or tracked by simulation engines	**Average Processing Time (per job)**	$Tp = t_p + t_r (P_f / 1 - P_f)$
	Turn Around Time per job	It is generated directly from simulation engine (the duration from job arrives till job completed)
	Average Turn Around Time (TA_{avg})	Total turn Around Time/ Number of Jobs
	Throughput (Th_{put})	Page Volume (V)/ Time taken to process all the jobs

Average Waiting Time W_{avg_t}	W_t = Time spent by each job in the queue. $$W_{avg\_t} = \sum_{i=1to}^{N} W_t / N$$
Service Time	It is the time spent by the services performing useful work. The Service time of the workflow = Total Turn around time – Total idle time of all the services
Utilization	Ratio between service time and total turn around time.

Process Management in Health Care: A System for Preventing Risks and Medical Errors

Massimo Ruffolo[1,3], Rosario Curia[1], and Lorenzo Gallucci[1,2]

[1] Exeura s.r.l.
[2] Department of Electronics, Information science and Systems (DEIS)
[3] Institute of High Performance Computing and Networking - Italian National
Research Council (ICAR-CNR),
Università della Calabria, 87036 Rende (CS), Italy
{ruffolo, curia, gallucci}@exeura.it
ruffolo@icar.cnr.it
http://www.exeura.com
Tel: +390984493094

Abstract. This work describes the architecture of a clinical processes management system aimed to support a process-centered vision of health care practices. At the heart of the system there is a formalism well suited for representation of both processes and related domain knowledge. This language allows the semantic description of clinical processes using ontology and workflow representation formalisms. The main goal of the system is to assist in executing the clinical processes by providing intelligence functionalities, based on workflow mining techniques, and in monitoring processes during their execution. Acquired process instances can be analyzed to identify main causes of medical errors and high costs and, potentially, to suggest clinical processes restructuring or improvement able to enhance cost control and patient safety.

1 Introduction

Nowadays health care costs and risks management is a high priority theme for health care professionals and providers. Across the world the whole issue of patient safety, medical errors prevention and adverse events reporting is a very challenging and widely studied research and development topic that stimulates a growing interest in the computer science researchers community.

A strong research effort has been taken, in the recent past, to provide an uniform representation of clinical knowledge useful in health care information systems. Interesting results have been obtained in the field of medical knowledge representation, where many ontologies, such as UMLS, MESH, ICD9-CM, Snomed, OpenGALEN [2,3,6,7,17,18] have been developed on different medical topics. The Evidence Based Medicine movement has stimulated the definition of many guideline representation formalisms such as GLIF, PROFORMA, EON based on different paradigms [1,4,19]; e.g. Proforma is a process description language grounded in a logical model whereas GLIF is a specification consisting of

W.M.P. van der Aalst et al. (Eds.): BPM 2005, LNCS 3649, pp. 334–343, 2005.

an object-oriented model. The promising young project DeGel [16] aims to construct a digital electronic guideline library extracted from descriptions available in textual documents.

Health care practices are characterized by complex clinical processes in which high risk activities take place. A clinical process can be seen as a particular workflow where medical (e.g. treatments, drugs administration, guidelines execution, medical examinations, etc.) and non-medical (e.g. patient enrollment, medical record instantiation, etc.) activities and events occur. A successful approach for reducing cost and risk and enhancing patient safety is a process-oriented vision of health care services and practices.

Systems providing clinical processes design, execution and analysis functionalities can change clinical practices and can help diffusion of a process and quality awareness in health care organizations. Both technologies at the state of the art of the health care information systems and knowledge representation field are required to build those systems able to support policies that ensure efforts of the health care organizations are mainly focused on improving quality of clinical governance, delivering high quality standards of care, service and patient safety.

This work describes the architecture of a clinical processes management system aimed to support a process-centered vision of health care practices. At the heart of the system there is a formalism well suited for representation of both process and related domain ontologies. The system is built upon the results coming from different fields of information technologies. Today, in fact, thanks to natural language processing techniques, ontology and workflow representation languages [8,9,10,15], data and workflow mining techniques [12,13], is possible to create a new generation of information technologies capable of capturing, recording, monitoring, classifying and analyzing events occurring in clinical processes. In particular, the system is based on ontologies and workflows representation mechanism enabling clinical processes and guidelines extraction, from textual documents, and their formal representation in machine manageable form. The system acquires and stores clinical processes and guidelines schema in a knowledge base and classify them w.r.t. the concepts contained in the ontologies. The main system goals are to assist in executing the clinical processes by providing intelligence functionalities, based on workflow mining techniques, and to assist in monitoring processes during their execution. Furthermore, acquired process instances can be analyzed to identify main causes of risks, to control costs and, potentially, to suggest clinical processes restructuring or improvement. Thanks to the system health care professionals have knowledge management and decision support functionalities able to enhance cost control and patient safety, reducing risks due to medical errors and adverse events.

2   System Description

The clinical processes management system is designed to follow the clinical processes life-cycle model shown in figure 1 based on three phases: design, execution and monitoring, intelligence. The architecture, depicted in figure 2, is organized

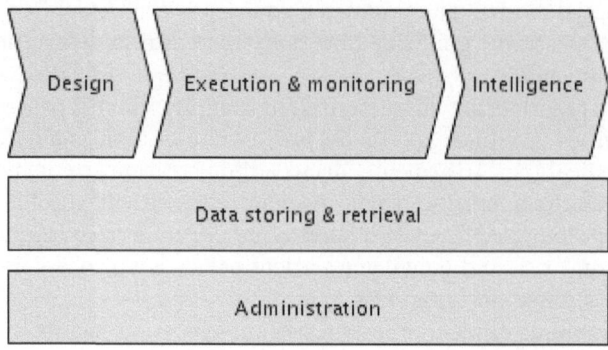

Fig. 1. Clinical Processes Life-Cycle

using a classical three-layer information system structure with interface, control and knowledge base layers. The packages *Design, Execution & Monitoring* and *Intelligence* contain software modules providing *main* functionalities for clinical processes life-cycle implementation that are based on *support* functionalities carried out by the packages *Data Storing & Retrieval* and *Administration*. In the following packages structure and functionalities are described in detail.

2.1   Clinical Processes Design

The *Design Package* contains software modules providing functionalities for the representation of clinical processes, medical guidelines and related ontologies obtained either by means of direct (on-screen) drawing and specification or using a text-extraction and annotation facility. The obtained ontologies and clinical processes models are stored in the repository contained in the knowledge base layer.

Design functionalities allow a formal and machine-readable specification of the clinical processes useful for their assisted and monitored execution. In particular, process activities and their parameters, decision steps, conditions, patient states, patient data etc. can be represented and semantically indexed using concepts contained in the ontologies. The indexing procedure allow the retrieval of process and sub-process (e.g. guidelines) using concept-based queries.

The system manages two kind of ontologies the *domain ontologies* containing the patients' data structure, medical records data, hospital and ward data and the *medical ontologies*. The latter concern a wide variety of concepts related to the medical domain and to specific clinical knowledge areas (e.g. diseases, drugs, medical examinations, medical procedures, laboratory terms, etc.) coming from standard medical thesaurus such as: ICD9-CM, Loinc, MeSH, Snomed. Ontologies are used for the semantic description of process activities and data.

Clinical processes, guidelines as well as domain ontologies are represented using a developing ad hoc formalism called CPML (Clinical Processes Modeling Language). The formalism is based on DLP+ [20] an ontology representation language founded on Disjunctive Logic Programming extended with object oriented capabilities and built upon the reasoning system DLV [21]. CPML allows the definition of clinical processes and ontologies through the joint representation of both ontology elements (i.e. concepts, attributes, taxonomic and non-taxonomic relations between concepts, instances, etc.) and workflow elements (i.e. sub-processes, activities, events, conditional forks, conditional joins, process participants, triggers, etc.). In the clinical processes ontologies sub-processes, activities and data are indexed and classified w.r.t. medical and domain ontologies concepts.

2.2   Clinical Processes Execution and Monitoring

The *Execution & Monitoring* package provide functionalities for the acquisition of the process instances and the assisted execution of clinical processes. Process instances are acquired following a schema already designed or selecting just-in-time the activities and/or sub-processes to perform. The process evolution is monitored during the execution using data and workflow mining techniques. The execution, performed through a graphical user interface constituted of web-based forms filled by doctors and nurses, is assisted in two different ways:

- through the execution of a clinical process schema as *workflow enactment* where actors can be humans or machines (e.g. legacy systems supplying results of medical examinations). During the workflow enactment the designed workflow schema is followed exactly producing clinical process instances where each of them contains the values of activities and events parameters related to a given cared patient;
- through a dynamic *workflow composition*. In this case, each activity or sub-process instance is acquired selecting the most appropriate one to execute in a given moment. The selection is supported by queries on the process ontologies contained in knowledge base. The queries syntax lets specification of the patient clinical data coming from anamnesis and medical examination, and each significant information available in the particular moment of the execution.

Adverse events and errors prevention, risks and costs reduction and patients safety enhancement need, also, a monitored execution of clinical processes. Monitoring lets the application of prediction models to running process instances, to identify exceptions, unusual or undesired behavior and to inform the user. Monitoring clinical process during their evolution, using workflow mining techniques, allows doctors and nurses to estimate the probability of errors or adverse events occurrence or find if an error or an adverse event has took place.

2.3   Clinical Processes Intelligence

The package *Intelligence* aims to allow analysis of the clinical processes instances after their execution. The software module *Process & Workflow Mining* provide

functionalities able to extract, from the workflow logs, relevant features charac-
terizing clinical process instances acquired during the execution and to represent
them in suitable datasets. This make available large amount of data on which
process intelligence [11], based on data and workflow mining techniques, can
be performed to discover patterns related to adverse events, errors and cost
dynamics, hidden in the structure of clinical processes, that are cause of risks
and of poor performances. When clinical process instances are obtained through
workflow composition of single activities or sub-processes the workflow mining
techniques are able to classify clinical process instances w.r.t. their behaviour
and, possibly, to suggest new schema able to reduce risks for patients and the
impact of errors to use as future reference.

The software module *Clinical Processes Evaluation* provide functionalities to
construct reports which can be customized defining performance indexes comparing process instances to a reference clinical process on the base of costs and risks.

The process intelligence functionalities support medical and managerial decision making about cost- and risk-effective intervention and enable *lesson learning*
about health care practices that suggest clinical processes restructuring or improvement criterions.

2.4 Administration and Knowledge Storing and Retrieving

The packages *Knowledge Storing and Retrieval* and *Administration* contain software modules providing support functionalities that assist clinical processes design, execution and intelligence. In particular:

- the *data storing and retrieving* services are built on top of high-level information representation techniques, such as ontologies (either domain and medical ontologies or process ontologies) and process workflows. The software
  components contained in the package are suited to represent, manipulate
  and query process ontologies, domain and medical ontologies. Repositories
  include Relational DB and semi-structured (XML) DB positioned in the Kn-
  woledge Base Layer; the access is mediated by the *ontomapper* component
  a library for ontology interface;
- the *administration* services are provided to properly co-ordinate simultaneous work from many users, by means of system management in terms of
  known users and their associated permissions. A dedicated web console lets
  administrative-role users to manage permissions, thus directing the whole
  working on knowledge base.

3   An Application to the Oncological Domain

This section describs an application of the system to a real case concerning
a clinical process for caring the mammary carcinoma. The clinical process is
referred to the practices carried out in the oncological ward of an italian hospital,
for this reason it is not general but specific for the domain of the considered ward.

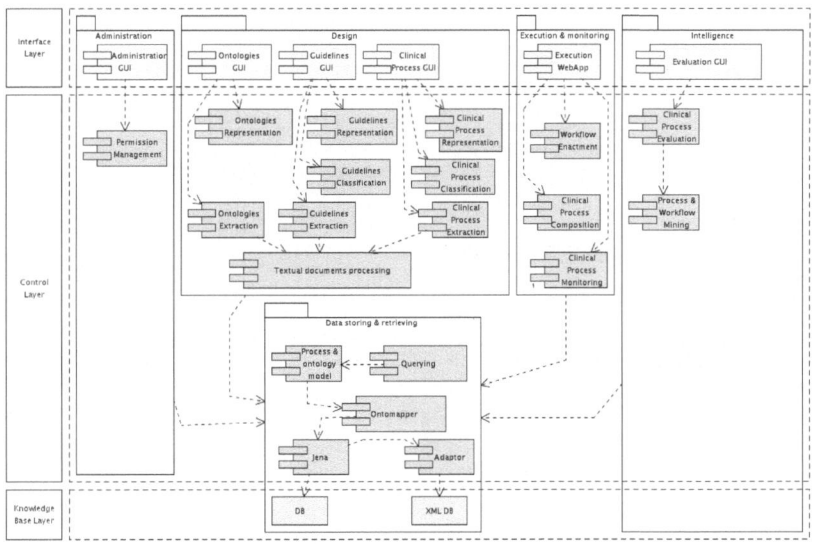

Fig. 2. System Architecture

The application suggests how the system can improve patient safety and cost control.

The clinical process which the patient is subjected to, depicted in figure 3, can be organized and represented using the following 10 activities and sub-process:

1. Patient enrollment (acceptance). The patient arrives to the ward with a previous clinical diagnosis of a mammary carcinoma; patient personal data are being collected and stored into a clinical records folder (CRF), initialized in this step.
2. Anamnesis. This sub-process is divided in three parts:
   - general anamnesis, in which general data, like physiological (allergies, intolerances, etc.) or personal data, are being collected;
   - remote pathological anamnesis, concerning past pathologies;
   - recent pathological anamnesis, in which each data or result derived from examinations concerning the current pathology (or pathologies) are acquired.

   Every information collected is stored, as in previous step, into CRF.
3. The patient is examined by an oncologist (Initial clinical evaluation). The results of the check-up are registered in the CRF.
4. Additional clinical tests (Other exams), if requested, are being conducted to find out general or particular conditions of the patient, if they are not fully deducible from the test results already available.
5. The physician picks a therapeutic strategy: the strategy itself depends upon actual pathology state as well as other patient data collected and is selected from one or more libraries among the guidelines available therein. The final

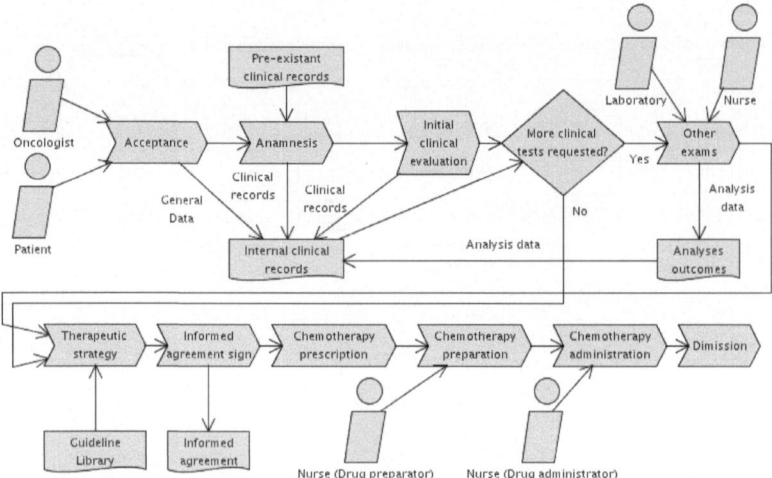

Fig. 3. A clinical process for caring the mammary carcinoma

choice among them descends from an agreement between the patient and the physician, given that different guidelines may cause variated side effects and have different associated recovery probabilities. Pathology stage and, in case, metastasis diffusion constitutes key information in this process, so there is an agreement at international level to code them strictly.

6. The patient is asked to sign one or more agreements, which may concern: understanding and acceptance of consequences (either side effects or benefits) which may derive from the therapy chosen (e.g. chemotherapy), privacy agreements, etc.
7. Prescription of drugs (chemotherapy prescription) as designed in the chosen guideline. This implies computation of doses, which may depend on patient's biomedical parameters, such as body's weight or skin's surface. Cross-checking doses is fundamental here, because if a wrong dose is given to the patient the outcome could be lethal;
8. Drug preparation (chemotherapy preparation), which must include an additional check on appropriateness of prescription.
9. The prepared drug is administered to the patient (chemotherapy administration), following the scheme (sub-process) outlined in the selected guideline.
10. discharging of patient from the ward (dimission).

Steps 8th and 9th are repeated till therapy's end; due to drug inherent toxicity, before each dose follows the preparation/giving chain, the oncologist must visit the patient to check for actual severity of side effects. At this point, he/she could decide changes in therapy, which may range from dose reduction to therapy suspension or termination. In the latter case, the guideline applied must be switched, possibly with subsequent changes in drugs needed for the therapy. The Systems described in this paper make viable:

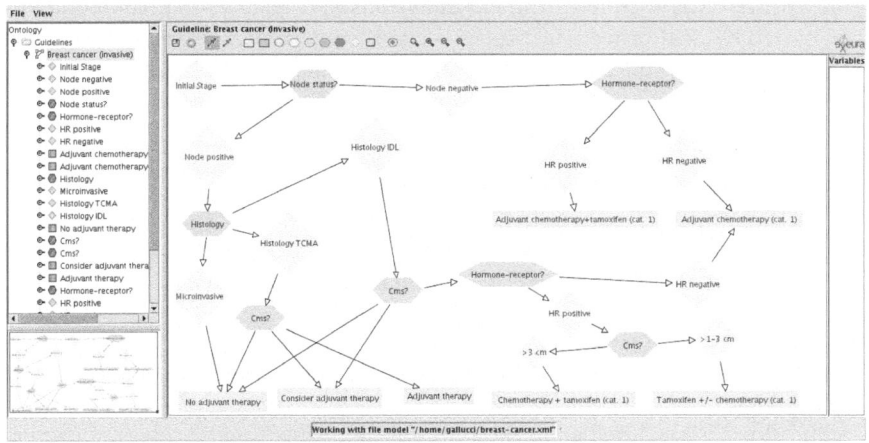

Fig. 4. An example of a computerized guideline

- the formal representation of the whole clinical process and its classification in an ontology containing all noteworthy processes that take place in the hospital;
- computer-aided choice of the applicable guidelines; the user, having patient and pathology data available in electronic form, can query the knowledge base which contains one or more guidelines' ontologies, getting as result a collection of every guideline applicable to the patient. Final selection pertains to the oncologist and the patient, because it depends on side effects level considered "bearable", but also wanted results. In this phase, availability of domain ontologies concerning drugs' interactions, side effects and contraindications for some patients could reduce dramatically the probability of fatal mistakes. The guidelines ontology is built through the extraction from natural language text sources, and the representation in structured form. In Figure 4 is depicted the system guidelines editor. This enable agile guidelines browsing and automatic clinical workflow enactment. The guidelines are also classified using different cost and risk evaluation criteria;
- assisted prescription of chemotherapy drugs; the doses, computed as mentioned, would be double checked against information extracted from drug ontologies, which could specify maximum absolute ratings for certain drug (e.g. "max -nnn- mg in a whole life") or other ratings relative to biomedical parameters of the patient. A change in a prescribed dose can be flagged immediately as dangerous, or even lethal, and the associated risk is notified to the oncologist;
- chemotherapy drugs preparation and dispensing; for a given patient, no ambiguity (e.g. due to handwriting) can arise on each dose to prepare or dispense;
- assisted acquisition and monitoring of clinical process instances; the system could also help in acquisition of personal data, anamnesis data as well as

any other data acquired from examinations and tests. When the number of acquired clinical process instances is significant clinical processes evaluation can take place and restructuring or improvement of the clinical processes can be suggested by the system to prevent risk and excessive cost.

4   Conclusions and Future Works

The practical application of the developed clinical processes management system show that is possible to improve risk prevention and cost control performing:

- the selection of the correct guideline to apply on a patient through queries on the guideline library based on data available about patient state and pathology;
- clinical processes intelligence activity for analyzing and discovering patterns and trends across all reported process instances, making possible to identify behaviors that pose risks for the patients;
- the benchmarking of processes and activities, obtaining comparative data to support the identification of practices that lead to good patient outcomes and the identification of further opportunities for improving patient safety;
- the identification and understanding of the human and system factors, which cause unintended harm to patients;
- the classification of errors and adverse events useful for the definition of a common understanding and the development of an agreed definition of terms relating to error, adverse event, risk and patient safety;
- the discovering of poor outcomes of care that are not just the result of adverse events and sub-optimal performances of the health service, but depend on a bad execution of related processes.

The main challenging research and development problems to approach in the future work are the development of an efficient query engine working on the conjunct representation of workflows and ontologies and the definition of further suitable ad hoc workflow mining algorithms.

References

1. Guide Line Interchange Format (GLIF). http://www.glif.org
2. Unified medical Language System (UMLS). http://www.nlm.nih.gov/research/umls
3. Medical Subject headings (MeSH). http://www.nlm.nih.gov/mesh/meshhome.html
4. Sutton D.R. and Fox J. The Syntax and Semantics of the PROforma guideline modelling language. JAMIA, Sep-Oct;10(5):433-43, 2003.
5. Logical Observation Identifiers Names and Codes (LOINC). http://www.loinc.org
6. World Health Organization (WHO). http://www.who.int/whosis/icd10
7. National Center for Health Statistic (NCHS). http://www.cdc.gov/nchs/icd9.htm
8. Workflow Management Coalition (WfMC). http://www.wfmc.org
9. Business Processe Management Initiative (BPMI). Business Process Modelling Language (BPML) specification. http://www.bpmi.org

10. Casati F. and Ming-Chien S. Semantic Analysis of Business Process Executions. Proceedings of EDBT'02, Prague, Czech Republic, March 2002.
11. Casati F., Dayal U., Sayal M., Shan MC.   Business Process Intelligence. http://www.hpl.hp.com/techreports/2002/HPL-2002-119.pdf
12. Greco G., Guzzo A., Manco G., Sacca' D. Mining Frequent Instances on Workflows. Proc. Seventh Pacific-Asia Conference on Knowledge Discovery and Data Mining (PAKDD), Seoul, South Korea, 2003.
13. Wil M.P., van der Aalst B.F., et al. Workflow mining: A survey of issues and approaches. Data Knowl. Eng. 47(2): 237-267, 2003.
14. Maruster L., van der Aalst W.M.P., et al. Automated Discovery of Workflow Models from Hospital Data. Proceedings of the ECAI Workshop on Knowledge Discovery and Spatial Data, pages 32-36, 2002.
15. Zamil K.Z. and Lee P.A. Taxonomy of Process Modeling Languages. CS-TR: 725, Department of Computing Science, University of Newcastle, 2001.
16. Shahar Y., Young O., Shalom E. et al. DeGeL: A Hybrid, Multiple-Ontology Framework for Specification and Retrieval of Clinical Guidelines. Proceedings of the 9th Conference on Artificial Intelligence in Medicine Europe (AIME) 03, Protaras, Cyprus, Springer-Verlag Heidelberg, pp. 122-131, Oct. 2003.
17. Rector A.L., Rogers J.E., Zanstra P.E., Van Der Haring E. OpenGALEN: Open Source Medical Terminology and Tools. Proc AMIA Symposium, 2003.
18. Systematized Nomenclature of Medicine (SNoMed). http://www.snomed.org
19. Tu S.W. and Musen M.A. From Guideline Modeling to Guideline Execution: Defining Guideline-Based Decision-Support Services. Proc. AMIA Symposium, Los Angeles, CA, 863-867, 2000.
20. Dell'Armi T., Leone N. Il Linguaggio DLP+ Exeura Internal Report, Cosenza, (June) 2004.
21. Faber W., Leone N. and Pfeifer G. The DLV homepage (www.dlvsystem.com) since 1996.

A Pathway for Process Improvement Activities in a Production Environment: A Case Study in a Rework Department

Onur Özkök, Fatma Pakdil, Fahri Buğra Çamlıca, Tolga Bektaş,
and İmdat Kara

Başkent University, Department of Industrial Engineering, 06530 Ankara, Turkey
{ozkok, camlica, fpakdil, tbektas, ikara}@baskent.edu.tr

Abstract. In this study, we describe our experience in process improvement in a study of one of the foremost manufacturing firms in Turkey. The firm is faced with a large increase in its inventory of returned products. To arrive at solutions to the problem and suggest improvements in the system under consideration, we offer a three-phase methodology which consists of identifying all the sub-problems that lead to the main problem and then solving those sub-problems. Although their solution involves the use of well-known techniques, the uniqueness of the approach lies in offering a systematic methodology that unifies these techniques from different disciplines in a sequential and integrated fashion.

1 Introduction

In this study, we describe a step-by-step business improvement experience in one of the foremost dishwasher production firms in Turkey. The firm is faced with a major problem of a huge amount of inventory of returned products due to a product takeback policy. For the solution to this problem, we propose a three-phase methodology. We also identify the tools that could or should be used in performing each step of the proposed methodology. Various studies, such as that performed by Rohleder and Silver [1], have focused on constructing a methodology for business improvement. However, none of them proposes any methodology as proposed in the present study. In very brief terms, our methodology includes three phases (see Figure 1). Phase 1 consists of *organizational issues*, such as describing and analyzing the system in order to find out the root and sub-causes of the problem, and their effects on the process. Connected to Phase 1, Phase 2 deals with the possible solutions regarding *workforce planning* related issues. The last phase constitutes a framework to address *material handling* related issues. Phase 3 is rather specific for this problem, whereas the first two phases are more general in the sense that they can be adapted to other firms. One important aspect of the methodology is that although it consists of well-known techniques belonging to different disciplines such as Total Quality Management (TQM), time and motion study, simulation, operations planning and material requirements planning, all the elements are combined in an integrated manner.

W.M.P. van der Aalst et al. (Eds.): BPM 2005, LNCS 3649, pp. 344–353, 2005.

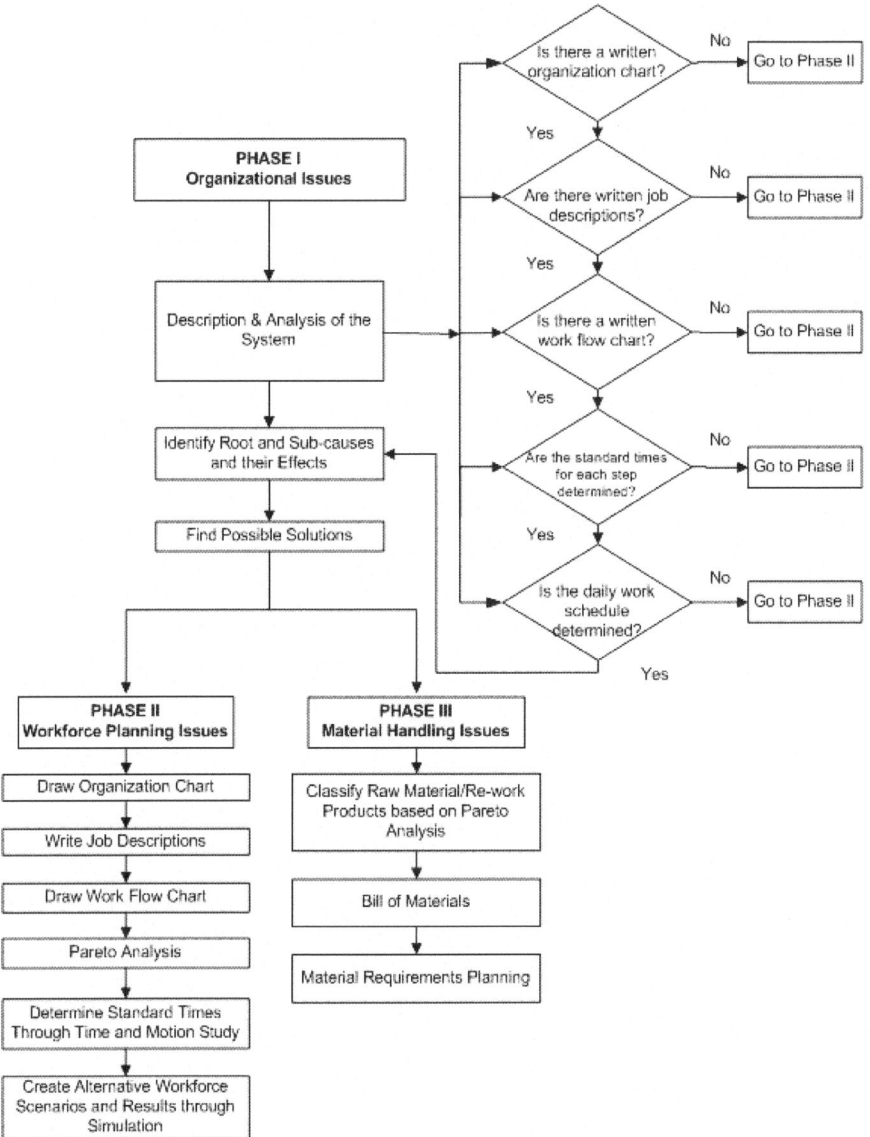

Fig. 1. Outline of the proposed methodology

Details on product takeback services can be found in Geyer and Van Wassenhove [2] and Klausner et al. [3]. More specifically, the former investigates product recovery strategies in product takeback schemes and the latter involves the integration of product takeback and technical service. Stevels et al. [4] report on the takeback and recycling costs of discarded television sets, and indicate that they can be brought down substantially by a combination of design improvements, technology improvements and by achieving economy of scale in the process. In what follows, we will describe in detail the system that is studied in this paper and the methodology proposed for the solution to the main problem considered.

2   Phase 1: Organizational Issues

This phase consists of system description, analysis of the current situation, identification of problems and possible solutions. Each are described in detail below. The heart of the study is in this phase. Any mistakes made at this phase cause further failures. It should be realized that if the problem and causes are not clear, further analysis cannot be performed.

2.1   Description and Analysis of the System

The firm in this study produces a single product, namely dishwashers. The firm has a specific policy to take back a sold product if the customer finds it to be defective or otherwise unsatisfactory. For such products, the *revision department* performs the re-work, repair and repacking operations, after which all of the products can be sold again. The current layout of the department is depicted in Figure 2. The revision department has three employees and its surface area is 189m^2. The department is divided into two sections. One section is used for repairing-packing and the other is for storing the products.

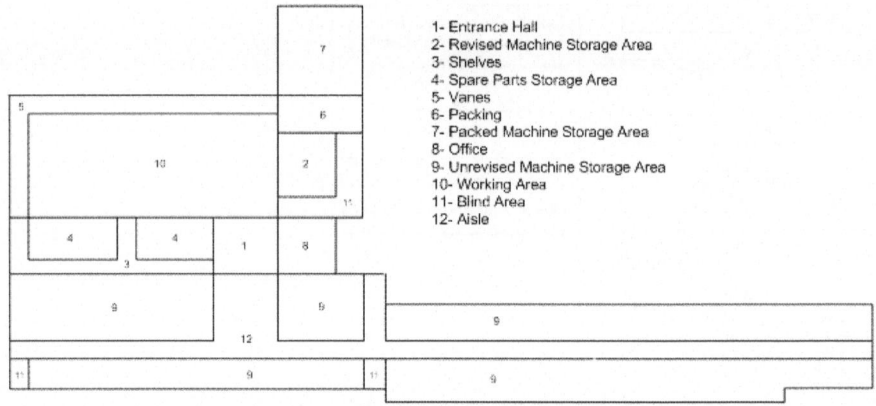

Fig. 2. Current layout of the rework department

The repairing-packing section consists of a working-tooling area, a spare part storage area, standard product elements and tool shelves, eight usable control vanes, a packing area, a packed product storage area and an office. The current workflow in the department can be roughly described as follows. The products that arrive at the department are transported from a storage area to the working area, which is in the middle of this section. After the employees identify the type of defect in each product, the product is subjected to a function test. There are eight vanes on which the function test can be performed. After the function test is completed and the related damage/defect is fixed, the products are stored for packing. Packing is performed in units of four, i.e. packing does not start until four products are revised. An outline of the workflow can be found in Figure 3.

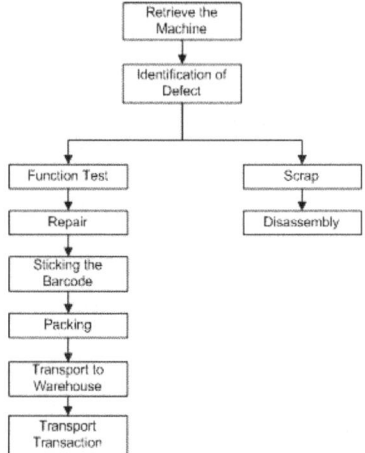

Fig. 3. Workflow diagram

The products that are returned to the rework department are classified with respect to their reason for refusal (either defective or found unsatisfactory). For this purpose, the firm uses 20 code groups, where each code group refers to a defect or damage found in the product. Each code group is further composed of specific subcategories, describing in detail the main source of the defect or damage. Table 1 shows all the code groups used in the firm, the number of subcategories included in each, and their explanations. For instance, the code group 140X refers to electronic defects. There are 5 subcategories of defects in this group, namely 1400, 1401, 1402, 1403 and 1409, which refer to general electrical, main card, control button, lamp and display card defects, respectively.

The rework department is now faced with a major problem of an unexpectedly large inventory of returned products. This inventory occupies a significant amount of workfloor space that is not allocated for it. Based on quality reports and a brainstorming session, the main problem was determined to be a lack of systematic workflow, which encompasses many specific and detailed subproblems in the revision department. Based on this specific problem, we first analyzed the organizational structure to determine whether there were any organizational deficiencies in terms of the organization chart, job descriptions, workflow chart, standard times for each sub-process, or daily operations planning. Previous documentation on these was limited and had not been implemented.

2.2 Identifying the Root Cause, Sub-causes, and Their Effects

By means of several comprehensive brainstorming sessions performed with the employees and engineers, various sub-problems, sub-causes and their effects on the whole process were determined. Based on this, we constructed the cause-and-effect (fish-bone) diagram given in Figure 4. The remaining part of this study was formed by analyzing and offering solutions to problems identified in the cause-and-effect diagram.

Table 1. Description of the code groups

Code Group	Explanation	Code Group	Explanation
10X	Button defects	20X	External damages
30X	Detergent, polish box defects	40X	Salt box defects
50X	Motor defects	60X	Cable defects
70X	Water spilling	80X	Water system defects
90X	External and internal trunk defects	110X	Program errors
100X	Heater, thermostat, controller defects	120X	Performance errors
130X	Furnace errors	140X	Electronic defects
150X	Dissatisfaction or misusage errors	160X	General machine defects
170X	Basket defects	180X	Emptying hose defects
190X	Door lock defects	9999	Unknown errors

2.3 Finding Possible Solutions

With a TQM approach in mind, we focused on using all possible techniques to eliminate the root cause and the main problem. At this step, all analysis results led to the rest of the study. As seen in Figure 1, we constituted this step in two phases, namely workforce planning issues and material handling issues, which seemed to cover the reasons for most of the problems.

3 Phase 2: Workforce Planning Issues

At this phase, we focused on re-organizing the revision department. This consisted of drawing the organization and the workflow charts, writing job descriptions (not already documented), performing a Pareto analysis to identify products that constitute a significant amount of inventory, determining standard times for such products through a time and motion study, and finally creating alternative scenarios through simulation for workforce and operations planning.

3.1 Organization Chart, Job Analysis, and Job Description

The organization chart and job descriptions are important parts of the organization structure. It should be remembered that a job description typically

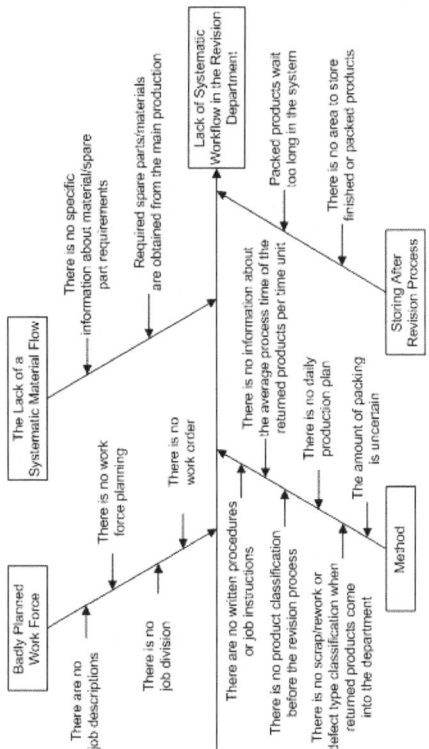

Fig. 4. Cause-and-effect diagram

describes job content, environment, and conditions of employment [5], [6]. In addition to this, the heart of the organization structure is the organization chart [7]. In analyzing the whole system, it appeared that there was no written organizational documentation, such as that mentioned in Phase 1 (section 2.1). As there were no job descriptions available, the employees were doing their jobs in individual ways which differed markedly from one employee to another in terms of process times. At this step, in order to eliminate this unorganized working system and pave the way for a standard process organization, we simultaneously monitored the whole process and the workflow, after which we performed a job analysis. Based on this, we structured an organization chart and a standard job description for the revision department employees.

3.2 Pareto Analysis

Each day, a number of products with varying defects and/or damages arrive at the revision department. As Table 1 indicates, the number of damages or defects is in the order of tens, and there are a total of 130 subcategories describing the defects, let alone those that cannot be appropriately identified. This makes the analysis too complicated to be carried out. Therefore, in order to properly distinguish the main sources of the problem and the bulk of the inventory sitting on the work floor, a Pareto analysis has been carried out with respect to the amount of products belonging to different code groups, and covering a period of 10 months (January - October 2004). The Pareto principle states that in general 80% of the errors in a system is caused by 20% of the problems [8]. According to the results of Pareto analysis, the main sources of the problem were found to be related to code groups 20X, 120X and 150X.

Table 2. Results of Pareto Analysis

Code	Explanation	Group %	Cumulative %
201	Packing has a defect	12.41	
202	Side trunk has a defect	49.64	14.63
204	External door has a defect	18.01	
1204	Product does not wash properly	44.50	
1205	Program does not work	15.81	11.68
1208	Product does not work	25.68	
1505	No problem, product working properly	43.65	
1508	Product change due to customer dissatisfaction	40.65	20.87
	Product is classified as scrap, no revision		24.62

These are presented in detail in Table 2, in which the column *Group %* indicates the percentage of the code in the code group (for instance, products with code 201 constitute 12.41% of all the products in group 20X). On the other hand, the last column labeled *Cumulative %* indicates the percentage of the code group among all the possible code groups (for instance, products in the code group 20X constitute 14.63% of all the products in the system). We found that products with defect/damage codes 201, 202, 204, 1204, 1205, 1208, 1505, or 1508 dominate the others in terms of their quantity on the workfloor. All the other code groups have a corresponding percentage of 6% or less. Therefore, in the ensuing analysis, we focused on these codes. Besides the previously mentioned defects, a significant amount of products are classified by the firm as scrap (24.62%). In this study we ignore this class since the decision is made by the firm.

3.3 Simulation Results

The solution to this problem should bear two aspects. Firstly, the amount of inventory must be reduced, and secondly a systematic workflow should be established so that a throughput, which will prevent an increase in the inventory, can be achieved. To get an idea of the state of the revision department, one needs to know its daily capacity. By comparing the daily capacity of the revision department with the rate of inflow, one can get an idea about changes in inventory. In order to determine the revision department's daily capacity, a simulation model was constructed with Arena 5.0 simulation software. Using such a model, one can estimate the department's average daily capacity for different scenarios with different workforces.

In order to develop the simulation model, the required standard times for each operation were determined with a time and motion study. The sample for the study consists of 15 observations for the packing operation and 10 observations for the rest of the process. We have also used this data to determine the probability distributions for each operation. In this way, the probabilistic nature of the revision process could be included in the capacity of the department. In the simulation study, two assumptions were made for all scenarios. The first related to working hours. The revision department works one 8-hour shift per day, with two 10-minute breaks and a 40-minute break for lunch. The second assumption concerned the packing operation, in which two employees work together and pack revised products in batches of 4.

From the simulation results, four workforce schedules, one of which was the current situation, were evaluated. The revision department currently has one part-time and two full-time employees. Full-time employees of the firm perform all operations of the revision process and the part-time employee only disassembles the returned products that are classified as scrap. In the first alternative the role of the part-time employee is changed to help the other employees pack the revised products, so that one full-time employee is sufficient for the packing operation and the other employee can continue the revision process. In the second alternative, there are two part-time employees and the packing operation is assigned exclusively to them. In the last alternative there are three full-time employees and one part-time employee who again helps pack the revised products. From the results of the simulation, daily capacities of the revision department were determined as in Table 3. The third alternative seems to be the most appropriate, as there is a huge inventory of returned products. However, after the amount of inventory is reduced to a reasonable level, the workforce schedule can be varied on the basis of the firm's decision.

Table 3. Comparison of alternatives via simulation

Scenario	Daily # of Reworked Products
Current Situation	32 products
Alternative 1 (2 full-time, 1 part-time employees)	45 products
Alternative 2 (2 full-time, 2 part-time employees)	68 products
Alternative 3 (3 full-time, 1 part-time employees)	65 products

4   Phase 3: Material Handling Issues

Even though simulation results indicate that 32 products can be repaired in the current situation, this quantity cannot be achieved in reality due to the lack of operations planning. That causes disorder in the revision department, as the employees do not know which machines are to be repaired in sequence. Also, spare parts are supplied by main production lines since there is no inventory for them. Both this disorder and the lack of spare parts result in inefficiency. To overcome these sub-problems, it is obvious that operations planning and an inventory system are required.

Knowing the department's daily capacity permits a daily operations plan to be made which includes information about the numbers and types of defects to be revised. With such a plan, one can easily see whether there is a problem in the process by comparing actual production to planned production. In terms of the inventory system, bills of materials were determined. Using these, material requirements were planned with software developed in MS Excel VBA. Using this software, material requirements can be determined according to the production plan, and the inventory in the revision department can be tracked. The number of products and defect types can differ according to the revision planning.

The other problem in material handling is that the products are placed randomly in the unrevised machine storage area. Random storage has major disadvantages and is used due to the fact that there is no planned storage policy. Products to be revised are also chosen randomly. These practices are not suitable for the proposed process, in which defect types to be repaired are determined before the retrieval of the products. To overcome these problems a dedicated storage policy [9] is proposed, where each defect type has a dedicated storage area. As we give higher priorities to products with the most frequent defects, they are placed as near as possible to the entrance of the working area. The placement of other products is less important, so they can be placed farther from the working area. In addition to this, to make material handling easier, the plan is to place the products on pallets and to use a hand forklift to transport them to the repair work area. In the current system two employees handle a product on the floor, due to its weight. However, with the use of pallets and a hand forklift, one employee could handle a product alone.

5   Conclusions

In this paper, we have described our experience in process improvement for a rework department in a production firm. We have proposed a methodology in which different techniques are integrated in a systematic manner. The application of the methodology results in an improved rework operation unit with a higher productivity and less work in process. The major contribution of this study lies in developing a specific methodology that improves a practical process of any firm. The steps of this methodology can be summarized as follows:

- Defining the main and sub-problems and root causes clearly,
- Establishing a well-documented process definition through a systematic approach (including job descriptions and organizational charts),
- Offering a well-built work-schedule in order to have a smooth workflow through Pareto analysis and extensive simulations, and
- Re-organizing the layout of the department and offering a storage policy that would result in a more productive and efficient working environment.

The generality of the proposed methodology, particularly that of Phases 1 and 2, shows that it can be applied to similar settings in many production environments. The success of Phases 1 and 2 affects the problem solving procedure's performance. What could be perceived as lessons derived from this study can be stated in two categories. Firstly, the problem and system should be comprehensively analyzed with an internal customer approach. In this specific case, the input rate of this department depends on the other relevant departments such as purchasing, manufacturing, quality control, and services network after sales. On this point, implementation of an internal customer performance measurement system seems to be an area for further research. The second lesson is that an analysis should be performed as to whether the system can reach its theoretical capacity or increase its actual capacity without further investments.

References

1. Rohleder, T.R., Silver, E.A.: A tutorial on business process improvement. Journal of Operations Management 15 (1997) 139-154
2. Geyer, R., Van Wassenhove, L.N.: Product take-back and component reuse, INSEAD Working Paper 2000/34/TM/CIMSO 12 (2000)
3. Klausner, M., Grimm, W.M., Horvath, A.: Integrating product takeback and technical service, Proceedings of the 1999 IEEE International Symposium on Electronics and the Environment, 1999, 48-53, Danvers, MA, USA
4. Stevels, A.L.N., A.A.P. Ram, Deckers, E.: Take-back of discarded consumer electronic products from the perspective of the producer Conditions for success. Journal of Cleaner Production 7 (1999) 383-389
5. Robbins, S.P. Coulter, M.: Management. 8th ed. Prentice Hall, New Jersey (2005)
6. Kreitner, R.: Management. Houghton Mifflin Co., Boston (1989)
7. Maynard, H.B.: Handbook of Business Administration. Mc-Graw Hill, New York (1970)
8. Kolarik, W.J.: Creating Quality. 3rd ed. McGraw-Hill, Berlin Heidelberg New York (1995)
9. Francis, R.L.: Facility layout and location : an analytical approach. 2nd ed. Prentice Hall, New Jersey (1992)

IT Support for Healthcare Processes

Richard Lenz[1] and Manfred Reichert[2]

[1] Institute for Medical Informatics, University of Marburg, Germany
`lenzr@mailer.uni-marburg.de`
[2] Information Systems Group, University of Twente, The Netherlands
`m.u.reichert@cs.utwente.nl`

Abstract. Patient treatment processes require the cooperation of different organizational units and medical disciplines. In such an environment an optimal process support becomes crucial. Though healthcare processes frequently change, and therefore the separation of the flow logic from the application code seems to be promising, workflow management technology has not yet been broadly used in healthcare environments. In this paper we discuss why it is difficult to adequately support patient treatment processes by IT systems and which challenges exist in this context. We identify different levels of process support and distinguish between generic process patterns and medical guidelines / pathways. While the former shall help to coordinate the healthcare process among different people and organizational units (e.g., the handling of a medical order), the latter are linked to medical treatment processes. Altogether there is a huge potential regarding the IT support of healthcare processes.

1 Introduction

Process-oriented information systems have been demanded for more than 20 years and terms like "continuity of care" have even been discussed for more than 50 years. Yet, healthcare (HC) organizations are still characterized by an increasing number of medical disciplines and specialized departments. The patient treatment process requires interdisciplinary cooperation and coordination. The recent trend towards HC networks and integrated care even increases the need to effectively support interdisciplinary cooperation along with the patient treatment process.

Healthcare heavily depends on both information and knowledge. Thus, information management plays an important role in the patient treatment process. Numerous studies have demonstrated positive effects when using IT systems in healthcare. In particular the preventability of adverse events in medicine has been in the focus of recent studies. Adverse events are defined as unintended injuries caused by medical management rather than the disease process [1]. It turned out that insufficient communication and missing information are among the major factors contributing to adverse events in medicine [2,3,4,5]. IT support for HC processes therefore has the potential to reduce the rate of adverse events by selectively providing accurate and timely information at the point of

W.M.P. van der Aalst et al. (Eds.): BPM 2005, LNCS 3649, pp. 354–363, 2005.

care [6]. Yet, there is a discrepancy between the potential and the actual usage of IT in healthcare. A recent IOM report even states that there is an "absence of real progress towards applying advances in information technology to improve administrative and clinical processes" [7].

Why is it so difficult to build IT systems that support a seamless flow of information along a patient's treatment process? In this paper we try to answer this question by identifying different levels of process support and by distinguishing between generic process patterns and the medical treatment process. Generic process patterns, such as medical order entry and result reporting, help to coordinate the HC process among different people and organizational units. Though clinical and administrative processes change over time, these generic patterns are a part of the fundamental processes of clinical practice, which basically remains the same for longer periods of time.

The specific patient treatment process, however, depends on medical knowledge and case specific decisions. Decisions are made by interpreting patient specific data according to medical knowledge. This decision process is very complex, as medical knowledge includes medical guidelines of various kinds and evidence levels, as well as individual experience of physicians. Moreover, medical knowledge continuously evolves over time. It is generally agreed that medical decision making cannot be automated. Yet, the patient treatment process can be improved by selectively providing medical knowledge in the context of the patient treatment process. The problem is to offer current knowledge, to only offer relevant knowledge according to the current context, to include the underlying evidence, and to support all of this in a way which seamlessly integrates with the physicians work practice.

In Section 2 we describe how traditional HC information systems support the fundamental processes in HC organizations and how standards contribute to integration. To find out how IT can support medical processes we will have a closer look on medical decision making and its implications for process-oriented IT architectures in HC environments in Sections 3– 5. Section 6 discusses demanding challenges with respect to the use of BPM technologies in the HC domain.

2   Generic Process Patterns in Healthcare

The architecture of typical hospital information systems is characterized by many different departmental systems, which are usually optimized for the support of different medical disciplines (e.g. radiology, cardiology, or pathology). The need to consolidate the data produced by these ancillary systems to a global patient-centered view and to support the cross-departmental processes has motivated the development of standards for data interchange in healthcare. These standards also play an important role when not only cross-departmental but also cross-organizational HC processes are to be supported. Today, HL7 is the leading standard for systems integration in healthcare. The name "Health Level 7" refers to the application layer in the OSI reference model [8].

HL7 is a message-based standard for the exchange of data among hospital computer applications. The standard defines which data items are to be interchanged when certain clinical trigger events occur (admission, discharge, or transfer of a patient are examples for such events). Since version 2.3 (1997) the standard has covered trigger events for patient administration, patient accounting, order entry, medical information management, clinical observation data, patient and resource scheduling, and others. The standard is continuously extended and newly arising topics, such as the integration of genomic data in Electronic Health Records, are handled in special interest groups (SIGs). Yet, the HL7 trigger events are intended to support standard communication patterns that will occur in any HC organization in basically the same way. Today's commercially available HC software usually only covers a relatively small portion of HL7, covering those communication patterns that are typically requested as essential basis for interconnecting disparate applications.

Despite well accepted standards for data integration (e.g., HL7, DICOM), HC applications are still far from plug and play compatibility (which is essential for realizing process-oriented clinical information systems). One reason is that existing standards do not address functional integration issues sufficiently. In order to avoid these difficulties common application frameworks are required which serve as a reference for programmers to create functionally compatible software components. Requirements for an application framework directed towards open systems in the HC domain are described in [9]. In general such a framework must provide specifications of interfaces and interaction protocols which are needed for embedding a software component into a system of cooperating components.

The best example for such a standard in the HC domain is the IHE initiative ("Integrating the Healthcare Enterprise") [10]. IHE does not develop new standards for data interchange but specifies integration profiles on the basis of HL7 and DICOM. Thereby actors and transactions are defined independently from any specific software product. An integration profile specifies how different actors interact via IHE transactions in order to perform a special task. These integration profiles serve as a semantic reference for application programmers, so that they can build software products that can be functionally integrated into an IHE conformant application framework. The core integration profile of IHE is called "Scheduled Workflow". The Scheduled Workflow Integration Profile establishes a seamless flow of information in a typical imaging encounter, by precisely specifying the actors and transactions that are involved in the process of image acquisition. By fixing the required workflow steps and the corresponding transactions, IHE ensures the consistency of patient information from registration through ordering, scheduling, imaging acquisition, storage, and viewing. This consistency is also important for subsequent workflow steps, such as reporting. However, this kind of workflow support has nothing to do with the traditional idea of workflow management systems: to separate the flow of control from application logic in order to keep the workflow maintainable [11]. The idea of these standards is to establish stable generic communication patterns that help to integrate autonomously developed IT components.

3   Medical Decision Making

The HC process is often called a diagnostic-therapeutic cycle comprising observation, reasoning, and action. Each pass of this cycle is aimed at decreasing the uncertainty about the patient's disease or the actual state of the disease process [12]. Thus, the observation stage always starts with the patient history (if it is available) and proceeds with diagnostic procedures which are selected based on available information. It is the job of an (Electronic) Patient Record to assist HC personnel in making informed decisions. Consequently, the system should present relevant information at the time of data acquisition and at the time of order entry. Thereby, an important question to be answered is how to determine what is relevant. Availability of relevant information is a precondition for decisions - medical knowledge guides these decisions. Medical knowledge, however, is not limited to what is found in medical textbooks. A large part of medical knowledge is not explicit but tacit, and tacit knowledge heavily influences information needs by care providers as well as the course of the care process [13,14]. Moreover, medical knowledge evolves over time. According to [15] knowledge is created and expanded through social interaction between tacit and explicit knowledge (cf. Fig. 1). This process, called "knowledge conversion", is a social process between individuals, rather than a process within an individual. Stefanelli describes this process of knowledge creation in [14]. In order to make medical knowledge broadly available, medical experts need to externalize their tacit knowledge. Thus, improving HC processes has a lot to do with stimulating and managing the knowledge conversion processes.

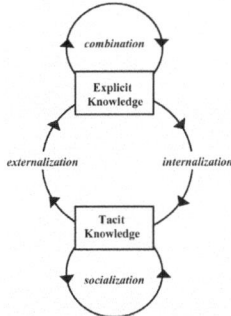

Fig. 1. The knowledge conversion process in a knowledge creating organization [15]

Supporting the HC process by bringing explicit medical knowledge to the point of care is closely related to developing and implementing medical practice guidelines. The MeSH (Medical Subject Headings) dictionary defines medical practice guidelines as "work consisting of a set of directions or principles to assist the health care practitioner with patient care decisions about appropriate diagnostic, therapeutic, or other clinical procedures for specific clinical circumstances". Guidelines are aimed at an evidence-based and economically reasonable medical treatment process, and at improving outcomes and decreasing the undesired variation of HC quality [16]. Developing guidelines is essentially a consensus process among medical experts. Yet, there is a gap between the information contained in published clinical practice guidelines and the knowledge and information that are necessary to implement them [16,17]. Methods for closing this gap by using information technology have been in the focus of medical informatics research for decades (e.g. [17,18,19]).

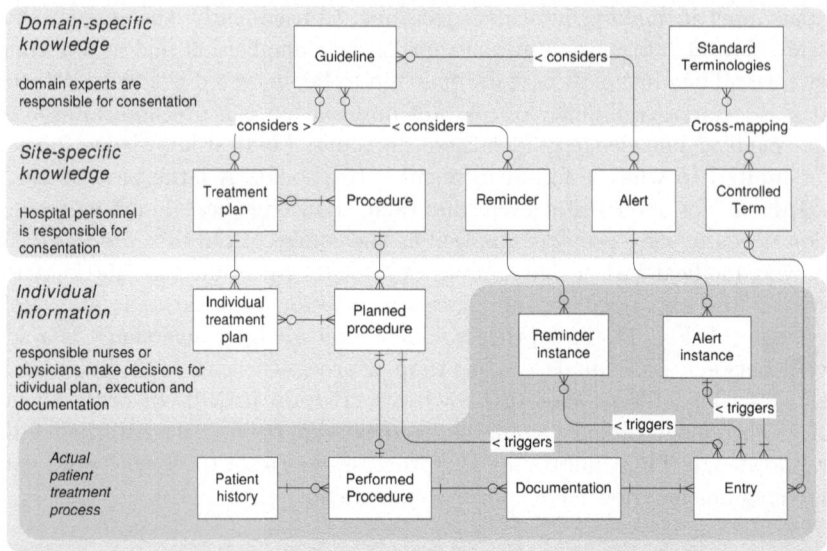

Fig. 2. Influence of explicit medical knowledge on the HC process

Medical pathways can be used as a basis for implementing guidelines [20] and sometimes they are confused with guidelines. In contrast to guidelines, though, pathways are related to a concrete setting and include a time component: Pathways are planned process patterns that are aimed at improving process quality and resource usage. Pathways are not standardized generic processes like those described within the IHE integration profiles (cf. Section 2). Pathways need a consensus process; they must be tailored to local and individual circumstances, which requires a cooperative initiative of clinical experts, process participants, and managers. Pathways can be used as a platform to implement guidelines (e.g., by routinely collecting the information required by a guideline). Selecting a guideline for implementation also requires an agreement of HC professionals and patients, because there are different kinds of guidelines with different origins and goals, and sometimes even conflicting recommendations. Likewise, to improve a patient treatment process across organizational borders, consensus on common practices is required in the first place. Once this consensus is achieved, the next question is how to implement it in practice. To be effective, a guideline must be easily accessible. Ideally, it should be embedded into the clinical work practice, and the physician should not need to explicitly look it up. Otherwise, there is always a risk of overlooking important information while the patient is in the office. Previous work has primarily demonstrated a positive influence of computer-generated alerts and reminders [21], which can be integrated into clinical work practice. Recent research indicates that this is exactly the major difficulty with implementing more complex multi-step guidelines: How to integrate them into the clinical workflow [19]?

The influence of different levels of explicit medical knowledge on the patient care process is illustrated in Fig. 2: Medical guidelines are distinguished from site specific treatment plans (e.g. clinical pathways). A treatment plan comprises multiple diagnostic or therapeutic steps (procedures). Instances of a treatment plan need to be adapted to the specific needs of an individual patient. The actual treatment process may still deviate from the individual treatment plan, because it is also led by tacit knowledge and not only by explicit knowledge. Yet, explicit medical knowledge can still be brought to the point of care: Documentation of performed or ordered procedures may trigger alerts or reminders. An alert (e.g., "Lab alert" for values out of bounds or about to evolve into dangerous areas) requires some kind of notification system to inform the physician. Reminders can be used to inform the person who enters data instantaneously if data are entered which are not plausible or if expected data entries have not been made.

4   Integrating Knowledge and Information Management

Medical pathways are one attempt to establish a platform for implementing complex guidelines. Thereby, predefined checklists that ask the right questions in the right context, predefined order sets, and well placed reminders are some of the techniques that can be used to improve process quality and reduce the required documentation overhead. All these techniques require the computer to be able to make use of the patient's clinical data. The first obstacle to achieving this is to represent guidelines in a computer-interpretable form, i.e., translating narrative guidelines into equivalent ones that use coded data. This task is cumbersome and also bares the risk of distorting the intent of the original guideline.

To overcome such problems numerous models have been developed to formally represent medical guidelines and medical knowledge (e.g., Arden Syntax [22], GLIF [23] PROforma [24], EON [25], Asbru [26]). Recent surveys have compared these approaches [27,28] One of the central goals is to define standard representation formats for medical knowledge in order to be able to share guidelines among different information systems in different organizations. In practice, however, it turned out that the main obstacle to be solved here is - once again - an integration problem: The data definitions in pre-defined formal guidelines may not map to the data available in an existing electronic health record system [29]. Typically, operational systems have to be modified and extended in order to acquire the necessary information needed for guideline implementation. Few guidelines have been successfully implemented into real clinical settings by using these systems and predefined, formally specified guidelines. Recent research has recognized these difficulties and focuses on process models for transforming text-based guidelines into clinical practice [17]. Standard formats for guideline representation do have their place in guideline implementation, but the integration problems to be solved are a matter of semantics rather than format. Guideline implementation requires a high level of data integration, because computerized reminders typically refer to both type and instance level semantics. More complex guidelines also need to refer to a formally established context comprising status information. The challenge to be solved for distributed HC networks is to establish a sufficient degree of integration as basis for guideline implementation, and to find practical solutions to cope with the evolving HC domain.

5   Implications for Process-Oriented IT Architectures

The adequate support of HC processes raises a number of requirements for process-oriented IT architectures. In particular, an integrated process support, information management, and knowledge management on different levels is needed.

In order to adequately support generic process patterns (cf. Section 2) and to provide the needed information at the point of care, responsive IT architectures must consider the cross-departmental nature of clinical processes. To avoid media breaks we either need highly integrated systems or semantically compatible application components. Semantic compatibility, in turn, subsumes functional integration. Besides application integration comprehensive process support is needed for coping with clinical and administrative processes. Process support functions should comprise both standard services (e.g., process enactment and monitoring, worklist management) and advanced features (e.g., ad-hoc changes of single process instances during runtime).

The handling of medical guidelines and pathways requires an approach which allows reaching an organization-specific consensus on them. Due to the evolving nature of guidelines and pathways, in addition, responsive IT infrastructures must enable their continuous extension and adaptation (cf. [11]). This should be accomplished under the control of the respective HC organization and its medical staff. In order to achieve this, we need sophisticated tools for (graphically) specifying the flow logic of guidelines and pathways at a high semantic level. Furthermore, patient treatment processes (and their monitoring) as well as patient information must be linked to the defined guidelines and pathways.

IT infrastructures, which support medical guidelines, should allow the explicit definition of medical knowledge and enable its combined use with patient-related information. This requires a minimum of semantic control. In order to avoid problems at the operational level (when linking guidelines with patient information), we need tools for defining guidelines based on the medical concepts and medical terminology already used within the operational systems. Doing so, again we must consider the evolving nature of the HC domain. In particular, we must support the evolution (and versioning) of ontologies and controlled vocabularies, to which the different guidelines refer, as well.

Current hospital information systems are far from having realized such process-oriented architectures. This has led to pragmatic solutions and workarounds in order to reduce the overall effort for integrating heterogeneous application components and to enable a requirements-driven system evolution.

6   Challenges for Process Management Technologies

Recently, we have seen an increasing adoption of BPM technologies and workflow management systems (WfMS) by enterprises. Respective technologies enable the definition, execution, and monitoring of operational processes. In connection with Web service technology, in addition, the benefits of process automation and optimization from within a single enterprise can be transferred to cross-organizational processes as well. In principle, WfMS offer promising perspectives for the support of HC processes as well. By separating the process flow logic from the application code, processes can be quicker implemented and adapted when compared to conventional approaches. Current WfMS, however, are far from being applicable to a broader range of HC processes. Existing WfMS are either too rigid or they do not meet the various requirements discussed above. In particular they are not able to cope with the dynamics and the evolving nature of HC processes. In any case, we need a more advanced process management technology, which enables the integrated support of medical processes, medical knowledge and patient-related information on different levels.

Though the scope of generic process patterns and medical pathways is different there are several commonalities. Graphical descriptions of the respective flow logic are useful and in both cases these descriptions must be linked with other components of the system architecture (e.g., application systems or patient-related information). At the process instance level, in addition, in both cases deviations from the pre-defined flow logic may become necessary and should therefore be supported. As mentioned, variations in the course of a disease or a treatment process are deeply inherent to medicine; the unforeseen event is to some degree a "normal" phenomenon. Medical personnel must be free to react and is trained to do so. However, respective deviations from the pre-planned process must not lead to errors or inconsistencies. Tools enabling them must be easy to handle, self-explaining and - most important - their use in exceptional situations should be not more cumbersome and time-consuming than simply handling the exception by a telephone call to the right person. Altogether we need adaptive process management technology which allows to rapidly set up new HC processes and to quickly adapt existing ones.

When deviations from predefined process patterns or medical pathways occur they should be documented and logged. A logical next step then is to continuously monitor, analyze, and mine this change log and to "learn" from it. Best case, based on this data necessary decisions can be made quickly and accurately to modify HC processes, to dynamically allocate resources, or to prioritize work.

Healthcare more and more changes from isolated patient treatment episodes towards continuous treatment involving multiple HC professionals and HC institutions. Therefore hospitals need to be linked with other HC organizations and general practioners, but also with insurance companies and governmental organizations, over wide area networks transporting sensitive patient data. The adequate support of distributed HC networks will result in novel workflow scenarios raising a number of challenging issues. A sufficient degree of process and information integration and the semantic interoperability of the different HC systems are crucial in this context. The same applies to privacy and security issues in connection with the exchange of patient data.

With advances in technology we can further observe that HC processes, which were previously confined to the hospital, will more and more be provided outside it. In this context technologies like mobile devices and wearable computing will be important drivers of change. Examples of upcoming application scenarios include the contactless monitoring of patients, the provision of smart agents collecting patient data during homecare, and the automatic detection of emergency situations. All these scenarios will demand further challenges with respect to the management of HC processes, ranging from the support of mobile and distributed processes to the seamless integration of different devices.

7   Summary

Based on many years of first-hand knowledge of the HC domain and our personal working experience in hospitals, based on the experiences made in numerous clinical projects, and having also deep insights into existing BPM technologies, we believe that the IT support of HC processes offers a huge potential. However, a number of challenges exist and new ones will arise in connection with novel technologies, which must be carefully understood and which require basic research before we can come to a complete solution approach. We believe that the realization of process-oriented IT architectures in HC is a great challenge for the BPM community - if not even the "killer application" for this type of technology.

References

1. Vincent, C., Neale, G., Woloshynowych, M.: Adverse events in british hospitals: preliminary retrospective record review. BMJ **322** (2001) 517–519
2. Brennan, T., Leape, L., Laird, N., Hebert, L., Localio, A., Lawthers, A.: Incidence of adverse events and negligence in hospitalized patients. results of the harvard medical practice study. N Engl J Med **324** (1991) 370–376
3. Kohn, L., Corrigan, J., Donaldson, M.: To Err Is Human. Building a Safer Health System. National Academy Press (2000)
4. Bhasale, A., Miller, G., Reid, S., Britt, H.: Analysing potential harm in australian general practice: an incident-monitoring study. Med J Aust **169** (1998) 73–76
5. Wilson, R., Runciman, W., Gibberd, R., Harrison, B., Newby, L., Hamilton, J.: The quality in australian health care study. Med J Aust **163** (1995) 458–471
6. McDonald, C., Hui, S., Smith, D., Tierney, W., Cohen, S., Weinberger, M.: Reminders to physicians from an introspective computer medical record. a two-year randomized trial. Ann Intern Med **100** (1984) 130–138

7. Committee on Quality of Health Care in America IoM: Crossing the Quality Chasm: A New Health System for the 21st Century. IOM (2001)
8. Tanenbaum, A.: Computer networks. Englewood Cliffs (1988)
9. Lenz, R., Huff, S., Geissbuehler, A.: Report of conference track 2: pathways to open architectures. Int J Med Inf **69** (2003)
10. Vegoda, P.: Introducing the ihe (integrating the healthcare enterprise) concept. J Healthcare Information Management **16** (2002) 22–24
11. Lenz, R., Kuhn, K.: Towards a continuous evolution and adaptation of information systems in healthcare. Int J Med Inf **73** (2004) 75–89
12. van Bemmel, J., Musen, M.: Handbook of Medical Informatics. Springer (1997)
13. Stefanelli, M.: Knowledge and process management in health care organizations. Methods Inf Med **43** (2004) 525–535
14. Stefanelli, M.: The socio-organizational age of artificial intelligence in medicine. Artif Intell Med **23** (2001) 25–47
15. Nonaka, I., Takeuchi, H.: The knowledge creating company. Oxford University Press (1995)
16. Gross, P., Greenfield, S., Cretin, S., Ferguson, J., Grimshaw, J., Grol, R.: Optimal methods for guideline implementation: conclusions from leeds castle meeting. Med Care **39** (2001) 85–92
17. Shiffman, R., Michel, G., Essaihi, A., Thornquist, E.: Bridging the guideline implementation gap: a systematic, document-centered approach to guideline implementation. J Am Med Inform Assoc **11** (2004) 418–426
18. Bates, D., Kuperman, G., Wang, S., Gandhi, T., Kittler, A., Volk, L.: Ten commandments for effective clinical decision support: making the practice of evidence-based medicine a reality. J Am Med Inform Assoc **10** (2003) 523–530
19. Maviglia, S., Zielstorff, R., M, M.P., Teich, J., Bates, D., Kuperman, G.: Automating complex guidelines for chronic disease: lessons learned. J Am Med Inform Assoc **10** (2003) 154–165
20. Schriefer, J.: The synergy of pathways and algorithms: two tools work better than one. Jt Comm J Qual Improv **20** (1994) 485–499
21. Shiffman, R., Liaw, Y., Brandt, C., Corb, G.: Computer-based guideline implementation systems: a systematic review of functionality and effectiveness. J Am Med Inform Assoc **6** (1999) 104–114
22. Jenders, R., Hripcsak, G., Sideli, R., DuMouchel, W., Cimino, J.: Medical decision support: experience with implementing the arden syntax at the columbia-presbyterian medical center. In: Proc. SCAMC. (1995) 169–173
23. Ohno-Machado, L., Gennari, J., Murphy, S., Jain, N., Tu, S., Oliver, D.: The guideline interchange format: a model for representing guidelines. J Am Med Inform Assoc **5** (1998) 357–372
24. Fox, J., Johns, N., Rahmanzadeh, A.: Disseminating medical knowledge: the proforma approach. Artif Intell Med **14** (1998)
25. Musen, M.: Domain ontologies in software engineering: use of protege with the eon architecture. Methods Inf Med **37** (1998)
26. Votruba, P., Miksch, S., Kosara, R.: acilitating knowledge maintenance of clinical guidelines and protocols. In: Proc. Medinfo. (1998) 57–61
27. De Clercq, P., Blom, B., Korsten, H., Hasman, A.: Approaches for creating computer-interpretable guidelines that facilitate decision support. Artif Intell Med **31** (2004) 1–27
28. Van de Velde, R., Degoulet, P.: Clinical Information Systems. Springer (2003)
29. Pryor, T., Hripcsak, G.: Sharing mlm's: an experiment between columbia-presbyterian and lds hospital. In: Proc. SCAMC'93. (1993) 399–403

From RosettaNet PIPs to BPEL Processes: A Three Level Approach for Business Protocols

Rania Khalaf

IBM TJ Watson Research Center, Hawthorne, NY 10532, USA
rkhalaf@us.ibm.com

Abstract. Business protocols in n-party interactions often require centralized protocol design but decentralized execution without the intervention of the designing party. In this paper, we tackle the problem for RosettaNet PIPs by creating a BPEL solution. We do so using a three–level approach, based on BPEL, for defining such multi–party protocols: templating for high–level patterns, specialization for particular protocols, and implementation for particular realizations of a protocol.

1   Introduction

Business protocols in n-party interactions often require the design of the protocol to be centralized but the execution to be done without the intervention of the designing party. In such cases it is desirable that instead of distributing one description from a global (neutral) point of view, or one description from a single point of view that mediates with all (hub-and-spokes), to distribute a package with an abstract business process for each party involved in the interaction and that defines its own required behavior. When plugged together, these processes show the complete behavior of all involved parties.

It is also desirable to easily derive compliant implementations at each party from the abstract definition. This lowers the bar for parties to participate, basically shifting the burden from the participating party to the protocol designer. In such a set-up, the latter's task now includes checking whether the created processes are in fact compatible while the participant, which may be a small business, has very little to do.

We present the mapping RosettaNet Partner Interface Processes (PIPs) [4], which match our target environment above, to BPEL processes using a three–level approach for creating such protocol definitions. For details on mapping the quality of service requirements of PIPs that are not at the business process level by using the appropriate additional Web services specifications, see [1].

2   Background

2.1   RosettaNet

RosettaNet (http://rosettanet.org) is a consortium of companies from the Electronics industry that defines an open e-business environment using an open–standards approach for the requirements and behavior of business partners. The

W.M.P. van der Aalst et al. (Eds.): BPM 2005, LNCS 3649, pp. 364–373, 2005.
© Springer-Verlag Berlin Heidelberg 2005

RosettaNet Implementation Framework (RNIF) represents requirements of a middleware infrastructure that supports the RosettaNet approach. Partner Interface Processes (PIPs) [4] are RosettaNet documents specifying two-party interactions to meet a business goal, such as processing a purchase order (PIP 3A4) or inquiring about a price (PIP 3A2). A PIP definition consists of a text document describing the expected requirements of the involved parties, including Message Sequence Charts (MSCs), DTD definitions of these messages, and quality of service requirements such as time-outs, security considerations, non-repudiation, and retries.

2.2   BPEL

The Business Process Execution Language for Web Services, BPEL for short [2], is an XML workflow-based composition language for Web services. A BPEL process is exposed as one or more Web services with WSDL portTypes (interfaces), with typed "partnerLinks" defining its connections with other parties. BPEL activities are of two types: Primitive activities with pre–defined behavior like invoking a Web service(invoke), receiving and replying to invocations(receive/reply), waiting (wait), throwing faults(throw); Structured activities impose control on activities nested within them, such as strict sequencing (sequence), or parallelism (flow). Conditional control links may impose additional ordering within a "flow". "Scopes" provide additional capabilities to their activities, such as data scoping, fault handling (detailed in [3]), compensation handling, and event handling. Instance management is done using "correlation sets", basically using fields in incoming messages to route the messages to one of possibly several running instances of the same process.

BPEL Processes can be either abstract and executable. A BPEL "abstract process" provides a specification of a service's behavior and may hide private information that is not relevant to the other participants in the interaction. For example, an abstract process may initialize a variable with an opaque value. The abstract process then only has the other parties it interacts with in the interaction protocol as partners. On the other hand, the executable variant provides a process definition with enough information to be interpreted.

3   Premise

We propose a "templating-specialization-implementation" approach for the design and implementation of processes that follow patterns, such PIPs:

1. Templating: Creating BPEL abstract "templates" that capture the message exchange and behavioral pattern of each party involved.
2. Specialization: Making use of such patterns by creating full valid *abstract BPEL* processes from these templates that represent a specialization of that pattern. For RosettaNet, these constitute the definition of a particular PIP.
3. Implementation: Creating executable artifacts from these abstract BPEL processes so that each party can have an implementation of the PIP with

minimal effort and low adoption barrier. Simple derivation rules are provided by the designing party for participants to follow.

4   Related Work

[6,14,5] provide "neutral" models of interaction protocols. BPSS[5] has been used to map PIPs. Drawbacks include: no tie–in to back–end systems, requiring a translation, limited fault handling, lack of separation of concerns where one party can get only its own requirements for participation. A hub–and–spokes model (single mediating process) is also clearly undesirable for our scenarios. Approaches combining a neutral definition with individual behavior definitions are in [12] and [7]. There does not seem to be a straight–forward mechanism to derive a fully executable processes from these abstractions. [9] use "RosettaNet Controls", Java entities that simplify process creation, to encode PIP patterns. They can exchange messages as prescribed by RNIF, but seem to be proprietary to BEA's products.

We presented a demo of part of this approach to the RosettaNet board and the BPEL Committee in Dec. 2003, focusing on combining multiple PIPs using a master proces. Each run kicked off 7 executable BPEL processes. These were manually created from abstract process, in turn manually created from the PIP definitions. It was a motivator for a RosettaNet working–group for PIPs over Web Services, and provided input to the BPEL standardization committee for abstract processes in V2.0. This paper adds to the demo and [1] in that they did not go into the details of deriving executables and did not use templates for process patterns.

5   Creating the Process Templates

A template consists of a single "pseudo-abstractBPEL" file for each partner involved. These are similar to abstract BPEL files, except for highlighting points of variability that a full abstract BPEL process would have to include. These are specified by omitting the BPEL elements/attributes required to be filled in. A schema validation of the process quickly shows where those points are, and could be done by a graphical editor. One should also provide any additional information needed in order to interpret hints about missing information, such as how to know which reply and receive activities match if the operation name is omitted.

Ongoing work such as [10] and [8] addresses the challenges of proving compatibility when starting with behavioral definitions of each partner separately.

5.1   Creating the Templates for Two–Action PIPs

PIPs can be separated into several patterns. We focus on "asynchronous two-action PIPs".In a two-action PIP party A sends a message to party B. Party B then sends a response back to party A. Business level acknowledgements for both

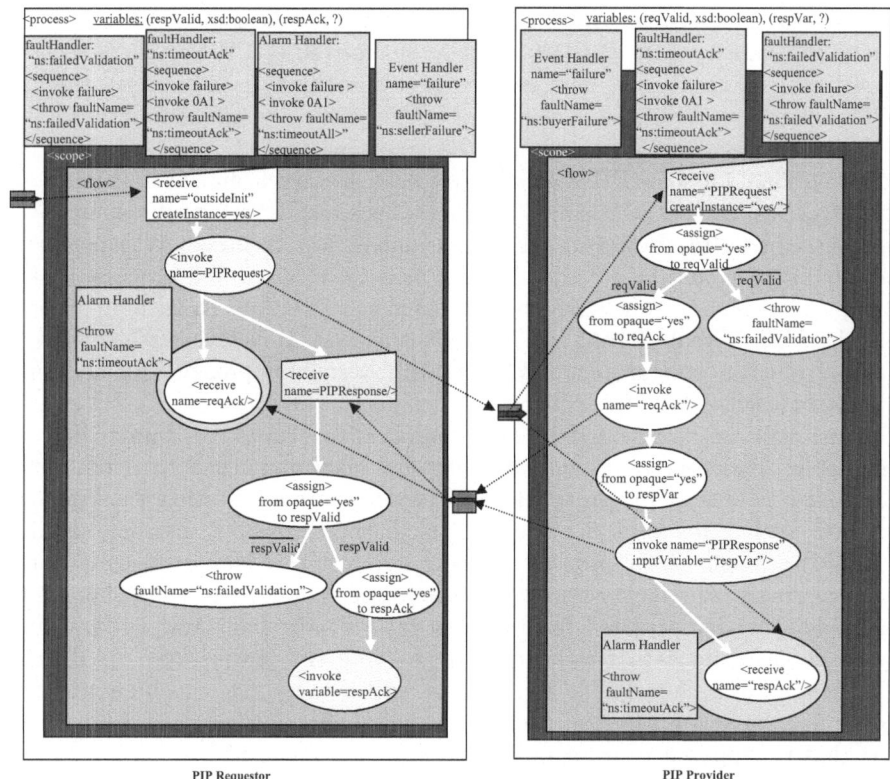

Fig. 1. Template of all asynchronous two-action PIPs

messages must also occur within a given time-frame. Based on this information and RosettaNet's message handling framework(RNIF), we can create a template for all asynchronous two-action PIPs and illustrated in Figure 1.

The points of variability between PIPs of this class, shown in the figure, are:

- *PartnerLinks* the process would be easier to read if these had names from the specific PIP definitions. Since PIPs are always two-party (except for 0A1, but the interactions with that are clear), it is straight–forward to know the activities with the other party.
- *Operation names* based on the portType and operation on the WSDL files created from the PIP.
- *Variables with Message Types* come from the mapping of the specific PIP's DTDs to schema and the resulting WSDL files.
- *Timer values* on the alarm handlers.
- *Correlation sets*. These will depend on the messages in the WSDL.

Since the names of the operations and portTypes are dependent on the specific PIP, we use the names of "receive" and "invoke" activities to hint that activities on the different processes must match. For example, the invoke named "PIPRequest" sends a message from the PIPRequestor to be received by PIPProvider's receive named "PIPRequest". The next section show how this is used to create the full BPEL abstract processes for each type of PIP from these templates.

Validation placeholders are seen at the conditions named *Valid. This validation is optional for some PIPs, going beyond the usual schema/XML validation.

Retries are not in the business processes, because PIP retries are at the network level in RNIF. However, acknowledgements are business level since they are used not for "message received" but "message received and processed", for example after having done an external dictionary validation. To faithfully represent the PIP, we require that any implementing system have a reliable delivery mechanism with configurable retry intervals. Alternate approaches for handling retries are discussed in [1].

Time-outs are represented as alarm handlers on scopes containing the corresponding receive activity, and throwing the relevant fault if time runs out. The fault is caught at the process level, sending failure notification to all parties involved so they don't block. The fault is then rethrown. A message handler is defined for each of the processes that can receive this fault message and react accordingly. The fault rethrow and the message handler provide plugpoints for implementations created from this to react to the fault and perform any necessary clean-up with back-end systems. Note the 'invokes' to "0A1", the "Notification of Failure PIP". This PIP is used if the partner to be notified of failure may have stopped executing. It receives a single error notification message that is supposed to come out of band, and is modeled here as a separate Web service.

6 Creating the Abstract BPEL Processes from a Template

In this step, one creates the complete abstract BPEL files that together encode the behavior of a subclass of the protocols represented by the template. The templates must be filled in using the information about these protocols. We will use the Purchase Order PIP, PIP3A4, to illustrate modeling of a two-action PIP from the template above. The resulting abstract processes are shown in Figure 2.

6.1 Abstract BPEL for PIP3A4 from Above Templates

The 3A4 has two parties: a buyer and a seller. The buyer sends a purchase order, expecting an acknowledgement and a confirmation that the order has been processed both within given time-frames. Upon receiving the confirmation, the buyer acknowledges its receipt.

One WSDL portType is created for each party, to exchange the PIP messages. Add two more portTypes: one to be used by the external initiator of the processes, and the other for the 0A1 PIP. For 3A4, we create a sellerPortType with three operations: purchaseOrderRequest, failure, and purchaseOrderAcknowledgment. On the other side, we have a buyerPortType that will have

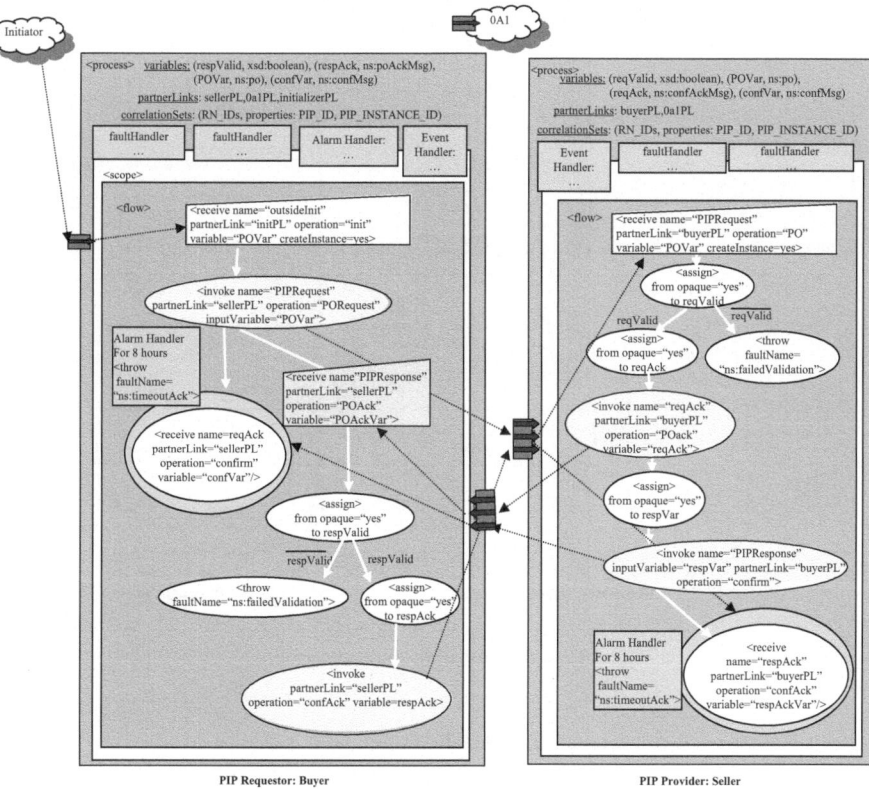

Fig. 2. Abstract processes of 3A4 PIP

another three operations: purchaseOrderInitiate, failure, and purchaseOrderConfirmation. These operations are all one–way. Now we have the information to define the BPEL partnerLinks in our templates. For each of the two process templates:

– Add a partnerLink for interacting with the other PIP participant, with a 'myRole' with the portType the process offers to the other participant, and a 'partnerRole' with the portType it requires from that participant.
  • Add this partnerLink, along with the corresponding portType, on all the activities that interact with the other participant. For example, in the buyer process add the partnerLink attribute with value "sellerPL" to the invoke named PIPRequest.
– Add a partnerLink on the PIP requestor with a myRole with the portType offered to the external initiator that will kick–off the process.
  • Add this partnerLink and its corresponding "myRole" portType to the receive named outsideInit.

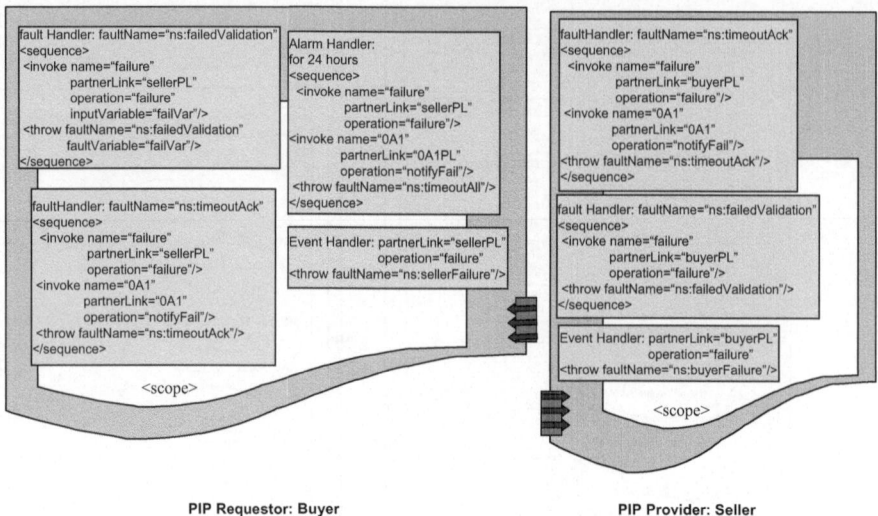

Fig. 3. Handlers on the scope

An example of a partnerLink for the seller is:

```
partnerLink name=''buyerPL'' partnerLinkType=''ns:PIP3A4plt'' myRole=''seller' partnerRole=''buyer''
```

Then, add the operations. The PIPRequest is the message with the main business document. The invoke on the PIPRequestor and the receive on the PIPProvider will therefore refer to the operations and messages that deal with the PurchaseOrder. Variables come next. Each message that will be exchanged results in one variable with that message type, with the corresponding interaction activity(ies) adding the variable attribute. Also, if the PIP does not require dictionary validation, we replace the assignment into one that copies "true" instead of an opaque value into the validation variable in each of the two processes.

Consider the wait times for the alarm handlers. The global handler gets its value directly from the PIP document (24 hours for the 3A4 buyer). The other time-outs are on the acknowledgements. Since we assume a reliable messaging infrastructure, these values are the result of multiplying the 'retry count' in the PIP document by the interval between retries (8 hours for 3A4). Such a failure would terminate the PIP, kicking off the 0A1.

Finally, we need to choose the correlation sets. In RosettaNet PIPs, two values uniquely identify a PIP instance: the PIP Code and the PIP Instance ID. We use these as our correlation tokens.initiator). Note that correlations and portType names are not in the figure to lessen clutter.

7   Creating Compliant Executable BPEL Processes

Up to here, the definitions are created by the designing party which is expected to have experienced business process designers. However, implementations are created by customers, who may be small and medium businesses whose core business is not technology and possibly lack expensive business modeling expertise or tools.

The abstract processes created above enable protocol implementers to generate compliant BPEL implementations simply by following an easy set of completion rules defined by the designers. This leverages the fact that BPEL abstract and executable processes share the same semantic and syntactic base. These compliant BPEL processes can be executed in any product supporting BPEL, thereby lowering the bar for an entity to participate.

Creating compliant executables is one of the main complexities of such approaches. By providing simple but strict rules for creating the executables from the abstract processes, we ensure that an entry level user can create compliant executables without much difficulty.

7.1   Creating the Executables for Our Class of PIPs

The real work is forwarded from the executable to the company's back–end systems, exposed as Web services. The partnerLink joining these systems to the process has the same type as that joining the two PIP processes together, but with roles reversed. The rules for creating the executables for our PIP are provided below. Adding anything else (ie: new links, lifecycle activities, ...) is prohibited.

- flip the abstractProcess attribute to "false".
- replace all opaque assign activities with a sequence activity containing either:
  - Interactions (invoke, invoke+receive) with backend systems for necessary data: In our asynchronous case, a one way invoke followed by a receive activity. The partnerLink on both these activities is that of the back–end systems.Added invokes are not allowed to throw faults. Or,
  - Any assignment activities to any necessary data adaptation or copying of values (except endpoint references), especially those to propagate correlation set fields.
- Add, in the handlers only, any assigns (in sequence) required for setting values of the variables used by activities in the handlers.
- Add any variables and correlation sets that are needed for the new activities and partnerLink, and for the activities in the handlers.
- Optionally add fault handlers at the top process level to do any required clean–up and notification of one's back–end systems in case of process errors.

7.2   Compliance of the Executables

Different notions of equivalence between artifacts [13] have been proposed based on the class of problems addressed. In this section, we show that the executable processes created by the steps above are equivalent to the abstract processes at hand according to a notion of conformance by Martens [11]. Martens states that the behavior of two processes is equivalent if an executable model could replace the abstract one without requiring changes to the environment in which the process operates. This is done by comparing "Communication Graphs" containing each process's externally visible behavior. It simple to show that this is true: The additions provided above are one–to–one replacements of existing activities. The new activities do not affect lifecycle since they cannot throw faults, and do not affect control because they are contained inside "sequence". Consider the externally visible behavior for the communication graph. The original (assign) activities did not have externally visible behavior. New invokes and receives may only interact with new partnerLinks not in the abstract process, and are therefore not visible to the environment(other PIP process). The communication graph of the executable is hence identical to that of the abstract making the processes equivalent with respect to the other PIP party.

8   Discussion and Future Work

BPEL abstract processes in V1.1 were created with 'observable behavior' in mind, leading to the current restrictions on syntax and semantics. The templates we define in section 5 are not valid abstract BPEL processes: They omit information essential for specifying full message exchanges. In response to requests for enabling a syntactic fill–in–the–blank templating usage, the BPEL committee is considering proposals for more 'opaque' tokens, and syntactic completion rules for the next version .

The RosettaNet case shows that there are legitimate, simple cases where one needs a mix between purely the externally observable behavior or pure fill–in–the–blank templating. For example, one must not change the timer values, but should still be allowed add new partnerlinks and fault handlers. Neither of these would be acceptable for this case.

Future work topics include wiring BPEL processes together (particularly relevant in more than 2 party interactions), looking into other classes of protocols for which this approach can be generalized, how each step can be generalized or parameterized, and expanding the approach for multiple parties.

9   Conclusion

This paper presents BPEL processes for RosettaNet PIPs, using a three–level approach that can generalize to similar environments with multi–party processes. We cover the creation from overall design, to specialization, to implementation. We highlight two main challenges in creating such bottom–up approaches, namely compatibility of the different processes and compliance between the abstract and executable artifacts. We focused on the usage of multiple levels of refinement using BPEL processes, levaraging a single language with wide–spread industry support (BPEL) due to the proliferation of tools and its abstract and executable variants with related semantics. We provide a real and practical illustrative example of using the BPEL abstract–executable continuum and show how it one can express various levels of abstractions through iterative refinement.

Acknowledgments. Paul Bunter and Sreedhar Janaswamy especially. Also Keeranoor Kumar, Ralph Hertlein, Peter Williams, Shishir Saxena for work on project. Francisco Curbera, Axel Martens and Frank Leymann for advice.

References

1. P. Bunter, R. Hertlein, R. Khalaf, and A. Nadalin. An approach to moving industry business messagung standards to web services. To appear online on IBM DeveloperWorks.
2. F. Curbera, Y. Goland, J. Klein, F. Leymann, D. Roller, S. Thatte, and S. Weerawarana. Business Process Execution Language for Web Service v1.1. Online at http://www.ibm.com/developerworks/library/ws-bpel, May 2003.
3. F. Curbera, R. Khalaf, F. Leymann, and S. Weerawarana. Exception handling in the BPEL4WS language. In *Proc. of BPM 2003*, LNCS 2678, Eindhoven, the Netherlands, June 2003. Springer.
4. S. Damodaran. B2B integration over the internet with XML: RosettaNet successes and challenges. In *Proc. of WWW 2004, Alternate Track Papers and Posters*, pages 188–195, New York, NY, May 2004. ACM Press.
5. ebXML. ebXML business process specification schema version 1.01. Online at http://www.ebxml.org/specs/ebBPSS.pdf, May 2001.
6. X. Fu, T. Bultan, and J. Su. Conversation specification: A new approach to design and analysis of e-service composition. In *Proc. WWW2003*, Budapest, Hungary, May 2003.
7. X. Fu, T. Bultan, and J. Su. A top-down approach to modeling global behaviors of web services. In *Requirements Engineering for Open Systems Workshop (REOS 2003)*, Monterey, California, sep 2003.
8. X. Fu, T. Bultan, and J. Su. Analysis of interacting bpel web services. In *Proc. of WWW 2004*, New York, NY, May 2004. ACM Press.
9. BEA Systems Inc. BEA weblogic workshop help online: Building RosettaNet participant business processes. Technical report, dec 2003.
10. A. Martens. On compatibility of web services. *Petri Net Newsletter*, 65:12–20, October 2003.
11. A. Martens. Consistency between executable and abstract processes. In *Proc. of the IEEE Conference on e-Technology, e-Commerce and e-Service (EEE 2005)*, Hong Kong, Mar 2005.
12. W.M.P. van der Aalst and M. Weske. The P2P approach to interorganizational workflow. In *Proc. of the Conference On Advanced Information Systems Engineering (CAiSE 2001)*, volume 2068 of *LNCS*, Berlin, Germany, 2001. Springer.
13. R. J. van Glabbeek. The linear time - branching time spectrum. In *Proc. of CONCUR 90*, number 458 in LNCS. Springer-Verlag, 1990.
14. W3C. Web Service Choreography Interface (WSCI) 1.0. Online at http://wwwo.sun.com/software/xml/developers/wsci/wsci-spec-10.pdf.

Using Software Quality Characteristics to Measure Business Process Quality

A. Selcuk Guceglioglu and Onur Demirors

Informatics Institute, Middle East Technical University, Inonu Bulvari,
06531, Ankara, Turkey
+90 312 210 3741
aselcuk@ieee.org, demirors@metu.edu.tr

Abstract. Organizations frequently use product based organizational performance models to measure the effects of information system (IS) on their organizations. This paper introduces a complementary process based approach that is founded on measuring business process quality attributes. These quality attributes are defined on the basis of ISO/IEC 9126 Software Product Quality Model. The new process quality attributes are applied in an experiment and results are discussed in the paper.

1 Introduction

IS capabilities have been advancing at a rapid rate and motivating organizations to accomplish much more investment. In 2002, $780 billion was spent for IS in the United States alone [1]. Although IS expenditures seem quite high, there are few systematic guidelines to measure the organizational impact of IS investments [2], [3]. The available studies on organizational impact of IS focus on the product based organizational performance models to manage IS investment. These studies provide organizations with guidelines for measuring cost and time related issues, but they have some constraints in identifying IS effects, isolating the contributions of IS effects from other contributors and using the performance measures in specific categories of organizations such as in public organizations.

In this paper, a complementary process-based approach is developed to measure the effects of IS on business process. This new approach focuses on the quality aspects of the processes. As it is known that business processes are one of the most fundamental assets of organizations, modifications performed on them whether in the way of improvements or innovations cause immediate effects on the success of the organizations. This approach therefore enables organizations to get early feedback about the potential IS investment. In the remaining chapters of the paper firstly, related search is summarized as a background to depict the relation of our model to the IS literature. Secondly, the new model is introduced and its measurement categories are given. Thirdly, implementation of the model and its results are summarized. Finally, conclusions and future works are stated.

2 Related Research

One of the most widely known models for measuring the effects of IS is DeLone and McLean IS Success Model [2]. The available studies in Organizational Impact

W.M.P. van der Aalst et al. (Eds.): BPM 2005, LNCS 3649, pp. 374–379, 2005.

dimension include organizational performance based models and measures. These studies concentrate on the effects of IS for creating organizational changes and relations of these changes with the firm level output measures such as productivity growth and market value [3]. In this circumstance, DeLone & McLean IS Success Model states that the studies in this dimension are at beginning stage and much works are required to be done in categorizing and measuring the changes in the organizations and work practices. Another well-known model is Seddon's IS Effectiveness Matrix [4]. In similar to the DeLone & McLean model, this model focuses on organizational performance based measures such as firm growth, return on assets, percent change in labor, and market share. There is also a process oriented study for assessing IS effects on organizations [5]. Although IS effects on business processes are dealt with in the study, it is not precisely defined to measure the effects on the process. The changes occurred in organizations due to IS effects are given in only conceptual level.

There are some factors which affect business processes, and IS is one of the most considerable of them [5]. IS affects both operational and managerial processes. IS influences operational processes by automating them with providing technologies of work flow systems, flexible manufacturing, data capture devices, and computer aided design tools (CAD). Similarly, IS influences managerial processes by providing electronic mail, database and decision support tools. These effects can be categorized. For instance, Davenport [6] concentrates on the effects of IS in the business process reengineering perspective and identifies nine opportunities for business process innovation through IS effects as automational, informational, sequential, tracking, analytical, geographical, integrative, intellectual, and disintermediating. In another categorization [5], IS can have three separate but complementary effects on the business processes. These are automational, informational and transformational effects.

3 A Process Based Model for Measuring IS Effects on Business Process Quality

The structure of the model that we have developed is based on the ISO/IEC 9126 Software Product Quality Model [7]. There are close relationships between software and business process [8]. For instance, both of them have logical structures with inputs, operations and outputs whether in the form of functions or activities. The "software product" logically matches with "business process", and "function" of software product with "activity" of business process. A similar relation between software product and function exists in the business process and activity.

After the evaluation of the ISO/IEC 9126, some appropriate software quality metrics are chosen. The business process quality attributes are defined according to these selected business process specific metrics and, guidelines of how they can be measured are detailed. The model is designed in four-leveled structure [9]. The first level is called as category. There is one category as "quality". The second level is called as characteristic. The quality category includes Functionality, Reliability, Usability and Maintainability characteristics. The third level is for subcharacteristics

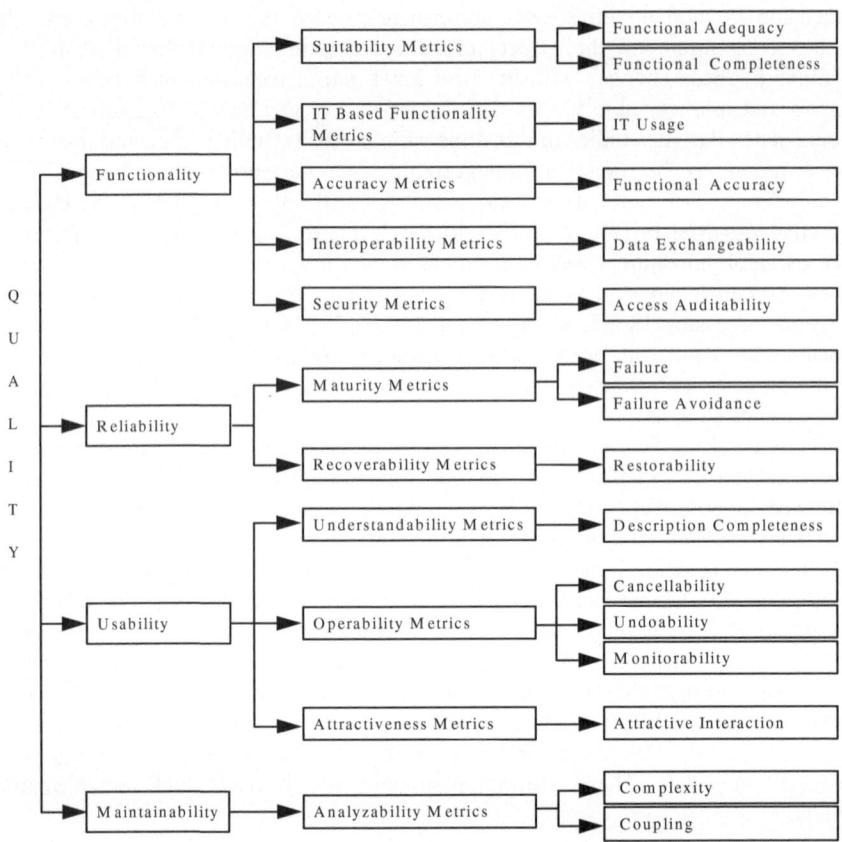

Fig. 1. Measurement categories and metrics of the model

and finally, fourth level is for metrics to measure the business process quality attributes. The quality category is given with its levels in Figure 1.

Functionality characteristic is defined for evaluating the capability of the process to provide functionality properties in the subcharacteristics of Suitability, Information Technology (IT) based Functionality, Accuracy, Interoperability and Security. Reliability characteristic is used for evaluating the capability of the process to provide reliability properties in the subcharacteristics of Maturity and Recoverability. Usability characteristic is used for evaluating the capability of the process to provide usability properties in the subcharacteristics of Understandability, Operability and Attractiveness. Maintainability characteristic is used for evaluating the capability of the process to provide maintainability properties in the subcharacteristic of Analyzability.

4   An Experiment for Measuring Process Quality Attributes

The implementation of the model is accomplished on a sample business process in an organization [9]. In the implementation, a business process, named as "Meeting

Material Request", is selected from Warehouse Department of the organization. While the departments are performing their tasks in the organization, they meet material needs from this department. The department is organized to meet these material requests and also purchase new material, repair and maintain existing material. It has approximately 80 staff and 7 basic business processes about material operations including Material Purchasing, Material Registration, Material Return, Material Repair and Maintenance. In the implementation of the model, static business process definitions are used. The present state (AS-IS) of the process is taken into consideration. This process has 29 activities. Each activity is clearly identified by explaining with actors who take part in, forms, tools and applications that are used in. Unified Modeling Language (UML) Activity Diagram is used for modeling the process.

Table 1. Results of the functionality characteristic

Subcharacteristic	Attribute	Formula	Result
Suitability	Functional Adequacy	$X = 1 - (6/29)$	0.793
	Functional Completeness	$X = 1 - (7/29)$	0.759
IT Based Functionality	IT usage	$X = 1 - (22/29)$	0.241
Accuracy	Functional Accuracy	$X = 1 - (14/29)$	0.518
Interoperability	Data Exchangeability	$X = 1 - (1/7)$	0.857
Security	Access auditability	$X = 1 - (2/29)$	0.931

The results of the first characteristic, functionality, are given in Table 1. The common desirable features of the functionality metrics are their closeness to the 1. The results of the functional characteristics can reveal some beneficial insights about the present state of the process. Access Auditability of the activities is near to 1 that is considered as satisfactory. The accesses of the users to the resources such as reading or updating inventory records and document record books are under the control. Unlike the Access Auditability, IT Usage is the most far away from 1. This low value shows improvement opportunities. On the other hand, another low value is about Functional Accuracy. It shows that process has critical functional accuracy problems and needs to be improved. The results of Functional Adequacy and Functional Completeness are close to each other and also to 1. It can be said that process activities are almost adequate and complete. The last result is for Data Exchangeability. Its value emphasizes that the process can be interoperable with other processes.

The results of the reliability characteristics are given in Table 2. Failure attribute shows the number of user based errors. According to the measurement, 23 Failures may be happened in the process (one activity may have more than one error). When the failures are investigated, it is recognized that most of the failures are user based failures such as writing incorrect material name and updating incorrect material number. The second attribute is Failure Avoidance. 6 Failure Avoidance mechanisms are detected in the present state of the process such as using the previous document template. The last attribute is about Restorability. There is 1 Restorability mechanisms as daily backups of inventory records to floppy disks.

Table 2. Results of the reliability characteristic

Subcharacteristic	Attribute	Result
Maturity	Failure	X = 23
	Failure Avoidance	X = 6
Recoverability	Restorability	X = 1

Table 3. Results of the usability characteristic

Subcharacteristic	Attribute	Formula	Result
Understandability	Description Completeness	X = 1 – (5/29)	0.828
Operability	Cancellability	X = 1 – (6/29)	0.793
	Undoability	X = 1 – (6/29)	0.793
	Monitorability	X = 1 – (25/29)	0.138
Attractiveness	Attractive Interaction	4 good,	4 good,
		4 very good	4 very good

Table 4. Results of the maintainability characteristic

Subcharacteristic	Attribute	Result
Analyzability	Complexity	X = 3
	Coupling	X = 2

The results of the third characteristic, usability, are given in Table 3. According to the results, Description Completeness attribute is near to 1 considered as understandable with its present definitions. This thought may be supported by Attractiveness Interaction attribute with its high value. The other attributes that are close to 1 are Cancellability and Undoability. These attributes show that the process activities can be undone or canceled before they are completed. On the other hand, Monitorability attribute has the lowest value. This indicates that status of the process activities cannot be monitored satisfactorily.

The results of the fourth characteristic, maintainability, are given in Table 4. Complexity attribute indicates the number of decision points as 3. The other attribute, Coupling, implies the number of business processes that are communicated as 2.

In order to give additional information about the process, cycle time and cost values are measured. Cycle time is calculated by adding the elapsed time in each activity. According to the result, the cycle time is 260 minutes. The other information is about cost. Although cost concept includes wide range coverage, we only calculate actors' salary-based cost. The actors' salaries and elapsed time in each activity are multiplied to find the cost. The cost is $25.340 for one cycle.

5 Conclusions

In this paper, a new process based model is proposed as a complementary to the available product based models. The model can be used with product based models to evaluate different IS investment alternatives. The product based measurements and results of the model can help the organizations for selecting the most suitable alternatives to their processes. When the effects of IS on processes are considered in process improvement (PI) scope, the implementation of the model shows that the model can be useful in PI purpose studies. The changes in the process quality attributes after implementation of a PI study demonstrate the impacts of the study.

As a prerequisite, organizations must model their business processes to apply the model. It may be thought as a possible restriction, but, today, organizations should already have modeling of their processes to follow and improve them. Another possible restriction may be high number of process. This makes difficult the implementation of the model. In this case, a sample business process set can be formed according to the criticality of the processes before applying the model.

In the future, further experiments will be performed to improve the model. These studies provide significant feedbacks to the model. By means of the feedbacks, the definitions of the attributes will be more clear and concrete. New measurement categories or attributes can be added to extent the scope of the model. The correlations between the attributes can also be examined and identified.

References

1. Jeffery, M., Leliveld, I., Best Practices in IT Portfolio Management, MIT Sloan Management Review (2004)
2. DeLone, W.H., McLean, E.R., The DeLone and McLean Model of Information Systems Success: A Ten-Year Update, Journal of Management Information Systems, Vol. 19, No. 4 (2003) 9-30
3. Brynjolfsson, E., Hitt L., The Three Faces of IT Value: Theory and Evidence, Proceedings of the Fifteenth International Conference on Information Systems, Vancouver, BC (1994) 263-276
4. Seddon P.B., Staples S., Patnayakuni R., Bowtell M., Dimensions of Information Systems Success, Communications of the Association for Information Systems, Vol.2 Article 20 (1999)
5. Mooney J.G., Gurbaxani V., Kraemer K.L., A Process Oriented Framework for Assessing the Business Value of Information Technology, The Data Base for Advances in Information Systems, Vol. 27, No. 2 (1996)
6. Davenport, T.H., Process innovation: reengineering work through information technology, Boston, Mass: Harvard Business School Press, 062117110523 (1993)
7. ISO/IEC FCD 9126-1.2: Information Technology - Software product quality -Part 1: Quality model
8. Osterweil, L., Software Processes are Software Too, Proceedings of the Ninth International Conference on Software Engineering, Monterey, CA (1987) 2-13
9. Demirors, O., Guceglioglu, A.S., A Model for Using Software Quality Characteristic to Measure Business Process Quality, Technical Report, METU/II-TR-2005-08, Department of Information System, University of METU (2005)

Business Process Modelling and Improvement Using TAD Methodology

Nadja Damij[1] and Talib Damij[2]

[1] Faculty of Economics, University of Ljubljana, Kardeljeva ploscad 17,
1000 Ljubljana, Slovenia
`nadja.damij@gmail.com`
[2] Faculty of Economics, University of Ljubljana, Kardeljeva ploscad 17,
1000 Ljubljana, Slovenia
`talib.damij@ef.uni-lj.si`

Abstract. This paper aims at carrying out business process modelling and business process improvement using TAD methodology. The methodology consists of six phases; the first three deal with business process modelling and improvement, and the last three phases continue with the implementation of the improved business processes by developing an information system, which covers the areas discussed. To make the methodology capable of carrying out business process modelling and improvement, the first three phases of the methodology are intended to solve this problem successfully. The problem of Sales_Claim is used as an example to show the implementation of the new concepts of the methodology. The results of this work are very encouraging.

1 Introduction

The aim of this work is to present the use of a methodology called TAD (Tabular Application Development) in the field of business process modelling and improvement. To make the methodology capable of carrying out business process modelling and improvement, the first three phases of the methodology are aimed at a complete solution of this problem. The first phase identifies the business processes of the enterprise discussed, the second phase presents a new idea of business process modelling, and the third phase shows an interesting way to improve business processes. The last three phases deal with implementing changes identified in the first three phases by developing the information system of the enterprise. The fourth phase develops a systems object model, the fifth phase designs the systems, and the last phase implements the system. The TAD methodology presents a new concept, which is simple and very different from the ideas used in other approaches. This paper discusses the first three phases in detail to cover the field of business process modelling and improvement.

2 TAD Methodology

TAD is an object-oriented methodology, which consists of six phases. The first three deal with problem definition, process modelling, and process improvement. And the

W.M.P. van der Aalst et al. (Eds.): BPM 2005, LNCS 3649, pp. 380–385, 2005.

last three concern object model development, design of the system, and implementation of the system.

The created tables describe the functioning of the enterprise by showing business processes, work processes and activities. These tables are then analysed in order to identify the necessary changes that have to be implemented to improve the functioning of the enterprise.

2.1 Problem Definition

In the first phase, the real world of the problem to be solved is identified and reduced to understandable terms. This is achieved by identifying the business processes of the system discussed. To do that, we have to conduct interviews with the management at strategic, business and operational levels. The purpose of these interviews is to define the strategic goals and objectives of the enterprise. After defining the goals and objectives at the enterprise level, we continue conducting interviews with the management at business and operational levels with the aim of identifying business processes, and the goals of the management at business level.

TAD methodology uses the term "entity" to define a user, group of users, a unit department or any source of information. An entity may be internal or external.

To connect the identified strategic and business goals we define one or more analyses related to the each goal or business process. These analyses are collected into a table called the analysis table. The analysis table is structured as follows: the columns of the table represent the entities, and rows of the table represent the analyses defined. An asterisk in any square(i,j) in the analysis table means that the entity defined in column j requires analysis defined in row i.

Sales_Claim. Many customers are not satisfied with the solution obtained considering their claim application, and the process is time consuming. Corresponding to the first phase of the methodology, interviews were conducted at different management levels and strategic, business, and operational goals identified.

Table 1. The Analysis Table of Sales_Claim

Entity / Analysis	Sales Management	Purchasing Management	Dispatch Management	Quality Coordinator
1. Analysis of claims by sales units	*	*		
2. Analysis of claims by customers	*	*		*
3. Analysis of claims by products g	*	*		*
4. Analysis of claims by suppliers	*	*		*
5. Information about causes of claims	*	*	*	*
6. Information about solution of claims	*	*	*	
7. Information about duration of claims	*	*		

2.2 Business Process Modelling

The second phase of TAD methodology deals with modelling the business processes of the enterprise for which we intend to carry out business process improvement or to develop an information system. To do that, we continue with organising interviews with the management at operational level corresponding to the plan of interviews developed in the first phase. The purpose of the interviews is to identify work processes.

A work process is the lowest-level group activity within the organisation [5]. A work process is a collection of activities followed in a determined order in carrying out distinguishable work to produce a certain output.

Activity Table. An effective way to carryout business process modelling is achieved by developing two tables called the activity table and the task table [4]. The activity table is organised as follows: the first column represents business processes, the second column shows work processes, the activities are listed in the rows of the third column, and the entities are introduced in the remaining columns of the table grouped by the departments to which they belong. Such organisation of the activity table enables us to create a clear and visible picture of every business process and its work processes, and also of each work process and its activities (see Table 2).

To make the activity table represent the real world, we link the activities horizontally and vertically. Horizontal linkage means that each activity must be connected with those entities in the columns which are involved in performing it. To indicate this, letters S and T are used. Letter S in square(i,j) means that entity(j) is a source entity for activity(i). This means that entity(j) performs activity(i), otherwise entity(j) starts the activity(i). For example, S_1 in square(2,1) means that internal entity(1) performs activity(2). Letter T in square(i,j) means that entity(j) is a target entity for activity(i). This entity(j) performs activity(i) if it is an internal entity, otherwise entity(j) accepts an output from other entity. For example, S_6 in square(1,6) and T_1 in square(1,1) mean that external entity(6) start activity(1) and internal entity(6) performs it; see Table 2. Any activity may have one or more source entities and also a number of target entities. For this reason the letters S and T are also indexed by the index of the source entity of the treated activity.

Vertical linkage is used to define the order in which the activities are performed. Vertical linkage is used only in connection with the internal entities. This is achieved by using the letters P and U to connect the activities. Letter P in square(i,j) means that activity(i) is a predecessor to some activity indicated by U. Letter U in square(i,j) means that activity(i) is a successor to another activity indicated by P.

Any activity may have one or more predecessors and also one or more successors. The letters P and U are indexed by the index of the predecessor activity. For example, to define that activity(1) is a predecessor to activity(2), we write P_1 in square(1,1) and U_1 in square(2,1); see Table 2. Furthermore, an activity may have several successors. In this case, an OR operator is used to indicate each of the alternative successors. An OR operator is represented by an asterisk written on the right side of letter U. An OR operator can also be used with letters S and T.

Sales_Claim. The first column of Table 2 shows the business process Sales_Claim and the second column shows its four work processes. Table 2 has 21 activities and 6 entities. The first 5 entities, belong to the Sales and Warehouse departments, are internal and the last one is external. The first activity "Receive Claim_Note" means the

Sales_Claim clerk receives a claim note from a customer. The second activity means that the Sales_Claim clerk registers the claim note. The third activity means that Sales_Claim clerk prints a claim form. Thus we write S_6 in square(1,6), T_6 in square(1,1) and S_1 in square(2,1) and square(3,1) to indicate the first, second and third activity. Furthermore, concerning the Sales_Claim clerk we find that the first activity is a predecessor to the second activity and third activity. For this reason we write P_1 in square(1,1), U_1 in square(2,1), and square(3,1).

Task Table. As we develop the activity table we simultaneously develop another table, the task table, which is very important in describing activities in detail. The task table is organised as follows: the activities are represented in the rows and the characteristics of the activities are defined in the columns. Activity characteristics are:

<u>Description:</u> this is used to write a short description of the activity defined in the current row of the task table.

<u>Time:</u> this is used to denote that the activity discussed needs a determined time to be accomplished.

<u>Business Rule:</u> Business rules are precise statements that describe, constrain and control the structure, operations and strategies of a business.

<u>Input/Output:</u> this is used to indicate which inputs and outputs are connected with the activity described.

<u>Cost:</u> this is sum of the costs of the resources needed to accomplish an activity.

Sales_Claim. Table 3 describes in detail all parameters determined in the above defined organisation of the task table. Because of space limitations, only the first 6 activities of the activity table (Table 2) are described in Table 3.

2.3 Business Process Improvement

The third phase of TAD methodology deals with carrying out business process improvement (BPI) in the enterprise concerned. More precisely, this phase deals with identifying and implementing changes to improve the functioning of the enterprise. The third phase has two steps. The first step deals with partitioning of the activity table. The second carries out BPI in the analysis, activity and task tables.

Business process improvement is achieved by precise analysis of the analysis and activity tables, suggesting changes and improvements, and giving solutions for existing problems. To make our work easier the activity table could be divided by using an approach called table partitioning. The concept of table partitioning is based on the idea of identifying basic work processes in the framework of each business process. A basic work process is a work process performed in all circumstances and related to different alternative work processes for performing alternative options. A non-basic work process is recognized in the activity table by an OR operator indicating the row of its first activity.

Thus the result of table partitioning is the creation of several parts (subtables) of the activity table. Each of these parts could be analysed separately. Therefore, the analyst analyses each part carefully in order to understand it completely.

Sales_Claim. Tables 1, 2 and 3 were analysed carefully. We found that the claim documentation is in a state of permanent movement between the three different entities which are involved in performing similar activities. For this reason, we suggested unifying these jobs by defining one entity; this is claim clerk, who takes care of the claim.

Table 2. Activity Table of Sales_Claim business process

Business Process	Work Process	Department / Entity / Activity	Sales		Warehouse			
			1. Sales Claim Clerk	2. Sales Clerk	3. Warehouse Claim Clerk	4. Manager of Dispatch D.	5. Stock Keeper	6. Customer
Sales_Claim	1. Reception of Sales_ Claim	1. Receive Claim_ Note	T_6 P_1					S_6
		2. Register Claim_Note	S_1 U_1					
		3. Print Claim_Form	S_1 U_1, P_3					
		4. Collect Claim Documentation	S_1 U_1, P_4					
		5. Send Claim Documentation	S_1 U_3, U_4, P_5	T_1				
		6. Determine Claim Solution Path		S_2 U_5, P_6				
		7. Return Claim Documentation	T_2	S_2 U_6, P_7				
	2. Under-Received Products	8. Forward Claim Documentation	S_1 U_7^*, P_8		T_1			
		9. Analyse Quantity of Products			S_3 U_8, P_9			
		10. Approve the Claim			S_3 U_9^*, P_{10}			
		11. Reject the Claim			S_3 U_9^*, P_{11}			
		12. Return Claim Documentation	T_3		S_3 U_{10}, U_{11}, P_{12}			
		13. Check Approval	S_1 P_{13}, U_{12}					
		14. Issu Credit_Note	S_1 U_{13}^*				T_1	
		15. Send Explantion	S_1 U_{13}^*				T_1	
	3. Over-Received Products	16. Create Addiional Invoice	S_1 U_7^*, P_{16}					
		17. Send Additional Invoice	S_1 U_{16}					T_1
	4. Warehouse Prod- uct Return	18. Require Tranport Products Back	S_1 U_7^*, P_{18}		T_1	T_1		
		19. Create Transport Schedule			S_3 U_{18}, P_{19}			
		20. Send Transport Schedule			S_3 U_{19}, P_{20}		T_1	
		21. Inform about Shipment Recepion			T_5, S_3	T_3	S_5 U_{20}	

Table 3. Task Table of Sales_Claim business process

Characteristic Activity	Description	Time	Business Rule	Input/ Output	Cost ($)
1. Receive Claim_Note	Sales claim clerk receives a Claim_Note from customer	10 m	Check Order and Shipment	Claim_Note, Order, Shipment	14
2. Register Claim_Note	Sales claim clerk registers customer's Claim_Note	10 m		Claim_Note	14
3. Print Claim_Note	Sales claim clerk prints a Claim_form	5 m		Claim_Form	7
4. Collect Claim Docs	Sales claim clerk collects the rest of claim documentation	30 m		Dispatch Order Invoice	42
5. Send Claim Docs	Sales claim clerk sends claim doc. to Sales clerk	5 m		Claim Documentation	7
6. Determine Solution Path	Sales clerk determines the cause of the claim	30-60m	Determine the claim cause	Claim Documentation	45-90

3   Conclusion

The main problem of business process modelling is the visibility of the model obtained and the follow-up of its activities, particularly when the model of the process contains hundreds of activities.

The use of TAD methodology contributes a great deal in solving the problem of process visibility and follow-up of its activities. This is achieved by using the partitioning approach to create subtables. Each of them shows an alternative path, where the follow-up of the activities of each path is clear and manageable. This is essential in inventing improvements and solving problems of the process discussed.

We found that using TAD methodology for modelling and improving business processes could be very successful and enabled us to comprehend the process discussed in an easy manner.

References

1. Aguilar-Saven R.: Business Process Modeling. Review and Framework. Internatinal Journal of Production Economics, Vol. 90, No. 2, (2003) 129–149
2. Anapindi R., Chopra S., Deshmukh S. D., van Mieghem J. A., Zemel E.: Managing Business Process Flows. Upper Saddle River, NJ: Pretence Hall, (1999)
3. Chan M.: A Framework to Develop an Enterprise Information Portal for Contract Manufacturing. International Journal of Production Economics, 75 (1–2), (2002) 113–126
4. Damij T.: An Object-Oriented Methodology for Information Systems Development and Business Process Reengineering, Jounal of Object-Oriented Programming, Vol. 13, No. 4, pp. 23-34
5. Watson H.G.: Business Systems Engineering. Managing Breakthrough Changes for Productivity and Profit. John Wiley & Sons, New York (1994)

On the Suitability of Correctness Criteria
for Business Process Models

Juliane Dehnert[1] and Armin Zimmermann[2]

[1] Fraunhofer ISST Berlin
juliane.dehnert@isst.fhg.de
[2] Technische Universität Berlin
azi@cs.tu-berlin.de

Abstract. A popular requirement for the validation of workflow models
is soundness. As soundness can not be easily seen on the model level, different correctness criteria have been proposed in the literature to bridge
the gap between the modeling process and a executable workflow model.
Well-structuredness and relaxed soundness are investigated in the paper.
Relationships between the properties are derived.

Keywords: Workflow, Validation of business process models, Petri nets.

1 Introduction

An increasing number of companies have adopted process-aware information
systems during the past years. By doing so, complex and distributed business
processes can be managed and improved easier. Standard ERP software tools
have been enhanced by a workflow module, while other examples like Staffware
are dedicated workflow management systems (WfMS).

The basis of any of these systems is a model of the company's business processes in a machine-readable manner: the workflow definition. Modeling workflow
requires a deep inside into the application context. Domain experts are often
put in charge of the modeling, although they do not necessarily are modeling
experts. The models describe the processes with the modelers view, and thus do
not necessarily adhere strictly to sound models.

Soundness [Aal98] guarantees that there are no faulty executions at run-
time, like deadlocks or processes that leave dangling documents when terminating. The soundness of a workflow definition can be checked, but is not easy
to see on the model level. To bridge that gap, different properties have been
proposed in the literature to assist non-expert modelers in creating sound business process models. Maybe the most commonly used property in this context is
well-structuredness [Aal98, CWBH+03, Ver04]. The advantage of this property is
that it can be checked easily on the structural level of the model. Well-structured
process descriptions are guaranteed to be sound if they are live. However, well-
structuredness is quite restrictive, and does not support all workflow patterns.

Relaxed soundness [DR01] has been proposed as a property which is claimed
to be better suited for this task. It is a weaker property than soundness, thus

W.M.P. van der Aalst et al. (Eds.): BPM 2005, LNCS 3649, pp. 386–391, 2005.

allowing more workflow structures. However, an additional step is required to achieve a sound WF-net. It has been shown recently how methods from Petri net controller synthesis can be adopted to automatically generate a sound model from a relaxed sound one [DZ04, DvdA04]. The question for workflow modelers as well as WfMS tool designers is now which one of the existing criteria — soundness, well-structuredness or relaxed soundness — should be used to guide the modeler in creating a sound workflow definition. This discussion is the main motivation for this paper. The main contribution is the theoretical background for the comparison. Proofs are provided that put the criteria in relation to each other. Based on these considerations, the usability of the properties is briefly compared from the modeler's point of view.

2   Soundness - Well-Structuredness - Relaxed Soundness

For the modeling of business processes we refer to WF-nets, cf. [Aal98]. WF-nets are a special class of Petri nets characterized by a source place ($^\bullet i = \emptyset$) and a sink place o ($o^\bullet = \emptyset$). Furthermore, short-circuiting the net by an additional transition t' the resulting net is strongly connected.

Figure 1 shows a WF-net modeling the business process initiated by incoming goods. The process is assigned to two departments, the accountancy and the storage, which may work in parallel. The thread of control is split accordingly in the beginning. In the accountancy (c.f. upper thread in the figure) the receipt of the goods is recorded. The incoming goods are checked and stored in the other department (lower thread). In case the check is negative, a notification is sent to the accountancy. Finally, the threads are joined again.

In the following we introduce some relevant process properties.

Well-structuredness is a property that has been proposed in the literature (e.g. [Aal98, Ver04]) to assist non-expert modelers in formalizing their business processes. Well-structuredness is a property requiring a strict block structuring of a process description. It is satisfied if every split (OR, AND etc.) is followed be a corresponding join of the same type. The restriction to well-structuredness is also present in UML v1.4 [UML02] activity diagrams, BPEL4WS [BEA03] and ADEPT [RD98][1]. In terms of Petri-net theory, well-structuredness is characterized by the absence of handles[2] [ES90].

Definition 1 (Well-structured). *A WF-net PN is well-structured if the short-circuited net $\overline{PN}$ does not contain any handles, i.e. $\overline{PN}$ is well-handled.*

The WF-net of Figure 1 is not well-structured. Examples for existing handles are the transition-place pair (t1, p5) and the place-transition pair (p2, t10).

---

[1] In the latter two, the strict block structuring conditions are relaxed by allowing control links (resp. synchronization edges) to synchronize tasks belonging to different parallel control flow paths.

[2] A handle is a pair of two different nodes (a place and a transition) that are connected via two elementary paths sharing only these two nodes.

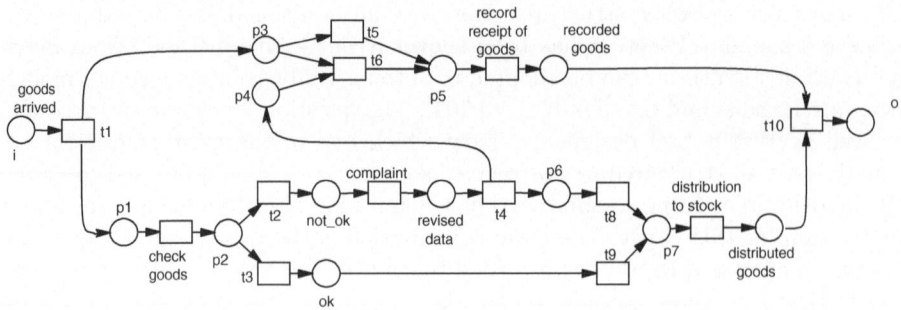

Fig. 1. WF-net "Processing of incoming goods"

Soundness has been introduced in [Aal98]. A WF-net is sound if termination is always possible and once terminated there are no residual tokens outside o. Furthermore, there are no dead transitions and neither deadlocks nor livelocks.

Definition 2 (Soundness). *A WF-system (PN, i) is sound iff:*

(i) *For every state M reachable from state i, there is a firing sequence leading from state M to state o: $\forall M : (i \xrightarrow{*} M) \Rightarrow (M \xrightarrow{*} o)$.*

(ii) *State o is the only state reachable from state i with at least one token in place o (proper termination)[3]: $\forall M : (i \xrightarrow{*} M \land M \geq o) \Rightarrow (M = o)$*

(iii) *There are no dead transitions in PN: $\forall t \in T \ \exists M, M' : (i \xrightarrow{*} M \xrightarrow{t} M')$*

The WF-net "Processing of incoming goods" (c.f. Figure 1) is not sound. There are firing sequences that do not terminate properly.

Relaxed soundness was introduced with the intention to represent a more pragmatic view of correctness. Relaxed soundness does not require the absence of residual tokens, livelocks or deadlocks. A process is relaxed sound if each task of the business process is part of a properly terminating sequence.

Definition 3 (Relaxed soundness). *A workflow system (PN, i) is relaxed sound iff each transition t of PN appears in some sound firing sequence σ: $\forall t \in T \ \ \exists \sigma : i \xrightarrow{\sigma} o \ \ with \ \ t \in \sigma$*

The process specification shown in Figure 1 is relaxed sound. There are enough sound firing sequences, i.e. all transitions are covered.

3   Relations Between the Properties

Clearly, soundness implies relaxed soundness. This can be derived directly from the definitions. Another proposition relates all three properties. We will show that a process description is sound if it is well-structured and relaxed sound.

---

[3] Note that this statement from the original definition already follows from requirement (i) [HSV04].

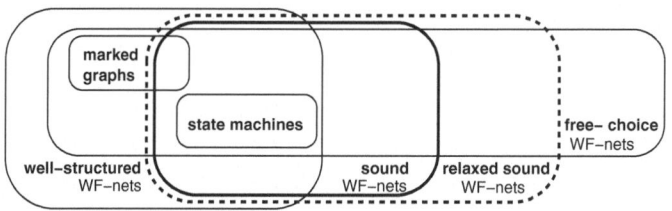

Fig. 2. Relations between different Petri net-properties

Proposition 1. *Let PN be a WF-net with input place i. If PN is well-structured and (PN, i) is relaxed sound, then (PN, i) is sound.*

The proof of this claim is provided in two steps. It is first proved for free-choice[4] WF-nets. The result is then applied to the unrestricted class of WF-nets.

Proof 1 (PN is free-choice): Because PN is well-structured, the short-circuited net $\overline{PN}$ is well-handled and strongly connected. With [ES90] (Theorem 3.1 & 3.2) we can conclude that $\overline{PN}$ is well-formed, i.e. structurally bounded and structurally live. Soundness of (PN, i) coincides with liveness and boundedness of $(\overline{PN}, i)$ [Aal98]. Boundedness of $(\overline{PN}, i)$ follows directly from the fact that $(\overline{PN}, i)$ is structurally bounded. It thus remains to prove that $(\overline{PN}, i)$ is live.

As (PN, i) is relaxed sound, there is an infinite firing sequence σ of $(\overline{PN}, i)$ which supports each transition. σ can be constructed, linking the set of sound firing sequences in (PN, i) via firing of transition t^*. This is done infinitely often: $\sigma = \sigma_1 t^* \sigma_2 t^* ... t^* \sigma_n t^* \sigma_1 t^* \sigma_2 t^* ... t^* \sigma_n ....$ With [ES90] (Theorem 3.2) we know that $\overline{PN}$ is covered by S-components[5]. The infinite firing sequence σ is enabled in i and contains all transitions. Since $\overline{PN}$ is strongly connected, every place and therefore every trap is marked during the occurrence of σ. Since marked traps remain marked, every trap and therefore every S-component is marked in i. With [DE95] (Theorem 5.8), it can be concluded that $\overline{PN}$ is live.   □

In order to transfer the result to the class of non-free-choice WF-nets we first establish some prerequisites. We recall a transformation rule ([DE95]) that transforms a non-free choice net PN into a free-choice net PN'. According to this rule, every arc $(p, t) \in F$ in PN is replaced by a sequence $(p, t')(t', p')(p', t) \in F'$. The sets P and T are extended appropriately. Note that this transformation preserves well-structuredness (I) and relaxed soundness (II), whereas it does not hold for liveness. However, the properties liveness and boundedness are preserved during the backward direction of the considered transformation (III).

Proof 2 (PN is not free-choice): We apply the free-choice transformation rule to PN and obtain a well-structured (I) and relaxed sound (II) WF-net (PN', i), which is additionally free-choice. We short-circuit PN' and obtain the strongly

---

[4] A Petri net $PN = (P, T, F)$ is a free-choice net (basically extended free-choice) iff $\forall t, t' \in T : {}^\bullet t \cap {}^\bullet t' = \emptyset \vee {}^\bullet t = {}^\bullet t'$.

[5] A Subnet $PN' = (P', T', F')$ is an S-component of the net $PN = (P, T, F)$ iff PN' is a strongly connected state machine and $\forall p \in P' : {}^\bullet p \cup p^\bullet \subseteq T'$.

connected, well-handled and free-choice net $\overline{PN'}$. Using again [ES90], we can conclude that $\overline{PN'}$ is well-formed. Exploiting the result of the first proof, we can infer that $(\overline{PN'}, i)$ is live and bounded. As the reverse direction of the free-choice transformation preserves these properties (III), we can conclude that $(\overline{PN}, i)$ is also live and bounded. Therefore, PN is sound. □

Figure 2 shows an Euler diagram depicting the established relations between Petri net classes considered in workflow modeling.

4   Usability of the Properties

Supporting the modeler, meaningful but possibly loose modeling restrictions should be posed, such that a wide range of process descriptions can be defined having a sensible/useful interpretation. However, this means especially that all process descriptions which are sound (c.f. [Aal98]) should satisfy the used criterion. On the other hand, all process descriptions not satisfying the property should clearly contain design faults.

Soundness is now widely accepted as a necessary requirement for any executable workflow model. However, soundness is not easily seen on the model level. The reason for this is that soundness requires complete knowledge of all possible behavior. As a consequence, the modeler is required to think about the "how" of the execution. This contradicts the argument that the specification of business processes should be as abstract as possible.

Demanding a strict hierarchical design, as done using **well-structuredness**, seems to be a valuable help in the modeling process. The modeler must only follow simple structural rules to get a correct process description. However, some business processes can hardly be matched by a well-structured process description. The demand for a strict hierarchical design ignores the need to assign the tasks according to their organizational assignment. Modeling in a well-structured manner requires overview of the whole process. This can hardly be assumed if the process to be described is spanning different organizational units of the company, involving various modelers. The mentioned shortcoming can be characterized also as follows: When modeling in a well-structured manner, some useful process descriptions are disregarded right from the beginning. Figure 2 shows that there are processes which are sound but not well-structured.

In previous publications (e.g. [DZ04]), it is argued that **relaxed soundness** meets the modeling capabilities of modelers, as it does not require high modeling knowledge but acknowledges the process view of domain experts. The main reason for that is that relaxed soundness does not impose operational semantics. Whereas the users point of view should be reflected, it is clear that formally correct process descriptions, such as described by the soundness criterion, should as well assessed to be correct. This holds for relaxed soundness. All sound process descriptions are relaxed sound by definition, c.f. Figure 2. Therefore, no sound process description is disregarded already at design time. On the other hand, WF-nets which are not relaxed sound are not sound either. Such process descriptions contain transitions which are not contained in any sound firing sequence. Such redundant transitions clearly constitute a design flaw. They relate to tasks that have been modeled but do not contribute to any proper terminating execution. It is hardly imaginable that such modeling is intended. Such WF-nets clearly needs revision.

5   Conclusion

In this paper the main criteria for Petri net workflow models have been put into relation. The formal part relates soundness, well-structuredness and relaxed soundness. It is shown that relaxed soundness and well-structuredness together imply soundness. The shown relations are briefly interpreted from the modeler's point of view.

References

[Aal98]   W.M.P. van der Aalst. The Application of Petri Nets to Workflow Management. *The Journal of Circuits, Systems and Computers*, 8(1):21–66, 1998.

[BEA03]   BEA Systems, IBM Corporation, Microsoft Corporation, SAP AG, Siebel Systems. *Business Process Execution Language for Web Services (Version 1.1*, 2003.

[CWBH+03]   P. Chrzastowski-Wachtel, B. Benatallah, R. Hamadi, M. O'Dell, and A. Susanto. A top-down petri net-based approach for dynamic workflow modeling. In W. van der Aalst, A. ter Hofstede, and M. Weske, editors, *Int. Conf. on Business Process Management*, volume 2678 of *LNCS*, pages 336–353. Springer, 2003.

[DE95]   J. Desel and J. Esparza. *Free Choice Petri Nets*. Cambridge University Press, 1995.

[DR01]   J. Dehnert and P. Rittgen. Relaxed Soundness of Business Processes. In K.L. Dittrich, A. Geppert, and M.C. Norrie, editors, *Advanced Information System Engineering, CAISE 2001*, volume 2068 of *LNCS*, pages 157–170. Springer, 2001.

[DvdA04]   J. Dehnert and W.M.P. van der Aalst. Bridging the Gap Between Business Models and Workflow Specifications. *Int. Journal of Cooperative Information Systems (IJCIS)*, 13(3):289–332, 2004.

[DZ04]   J. Dehnert and A. Zimmermann. Making Workflow Models Sound Using Petri Net Controller Synthesis. In R. Meersman and Z. Tari et.al., editors, *Int. Conf. Cooperative Information Systems (CoopIS) 2004*, volume 3290 of *LNCS*, pages 139–154, Cyprus, 2004.

[ES90]   J. Esparza and M. Silva. Circuits, Handles, Bridges and Nets. In G. Rozenberg, editor, *Advances in Petri Nets 1990*, volume 483 of *LNCS*, pages 210–242. Springer, 1990.

[HSV04]   K. van Hee, N. Sidorova, and M. Voorhoeve. Generalised Soundness of Workflow Nets is Decidable. In W. Reisig J. Cortadella, editor, *Int. Conf. on Application and Theory of Petri Nets*, volume 3099 of *LNCS*, pages 197–216. Springer, 2004.

[RD98]   M. Reichert and P. Dadam. ADEPTflex: Supporting Dynamic Changes of Workflow without Loosing Control. *Journal of Intelligent Information Systems*, 10(2):93–129, 1998.

[UML02]   Unified Modeling Language: version 1.4.2, ISO, 2002.

[Ver04]   Eric Verbeek. *Verification of WF-nets*. PhD thesis, TU Eindhoven, 2004.

Service Retrieval Based on Behavioral Specifications and Quality Requirements

Daniela Grigori, Verónika Peralta, and Mokrane Bouzeghoub

Laboratoire PRiSM, Université de Versailles
45, avenue des Etats-Unis, 78035 Versailles Cedex, France
{Daniela.Grigori, Veronika.Peralta,
Mokrane.Bouzeghoub}@prism.uvsq.fr

Abstract. The capability to easily find useful services becomes increasingly critical in several fields. In this paper we argue that, in many situations, the service discovery process should be based on both behavior specification (that is the process model which describes each composite service) and quality features of services. The idea behind is to develop matching techniques that operate on process models and allow delivery of partial matches and evaluation of semantic distance between these matches and the user requirements. To do so, we reduce the problem of service behavioral matching to a graph matching problem and we adapt existing algorithms for this purpose. The matching algorithm is extended by a flexible quality evaluation procedure which checks whether a given service is worth to be delivered or not.

1 Introduction

The capability to easily find useful services becomes increasingly critical in several fields. Examples of such services are numerous: (i) software applications as web services which can be invoked remotely by users or programs; (ii) programs and scientific computations which are important resources in the context of the Grid; (iii) software components which can be downloaded to create a new application.

In all these cases, users are interested in finding suitable services in a library or in a catalog of services. Service retrieval may be based on their inputs/outputs and their constraints (pre and post conditions) or on some high level attributes which characterize, at some extent, their functional semantics. Recent publications have demonstrated that this approach is not sufficient to discover relevant services [2]. Many services with different semantics may have the same inputs/outputs or the same constraints, therefore leading to a lack of relevance of the retrieval process. To go beyond the limits of this approach, a substantial effort has been done by different works on semantic web and ontologies by exploiting more knowledge on the semantics of the services [1][7]. High level conceptual graphs and assertions intend to capture more meaning of supplied services. However, this effort still remains insufficient and does not fulfill user needs as many functional or quality aspects are hidden within the specification of services behavior.

In this paper we argue that, in many situations, the process of service discovering should be based on both behavior specification and quality features of services.

W.M.P. van der Aalst et al. (Eds.): BPM 2005, LNCS 3649, pp. 392–397, 2005.

Behavior specification abstracts the functional semantics of the components while quality features describes non-functional aspects, i.e. their added-value, constraints in terms of performance, reliability, data consistency, data freshness, etc. The idea behind is to develop matching techniques that operate on process models as well as the associated quality features and allow delivery of partial matches and evaluation of semantic distance between these matches and the user requirements. Consequently, even if a service satisfying exactly the user requirements does not exist, the most similar ones will be retrieved and proposed for reuse by extension or modification.

In the approach presented in this paper, we reduce the problem of service behavioral matching to an adorned graph matching problem, where the graph represents the functional process of a service and the adornment represents quality constraints. Our approach is based on existing algorithms on graph matching [6] which are adapted to workflow diagrams. The matching algorithm is extended by a flexible quality evaluation procedure which checks whether a given service is worth to be delivered or not. A set of similarity metrics based on functional and non-functional requirements are defined.

The rest of the paper is organized as follows: Section 2 presents a graph representation of the process models, shows how the behavioral matching problem can be reduced to a graph matching problem and defines a structural similarity measure. Section 3 presents the data quality evaluation principle and defines a qualitative similarity measure that is added to the matching approach. Finally, section 4 presents ongoing work and conclusions.

2 Behavioral Matching

In this section, we present our approach to service retrieval based on their behavioral models. Behavioral models are process models that describe user requirements as well as provided services. We assume that the formal models are workflow models although the approach can easily be adapted to other formal models such as Petri nets or state chart diagrams, provided that it is a uniform model. After giving the preliminary definitions, we describe the principles of service matching algorithm.

2.1 Background

Most of existing proposals (standard and research models) for process specification are graph based. For this reason, we choose to base our service retrieval approach on process graphs. A process is represented as a directed graph, whose vertices are activities or data repositories. Activities associated to web services consume input data elements and produce output data elements which may persist in repositories. There are two types of edges: (i) *control edges* that have associated transition conditions expressing the control flow dependencies between activities, and (ii) *data edges* representing data flow between activities.

Using graphs as representation formalism for both user requirements and service models, the service retrieval problem turns into a graph matching problem. We want to compare the process graph representing user requirements with the process graphs in the library. The matching process can be formulated as a search for graph or subgraph isomorphism. However, it is possible that it does not exist a process model such that an exact graph or subgraph isomorphism can be defined. Thus, we are interested in finding process models that have similar structure, if models that have identical structure do not exist. The *error-correcting graph matching* integrates the concept of error correction (or inexact matching) into the matching process [6].

In order to compare the graphs in the library (that will be called *catalog graphs* in the following) to the graph expressing user requirements (called *query graph*) and decide which model is more similar to the latter, it is necessary to define a distance measure for graphs. Similar to the string matching problem where edit operations are used to define the string edit distance, the subgraph edit distance is based on the idea of edit operations that are applied to the catalog graph altering it until there exists a subgraph isomorphism to the query graph. A certain cost is assigned to each graph edit operation. The subgraph edit distance from a model to an input graph is then defined to be the minimum cost taken over all sequences of edit operations that are necessary to obtain a subgraph isomorphism. It can be concluded that the smaller the subgraph edit distance between a model and an input graph, the more similar they are. The subgraph isomorphism detection is based on a state-space search by means of an algorithm similar to A*. Different algorithms have been proposed for error correcting graph matching in order to reduce the computation complexity (see for example [6]).

2.2 Service Retrieval Based on Behavioral Matching

In this section we begin by showing how the error-correcting subgraph isomorphism algorithm can be used for behavioral matchmaking. Then we define a similarity measure based on the subgraph edit distance allowing ranking models in the library.

Suppose that a user needs to develop a new composite web service. He specifies his composition model as a query graph that he submits to the service retrieval system to find in the library similar web services or fragments that could be composed to develop his application or a new web service. If we assume that existing services are represented as graphs (e.g. workflows), the problem of service matchmaking is transformed into a problem of graph matching. If user defines input/output parameters and operation name for the new composite web service, then a first filter could be applied for components retrieval based on these properties. The behavioral matching will be applied either to the set of services retrieved in the first step or independently.

We use the algorithm for error correcting subgraph isomorphism to retrieve the most similar models. For the error correcting algorithm, the cost function for each graph operation has to be defined. The costs are application dependent and reflect the likelihood of graph distortions. The more likely a certain distortion is to occur the smaller is its cost.

The cost for deletion/insertion of an edge and a vertex can be set to constants. The cost for substituting a label and its attributes is defined as follows. If the labels are not identical then the substitution cost is set to the corresponding constant. If they are identical and they have the same number of attributes, the substitution cost is defined to be the weighted mean of distances between the corresponding attributes. For each attribute of a service, the cost function of substituting an attribute value has to be defined. In [4], we showed how this cost can be defined if service attributes are annotated with ontological concepts. Attributes being associated with concepts in the ontology, the cost function is the distance between these concepts in the ontology.

In order to rank the catalog graphs, a similarity measure has to be defined. The total distance between the two graphs can be defined as the sum of the subgraph edit distance and the cost of inserting the vertices of the query graph not covered by the error correcting subgraph. The similarity measure is the inverse of this distance. For more details on behavioral matching see [4].

3 Data Quality Evaluation

The relevance of the service retrieval process can be enhanced by adorning service behavior with quality features. Section 2 has shown how the retrieval process is reformulated as a graph matching problem. However, if the matching algorithm is only based on functional requirements, the resulting services may not necessarily fulfill user expectations in terms of performance, reliability and consistency. Additionally, if some of these services are data providers, freshness and accuracy of this data could be of high importance to their users. This section goes further to improve such matching by introducing quality measures which help to discriminate between several matches.

The idea behind this approach is to adorn catalog graphs with quality features (e.g. costs, delays, data freshness, data accuracy) which allow estimating the quality of the data that can be produced by the graphs. The calculated quality values are compared to the quality constraints of the query graph (also used as adornments).

In this section we present an example that illustrates how we utilize these adornments for evaluating a concrete quality factor: *data freshness* and then we describe the matching approach using a similarity metric based on data quality.

3.1 Evaluating Freshness Within the Adorned Graph

Data freshness is a quality factor which is very important in many data centric applications. Decision making may not be relevant if it is based on stale data. Data freshness can be measured as the time passed since the creation of data or as the time elapsed since the last delivery of data; an extensive study of this has been done in [3].

The freshness of result data depends on the freshness of input data but also on the amount of time the service needs for executing its activities. The latter depends on the processing cost of activities and on the synchronization delays that can exist between their executions due to control flow constraints. Then, the quality features used as graph adornments mainly are:

- *Processing cost*: It is the amount of time, in the worst case, that an activity needs for reading input data, executing and building output data.
- *Synchronization delay*: It is the amount of time passed between the executions of two consecutive activities.
- *Input actual freshness*: It is a measure of the actual freshness of source data.

We propose an evaluation algorithm that estimates the freshness of result data based on the graph adornments. The algorithm traverses the graph following the sense of the edges, calculating the freshness of the data resulting from each activity. The principle is the following (algorithm pseudocode can be read in [5]):

- If an activity A reads external input data, result data freshness is calculated adding the actual freshness of input data and the processing cost of the activity.
- If an activity A has one predecessor B, result data freshness is calculated adding the freshness of the data produced by B, the synchronization delay between B and A and the processing cost of A.
- In the general case, if activity A has several predecessors, the freshness of data coming from each predecessor (plus the corresponding synchronization delay) should be combined and added to the processing cost of the activity A. If activity A also reads external data, actual freshness of input data should also be considered and combined. Typical combination functions are maximum, average or weighted average, but other user-specific functions can be considered, for example, for ignoring some predecessor because its data is stable (ex. country names).

3.2 Using Quality Requirements in Behavioral Service Retrieval

In this section we extend the behavioral matching procedure in order to take into account quality requirements. We distinguish two scenarios: (i) the query requirements express constraints, i.e. the catalog graphs that do not fulfill quality constraints should not be returned, and (ii) the quality requirements express expectations, i.e. a catalog graph that does not fulfill quality expectations but offers the most similar values can be returned. In the former, the user prefers retrieving a graph that is structurally less similar (and then having more development cost) but satisfying his quality requirements. In the latter, quality expectations are more flexible, and can be balanced with structural similarity. In the following we discuss both approaches.

Matching under Quality Requirements: In this scenario, quality requirements are expressed as quality thresholds that catalog graphs must verify to be retrieved. The retrieval steps are: (i) match the catalog graphs in order to obtain the isomorphic subgraphs and calculate the structural similarity measure; (ii) evaluate data freshness and eliminate the candidates that do not achieve freshness requirements; and (iii) rank the candidate graphs according to their structural similarity and retrieve the best one.

Matching with Quality Expectations: In this scenario, the model graphs can be ranked according to a similarity measure that takes into account both structural similarity and freshness expectations. A qualitative similarity measure can be defined in order to express the degree in which freshness expectations are achieved. Depending on the application, the similarity measure can be defined in different ways.

An example of such a measure is given in the following formula:

$$S_Q = (ExpectedFreshness - ActualFreshness) / ExpectedFreshness \qquad (1)$$

The formula calculates the difference between expected and actual freshness values and normalizes it dividing by the expected values. When freshness expectations are achieved, the similarity is a positive value that will act positively in the global similarity measure. When freshness expectations are not achieved the similarity is a negative value having the opposite effect.

Having defined a qualitative similarity measure, we can build a global similarity measure that combines structural and qualitative ones in a weighted sum. The weights should indicate user preference for structural over quality criteria. The retrieval steps are: (i) match the catalog graphs in order to obtain the isomorphic subgraphs and calculate the structural similarity measure; (ii) evaluate data freshness and calculate the qualitative similarity measure; and (iii) rank the candidate graphs according to their global similarity and retrieve the best one.

4 Conclusion

In this paper we proposed a solution for service retrieval based on behavioral specification and quality requirements. First, we proposed to use a graph error correcting matching algorithm in order to allow an inexact matching. Then, we showed how the quality factors can be used in the matchmaking process.

We implemented the behavioral matchmaking as a web service. The prototype takes as input the graph representations of two services and calculates the degree of similarity between them, also returning the sequence of transformations needed to transform one service into the other. We also implemented a prototype of a quality evaluation tool, which takes as input the graph representation of a service and returns a measure of its data quality. Current work addresses the integration of the two functionalities in the perspective of the ideas presented in this paper.

While in this paper we dealt with the semantics aspects of behavioral matchmaking, we did not address the operational aspects. The graph matching computation is a NP-complete problem. In our future work we will try to apply constraints and heuristics to cut down the computational effort to a manageable size.

References

1. Benatallah, B., Hacid, M.S., Rey, C., Toumani, F.: Semantic Reasoning for Web Services Discovery. In Proc. of the Workshop on E-Services and the Semantic Web (ESSW), Hungary (2003)
2. Berstein, A., Klein, M.: Towards High-Precision Service Retrieval. In Proc. of the 1st Int. Semantic Web Conference (ISWC), Italy (2002)
3. Bouzeghoub, M., Peralta, V.: A Framework for Analysis of Data Freshness. In Proc. of the 1st Int. Workshop on Information Quality in Information Systems (IQIS), France (2004)
4. Grigori, D., Bouzeghoub, M.: Service Retrieval Based on Behavioral Specifications. In proc. of Int. Conf. Of Service Computing, USA (2005)
5. Grigori, D., Peralta, V., Bouzeghoub, M.: Service Retrieval Based on Behavioral Specifications and Quality Constraints. Technical report, University of Versailles (2005)
6. Messmer, B.: Graph Matching Algorithms and Applications. PhD Thesis, University of Bern (1995)
7. Paolucci, M., Kawamura, T., Payne, T.R., Sycara, K.: Semantic Matching of Web Services Capabilities. In Proc. of the 1st Int. Semantic Web Conference (ISWC), Italy (2002)

On the Semantics of EPCs:
Efficient Calculation and Simulation
Extended Abstract

Nicolas Cuntz[1] and Ekkart Kindler[2]

[1] University of Siegen, Computer Graphics and Multimedia Systems Group
nicolas.cuntz@uni-siegen.de
[2] University of Paderborn, Software Engineering Group
kindler@upb.de

Abstract. Recently, we have defined a formal semantics of *Event driven Process Chains* (EPCs) that, for the first time, faithfully captures the non-local behaviour of the XOR- and OR-join connectors. This fixed-point characterisation of the semantics of EPCs, however, does not provide an efficient algorithm for calculating the semantics of an EPC and for simulating it.

In this paper, we will show how to calculate this semantics of an EPC in an efficient way by employing Kleene's fixed-point theorem and different techniques from symbolic model checking. These algorithms have been implemented in an open source tool for simulating and analysing EPCs: *EPC Tools*.

1 Introduction

Event driven Process Chains (EPCs) have been introduced in the early 90ties for modelling business processes [1]. Initially, EPCs have been used informally only, without a fixed formal semantics. For easing the modelling of business processes with EPCs, the informal semantics proposed for the OR-join and the XOR-join connectors of EPCs was *non-local*. This non-local semantics, however, results in severe problems when it comes to a formalisation of the semantics of EPCs and, recurrently, resulted in a debate on the semantics of EPCs [2,3]. It turned out that these problems are inherent to the informal non-local semantics of EPCs. In [4], we pin-pointed these arguments, which render a formal semantics that exactly captures the non-local semantics of an EPC in terms of a single transition relation impossible. But, we could define a semantics for an EPC that consists of a pair of two correlated transition relations by using fixed-point theory [5].

Due to the non-local semantics, an EPCs cannot be simulated by looking at its current state only; rather it requires calculating the transition relations of the EPC beforehand. In principle, the two transition relations defined as the semantics of an EPC can be calculated by fixed-point iteration. The problem, however, is that the calculation of the two transition relations by naive fixed-point iteration is very inefficient and intractable in practise. In this paper, we

W.M.P. van der Aalst et al. (Eds.): BPM 2005, LNCS 3649, pp. 398–403, 2005.

sketch the main idea how techniques from *symbolic model checking* [6] can be used for calculating the semantics of an EPC in a more efficient way. For details, we refer to [7,8].

2   Syntax and Semantics of EPCs

In this section, we informally introduce the syntax and the semantics of EPCs as formalised in [5], which is a formalisation of the informal ideas as presented in [1,9].

2.1   Syntax

Figure 1 shows an example of an EPC. It consists of three kinds of *nodes*: *events*, which are graphically represented as hexagons, *functions*, which are represented as rounded boxes, and *connectors*, which are represented as circles. The dashed arcs between the different nodes represent the *control flow*. The two black circles do not belong to the EPC itself; they represent a *state* of the EPC. A state, basically, assigns a number of *process folders* to each arc of the EPC. Each black circle represents a process folder at the corresponding arc.

Mathematically, the nodes are represented by three pairwise disjoint sets E, F, and C, which represent the events, functions, and connectors, respectively. We denote the set of all nodes by $N = E \cup F \cup C$. The type of each connector is defined by a mapping $l : C \rightarrow \{and, or, xor\}$. The control flow arcs are a subset $A \subseteq N \times N$. Note that there are some syntactical restrictions on EPCs. But, we omit these restrictions for lack of space, since they are not so important for our semantical considerations.

A *state* of an EPC assigns zero or one *process folders* to each arc of the EPC. So a state σ is a mapping $\sigma : A \rightarrow \{0, 1\}$. The set of all states of an EPC will be denoted by Σ.

2.2   Semantics

The semantics of an EPC defines how process folders are propagated through an EPC. This can be formalised by a *transition relation* $R \subseteq \Sigma \times N \times \Sigma$, where the first component denotes the source state, the third component denotes the target state, and the middle component denotes the involved node.

For events and functions, a process folder is simply propagated from the ingoing arc to the outgoing arc as shown in Fig. 2 a. and b. The semantics of the other nodes is shown in Fig. 2, too. For lack of space, we discuss only the XOR-join connector (case h.): An XOR-join connector waits for a folder on one ingoing arc, which is then propagated to the outgoing arc. But, there is one additional condition: The XOR-join must not propagate the folder, if there *is* or there *could arrive* a folder on the other ingoing arc. This additional condition is graphically indicated by the label #• at the other arc. Note that this condition cannot be checked locally in the current state: whether a folder could arrive on

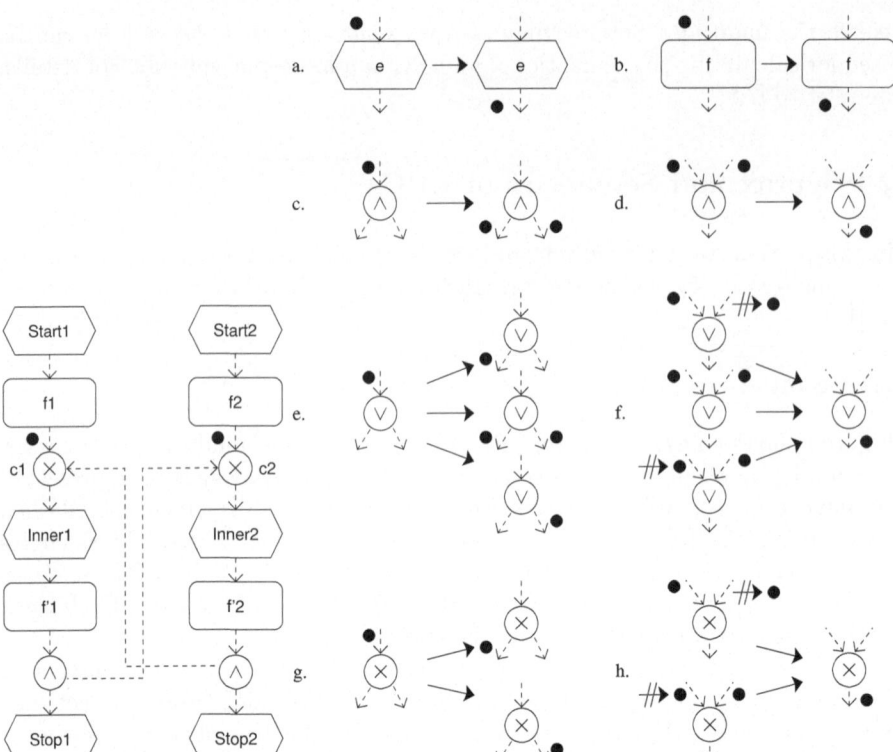

Fig. 1. An EPC   **Fig. 2.** The transition relations for the different nodes

the other arc depends on the overall behaviour of the EPC. Therefore, we call the semantics of the XOR-join connector *non-local*.

Note that, in this informal definition of the *transition relation*, we refer to the transition relation itself when we require that no folders can arrive at some arc according to the transition relation. Therefore, we cannot immediately translate this informal definition into a mathematically sound definition. In order to resolve this problem, we assume that some transition relation P is given already, and whenever we refer to the non-local condition, we refer to this transition relation P. Thus, Fig. 2 defines a mapping $R(P)$: for some given transition relation P, it defines the transition relation $R(P)$.

The most important property of $R(P)$ is that it is monotonously decreasing in P. This property guarantees that there exists a least transition relation P and a greatest transition relation Q such that $R(Q) = P$ and $R(P) = Q$, where P is called the *pessimistic transition relation* and Q is called the *optimistic transition relation* of the EPC. The pair of these two transition relations (P, Q) was defined as the semantics of the EPC. In most cases, we have $P = Q$, which means that P is a fixed-point of R. If P and Q are different, there are some ambiguities in the interpretation of the EPC (see [5] for details). Therefore, we call an EPC *unclean* if P and Q are different, and we call it *clean* if P and Q are equal.

3   Calculating the Transition Relations

In [5], we have shown that the pair (P, Q) exist and is uniquely defined for each EPC; but it was not clear how to calculate it. In this section, we will show how to actually calculate P and Q.

3.1   Iterative Fixed-Point Characterisation

A first idea for calculating P and Q comes from Kleene's fixed-point theorem, which gives us an iterative characterisation of P and Q: Let $P_0 = \emptyset$ and $Q_0 = \Sigma \times N \times \Sigma$. For each $i \in \mathbb{N}$, we define $P_{i+1} = R(Q_i)$ and $Q_{i+1} = R(P_i)$. Since $R(P)$ is a monotonously decreasing function and since the set of possible states is finite, eventually, we will have $P_{i+1} = P_i$ and $Q_{i+1} = Q_i$. And it turns out that this P_i and Q_i are the pessimistic and the optimistic transition relations of the EPC, respectively.

Unfortunately, an explicit representation of the transition relations P_i and Q_i and an explicit calculation of $P_{i+1} = R(Q_i)$ and $Q_{i+1} = R(P_i)$ is extremely inefficient. For realistic EPCs, there are millions of potential states Σ and billions[1] of potential arcs in the transition relation. Moreover, an explicit calculation of $R(P)$ involves a reachability analysis on P. So a naive explicit implementation does not work in practise.

3.2   Symbolic Calculation

Therefore, we use techniques from symbolic model checking for calculating the relations P and Q. The idea is that the transition relation can be represented and calculated symbolically. For example, the transition relation for an AND-split connector with ingoing arc i and outgoing arcs o_1 and o_2, can be represented by the following formula (cf. Fig. 2 (c)):

$$i \wedge \neg o_1 \wedge \neg o_2 \wedge \neg i' \wedge o_1' \wedge o_2'$$

In this formula, i and o_1 and o_2 are boolean variables. The values of these variables represent the state before the transition, where value *true* means that there is a process folder on the corresponding arc, and value *false* means that there is no process folder. The primed variables i', o_1', and o_2' represent the state after the transition. Altogether, this formula represents the state change of an AND-split connector, where the use of primed variables is a standard technique for representing transition relations by formulas. The transition relations for all nodes with a local semantics can be expressed in this way.

For the connectors with non-local semantics, however, we need a new idea. We must formalise that no folder can reach some arc i (with respect to some transition relation). This can be expressed by a temporal formula[2] $\neg EF\, i$. With this idea, the transition relation for an XOR-join connector with ingoing arcs

---

[1] Remember that the iteration starts with $Q_0 = \Sigma \times N \times \Sigma$.
[2] To be precise, EF is a temporal operator from CTL.

i_1 and i_2 and outgoing arc o can be formalised by the following formula (cf. Fig. 2 (h)):

$$((i_1 \wedge \neg EF\, i_2) \vee (\neg EF\, i_1 \wedge i_2)) \wedge \neg o \wedge \neg i_1' \wedge \neg i_2' \wedge o'$$

The formulas $\neg EF\, i_1$ resp. $\neg EF\, i_2$ express that a transition is only possible when no folder can arrive on the other arc, respectively. Note that this exactly captures the graphical notation ⇸• in Fig. 2. The transition relation for the OR-join connector can be expressed in a similar way (see [8] for details).

By interpreting these formulas on transition relation Q_i, we can calculate the next transition relation P_{i+1} in a symbolic way; by interpreting them on P_i, we can calculate the next transition relation Q_{i+1} in a symbolic way. The interpretation of a temporal formula in some transition relation is usually calculated by a model checker; we implemented it by the help of our own MCiE model checker [10].

3.3 Optimisation

For lack of space, we cannot go into the details of this implementation. In order to make the calculation more efficient, we employed two different kinds of optimisations. The first replaces chains of functions and events by a single node, which reduces the number of arcs of the EPC. Since each arc will be a variable in the formulas, the reduction of arcs reduces the number of variables in the formulas, which in turn reduces the computation time. Unfortunately, this technique is not always applicable (see [8] for a detailed discussion).

The other optimisation used standard techniques from model checking. We investigated several schemes for optimising the variable order in the underlying ROBDDs and we partitioned the set of transitions in order to reduce the size of the ROBDDs representing the transition relations.

We implemented the algorithm with all of the above optimisations, which could calculate the complete transition relation of a medium sized EPCs within seconds. With the transition relation calculated, the EPC can be simulated in virtually no time. Precise figures are discussed in [8].

4   Conclusion

In this paper, we have sketched the idea of an algorithm for calculating the semantics of an EPC. With the presented optimisations, the simulation of medium size EPCs works quite well and is practically feasible.

The presented algorithm is implemented in a new Eclipse based tool for simulating EPCs: *EPC Tools*. This tool comes with a graphical editor and visually simulates the EPC. EPC Tools also provides some simple analysis features. For example, it checks whether the EPC is clean. EPC Tools is open source and published under the GNU Public License, which might make it a good starting point for an open source tool for EPCs. It can be downloaded from [11].

Once the semantics of an EPC is calculated, it would be easy to use it for verifying all kinds of properties by model checking (which would take less time than calculating the semantics itself). The only question is what kind of properties should be verified for EPCs; this needs to be discussed with people using EPCs in practise.

References

1. Keller, G., Nüttgens, M., Scheer, A.W.: Semantische Prozessmodellierung auf der Grundlage Ereignisgesteuerter Prozessketten (EPK). Technical Report Veröffentlichungen des Instituts für Wirtschaftsinformatik (IWi), Heft 89, Universität des Saarlandes (1992)
2. Langner, P., Schneider, C., Wehler, J.: Petri Net Based Certification of Event driven Process Chains. In Desel, J., Silva, M., eds.: Application and Theory of Petri Nets 1998. LNCS 1420. Springer (1998) 286–305
3. Rittgen, P.: Quo vadis EPK in ARIS? Wirtschaftsinformatik **42** (2000) 27–35
4. van der Aalst, W., Desel, J., Kindler, E.: On the semantics of EPCs: A vicious circle. In Nüttgens, M., Rump, F.J., eds.: EPK 2002, Geschäftsprozessmanagement mit Ereignisgesteuerten Prozessketten. (2002) 71–79
5. Kindler, E.: On the semantics of EPCs: Resolving the vicious circle. In Desel, J., Pernici, B., Weske, M., eds.: Business Process Management, Second International Conference, BPM 2004. LNCS 3080. Springer (2004) 82–97
6. Burch, J., Clarke, E., McMillan, K., Dill, D., Hwang, L.: Symbolic model checking: 10^{20} states and beyond. Information and Computation **98** (1992) 142–170
7. Cuntz, N.: Über die effiziente Simulation von Ereignisgesteuerten Prozessketten. Master's thesis, University of Paderborn, Department of Computer Science (2004)
8. Cuntz, N., Kindler, E.: On the semantics of EPCs: Efficient calculation and simulation. In Nüttgens, M., Rump, F.J., eds.: EPK 2004, Geschäftsprozessmanagement mit Ereignisgesteuerten Prozessketten. (2004) 7–26
9. Nüttgens, M., Rump, F.J.: Syntax und Semantik Ereignisgesteuerter Prozessketten (EPK). In: PROMISE 2002, Prozessorientierte Methoden und Werkzeuge für die Entwicklung von Informationssystemen. GI Lecture Notes in Informatics, Vol. P-21. Gesellschaft für Informatik (2002) 64–77
10. Kindler, E.: The Model Checking in Education (MCiE) project: Home page. http://www.upb.de/cs/kindler/Lehre/MCiE (2004)
11. Cuntz, N., Kindler, E.: The EPC Tools project: Home page. http://www.upb.de/cs/kindler/research/EPCTools (2004)

Towards Integrating Business Policies with Business Processes

Zoran Milosevic

CRC for Enterprise Distributed Systems Technology (DSTC)
The University of Queensland, Brisbane, Q 4072, Australia
zoran@dstc.edu.au

Abstract. We present a framework for augmenting business process specifications with policy expressions such as obligations, permissions and prohibitions. One use of such a combined model is to support monitoring of participants' behaviour against agreed policies as in business contracts.

1 Introduction

One limitation of current business process (BP) initiatives, e.g. [1, 2, 3], is their lack of positioning of BP models within a broader enterprise model covering organisational structures, policies and contracts. This paper partly addresses this limitation by applying our community model [4] to a behavioural style typical of BP approaches. The aim of the paper is to augment BPs with policy expressions to support monitoring of participants' behaviour against agreed policies. Section 2 introduces key concepts from ebXML's BP specification (ebBP) [1] and BP modelling notation (BPMN) [3] of relevance for business policies, and illustrate them with a simple example (Fig. 1). Section 3 introduces our policy concepts and maps them onto the relevant BPMN and ebBP concepts. Section 4 discusses open issues and future work.

2 Key Collaborative Business Process Concepts

An end-to-end BP model can cover both *private* (internal) and *collaborative* (global) sub-models [3]. We outline key *collaborative* concepts in the ebBP and BPMN standards to show the role of business policies. In ebBP, a *business collaboration* defines a set of roles and a set of business transactions between participants filling these roles. We use BPMN *pool* (shown as a rectangle, Fig. 1) to represent the participants. An ebBP *business transaction* (BT) is a basic unit of work in business collaboration. It specifies how business documents are exchanged between the participants. We use a BPMN *event* to denote the start and end of the collaboration and the arrival of business documents [3]. Each BT has a requesting document flow from the requesting to responding activity and can have a response document flow in the opposite direction. A BT may involve exchange of one or more *business signals* that support synchronisation of the business states between parties, e.g. to indicate that a document was successfully delivered but cannot be processed by the receiving application because of an invalid document schema. Business signals are separate from lower protocol and

W.M.P. van der Aalst et al. (Eds.): BPM 2005, LNCS 3649, pp. 404–409, 2005.

transport infrastructure. We use BPMN *message flow* to show ebBP messages and signals (dashed line arrows, Fig.1). A *business transaction activity* (BTA) denotes the use of a BT within the collaboration. The same BT can be performed by multiple BTAs in the same or multiple collaborations. A BTA may specify that a document interchange has a *legal intent* indicating a commitment by the trading partners. The ebBP standard does not provide any recommendation as to how it should be interpreted or enforced, leaving it as a concern of an external contractual framework such as the one proposed in [8]. *Choreography* is the ordering of BTAs, defining the expected flow of business documents and signals and can be shown using UML activity diagrams or other notations like BPMN, as in the ebBP specification.

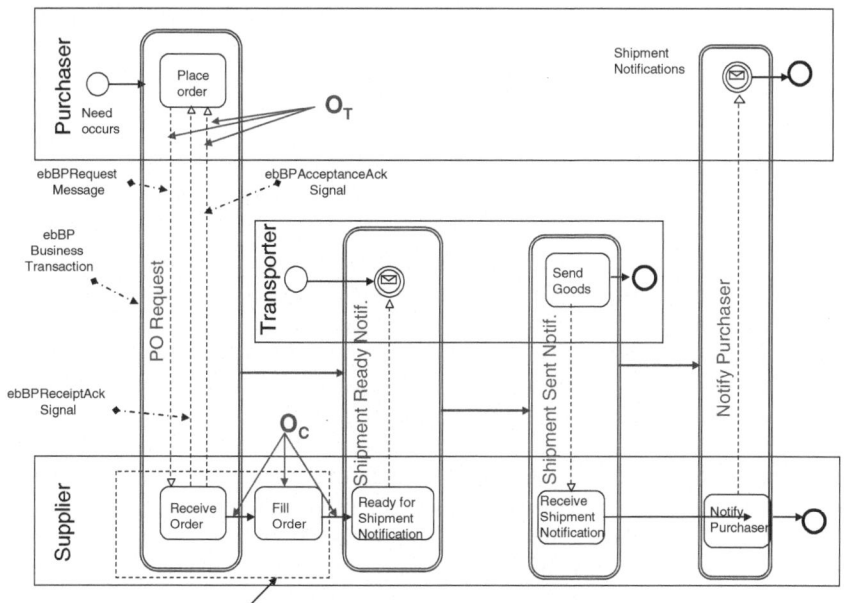

Fig. 1. End-to-end business process model

Our example depicts four BTs with their requesting and responding activities. The PlaceOrder activity initiated by the purchaser sends an ebBPRequest message carrying the purchase order (PO) to the supplier. The RecieveOrder activity carried out by the supplier, validates the PO schema and logs this request, if valid, for subsequent processing. This BT involves ReceiptAcknowledgment and AcceptanceAcknowledgment signals. The former acknowledges that the PO is received, while the latter acknowledges PO validity. The signals provide assurance to the purchaser of the successful delivery and acceptance of the PO by the supplier. After the supplier has processed the PO and is ready to supply goods, it sends a Shipment Ready Notification to the transporter. The BPMN sequence flow shows the ordering between these two BTAs and between subsequent transactions. The figure also shows the FillOrder private activity for the supplier and includes a *business policy*, i.e. *Supplier is obliged to fill the PO within one day of the PO request* (BMPN text annotation). This places an additional constraint on the supplier's FillOrder activity, triggered by successful validation of the PO. The completion of the FillOrder triggers another BTA, between the supplier and transporter.

3   Linking Policies and Business Processes

BPMN and ebBP standards allow the specification of the 'normal' flow of control and data between business activities, i.e. a computationally complete process where points of failure are identified in advance and explicitly represented in the collaboration. Typically, these are network or application failures. In addition, ebBP business signals discussed previously, support the expression of *business failures*. For a successful BT, both its network/application and business aspects must be successful. Although business signals support predictability of interactions when crossing technology and organisational boundaries, they are not sufficient to specify whether business failures are caused by participants not fulfilling their commitments as agreed in the collaboration. Namely, there is a weak link between the specification of business policies that apply to the participants and the 'normal' flows in the collaboration. This is a limitation because a complete specification of the collaboration needs to explicitly state policies that apply to the constituent roles. Consider a situation where a purchaser has received both ReceiptAcknowledgment and AcceptanceAcknowledgment signals. A success of this first BTA only provides guarantees to the purchaser that the supplier is in a position to start executing its own process. However, the purchaser cannot determine whether the supplier has fulfilled their PO and thus satisfied the constraint from the agreement, e.g. 'supplier is obliged to fill the PO within one day of the PO request'. Thus, an added mechanism is needed to deal with unpredictability of behaviour of participants and to check (unintentional or deliberate) violation of their policies. Although ebBP's labelling of BTAs with a 'legal intent' attribute gives special weight to such BTAs, the policy conditions in a collaboration are typically more complex and involve internal actions of parties. In our example, the obligation applies to the supplier (although it is triggered by the purchaser) and it is the monitoring of this policy that determines correctness of their behaviour. In what follows we show how our policy language [4, 8] can be applied to the ebBP concepts to provide a model for the specification of both the normal behaviour and policy constraints.

3.1   Specifying Policy Constraints

A business *policy* specifies constraints on behaviour of a participant in an organisational context such as an ebBP collaboration. In our unified model [4] this context is called *community*. Community specifies the *roles* involved, their relationships and *basic behaviour* constraints, e.g. control and data flow between participants and/or business steps in a BP. A community also supports the expression of policies applying to the roles, e.g. obligation constraints as mentioned before, or a permission constraint such as 'The supplier is permitted to provide an invoice immediately after goods delivery'. A more complex policy expression involving the concept of state is: 'The purchaser has a credit limit with the supplier, which is a maximum outstanding amount with no particular time limit. The purchaser is not permitted to exceed the credit limit' [5]. Basic behavioural constraints and policy constraints are specified using the concept of events and their relationships. An *event* can represent the actions of participants in the collaboration, either their internal actions or their interactions with other parties, or any other occurrence of interest, e.g. the events from the environment or timeouts. Event relationships are called *event patterns* and they have many similarities with the complex event processing ideas [6]. The main role of event patterns is to support event-based monitoring of activities of relevance to policies. An event pattern is evaluated as events come into the system and its progressive evaluation is finished when it's condition is matched. Event patterns range from simple relationships, e.g. sequence of events and logical event relationships to more complex events, e.g. quorum, event causality and temporally-oriented constraints like a sliding time window [4, 8]. Using a simplified version of our policy language, the last policy is:

Policy: CreditLimitForPurchaser
  Role: Purchaser
  Modality: Not Permitted
   Condition: PO (OutstandingDebt + PO.value > CreditLimit)

Policy is defined in terms of a name, a role to which it applies, modality and event pattern condition (a singleton event of a PO type in this example). Its value is used as a parameter in the condition for checking the value of the *OutstandingDebt* state.

State: OutstandingDebt
  CalculationExpression
   UpdateOn: Payment
   UpdateSpecification:
     return (this - Payment.amount)
  CalculationExpression
   UpdateOn: InvoicePurchaser
   UpdateSpecification:
     return (this + InvoicePurchaser.amount)

The *OutstandingDebt* state value updates are triggered by the events that affect this state. The concept of state is significant for run-time monitoring of a contract since state variables can be embedded in policy checking expressions as above.

3.2 Applying Policies to Business Collaborations

One motivation for applying policies to the ebBP collaboration is to explicitly associate responsibilities, authorisations, permissions and other policies with collaboration roles which is, for example, needed when integrating contract conditions with the BPs governed by contracts. Another motivation is to support run-time monitoring of participants' behaviour to detect existing or potential violations of the agreed behaviour, as presented in [8] and which is of increasing importance for meeting compliance requirements, e.g. [7]. A further value in separating policies from basic behaviour is the ability to change policies while preserving the fundamental properties of a BP.

The first step in applying policies is to identify events in the collaboration that, via event patterns, are part of policies that apply to collaboration roles, in particular those that are involved in BTs labelled with 'legal intent'. Such events can be part of ebBP transactions, shown as O_T symbols in Fig.1 (denoting observation points in a transaction), e.g. the sending of Request and Response messages and the generation of ReceiptAcknowledgment and AcceptanceAcknowledgment signals. Each of these can have an associated timeout and their occurrences (e.g. as generated by an ebBP engine) can be modelled as deadline events. These can then be used as input to the business monitoring engine for subsequent management actions, e.g. generation of human readable notifications. The events in a collaboration can, on the other hand, correspond to the transition between ebBP business states in an ebBP choreography. These events are as shown as O_c symbols (observation points in a choreography). Considering this, a possible implementation of the third policy from section 3.1 is:

Policy *PromptOrderFulfillment*
    Role: *Supplier*
    Modality: *Obliged*
    Condition:
        OrderFilled before (ReceiveOrder + 1 days)

The successful receipt of the PO triggers an obligation for the supplier to fulfil the remaining part of the document. The RecieveOrder triggering event can be defined to be: *i)* the AcceptanceAcknowledgment signal (left arrow next to the O_c symbol) or *ii)* state transition at the sequence flow from the RecieveOrder activity to the FillOrder activity (middle arrow next to the O_c symbol). The OrderFilled event (right arrow next to the O_c symbol), if occurring within one day of the RecieveOrder event, signifies the fulfilment of this obligation, otherwise, the violation occurs. In this policy the trigger event was generated by the purchaser and the obligation is on the execution of OrderFilled event by the supplier. Thus, the policy involved the events occurring within two roles in this cross-organisational BP and one is within the scope of the global BP, namely the ebBP transaction (i.e. RecieveOrder event) while the second is within the scope of a private BP (OrderFilled event).

This analysis suggests that our policy language can be used to define additional constraints on the behaviour of trading partners in cross-organisational processes, provided all the events of relevance for the policies are available to the policy specifier. However, the current ebBP standard only allows the specification of the choreography of *collaborative* business activities. This means that the important events in a policy's event pattern that correspond to *private* processes need to have their global counterparts, i.e. these need to be defined as part of ebBP BTAs. An alternative would be to extend ebBP semantics to provide integration points with private processes, or allow for the specification of asynchronous events made visible to trusted third parties for monitoring purposes.

4 Open Issues and Future Work

If policies are to be applied to a cross-organisational BP then the private activities and their integration with the global activities need to be made visible (at least to process designers, not necessarily to end-users), because policy specifications often require the expression of event patterns that include events associated with *either* of the activity types. This is currently not possible in the ebBP models. Further, policy monitoring requires: *i)* the concept of state in ebBP and BPMN and *ii)* definition of a concept of event in ebBP so that ebBP message arrivals, signals and timeouts can be treated as a special kind of that event. BPMN provides a rich set of events and these can be used in ebBP models. Finally, the full power of the community model can further augment cross-organisational BP with other enterprise modelling constructs and provide input to the development of ebBP, BPMN and BPEL standards.

In future we plan to develop a stronger link between ebBP and contract frameworks to enable richer support for contract monitoring, as initially proposed in [9] and contribute to aligning ebXML and legalXML e-contracts [10] standards. We also plan to apply model-driven design for the mapping of policy language to ebBP.

Acknowledgements

The work reported in this paper has been funded in part by the Co-operative Research Centre for Enterprise Distributed Systems Technology (DSTC) through the Australian Federal Government's CRC Programme (Department of Education, Science, and Training). The author would also like to thank Andrew Berry and Andy Bond for their comments on an earlier version of this paper.

References

1. ebXML Business Process Specification Schema 3, v2.0 4 Working Draft 10, 23 February 2005 (pre-notification Committee Draft)
2. Business process execution language for web services, May 2003. http://www.ibm.com/developerworks/library/wsbpel/.
3. Business process modelling notation, 2004. http://www.bpmn.org/.
4. P. Linington, Z. Milosevic, J. Cole, S. Gibson, S. Kulkarni, S. Neal, *A unified behavioural model and a contract language for extended enterprise*, Data Knowledge and Engineering Journal, Elsevier Science, October 2004
5. A. Berry, Z. Milosevic, *Extending choreography with business contract*, special issue of the IJCIS journal on contract architecture and languages, to appear.
6. D. Luckham, *The Power of Events*, Addison-Wesley, 2002
7. http://www.sarbanes-oxley.com/
8. Z. Milosevic, S. Gibson, P. F. Linington, J. Cole, S. Kulkarni, *On design and implementation of a contract monitoring facility*, Proc. the 1st IEEE Workshop on e-contracting, July 2004.
9. J. Cole, Z. Milosevic, *Extending Support for Contracts in ebXML*, ITVE workshop, Australian Computer Science Week, Jan 2001.
10. www.oasis-open.org/committees/legalxml-econtracts/charter.php

A Contract Layered Architecture for Regulating Cross-Organisational Business Processes

Mohsen Rouached, Olivier Perrin, and Claude Godart

LORIA-INRIA-UMR 7503
BP 239, F-54506 Vandœuvre-lès-Nancy Cedex, France
{mohsen.rouached, olivier.perrin, claude.godart}@loria.fr

Abstract. As technology infrastructure becomes available for electronic exchange of contracts, the IT community is becoming more interested in modeling of contracts as governance structures for inter-organisational interactions and business processes. This paper investigates e-contract modeling and monitoring. Subsquently, we propose a contract layered model that allows for the convenient monitoring of multi-party contracts during contract fulfillment and reduces complexity of interrelationships. Communication between contract parties relies on a event-based mechanism which extends the scope and flexibility of our model.

Keywords: e-contract modeling and analysis, business process management, event-based monitoring.

1 Introduction

Nowadays, there is a renewed interest for modeling and orchestrating cross-organisational and cooperative processes using business contracts. This is motivated by the fact that enterprises increasingly use the Internet for communication with their partners and would like to leverage this technology in order to gain efficiency in contracting processes. Moreover, contracts are important in the context of loosely coupled structures (supply chains for instance). In fact, there is no central authority that coordinates activities of independent entities making up a supply chain, each entity being responsible to arrange a contract with their partner for the collaboration to which they belong.

Usually, contracts define rights and obligations of parties as well as conditions under which they arise and become discharged. The rights and obligations concern either states of the affairs or actions that should be carried out. Often contracts also specify secondary obligations (reparation) that come into force when a party does not carry out an obligation. An e-contract is a contract regulating cross-organisational business processes over the Internet.

Problems in analysing contracts generally arise from ambiguity and fuzyness of natural langages, the autonomous nature of individual organisations, and the complexity due to the richness of the structures in business organisations. Events that need to be monitored often come from counter parties in other organisations, and might not be monitorable. Thus, cooperation and trust should be developed

W.M.P. van der Aalst et al. (Eds.): BPM 2005, LNCS 3649, pp. 410–415, 2005.

among trade partners to alleviate this problem. In general, this improves the transparency of operations, services, and is therefore vital in contemporary e-service providers under strong competitions. It may be the case in SOA (Service Oriented Architecture) or BPM (Business Processes Management).

In this paper, we present a model and a platform to support contracts. We adopt a layered and distributed event-based architecture for modeling and executing electronic contracts. It is worth noting that our approach is completely different from the workflow aspect because we do not specify *how* to manage the business process but we concentrate on regulating cross-organisational business processes over the Internet and determining the responsibility of each partner to respect contract clauses. Our contract approach for coordinating business processes is interesting because it allows the separation of the rules that govern the behaviour of the overall process from the internal processes in the organisations. This feature is important as it allows to ensure the autonomy of the partners, and also to respect their privacy. An other benefit of this approach is the dynamic adaptation, this means that it is possible to adapt the behaviour of the business process without the need to fully reconfigure it, and changes can be applied without stopping the execution. Then, such an architecture for supporting contract is valuable as it permits to garantee several criteria required for business processes such as expressivity, flexibility, reusability and completeness.

The rest of the paper is organized as follows. In section 2, we present our contract model, while section 3 details contract events. In section 4, related work will be discussed, and section 5 concludes this paper.

2   A Layered Contract Model

To reduce the degree of the complexity and alleviate problems introduced so far, we propose the following contract model, illustrated in figure 1. It is based on a three-layers architecture. This architecture is different from the one proposed by Chiu & al. [CCT02] in the sense that we are not interested in contract negotiation

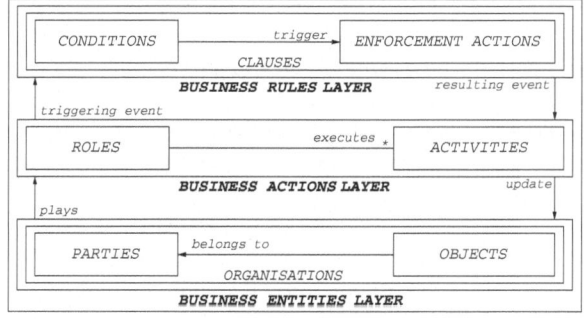

Fig. 1. Contract Model

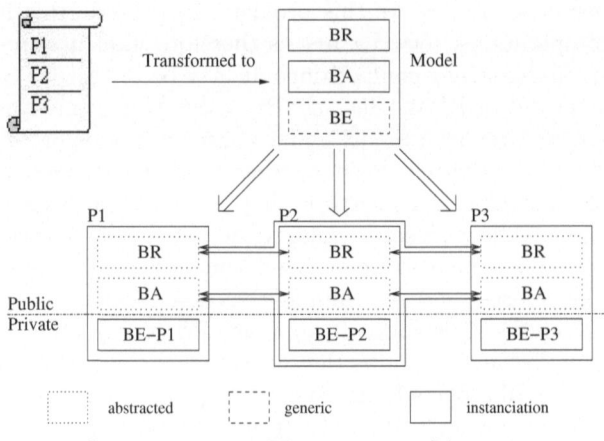

Fig. 2. Global View

nor contract automatic writing. In fact, we suppose these steps are already done. We are rather interested in the instantiation of the execution infrastructure. Our architecture consists of a *business entities* layer, a *business actions* layer, and a *business rules* layer. Those three layers are coordinated by an event-based interaction mechanism entailing a dispatching and coordination paradigm, which offers the advantage of a complete separation of the coordination aspects and functionality aspects (see section 3). Let us now detail each layer.

The *business entities* layer specifies the organisations involved in the contract. An organisation consists of one or more parties and objects belonging to the parties that are relevant to the e-contract. In the same organisation, each party can participate in several contracts. Thus, our model supports a multi-party contract, avoiding the need to break it down into a number of bilateral contracts. The *business actions* layer captures the details of the actions required in the contract, including the set of roles involved in each action and the set of partners's activities executed by each role. These activities are only those seen from outside during the e-contract execution. We are not interested in internal activities for each party. The *business rules* layer specifies the clauses stipulated in the e-contrat. It consists of two parts: the conditions and the enforcement actions that should be executed under these conditions. The evaluation of the conditions is triggered by a generated event resulting from the execution of an activity in the business actions layer or by an external event.

As such, the layered architecture allows an e-contract to be seamlessly defined and enacted by considering an e-contract as an "abstracted" business process. Then, to map the contract document into electronic format allowing automated management, we carry out two operations, viz., *instantiation* and *execution*. The first operation consists of determining elements of each layer. Then during the execution, the parties start communicating and interacting.

Our approach of sharing the layers between contract parties, illustrated in figure 2, consists of attributing only the business entities layer to each party whereas the other layers are shared between all parties involved. In this figure, a contract is established between three parties $P1$, $P2$, and $P3$. The term *generic* precises that the layer is instantiated by each party whereas the term *inherited* indiquates that the layer is shared by all parties. Thus, each party has its own objects and only those necessary for enacting and enforcing the e-contract are communicated to the others. At the same time all parties collaborate to apply contract clauses and execute necessary activities to accomplish the desired service. As such, we alleviate the problem of using a Third Trusted Party which often could not have only external knowledge about parties which it supervises and thus it could not apply relevant corrective actions. Moreover, this approach permits facilitate update operations for each party. It allows contract parties to be autonomous entities that encapsulate the integrality of their behaviors without any centralized control.

3   An Event-Based Architecture to Support Contracts

Event-based communication is an interesting paradigm for building large-scale distributed systems. It has the advantages of loosely coupling communication partners, being extremely scalable. To synchronize the three levels (Figure 1), we consider an event as a significant occurrence in time or instantaneous (punctual). Events are relevant to roles within a context determined by the contract. This context has attributes which can be *repetition* operators, *detection* mode, *composition* operators, *counting* operators, *negation* operators, and *temporal* management. We have also the possibility to express conditions on these operators.

From an abstract point of view, an e-contract execution can be described by the types and relative order of events occuring in each party. Therefore, after defining events, it is necessary to study their relationships in order to ensure the synchronization of the layered architecture and enable the communication among contract parties.

3.1   Event-Driven Causality

Let E_i denote the set of events occurring in a party P_i, and let $E = \cup_{i=1,...,N} E_i$ denote the set of all events in the N e-contract parties. These event sets are evolving dynamically during the e-contract execution. The causality relation $\prec$ onto $E \times E$ is the smallest transitive relation satisfying: (1) if e_{ij}, $e_{ik} \in E_i$ occur in the same party P_i and $j < k$, then $e_{ij} \prec e_{ik}$, (2) if $s \in E_i$ is a sent event and $r \in E_j$ is the corresponding received event, then $s \prec r$.

Given two events $e1$ and $e2$, if neither $e1 \prec e2$, nor $e2 \prec e1$ holds, they are said to be concurrent. The concurrency relation $\parallel$ onto $E \times E$ is defined as $e1 \parallel e2 \equiv \neg((e1 \prec e2) \vee (e2 \prec e1))$. In general, an unspecified pair of events always satisfies one and only one of the following relations $\forall e1, e2 : e1 \prec e2 \bigoplus e2 \prec e1 \bigoplus e1 \parallel e2$.

3.2 Events Contract Model

For reliability and efficiency, our event-driven mechanism consists of two meta-models. An **Event Types Meta-model** offers a grammar to describe event types and formal tools to specify composition operators semantics. We specified an event as $(instant, type, validity, cond, mask)$. The *instant* expresses the observation granularity of a special situation. The *type* identifies primitive or composite events defined by applying event operators the primitive ones. The validity interval *validity* indicates the begining and the end moments of the event effect. The *cond* contains information which informs about conditions under which event occurs. The *mask* is a predicate expressing *cond* constraints and temporal expressions that events must satisfy. An **Event Management Meta-Model** describes how events are recognized and notified. The context mentioned so far has several properties which include temporal characteristics, semantic characteristics, space characteristics, and state characteristics. Moreover, event relationships are primarily based on concepts of causality and events composition. This proves that the business process automation requires **semantic** level monitoring, rather than **system** level monitoring. Therefore, we focus on relationships between events to deal with monitoring issues, which makes it possible to achieve the pro-active monitoring goal.

4 Related Work

There has been an important numbers of researches concerning the representation of contracts for the purpose of reasoning over, and monitoring, them at run-time. In [Gro99], Grosof introduced a declarative approach to business rules in e-commerce contracts by combining Courteous Logic Program and XML. Marjanovic et Milosevic [MM01] modeled a contract with deontic logic, based on obligation, permission and prohibition. Business Contract Architecture (BCA) [AZAK95] does not provide generic monitoring facilities, expecting each application to develop its own monitoring code to detect and signal non-conformance to the contract monitor. In Seco (Secure electronic contracts) [MK00], the monitoring services allow events to be triggered according to the current state of the contract and informs enforcement service to initiate an enforcement activity. In paper [GLA02], the authors present a three-level process framework for dynamic contract-based service outsourcing and discuss an abstract architecture for dynamic service outsourcing. Comparing with our architecture, this paper did a vertical level research which is involved with workflow system details. On the other hand, our contribution is a horizontal level which interested in interactions among several contractual parties in terms of complex events.

5 Conclusion and Future Work

This paper presents an approach to formalize electronic contracts into a meta-model that enables automatic monitoring. We have detailed a pragmatic architecture for cross-organisational e-contract enforcement and enactement comprising three layers. We have detailed elements contained in each layer. We have also developed an event-based paradigm to facilitate the executable specification of e-contracting applications.

At the same time, we are working on further details for complex events management and their impact on electronic contracts monitoring. We are also implementing the suggested model using Jena which is a Semantic Web framework containing a reasoner subsystem for building Semantic Web applications, allowing both backward and forward chaining.

References

[AZAK95]   Bond A., Milosevic Z., Berry A., and Raymond K. Supporting business contracts in open distributed systems. In *2nd International Workshop on Services in distributed and Network Environments, (SDNE'95) Whister, Canada*, 1995.

[CCT02]   Dickson K.W. Chiu, S.C. Cheung, and Sven Till. A three-layer architecture for e-contract enforcement in an e-service environment. In *Proceedings of the 36th Hawaii International Conference on System Sciences (HICSS'03)*, 2002.

[GLA02]   P. Grefen, H. Ludwing, and S. Angelov. A framework for e-services: A three-level approach towards process and data management. Technical report, IBM Research Report RC22378, University of Twente, 2002.

[Gro99]   B. N. Grosof. A declarative approach to business rules in contracts: Courteous logic programs in xml. In *Proceedings of the 1st ACM Conference on Electronic Commerce (EC99), USA*, pages 68–77, November 1999.

[MK00]   Schopp B Greunz M and Stanoevska-Slabeva K. Supporting market transactions through xml contracting containers. In *Proceedings of the Sixth Americas Conference on Information Systems (AMCIS 2000). Long Beach, CA*, 2000.

[MM01]   O. Marjanovic and Z. Milosevic. Towards formal modeling of e-contracts. In *Fifth IEEE International Enterprise Distributed Object Computing Conference, Seattle, USA*, pages 59–68, September 2001.

An Effective Content Management Methodology for Business Process Management

Young Gil Kim[1], Sang Chan Park[1], Chul Young Kim[1], and Jin Ho Kim[2]

[1] Dept. of Industrial Engineering, Korea Advanced Institute of Science and Technology,
373-1 Gusung-Dong, Yusong-Gu, Daejeon, Korea 305-701
{ttaldul, sangchanpark, fezero}@kaist.ac.kr
[2] 140-19 Samsong Bldg., Samsong-Dong, Kangnam-Gu, Seoul, Korea 135-090
jhkim@dsrgroup.co.kr

Abstract. BPM is considered as the suitable framework for today's process-centric trends because it addresses the interplay of people and organizations on the one hand and process-aware software on the other hand. In such an environment, it is very important for each business participant to trace contents of total business processes. However, it has been a difficult problem for BPMS to support management of content usages because of limited storage, complex and dynamic process change, and absence of formal usage model. Content management technologies play an important role in e-business environment because they enable the seamless flow of information among business participants. Using these, we present a methodology that manages various process-related contents for implementing business process management system. The proposed framework can enable real-time content integration between user's workflow information, processing knowledge and various enterprise applications.

1 Introduction

In recent years, a lot of new technologies have become available for adopting the process-driven approaches. To gain a competitive advantage, many companies emphasis on their processes and process-driven approach. One of these approaches, the Business Process Management (BPM) is considered as the suitable framework for the process-centric methods because it addresses the interplay of people and organizations on the one hand and process-aware software on the other hand. Many people consider the BPM to be the "next step" after the workflow systems [8], and we think there is no doubt about the promise of BPM.

However, there are two questions for BPM implementation: how to support the first step process setup of immature companies and how to support companies' content reusability for other service processes. These two questions are basically how to support effectively management of content usage in BPMS. In this manner, we consider the method for coping with these problems by incorporating content management modules into BPMS.

For companies to manage their processes effectively, they should identify and visualize their processes and collect overall process-related information. Content Management (CM) utilizing Web services could enable these activities. More specifi-

W.M.P. van der Aalst et al. (Eds.): BPM 2005, LNCS 3649, pp. 416–421, 2005.

cally, the information acquisition, storage and distribution activities in content management enable the dynamic management and maintenance of process management activities.

In this paper, we present a methodology for business process management system coupled with content management and apply to a market research service company for demonstrating the feasibility of the proposed methodology. The proposed system can enable real-time content integration among user's workflow information, processing knowledge, and various enterprise applications.

2 Literatures Review

2.1 Business Process Management (BPM)

Aalst [1] defined Business Process Management as follow: Supporting business processes using methods, techniques, and software to design, enact, control, and analyze operational processes involving humans, organizations, applications, documents and other sources of information. The BPM can be used to integrate existing applications and support process change by merely changing the process diagram. As Lu and Chen introduced in their paper on web-based information system, the Web and Internet tools can support these aspects effectively.

Recently, some useful tools for building BPMS are introduced. BPML [4], XPDL [10] and BPEL4WS [3] are XML-based process definition languages for this purpose. They provide a formal model for expressing executable processes that addresses all aspects of enterprise business processes, but they are based on significantly different paradigms.

All of these tools utilize *activity* as the basic component of process definition. In each, activities are always part of some particular process. All Each has *instance-relevant data*, property for BPML, workflow-relevant data (data fields) for XPDL and Containers for BPEL4WS, which can be referred to in routing logic and expressions. These give some chances to leverage the management of contents produced by the business process management systems, if we can manipulate the process activities and their properties as integrated content objects using a mediating facility.

2.2 Content Management (CM)

Content management (CM) is generally a term that describes the issues around creating, versioning, storing and disseminating semi-structured and unstructured information owned by an enterprise, or it is often equated a repository based facility to store web content with some metadata management.

Trappey et al. [9] proposed a global content management platform and Najmi [7] introduced content management using ebXML Registry standard. In other approaches, Fernández-García et al. [6] presented architecture for the management of distributed contents. While most researches are focused on managing transactions [9] [7] or managing enterprise documents [6], we concentrate on content management to control the company's total process-related information for supporting business process management. A content management itself is not the source of knowledge, but it can be a very valuable enabler in knowledge-capturing processes.

Although originally applied to data resources, metadata may refer to any kind of resource, such as applications or processes or people. The meaningful definition of metadata corresponding to our framework is the descriptive information about the structure and meaning of a resource [5]. Metadata used in content managements can function as follows: tracking content owners, capturing relationship & link, capturing classification information (keywords), tracking a range of business-specific information. However, for applying metadata to process-aware content management, these are lacking in information about the structure and relationship of contents (dynamic aspects of contents).

Although process definition languages for building BPMS are based on different paradigms, content management technologies give useful chances for coping with these gaps and so applying various environments, if we can incorporate content aggregation model having metadata into above process definition languages using "*activity*" concept as the basic components of process definition.

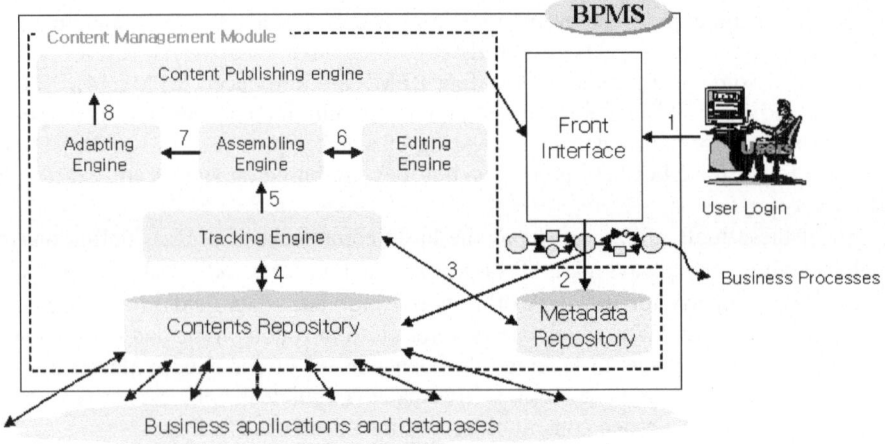

Fig. 1. Framework of Content Management Module incorporating into BPMS

3 Methodology

The BPMS is a single, unified modeling, integration, and execution environment. BPM can model not only computer-based processes, but also manual, abstract and real world processes. In order to implement BPM in process set-up of immature or dynamic change companies, it should be required that the companies capture, store, share, use and reuse their contents form present status to next improvement. Figure 1 depicts the proposed content management framework within BPMS.

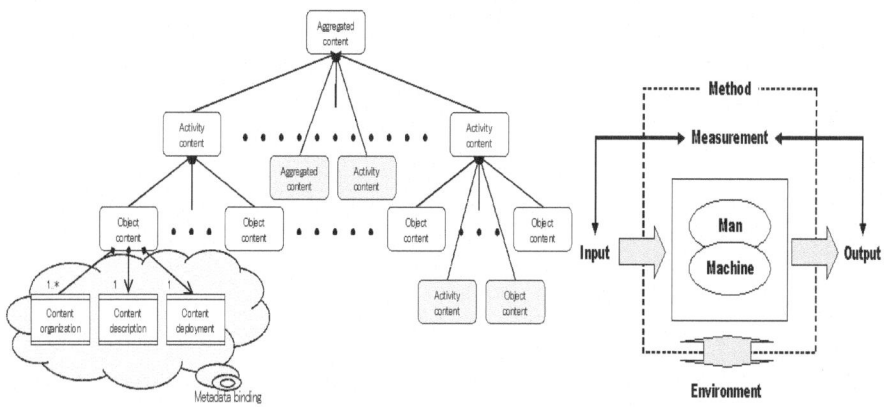

Fig. 2. Content aggregation model for BPM and process-entity for dynamic metadata

The primary role of the Content Management Module is to bind contents with synthesized metadata and thus to acquire, manage, reuse, and service the various types of process-related contents. The proposed module has five major engines and two repositories for managing contents.

When a user logs on the system and uploads his outputs, the Tracking engine searches appropriate contents form the two repositories. Then, the Adapting engine is used to infer the suitable rules for manipulating the retrieved contents. Also, the Tracking engine saves content usage data and its structures of the user, and the Adapting engine is used to support the user to a suitable interface and content through the analysis of the usage history. The Assembling engine packages contents retrieved content objects. The Publishing engine provides with packaged contents which can have various formats. The Editing engine supports to edit the content or content organizations.

In this framework, the content aggregation can be used to aggregate information assets into a cohesive unit of work (e.g. process, activity, operator, data, etc.), and apply process structures and processing taxonomies. Content aggregation and content packaging concepts of SCORM specification [2] can be used in overcoming and expanding the shortage of traditional metadata techniques. Figure 2 (left) shows how to package the process processing contents in our framework. We define object content, activity content and aggregated content for content packaging. The object content is the information asset of process-related information like as man, machine, material, method and environment (4M1E) depicted in figure 2 (right).

Metadata should be used to describe the various parts that is managed by BPMS, such as available applications, service interfaces and their binding points, batch queuing systems, hardware profiles, and so on. For streamlined and effective supports for building mature processes, our idea is firstly to embed the structure of information items into metadata template, which are then filled semi- or automatically on the fly doing daily work.

Figure 3 (left) shows a web interface of metadata template. In this figure, the elements for applying the metadata to content objects in the proposed framework consist

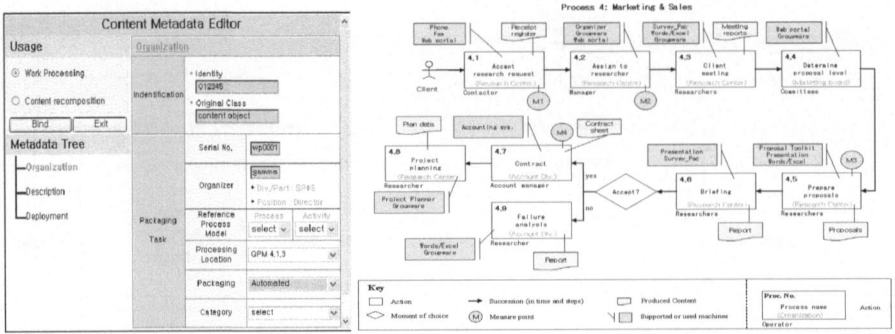

Fig. 3. Web interface for metadata capturing and an example of extended process map

of three major components: organization for content structure, description for static info and deployment for dynamic info. Using this capabilities, a company can accumulate process-aware contents, so can achieve more extended process map for toward next step in improvement as depicted in figure 3 (right).

4   Implementation

In order to demonstrate the feasibility of the proposed framework, we design and implement an application for marketing research processes. A typical service process for a marketing research company consists of four consecutive stages: Marketing & sales, Planning, Fieldwork, and Reporting.

For example, the typical marketing & sales process in a quantitative research project begins with a contactor. When an opportunity is identified or a proposal sought by potential or existing clients is received, the contactor registers the receipt information in the register. The manager of research center assigns the received information to a researcher, and the assigned researcher meets the client and reports the meeting results. After the proposal level is decided by the company's marketing board, the researcher prepares a proposal and presents the proposal (see figure 3 right).

Fig. 4. Upload (register), recompose content objects and process monitoring in the application

In the application, users are allowed to upload their process-related contents and to verify the metadata according to their authority. They also retrieve appropriate contents and re-compose the contents for other purposes. Furthermore, for automated process control and decision support, the system can extract process processing information through the deployment part of the process management metadata and present research status graphs and tables depicted in figure 4.

5 Conclusions

In this paper, we proposed the methodology that manages various process-related contents for implementing BPMS. In the proposed framework, the metadata and content aggregation methods could be used for better understanding a company's processes and easily constructing a business process management system. Also, the metadata consisting of organization, description and deployment elements could be used for process control and content recomposing and reusing. The proposed methodology enables real-time content integration among user's workflow information, processing knowledge, and various enterprise applications.

With the growing technologies, such as semantic web and web services, the constitution of ontologies for content management like as ontology about processes, knowledge and content usage will be useful facilities of the e-business environment.

References

1. van der Aalst, W.M.P.: Business Process Management: A personal view. BPMJ, Vol. 10. No. 2 (2004) 248-253
2. Advanced Distributed Learning: Sharable Content Object Reference Model (SCORM) ver. 1.2 The SCORM Content Aggregation Model. ADLnet (2001) http://www.adlnet.org
3. Andrews, T., Curbera, F., Dholakia, H., Goland, Y., Klein, J., Leymann, f., Liu, K., Roller, D., Smith, D., Thatte, S., Trickovic, I., Weerawarana, S.: Business Process Execution Language for Web Services Flow specification ver. 1.1 (2003)
4. Arkin, A.: Business Process Modeling Language (BPML) specification. BPMI.org (2001)
5. Calvalcanti, M. C., Targino, R., Baião, F., Rössle, S. C., Bisch, P. M., Pires, P. F., Campos, M. L. M., Mattoso, M.: Managing structural genomic workflow using Web services. Data & Knowledge Engineering, Vol. 53 (2005) 45-74
6. Fernández-García, N., Sánchez-Fernández, L., Villamor-Lugo, J.: Next Generation Web Technologies in Content Management. Proceedings of the 13th international W3C conference (2004)
7. Najmi, F.: Web Content Management Using the OASIS ebXML Registry Standard. XML 2004, Amsterdam, the Netherlands (2004)
8. Smith, H., Fingar, P.: Business Process Management – the third wave. 1st ed. Meghan-Kiffer Press (2002)
9. Trappey, A. J.C., Trappey, C. V.: Global content management services for product providers and purchasers. Computers in Industry, Vol. 53 (2004) 39-58
10. Workflow Management Coalition (WfMC): Workflow Process Definition Interface – XML Process Definition Language (XPDL) ver. 1.0 (2002)

Specification and Management of Policies in Service Oriented Business Collaboration

Bart Orriëns[1] and Jian Yang[2]

[1] Dept. of Information Management, Tilburg University,
PO Box 90153, 5000 LE, Tilburg, Netherlands
`b.orriens@uvt.nl`
[2] Dept. of Computing, Macquarie University,
Sydney, NSW, 2109, Australia
`jian@comp.mq.edu.au`

Abstract. Current composite web service development and management solutions, e.g. BPEL, do not cater for developing adaptive business collaborations while adhering to the requirements imposed by the business environment. In this paper we introduce the Business Collaboration Design Framework which uses a blend of design perspectives, facets and aspects to provide designers with the means to develop and deliver business collaborations in a effective and manageable manner. We explain how policies can be defined in the BCDF to specify a wide range of requirements. We also conceptually introduce policy management mechanisms, which facilitate management of policy conformance and consistency, alignment and compatibility.

Keywords: Business process modeling and analysis, processes and service composition, process verification and validation, intra-organizational process support.

1 Introduction

Recently there has been increasing focus on service oriented computing [12] to deliver adaptive corporate business services by utilizing existing services cross organizational boundaries, i.e. via *business collaboration*. In order to realize this the specifics of business collaborations and their associated policies must be properly captured and modeled. Unfortunately, current composite web service development and management solutions including the defacto standard BPEL4WS [5] do not support specification and alignment of high (abstract) level requirements and technical demands, definition and management of policies and their compatibility among collaboration participants, and adaptiveness to environmental changes. In this paper we introduce a **Business Collaboration Design Framework (BCDF)** for designing business collaborations, which provides a systematic way for analyzing the requirements and modeling the activities involved in business

W.M.P. van der Aalst et al. (Eds.): BPM 2005, LNCS 3649, pp. 422–427, 2005.

collaboration development. The ideas presented are illustrated using a complex multi-party, insurance claim handling scenario [9]. [1]

2   Business Collaboration Design Framework

This section introduces the framework for design of business collaborations, being the **Business Collaboration Design Framework (BCDF)**. The BCDF, displayed in Fig.1, employs a multi-layered approach where collaboration design is facilitated through the usage of perspectives, facets and aspects.

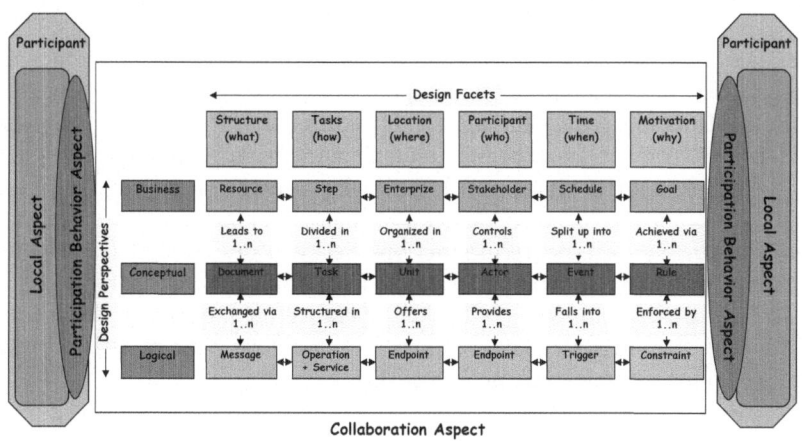

Fig. 1. Business Collaboration Design Framework (BCDF)

The *Design Perspectives* represent different levels of abstraction in a business collaboration development, supporting 'separation of concern': 1) *business perspective*: from which the purpose and the requirements are modeled and specified in terms of **Goal, Schedule, Resource, Enterprize, Stakeholder, Step**. 2) *conceptual perspective*: from which a computational independent conceptual model is generated that depicts the business activities in terms of **Rule, Event, Document, Unit, Actor, Task**. 3) *logical perspective*: from which a service-oriented business collaboration is modeled in terms of **Constraint, Trigger, Message, Endpoint, Service, Operation**.

Design Facets express different, but equally valid points of view at a particular perspective that emphasize the specification of different business description elements (as described in e.g. [14]: *what, how, where, who, when, why*). The *what* facet emphasizes the informational view. The functional standpoint is taken in the *how* facet. The geographical facet is expressed in the *where* facet, whereas

---

[1] This work was partially funded with an UNSW Australian ARC Discovery Research Grant.

the *who* facet captures participants. The temporal aspect covers in the *when* facet. The *why* facet describes rationale.

Design aspects accommodate different design considerations from different viewpoints: 1) *collaboration aspect:* describes the externally visible behavior between participants in a business collaboration, specifying how its participants are expected to behave in the collaboration, 2) *participant behavior aspect*: describes how an individual participant can behave in a business collaboration, i.e. its externally observable (public) behavior, and 3) *local aspect* describes the internal (private) behavior which is only of the interest of a particular participant.

3   Modeling in the BCDF Framework

The BCDF uses two types of models to accommodate modeling of business collaborations: meta models and instance models, both of which are defined for individual perspectives. The meta models are generic which provide design guidelines; while the models represent a particular design of an application, which have to conform to the meta-model. Each meta-model describes the relevant elements of (six) design facets as classes and their relationships. The relationships connect the classes to indicate interactions between design facets. Meta-model variants for each design aspect are defined by adding constraints to a meta-model. Fig. 2 displays an example business perspective model represented based on UML conventions. (Note: due to space limitations meta models as well as additional example models are not shown).

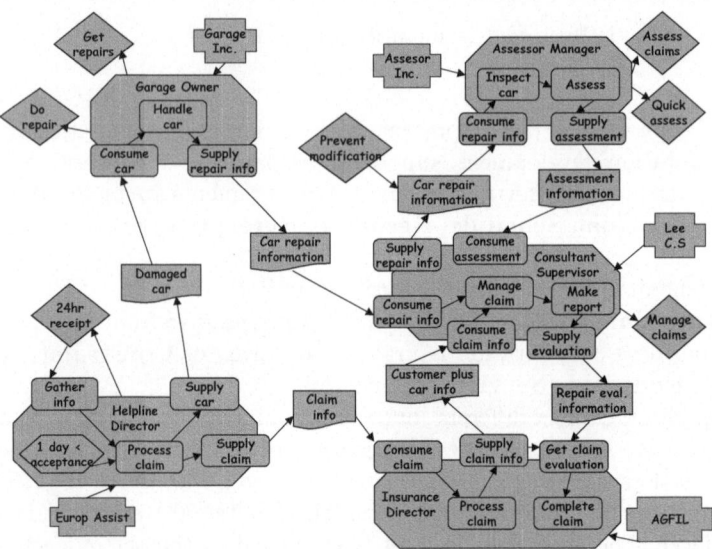

Fig. 2. AGFIL Business Model (AGFIL-BM)

To illustrate the working of the models look for a moment at the business model in Fig. 2. A main class instantiated is the *Goal* class, representing concrete goals such as 24hr receipt. Goals are pursued by stake holders (e.g. insurance director) who serve for participating enterprizes. Information about an enterprize is captured in the *Enterprize* class like AGFIL. Stakeholders exchange resources such as customer plus car, which are based on the *Resource* class.

Goals are achieved through steps representing high level functions such as process claim. Steps can be dependent on one another or contain other steps. Steps are of type 'internal' or 'exchanged'. Internal steps like process claim are enterprize specific and are not observable by others. Exchange steps provide a definition mechanism for resource supply and/or consumption (e.g. consume claim information).

Steps can have schedules linked to them like 1 day > performance. Schedules for different steps must be synchronized to avoid conflicts, so that, for example, the schedule of process claim does not interfere with that of supply claim information. Conceptual and logical models can be interpreted in a similar manner; however, naturally the semantics of the used elements will be different as they describe different perspectives.

4 Policies in the BCDF

Policies facilitate definition of the conditions under which involved enterprizes cooperate with one another. Policies are associated with the modeling description elements in the BCDF models, constraining the manner in which they can be connected to other description elements. As such, when models adhere to these different policies, this means that the described business collaborations are compliant with the set out conditions. In this manner policy specification gives empowerment with regard to managing and controlling collaborations.

4.1 Specification

In the BCDF different policies exist varying both in terms of design perspective and aspect. To accommodate their specification BCDF offers a generic policy language where policies are defined as sets of exclusive alternatives (each covering a possible collaboration scenario). Each alternative constitutes a group of rules, combinable using standard logical operators. Rules are defined using first order predicate logic. This accommodates definition of a wide variety of rules in terms of functionality ([10] [13]), optionality, genericity, and, perspective, facet and aspect. The atoms referred to in the rules constitute the modeling description elements used in BCDF to express business, conceptual and logical models, i.e. elements, properties and links like 'resource', 'modification' and 'consumes'. By constraining their existence and/or value, the rules can thus drive development of business collaboration models compliant with the set out conditions.

To exemplify look for a moment at the policy of claim information (exchanged between helpline director and insurance director), which is defined as $P_{ClaimInfoTrade}$: $\{PA_{HighClaim} \oslash PA_{LowClaim}\}$. This policy contains alternatives 'HighClaim' and 'LowClaim', where the first is applicable to high-valued claims and the second one to low-valued ones. $\text{PA}_{HighClaim}$ is defined as $\{R_{modification} \wedge R_{legitimation}\}$, depicting that $\text{PA}_{HighClaim}$ contains the security oriented rules $R_{modification}$ and $R_{legitimation}$. $R_{modification}$ is itself defined as the tuple (Content, '5','Mandatory') indicating that its application is mandatory and has the highest priority. Its content is defined as "resource('ClaimInformation') $\wedge$ value('ClaimInformation',_), '$\geq$', '1000') $\rightarrow$ (modification('ClaimInformation'), '=', 'true')"; depicting that all claims with value greater than $1000 must be sent in such a way that any modification will always be detected.

4.2 Policy Management

Policy management in the BCDF is facilitated in three ways: 1) policy conformance to verify whether a model is valid with regard to a policy; and policy consistency verification which in turn allows determining whether the model resulting from the application of these rules is consistent. 2) policy alignment by guarding the validness of mappings between perspectives, which express dependencies among the classes in different meta-models and instance models at different perspectives (also shown in Fig.1 in section 2). 3) policy compatibility for maintaining consistency among policies from different design aspects, i.e. between internal and behavioral policies, and between behavioral and collaboration policies.

To illustrate, suppose that claim information with value $1500 is sent without modification protection. This is not conform to the policy of `claim information`, since protection is mandated for all claims valued higher than $1000. In addition assume that we have $R_{confidentiality}$ defined as "(confidentiality('ClaimFile',_), '=', 'true') $\leftarrow$ document('ClaimFile') $\wedge$ has('ClaimFile', 'ClaimValuePart') $\wedge$ (documentPart('ClaimValuePart'), '>', '750')". These rules are not properly aligned as $R_{confidentiality}$ is too strict in comparison to $R_{modification}$. Lastly, assume that $R_{modification}$ constrains the potential behavior of `AGFIL`, and $R_{modification_{c}ol}$ the agreed collaboration stating "resource('ClaimInformation') $\wedge$ value('ClaimInformation',_), '$\geq$', '500') $\rightarrow$ (modification('ClaimInformation'), '=', 'true')". These rules are not compatible since `AGFIL` supports modification protection only for claims above $1000, whereas this is required for those at $500 already.

5 Conclusions

Current industry standards in business collaboration design, such as BPEL and WS-Policy [5,2] and ebXML [6], as well as approaches proposed from academia like [3,4,1,7,16,8,15] are not suitable for dealing with the complex and dynamic nature of developing and managing business collaborations and their policies.

In this paper we have presented the Business Collaboration Design Framework (BCDF), a framework that utilizes a three dimension approach (i.e. perspective, facet, aspect) to business collaboration design. The work presented gives a very brief overview of the framework and its relevance for policy specification and management. For more information the reader is referred to [11].

References

1. Business Process Management: A Survey, W. van der Aalst, A. ter Hofstede, M. Weske, *Proceedings of the International Conference on Business Process Management, 2003*
2. S. Bajaj, D. Box, D. Chappell, et all, Web Services Policy Framework (WS-Policy), *http://www-106.ibm.com/developerworks/library/specification/ws-polfram/*, *September 2004*
3. P. Bresciani, A. Perini, P. Giorgini, et all, Tropos: An Agent-Oriented Software Development Methodology, *Autonomous Agents and Multi-Agent Sytems, Vol. 8, No. 3, pp. 203236, 2004*
4. F. Casati, E. Shan, U. Dayal, et all, Business-Oriented Management of Web Services, *Communications of the ACM, Vol. 46, No. 10, pp. 55-60, 2003*
5. F. Curbera, Y. Goland, J. Klein, et all, Business Process Execution Language for Web Services, *http://www-106.ibm.com/developerworks/webservices/library/ws-bpel/, July 31, 2002*
6. ebXML, *http://www.ebxml.org*
7. D. Fensel, C.Bussler, The Web Service Modeling Framework WSMF, *Electronic Commerce Research and Applications, Vol. 1, No. 2, pp. 113-137, 2002*
8. D. Georgakopoulos, H. Schuster, D. Baker, et all, Managing Escalation of Collaboration Processes in Crisis Mitigation Situations, *Proceedings of the 16th International Conference on Data Engineering, San Diego, CA, USA, 2000*
9. P. Grefen, K. Aberer, Y. Hoffner, et all, CrossFlow: Cross-Organizational Workflow Management in Dynamic Virtual Enterprises, *International Journal of Computer Systems Science & Engineering, Vol. 15, No. 5, pp. 277-290, 2000*
10. B. von Halle, Business rules applied: Building Better Systems Using the Business Rule Approach, *Wiley & Sons, 2002*
11. B. Orriens, J. Yang, Establishing and Maintaining Compatibility in Service Oriented Business Collaboration, *To appear in Proceedings of the 7th International Conference on Electronic Commerce, Xi'an, China, August 2005*
12. M. Papazoglou, J. Dubray, A survey of web service technologies, *Technical Report DIT-04-058, Informatica e Telecomunicazioni, University of Trento, 2004*
13. R. Ross, Principles of the Business Rule Approach, *Addison-Wesley, 2003*
14. A. Scheer, Architecture for Integrated Information Systems - Foundations of Enterprise Modeling, *Springer-Verlag New York, Secaucus, NJ, USA, 1992*
15. P. Traverso, M. Pistore, M. Roveri, et all, Supporting the Negotiation between Global and Local Business Requirements in Service Oriented Development, *Proceedings of the 2d International Conference on Service Oriented Computing, New York, USA, 2004*
16. L. Zeng, B. Benatallah, H. Lei, et all, Flexible Composition of Enterprise Web Services, *Electronic Markets - The International Journal of Electronic Commerce and Business Media, Vol. 13, No. 2, pp. 141-152, 2003*

Yet Another Event-Driven Process Chain

Jan Mendling[1], Gustaf Neumann[1], and Markus Nüttgens[2]

[1] Vienna University of Economics and Business Administration, Austria
{firstname.lastname}@wu-wien.ac.at
[2] University of Hamburg, Germany
nuettgens@hwp-hamburg.de

Abstract. The 20 workflow patterns proposed by Van der Aalst et al. provide a comprehensive benchmark for comparing control flow aspects of process modelling languages. In this paper, we present a novel class of Event-Driven Process Chains (EPCs) that is able to capture all of these patterns. This class is called "yet another" EPC as a tribute to YAWL that inspired this research. yEPCs extend EPCs by the introduction of the so-called empty connector; inclusion of multiple instantiation concepts; and a cancellation construct. Furthermore, we illustrate how yEPCs can be used to model some of the workflow patterns.

1 Introduction

The 20 workflow patterns gathered by Van der Aalst, ter Hofstede, Kiepuszewski and Barros [1] are well suited for analyzing different workflow languages: workflow researchers can refer to these patterns in order to compare different process modelling techniques. This is of special importance considering the heterogeneity of process modelling languages (see e.g. [2]). Building on the insight that no language provides support for all patterns, Van der Aalst and ter Hofstede have defined a new workflow language called YAWL [3]. YAWL takes workflow nets as a starting point and adds non-petri-nets constructs in order to support each pattern in an intuitive manner (except implicit termination).

Besides Petri nets, Event-Driven Process Chains (EPC) [4] are another popular technique for business process modelling. Yet, their focus is rather related to semi-formal process documentation than formal process specification. The debate on EPC semantics has recently inspired the definition of a mathematical framework for a formalization of EPCs in [5]. As a consequence, we argue that workflow pattern support can also be achieved by starting with EPCs instead of Petri nets. This paper presents an extension to EPCs that is called yEPCs. In Section 2 we introduce EPCs and yEPCs. yEPCs include three extensions to EPCs that are sufficient to provide for direct support of the 20 workflow patterns reported in [1]. In Section 3 we discuss in detail how workflow patterns can be expressed with yEPCs. In particular, we highlight the non-local semantics of the XOR join, and its implications for workflow pattern support. After a survey on related work (Section 4), we give a conclusion and an outlook on future research (Section 5). An extended version of this paper is available as [6].

W.M.P. van der Aalst et al. (Eds.): BPM 2005, LNCS 3649, pp. 428–433, 2005.

2   Yet Another Event-Driven Process Chain (yEPC)

EPCs are introduced as a modelling concept to represent temporal and logical dependencies in business processes [4]. Elements of EPCs may be of *function type* (active elements), *event type* (passive elements), or of one of the three *connector types* AND, OR, or XOR. These objects are linked via control flow arcs. Connectors may be split or join operators, starting either with function(s) or event(s). In EPCs both OR join and XOR join have non-local semantics (cf. [5,7]). Concerning the XOR join, this implies that it blocks when there is one incoming branch finished and another still active. For a formal discussion of these semantics refer to Kindler [5]. Furthermore, *process interfaces* and *hierarchical functions* (see e.g. [7,8]) can be used to link different EPC models. A hierarchical function can be regarded as a synchronous call to a sub-process. After the sub-process has completed, navigation continues with the next function subsequent to the hierarchical function. The process interface can be regarded as an asynchronous spawning off of a sub-process. There is no later synchronization when a sub-process completes. For more on EPC sub-processes refer to [7].

Figure 1 illustrates the syntax elements of Yet Another Event-Driven Process Chain (yEPC). This extension of EPCs is motivated by incomplete workflow pattern support of EPCs. yEPCs reflect three measures that suffice to provide for direct support of all workflow patterns. These measures include the introduction of the so-called empty connector; an inclusion of a general multiple instantiation concept; and the introduction of a cancellation concept. The EPC extensions differ from Petri net extensions that were needed to define YAWL: Petri nets also had to be extended with multiple instantiation and cancellation concepts, but they lacked advanced synchronization patterns. EPCs, in contrast, miss support for state-based patterns. It should be mentioned that yEPC extensions have no impact on the validity of existing EPC models: this means valid EPCs according to the definitions in [7] are still valid with respect to this new class of EPCs.

As mentioned above, EPCs cannot explicitly represent state-based workflow patterns. This shortcoming can be resolved by introducing a new connector type that we refer to as the *empty connector*. This connector is represented by a cycle, just like the other connectors, but without any symbol inside. Also the same syntax rules as for other connectors hold. We follow control flow semantics as defined by Kindler [5], this means process folders (the EPC analogue to tokens

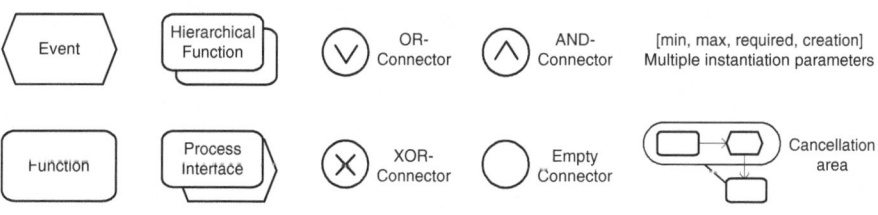

Fig. 1. yEPC Symbols

of Petri nets) are placed on arcs. The empty split then has to be interpreted as a hyperarc e.g. from the event before the empty split to the functions subsequent to it; the empty join analogously as a hyperarc from e.g. multiple functions before it to its subsequent event. Consider an event that is followed by an empty split linking to multiple functions. The empty split allows all subsequent functions to pick up the event. As a consequence, there is a run between the functions: the first function to consume the event causes the other functions to be no more active. This split semantics match the deferred choice pattern. Consider the other case of an empty join with multiple input events. The subsequent function is activated when one of these events has been reached. This behavior matched the multiple merge pattern. We will explain in Section 3 why such semantics are needed as an EPC extension.

The lack of EPC support for *multiple instantiation* has been discussed before (see e.g. [9]). In yEPCs we stick to multiple instantiation as defined for YAWL. YAWL defines a quadruple of parameters that control multiple instantiation. The parameters min and max define the minimum and maximum cardinality of instances that may be created. The required parameter specifies an integer number of instances that need to have finished in order to complete multiple instantiation. The creation parameter may take the values static or dynamic which specify whether further instances may be created at run-time (dynamic) or not (static). In the context of multiple instantiation, it is helpful to define sub-processes in order to model complex blocks of activities that can be executed multiple times as a whole. Accordingly, multiple instantiation parameters can be specified for functions as well as for hierarchical functions and process interfaces.

Cancellation patterns have not yet been discussed for EPCs. We adopt the concept of YAWL. Cancellation areas (symbolized by a lariat) may include functions and events. The end of the lariat has to be connected to a function. When this function completes, all functions and events in the lariat are cancelled.

3   Workflow Pattern Analysis of EPCs

In this section we will consider the EPC control flow semantics of Kindler [5] which reflect the ideas of [4,7]. For multiple instantiation and cancellation the concepts from YAWL are adopted. In the following we illustrate workflow patterns (WP) 4,5, and 17 and their yEPC representation. A full workflow pattern analysis can be found in [10]. We will speak of EPCs each time we make a statement that holds for both yEPCs and EPCs. Otherwise, we will explicitly refer to yEPCs when we present concepts that are not included in EPCs.

WP 4 (Exclusive Choice) and 5 (Simple Merge): WP 4 describes a point in a process where a decision is made to continue with one of multiple alternative branches. This situation can be modelled with the XOR split connector of EPCs. There has been a debate on the non-local semantics of the XOR join. While Rittgen [11] and Van der Aalst [12] proposes a local interpretation, recent research agrees upon non-local semantics (see e.g. [5,7]). This means that the XOR join is only allowed to continue if exactly one of the preceding functions

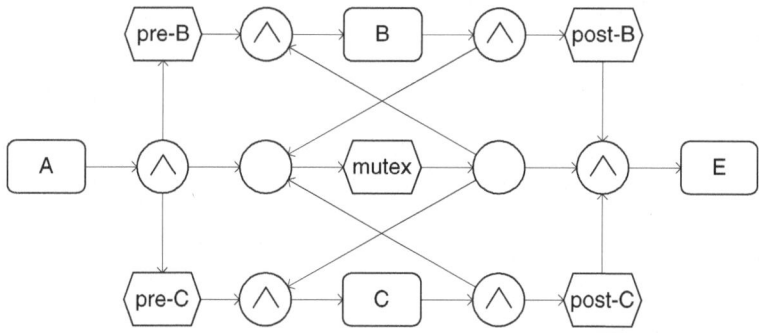

Fig. 2. yEPC Model for WP 17 Interleaved Parallel Routing

have finished, and it is not possible that the other functions will ever be executed. Accordingly, EPC's XOR join works perfect when used in an XOR block started with an XOR split, but may block e.g. when used after an OR split depending on whether more than one branch has been activated. Due to these non-local semantics it is similar to a synchronizing merge but with the difference that it blocks when further process folders may be propagated to the XOR join.

In contrast to this, WP 5 defines a simple merge without synchronization, but building on the assumption that the joined branches are mutually exclusive. The XOR join in YAWL [3] can implement such behavior with local semantics: when one of parallel activities is completed the next activity after the XOR join is started. But when the assumption does not hold, i.e., when another of the parallel activities has finished the activity after the XOR join is activated another time, and so forth. This observation allows two conclusions. First, there is a fundamental difference between the semantics of the XOR join in EPCs and YAWL: the XOR join in EPCs has non-local semantics and blocks if there are multiple paths activated; the XOR join in YAWL has local semantics and propagates each incoming process token without ever blocking. Accordingly, the YAWL XOR join can also be used to implement WP 8 (multiple merge). Second, as the XOR join in EPCs has non-local semantics, it cannot be used to model WP 8. Hence, yEPCs use the empty connector for WP 8.

WP 17 (Interleaved Parallel Routing): Empty connectors can be used for state-based patterns. Figure 2 shows the process model of WP 17 following the ideas presented in [1]. The event at the center of the model manages the sequential execution of functions *B* and *C* in arbitrary order. It corresponds to the "mutual exclusion place (*mutex*)" introduced in [1]. The AND split after function *A* adds a folder to this *mutex* event via an empty connector. The AND joins before the functions *B* and *C* consume this folder and put it back to the mutex event afterwards. Furthermore, they consume the individual folders in *pre-B* and *pre-C*, respectively. These events control that each function of *B* and *C* is executed only once. After both have been executed, there are folders in *post-B*, *post-C*, and *mutex*. Accordingly, *E* can be started. In [13] sequential split and join operators are proposed to describe control flow behavior of WP 17. Yet, it is no clear what the semantics of these operators are when not used pairwise.

Altogether, WP 1 to 7, 10, and 11 are supported by EPCs. In contrast, yEPCs provides additional modelling support of WP 8 (multiple merge), 9 (discriminator), 12-15 (multiple instantiation), 16 (deferred choice), 17 (interleaved parallel routing), 18 (milestone), and 19-20 (cancellation). As a consequence, business processes including control flow behavior that is related to previously unsupported workflow patterns can now be represented appropriately using yEPCs.

4   Related Work

The workflow patterns proposed by [1] provide a comprehensive benchmark for comparing different process modelling languages. A short workflow pattern analysis of EPCs is also reported in [3], yet it does not discuss the non-local semantics of EPCs XOR join. In this paper, we highlighted these semantics as a major difference between YAWL and EPCs. Accordingly, we propose the introduction of the empty connector in order to capture workflow pattern 8 (multiple merge). There is further research discussing notational extensions to EPCs. In Rittgen [11] a so-called XORUND connector is proposed to partially resolve semantical problems of the XOR join connector. Motivated by space limitations of book pages and printouts, Keller and Teufel introduce process interfaces to link EPC models on different pages [8]. We adopt process interfaces in this paper to model spawning off of sub-processes. Rosemann [13] proposes the introduction of sequential split and join operators in order to capture the semantics of workflow pattern 17 (interleaved parallel routing). While the informal meaning of a pair of sequential split and join operators is clear, the formal semantics of each single operator is far from intuitive. As a consequence, we propose a state-based representation of interleaved parallel routing inspired by Petri nets. Furthermore, Rosemann introduces a connector that explicitly models a decision table and a so-called OR_1 connector to mark branches that are always executed [13]. Rodenhagen presents multiple instantiation as a missing feature of EPCs [9]. He proposes dedicated begin and end symbols to model that a branch of a process may be executed multiple times. Yet, this notation does not enforce that a begin symbol is followed by a matching end symbol. As a consequence, we adopt the concept of YAWL that permits multiple instantiation only for single functions or sub-processes, but not for arbitrary branches of the process model.

5   Conclusion and Future Work

In this paper, we presented a novel class of EPCs called yEPCs that is able to capture all 20 workflow patterns. Basically, yEPCs introduce three extensions to EPCs: the introduction of the empty connector; the inclusion of a multiple instantiation concept; and the inclusion of a cancellation concept. These extensions permit some conclusions on the relation of Petri nets and EPCs in general.

Towards workflow pattern support, both include extensions for multiple instantiation and cancellation. In addition, Petri nets had to be extended with advanced synchronization concepts. On the other hand, EPCs had to be modified to address the state-based patterns. As a consequence, yEPCs and YAWL are quite similar concerning their modelling primitives. The XOR join is the major difference between both. In future research, we aim to implement a transformation between yEPCs available in EPML format and the interchange format of YAWL.

References

1. van der Aalst, W.M.P., ter Hofstede, A.H.M., Kiepuszewski, B., Barros, A.P.: Workflow Patterns. Distributed and Parallel Databases **14** (2003) 5–51
2. Mendling, J., Neumann, G., Nüttgens, M.: A Comparison of XML Interchange Formats for Business Process Modelling. In Feltz, F., Oberweis, A., Otjacques, B., eds.: Proceedings of EMISA 2004 - Information Systems in E-Business and E-Government. Volume 56 of Lecture Notes in Informatics. (2004)
3. van der Aalst, W.M.P., ter Hofstede, A.H.M.: YAWL: Yet Another Workflow Language. Information Systems **30** (2005) 245–275
4. Keller, G., Nüttgens, M., Scheer, A.W.: Semantische Prozessmodellierung auf der Grundlage "Ereignisgesteuerter Prozessketten (EPK)". Heft 89, Institut für Wirtschaftsinformatik, Saarbrücken, Germany (1992)
5. Kindler, E.: On the semantics of EPCs: Resolving the vicious circle. In J. Desel and B. Pernici and M. Weske, ed.: Business Process Management, 2nd International Conference, BPM 2004. Volume 3080 of Lecture Notes in Computer Science., Springer Verlag (2004) 82–97
6. Mendling, J., Neumann, G., Nüttgens, M.: Yet Another Event-Driven Process Chain (Extended Version). Technical Report JM-2005-05-27, Vienna University of Economics and Business Administration, Austria (2005)
7. Nüttgens, M., Rump, F.J.: Syntax und Semantik Ereignisgesteuerter Prozessketten (EPK). In J. Desel and M. Weske, ed.: Proceedings of Promise 2002, Potsdam, Germany. Volume 21 of Lecture Notes in Informatics. (2002) 64–77
8. Keller, G., Teufel, T.: SAP(R) R/3 Process Oriented Implementation: Iterative Process Prototyping. Addison-Wesley (1998)
9. Rodenhagen, J.: Ereignisgesteuerte Prozessketten - Multi-Instantiierungsfähigkeit und referentielle Persistenz. In: Proceedings of the 1st GI Workshop on Business Process Management with Event-Driven Process Chains. (2002) 95–107
10. Mendling, J., Neumann, G., Nüttgens, M.: Towards Workflow Pattern Support of Event-Driven Process Chains (EPC). In M. Nüttgens and J. Mendling, ed.: Proc. of the 2nd Workshop XML4BPM 2005, Karlsruhe, Germany. (2005) 23–38
11. Rittgen, P.: Quo vadis EPK in ARIS? Ansätze zu syntaktischen Erweiterungen und einer formalen Semantik. WIRTSCHAFTSINFORMATIK **42** (2000) 27–35
12. van der Aalst, W.M.P.: Formalization and Verification of Event-driven Process Chains. Information and Software Technology **41** (1999) 639–650
13. Rosemann, M.: Erstellung und Integration von Prozeßmodellen - Methodenspezifische Gestaltungsempfehlungen für die Informationsmodellierung. PhD thesis, Westfälische Wilhelms-Universität Münster (1995)

Comparing the Control-Flow of EPC and Petri Net from the End-User Perspective

Kamyar Sarshar and Peter Loos

Johannes Gutenberg-University Mainz,
Lehrstuhl für Wirtschaftsinformatik und BWL,
ISYM – Information Systems & Management,
D-55099 Mainz, Germany
{sarshar, loos}@isym.bwl.uni-mainz.de

Abstract. This contribution describes the results of a laboratory experiment which compares the Event-driven Process Chain (EPC) and Petri net (C/E net) regarding their approaches to represent the control-flow of processes. The outcome of the experiment indicates that from end-user perspective the EPC approach of applying connectors is superior to the token game. However, the non-local semantic of the EPC OR-connector clearly had a negative impact on end-user comprehension. The experiment also illustrates that the perceived ease-of-use and the intention to use the EPC notation is higher than C/E net.

1 Introduction

Two notations have been widely used for business process modeling. The Event-driven Process Chain (EPC) [13] which has been constructed on the basis of C/E net ([23], p. 47; [24], p. 428) gained broad acceptance in commercial projects especially in the context of SAP R/3 [14] and ARIS-Toolset [7]. However, the EPC is criticized since it suffers from ambiguity [31], and as a consequence, lack of the possibility to be analyzed and simulated properly. Petri nets have been investigated extensively within the scientific community during the last four decades and have successfully been applied to the concept of process modeling and analysis [20, 30]. Despite a strong agreement on applying Petri nets for process analysis, there are some doubts on whether or not formal Petri nets are appropriate for the development of conceptual process models [9].

A major difference between the two notations is their representation of the control-flow of processes. The control-flow deals with the order in which activities have to be executed. Petri nets use states and transitions in combination with the token game to represent the control-flow. The EPC in contrast applies different connectors as distinctive language primitives. This research aims to investigate which of these two approaches are more appropriate for conceptual modeling where processes are developed in cooperation with end-users [19]. Since EPC has been constructed on the basis of C/E net, it makes sense to compare the EPC notation with this Petri net type.

W.M.P. van der Aalst et al. (Eds.): BPM 2005, LNCS 3649, pp. 434–439, 2005.

Among a variety of empirical and non-empirical approaches that could potentially be used to evaluate the notations [25], we have chosen the laboratory experiment to investigate the following hypotheses:

H1: The EPC connectors are better comprehended by end-user than C/E net representations with respect to AND-type situation (H1a) and XOR-type situation (H1b).

H2: The EPC multi-level AND/XOR-connectors are comprehended by end-users as good as AND-connectors and XOR-connectors.

H3: The EPC OR-connectors are comprehended by end-users as good as AND-connectors and XOR-connectors.

H4: The overall end-user comprehension of the control-flow of an EPC model is higher than an equivalent C/E net model.

H5: The perceived ease-of-use of the EPC control-flow is higher than C/E net.

H6: The perceived ease-of-use of the EPC notation is higher than C/E net.

H7: The intention of end-user to use the EPC is higher than C/E net.

The remainder of this paper is organized as follows. After this introduction, the next two sections describe previous research and the research methodology. A summery of finding and limitations are presented briefly in section 4. We refer to [22] for further details regarding the experiment. A number of suggested areas for future research will be presented in the final section.

2   Previous Research

While aspects of conceptual data modeling have been explored reasonably by laboratory experiments [1-3, 11, 12, 15, 26, 32], the weakness within the existing literature is the lack of equivalent studies on process modeling notations. Existing experiments on Petri nets focus on their application as a visual programming language and for software specification rather than on business process modeling. Moher et al. [17] compared Petri nets with textual programming representations in a laboratory experiment with faculty members and graduates of computer science. In Boehm-Davis' and Fregl's study [4] professional programmers performed software modification tasks based on a Petri net documentation of the software system. Swigger and Brazile examined the effect of ERM and Petri net based documentation as well as the absence of documentation on the performance of programmer carrying out software modification tasks in two separate studies [27, 28]. Empirical contributions on the EPC notation include the study of Davies et al. [6], which is based on interviews with ARIS modelers from different organization and institutions throughout Australia and is founded on an analysis of ARIS against BWW constructs. Green & Rosemann [10] conduct a questionnaire based test with university students in order to investigate propositions discussed in [8]. Additionally, a number of non-empirical contributions discuss formalization issues of the EPC based on the Petri net semantic [5, 16, 21, 29].

As far as we know, this research is the first laboratory experiment comparing Petri net and EPC notation. Hence, it extends the literature on the issue of usability of modeling notations to the domain of business process modeling. It is complementary to the experiments on end-user data modeling conducted in [11, 12] where record-based and conceptual data modeling techniques were compared. Additionally, the study extends the experiments on Petri nets since it investigates the application of Petri nets in the context of business process modeling and from the perspective of the end-user.

3   Research Methodology

To test the hypotheses, a laboratory experiment was conducted with a two-group, post-test only experimental design which was adapted and modified from [18].

Independent variables: in this study the type of process modeling notation was regarded as the only independent variable.

Controlled variables: in this study task complexity, prior user experience and training were regarded as controlled variables.

Dependent variable: according to the hypothesis, the dependent variables were model comprehension and perceived ease-of-use. Model comprehension was measured by the number of correct answers to process related questions in the questionnaire. The perceived ease-of-use of the notation was assessed by various process independent questions of the questionnaire whereby the subjective estimation of the participants regarding the notation was asked for.

After recruiting 50 students with business and economy background with no or very limited prior knowledge on the notations, they were randomly separated into an EPC group and a C/E net group. Participants of each group got an experimental treatment consisting of general instruction for one of the modeling notations. During this introduction, the main elements of the each notation were presented to the participants. Afterwards, a business process model which was represented respectively for one group in EPC notation and for the other in C/E net notation was handed out to the participants. Additionally, each participant got a multiple-choice questionnaire which consisted of anonymous questions, questions based on situations within the supplied business process and post-task questions. The experiment took place in a regular lecture room where participants of one group simultaneously conducted the experiment. The material and the procedure were tested prior to the experiment with four students. The distributed business process was taken from ([23], p. 433). The chosen model was a sales process modeled in EPC notation. The model has been extended in order to make it more complex. The transformation to an equivalent C/E net was based on [29] in term of events, functions, AND and XOR joint and split connector. All EPC OR-connectors were replaced by XOR-type situation within the C/E net model.

4   Summary of Findings and Limitations

The findings of the study are summarized in table 1.

Table 1. Summary of the findings

	Results of the experiment	
H1a	AND-type situation where equally well comprehended	No
H1b	XOR-type situation where significantly better comprehended within the EPC group	Yes
H2	EPC multi-level AND/XOR-connectors where as good comprehended by end-users as AND-connectors and XOR-connectors	Yes
H3	OR-connectors where significantly less comprehended than AND-connectors and XOR-connectors	No
H4	The overall comprehension of the control-flow of the EPC group was significantly better than the C/E net group	Yes
H5	No definitive result	-
H6	There is a tendency that the perceived ease-of-use of the EPC notation is higher than C/E net	Yes
H7	There is a tendency that the intention of end-user to use the EPC is higher then C/E net	Yes

The results of this empirical study have to be treated cautiously and considered in the context of several limitations regarding external and internal validity. Regarding external validity, the use of students as subjects clearly limits the generalisability of the results. The behavior of students, their learning style and motivation might not be representative for the population of end-users in practice even though the use of student as subjects is a well-established practice of experimental studies. A further aspect is that the experimental task consisted of only one business process of a given complexity limits the generalization of the results to more or less complex models. Additionally, the sample size used for the experiments can be considered as small. In order to ensure internal validity of the results, all variables except the independent variable had to be held constant between the groups. Even participants were separated randomly it was unavoidable that some individual difference between the participants of the two groups occurred. Furthermore, the elimination of the OR-type situation in the C/E net representation of the business process indicates that the models distributed to the participants of the two groups were not identical.

5 Future Research

This research was a step towards empirical evaluation of process modeling notations regarding end-user comprehension. The outcomes indicate that obviously there is an essential need for particular notations for conceptual process modeling where end-user comprehension and formal semantic are emphasized equally. The construction of such notations and the gradual transformation of their models to the analysis stage is not understood well and needs more investigation. Future research might be conducted to overcome the limitations of this study which were mentioned in the previous section. Such research could seek to evaluate other notations or to provide more insight on the comprehension of more experienced end-users and professionals.

Furthermore, the ability of the end-user to model rather than to comprehend business processes could be addressed in future research.

References

1. Batra, D. and Davis, J.G.: Conceptual Data Modelling in Database Design: Similarities and Differences between Expert and Novice Designers. International Journal of Man-Machine Studies, Vol. 37, No. 1 (1992) 83-101
2. Bock, D.B. and Ryan, T.: Accuracy in modeling with extended entity relationship and object oriented data models. Journal of Database Management, Vol. 4, No. 4 (1993) 30-40
3. Bodart, F., et al.: Should optional properties be used in conceptual modelling? A theory and three empirical tests. Information Systems Research, Vol. 12, No. 4 (2001) 384-405
4. Boehm-Davis, D. and Fregly, A.: Documentation of Concurrent Programs. Human Factors, Vol. 27, No. (1985) 423-432
5. Chen, R. and Scheer, A.-W.: Modellierung von Prozeßketten mittels Petri-Netz-Theorie, in Veröffentlichungen des Instituts für Wirtschaftsinformatik, Scheer, A.-W., (ed.): Saarbrücken (1994)
6. Davies, I., Green, P., and Rosemann:, M.: Exploring Proposed Ontological Issues of ARIS with Four Different Types of Modellers. In: Proc. Proceedings of the Australasian Conference on Information Systems (ACIS 2004). Hobart. (2004)
7. Davis, R.: Business process modelling with ARIS : a practical guide. Springer, London et al. (2001)
8. Green, P. and Rosemann, M.: Integrated Process Modeling: An Ontological Evaluation. Information Systems, Vol. 25, No. 2 (2000) 73-87
9. Green, P. and Rosemann, M.: An Ontological Analysis of Integrated Process Modelling. In: Proc. Advanced Information Systems Engineering, 11th International Conference CAiSE'99, Proceedings. Heidelberg, Germany. Lecture Notes in Computer Science 1626 (1999) 225-240
10. Green, P. and Rosemann, M.: Perceived Ontological Weaknesses of Process Modeling Techniques: Further Evidence. In: Proc. Proceedings of the 10th European Conference on Information Systems (ECIS 2002). Gdansk, Poland. (2002) 312-321
11. Jarvenpaa, S.L. and Machesky, J.J.: Data Analysis and Learning: An Experimental Study of Data Modeling Tools. International Journal of Man-Machine Studies, Vol. 31, No. (1989) 367-391
12. Juhn, S.H. and Naumann, J.D.: The effectiveness of data representation characteristics on user validation. In: Proc. Proceedings of the Sixth International Conference on Information Systems. Indianapolis, Indiana. (1985) 212-226
13. Keller, G., Nüttgens, M., and Scheer, A.-W.: Semantische Prozeßmodellierung auf der Grundlage "Ereignisgesteuerter Prozeßketten (EPK)", in Veröffentlichungen des Instituts für Wirtschaftsinformatik, Scheer, A.-W., (ed.): Saarbrücken (1992)
14. Keller, G. and Teufel, T.: SAP R/3 Process Oriented Implementation: Iterative Process Prototyping. Addison-Wesley (1998)
15. Kim, Y.-G. and March, S.T.: Comparing data modeling formalisms. Communications of the ACM, Vol. 38, No. 6 (1995) 103-115
16. Langner, P., Schneider, C., and Wehler, J.: Petri Net Based Certification of Event-Driven Process Chains, in Application and Theory of Petri Nets 1998: 19th International Conference, ICATPN'98, Lisbon, Portugal, June 1998. Proceedings, Desel, J. and Silva, M., (eds.), Springer: Berlin (1998)

17. Moher, T., et al.: Comparing the Comprehensibility of Textual and Graphical Programs: The Case of Petri Nets. In: Proc. Empirical Studies of Programmers: Fifth Workshop. (1993) 137-161

18. Moody, D.L.: Comparative Evaluation of Large Data Model Representation Methods: The Analyst.s Perspective. In: Proc. Conceptual Modeling - ER 2002, 21st International Conference on Conceptual Modeling. Tampere, Finland. LNCS 2503 Springer-Verlag (2002) 214-231

19. Mylopoulos, J.: Information modeling in the time of the revolution. Information Systems, Vol. 23, No. 3-4 (1998) 127-155

20. Oberweis, A., et al.: INCOME/WF A Petri Net Based Approach to Workflow Management. In: Proc. Wirtschaftsinformatik '97. Internationale Geschäftstätigkeit auf der Basis flexibler Organisationsstrukturen und leistungsfähiger Informationssysteme. Physica-Verlag Heidelberg (1997) 557-580

21. Rodenhagen, J.: Darstellung ereignisgesteuerter Prozeßketten (EPK) mit Hilfe von Petrinetzen. Masterthesis. Universität Hamburg, Hamburg (1997)

22. Sarshar, K., Dominitzki, P., and Loos, P.: Comparing the Control-Flow of EPC and Petri Net from the End-User Perspective - Statistical Results of a Laboratory Experiment, In Working Papers of the Research Group Information Systems & Management Nr. 25 (www.isym.de): Mainz (2005)

23. Scheer, A.-W.: Business process engineering : reference models for industrial enterprises. Springer, Berlin et al. (1998)

24. Scheer, A.-W., Nüttgens, M., and Zimmermann, V.: Rahmenkonzept für ein integriertes Geschäftsprozeßmanagement. Wirtschaftsinformatik, Vol. 37, No. 5 (1995) 426-434

25. Siau, K. and Rossi, M.: Evaluation of Information Modeling Methods - A Review. In: Proc. Proceedings of the Thirty-First Annual Hawaii International Conference on System Sciences (HICSS). Kohala Coast, Hawaii, USA. (1998) 314-322

26. Sinha, A.P. and Vessey, I.: An Empirical Investigation of Entity-based and Object-oriented Data Modeling: A Development Life Cycle Approach. In: Proc. International Conference on Information Systems. Charlotte, North Carolina. (1999) 229-244

27. Swigger, K. and Brazile, R.: Experimental Comparisons of Design/Documentation Formats for Expert Systems. International Journal of Man-Machine Studies, Vol. 31, No. (1989) 47-60

28. Swigger, K.M. and Brazile, R.P.: An empirical study of the effects of design/documentation formats on expert system modifiability. In: Proc. Empirical Studies of Programmers: Fourth Workshop. Norwood, NJ. (1991) 210-226

29. van der Aalst, W.M.P.: Formalization and verification of event-driven process chains. Information and Software Technology, Vol. 41, No. 10 (1999) 639-650

30. van der Aalst, W.M.P.: Modelling and analysing workflow using a Petri-net based approach. In: Proc. 2nd Workshop on Computer-Supported Cooperative Work, Petri nets and related formalisms, Proceedings. Zaragoza, Spain. (1994) 31-50

31. van der Aalst, W.M.P., Desel, J., and Kindler, E.: On the semantics of EPCs: A vicious circle. In: Proc. EPK 2002 - Geschäftsprozessmanagement mit Ereignisgesteuerten Prozessketten, Proceedings des GI-Workshops und Arbeitskreistreffens. Trier. (2002) 71-79

32. Weber, R.: Are Attributes Entities? A Study of Database Designer's Memory Structures. Information Systems Research, Vol. 7, No. 2 (1996) 137-162

Overview of Transactional Patterns: Combining Workflow Flexibility and Transactional Reliability for Composite Web Services

Sami Bhiri, Khaled Gaaloul, Olivier Perrin, and Claude Godart

LORIA - INRIA - CNRS - UMR 7503
BP 239, F-54506 Vandœuvre-lès-Nancy Cedex, France
{bhiri, kgaaloul, operrin, godart}@loria.fr

Abstract. In this paper, we present an approach to easily define flexible and reliable services compositions. We introduce a new concept called *transactional patterns* to specify flexible and reliable composite Web services. A *transactional pattern* is a convergence concept between workflow patterns and advanced transactional models. It can be seen as a coordination pattern and as a structured transaction. Thus, it combines workflow flexibility and transactional processing reliability. Designers can simply connect together a set of *transactional patterns* to define a composite Web service. We use a set of techniques to ensure control and transactional coherence between patterns inside a composition of services.

Keywords: Web services compositions, Workflow patterns, transactional processing.

1 Introduction

Web services are a great technology for dealing with B2B business processes, such as e-procurement for instance, but handling failures using the traditional transactional model for long running, asynchronous, and decentralized activities has been proven to be unsuitable. Advanced Transaction Models (ATMs) [1] have been proposed to manage failures, but, although powerful and providing a nice theoretical framework, ATMs are too database-centric, limiting their possibilities and scope [2] in this context (e.g. their inflexibility to incorporate different transactional semantics as well as different behavioral patterns into the same structured transaction [3]).

In the same time, workflow [4] has became gradually a key technology for business process automation [5], providing a great support for organizational aspects, user interface, monitoring, accounting, simulation, distribution, and heterogeneity [2].

In this paper, we propose an approach to specify and orchestrate flexible and reliable Web services compositions based on the concept of *transactional patterns*.which combines workflow flexibility and transactional reliability.

Section 2 presents a motivating example. Section 3 and 4 specify the transactional composite Web services and workflow patterns concepts. Section 5 introduces the idea of transactional patterns, and shows how to use them to specify composite Web services, while guaranteeing the consistency. Section 6 concludes.

W.M.P. van der Aalst et al. (Eds.): BPM 2005, LNCS 3649, pp. 440–445, 2005.

2 Motivating Example

We consider an application for online travel arrangement, carried out by a composite service as illustrated in figure 1. The customer specifies its requirements for destination and hotels. The composite service launches in parallel hotel and flight booking. Then, the customer is requested to pay online. Once this is done, travel documents are sent to the customer. To avoid failures, the designers of the composite service may want to augment the control flow described above with a set of transactional requirements. For instance, they may require the services FB and TDU to be sure to complete, but also the service FB to be compensatable (when an hotel booking is cancelled, or when it fails). They may also specify that service TDU is an alternative for $TDFE$.

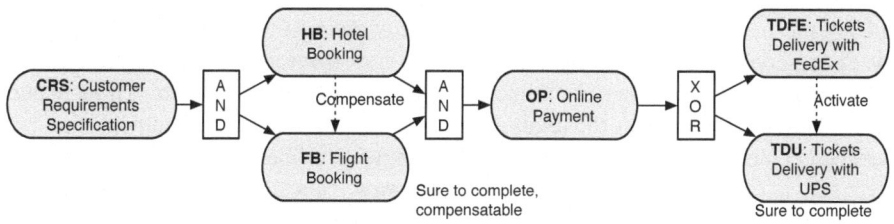

Fig. 1. A composite service for online travel arrangement

Modeling this example with ATM or workflow systems is not easy because ATM are too rigid to enable a such control structure, and they do not support bottom-up applications design, starting from predefined business process and using pre-existing systems or services with diverse semantics [3]. On the other hand, workflow systems lack functionalities to assess that the specified transactional behavior ensure the required reliability. In our example, if the service OP fails, causing the travel arrangement abortion, flight and hotel booking should be undone.

3 Transactional Composite Web Services

3.1 Transactional Web Service

A Web service is a self-contained modular program built with XML, SOAP, WSDL and UDDI specifications that can be discovered and invoked across the Internet ([6,7]). A *transactional Web service* is a Web service that emphasizes transactional properties for its characterization and correct usage.

The main transactional properties of a Web service we are considering are *retriable, compensatable, pivot* [8]. A service s is said to be **retriable** (s^r) if it is sure to complete after a finite number of activations. s is said to be **compensatable** (s^{cp}) if it offers compensation policies to semantically undo its effects. Then, s is said to be **pivot** (s^p) if once it successfully completes, its effects remains for ever and cannot be undone. A service can combine properties; the set of all possible combinations is $\{\emptyset; r; cp; p; (r, cp); (r, p)\}$.

Given the transactional properties of a service, a set of operations is available. For instance, a pivot service has a minimal set *abort(), activate(), cancel(), fail(), terminate()* allowing respectively its abortion before activation, its activation, its cancellation during its execution, its failure and its successful termination. A compensatable service has in addition a *compensate()* operation for its compensation. A retriable service has a *retry()* operation allowing to activate it after each failure.

3.2   Transactional Composite Web Service

A composite Web service is a conglomeration of existing Web services working in tandem to offer a new value-added service [5]. It coordinates a set of services as a cohesive unit of work to achieve common goals. A *Transactional Composite (Web) Service* (TCS) emphasizes transactional properties for composition and synchronization of component Web services. It takes advantage of services transactional properties to specify mechanisms for failure handling and recovery. A TCS defines orchestration between its services using dependencies to specify how services are coupled and how the behavior of some given services influences the behavior of some others. These dependencies are used to express the relationships (sequence, alternative, compensation,...) between component services. In our proposition, we consider the following dependencies: activation, alternative, abortion, compensation, cancellation, and we distinguish between activation dependencies and transactional dependencies (compensation, cancellation and alternative). More details on dependencies can be found in [9].

3.3   Control Flow and Transactional Flow of a TCS

Distinguishing two classes of dependencies, we separate the TCS control flow and the TCS transactional flow. The **control flow** specifies the partial ordering of component services activations, and it is defined as the set of the TCS activation dependencies.

The **Transactional Flow** specifies interactions for failures handling and recovery, and it is defined as the set of its transactional dependencies (compensation, cancellation and alternative).

Of course, transactional dependencies depend on activation dependencies semantics. Thus, a *transactional flow* is always defined according to a given *control flow*.

4   Workflow Patterns

As defined in [10], a pattern "is the abstraction from a concrete form which keeps recurring in specific non arbitrary contexts". A workflow pattern [11] can be seen as an abstract description of a recurrent class of interactions based on activation dependencies. For example, the *AND-join* pattern (see figure 2.b) describes an abstract services orchestration by specifying services interactions as follows: *a service is activated after the completion of several other services*. Thus, a pattern explicitly defines activation dependencies (i.e. the control flow) of a given set of services.

In this paper, we put emphasis on the following three patterns: *AND-split*, *AND-join* and *XOR-split*[1]. Figure 1 illustrates the patterns *AND-split* applied to *(CRS, HB, FB)*, *AND-join* applied to *(HB, FB, OP)*, and *XOR-split* applied to *(OP, TDFE, TDU)*.

---

[1] Our approach also considers the following list of patterns: sequence, *AND-split*, *OR-split*, *XOR-split*, *AND-join*, *OR-join*, *XOR-join* and *m-out-of-n* [11].

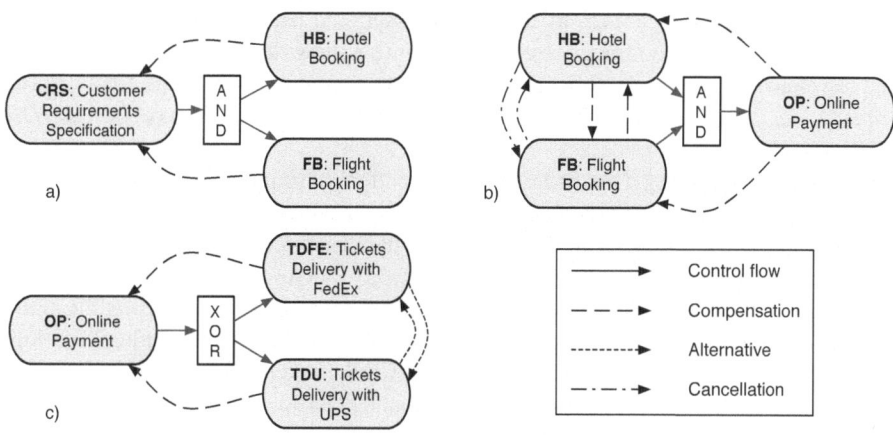

Fig. 2. AND-split, AND-join, and XOR-split patterns and their corresponding potential dependencies

We argue that once defined, a workflow pattern implicitly defines a new class of dependencies, called potential transactional dependencies, i.e. a set of dependencies not initially defined by the pattern, and that can be used/added in order to tailor (or modify) the control flow (see figure 2). In fact, these dependencies are directly related to the semantics of the activation dependencies of the pattern.

5   Web Services Composition Using Transactional Patterns

5.1   Transactional Patterns

Given the transactional properties of a service, and a workflow pattern for a composite service, we are able to deduce a new pattern, called a *transactional pattern*, that will be used to specify both the control and transactional flows. The control flow is inherited from the workflow pattern (i.e. the activation dependencies), while the transactional flow is specified using a set of transactional dependencies for managing alternatives, compensation, or cancellation.

From a transactional pattern, one can define several *transactional pattern instances* which are the application of a transactional pattern to a given set of services, and where the transactional dependencies of the instance is a subset of the set of transactional dependencies of the transactional pattern. For instance, on figure 2.c, a designer may choose to keep the alternative dependency for the delivery, while another may choose the compensation dependency. The choice depends not only on the designer, but also on the transactional properties of the compenent services.

5.2   Composition

Transactional patterns are interesting to compose a set of Web services to obtain a transactional composite service (TCS) which is reliable from an execution point of view. Thus, we specify a TCS as a set of transactional patterns instances connected together (sharing some component services). Figure 3 shows how we can specify the on-line travel arrangement service using the following transactional patterns composition: $Trans_{AND-split}(CRS, HB, FB)$, $Trans_{AND-join}(HB, FB, OP)$, $Trans_{XOR-split}(OP, TDFE, TDU)$.

However, connecting a set of transactional patterns instances can lead to a control flow and/or a transactional flow inconsistencies. For instance, control consistency problem can raise when instances are disjoined (no shared services allowing to connect the instances) or when an *XOR-split* instance is followed by an *AND-join* instance. Likewise, transactional inconsistency can raise when a component service fails, causing the entire TCS abortion, with remaining effects of the partial execution. For example, if we suppose that OP is not retriable (it can fail) in the TCS defined in figure 3, this means that FB should be compensated in order to be sure that it does not exist a remaining effect (a flight is booked) after the abortion of the TCS. This implies that it exists the compensate transactional dependency between OP and FB, and that FB is compensatable.

To manage these problems, we define a TCS as *valid* if it ensures both the control flow consistency and the transactional flow consistency. In order to guarantee the reliability of the TCS, we are using a set of rules to check both the control flow and the transactional flow consistency. These rules are described in [9]. Briefly summarized, the algorithm we are using is as follows:

1. after a component service failure, we are looking for an alternative dependency, if it exists,
2. after a composite service failure, we try to compensate what can be compensated given the transactional properties of the component services,
3. after a composite service failure, we cancel all the current executions of the TCS.

In order to compute the transactional consistency, we need not only these rules, but also the transactional properties of each service introduced in 3.1. Thus, there are two ways to use our approach. The first one is to fix the transactional dependencies for the TCS, and to compute what are the possible transactional properties of the component services. The second one is to choose a set of services and the transactional patterns, and to detect the transactional inconsistencies.

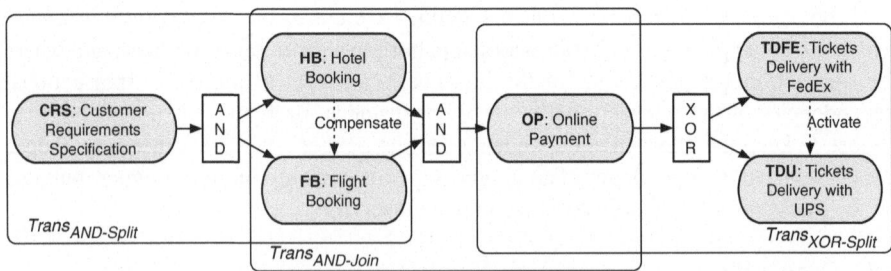

Fig. 3. A TCS is defined as a composition of a set of transactional patterns instances

6   Conclusion

In this paper, we propose a solution to ensure reliable and flexible a Web service composition. The main idea of our approach is to combine workflow flexibility and transactional processing reliability. We introduce an extension of workflow patterns, the *transactional patterns*, which can be seen as a convergence concept between workflow systems and transactional models to easily define flexible and reliable composite Web services. Then, we propose a set of rules in order to avoid inconsistencies which can result from the composition of the patterns.

Acknowledgments. The authors would like to thank Walid Gaaloul for the fruitful discussions we had during the writing of this paper.

References

1. A. Elmagarmid. *Transaction Models for Advanced Database Applications*. Morgan-Kaufmann, 1992.
2. G. Alonso, D. Agrawal, and A. El Abbadi. Process Synchronisation in Workflow Management Systems. In *8th IEEE Symposium on Parallel and Distributed Processing (SPDS'97)*, New Orleans, Louisiana, October 1996.
3. Nektarios Gioldasis and Stavros Christodoulakis. Utml: Unified transaction modeling language. In *Proceedings of the 3rd International Conference on Web Information Systems Engineering*, pages 115–126. IEEE Computer Society, 2002.
4. W. M. P. van der Aalst and K. M. van Hee. *Workflow Management: models, methods and tools*. Cooperative Information Systems. MIT Press, 2002.
5. B. Medjahed, B. Benatallah, A. Bouguettaya, A. H. H. Ngu, and A. K. Elmagarmid. Business-to-business interactions: issues and enabling technologies. *The VLDB Journal*, 12(1):59–85, 2003.
6. Francisco Curbera, Matthew Duftler, Rania Khalaf, William Nagy, Nirmal Mukhi, and Sanjiva Weerawarana. Unraveling the web services web: An introduction to soap, wsdl, and uddi. *IEEE Internet Computing*, 6(2):86–93, 2002.
7. Paulo F. Pires, Mario R. F. Benevides, and Marta Mattoso. Building reliable web services compositions. In *Web, Web-Services, and Database Systems*, pages 59–72, 2002.
8. Sharad Mehrotra, Rajeev Rastogi, Henry F. Korth, and Abraham Silberschatz. A transaction model for multidatabase systems. In *ICDCS*, pages 56–63, 1992.
9. Sami Bhiri, Olivier Perrin, and Claude Godart. Ensuring required failure atomicity of composite web services. In *14th International World Wide Web Conference, Japan*, May 2005.
10. Erich Gamma, Richard Helm, Ralph Johnson, and John Vlisside. *Design Patterns: Elements of Reusable Object-Oriented Software*. Addison-Wesley, Reading, Massachusetts, 1995.
11. W. M. P. van der Aalst, P. Barthelmess, C.A. Ellis, and J. Wainer. Workflow Modeling using Proclets. In O. Etzion and Peter Scheuermann, editors, *5th IFCIS Int. Conf. on Cooperative Information Systems (CoopIS'00)*, number 1901 in LNCS, pages 198–209, Eilat, Israel, September 6-8, 2000. Springer-Verlag.

Accelerated Enterprise Process Modeling Through a Formalized Functional Typology

Avi Wasser, Maya Lincoln, and Reuven Karni

ProcessGene Ltd.
15303 Ventura Boulevard, Sherman Oaks, CA 91403
{avi.wasser, maya.lincoln, reuven.karni}@processgene.com
www.processgene.com

Abstract. An enterprise process model encompasses a set of business processes implemented or to be implemented in the enterprise. As such, it expresses the requirements of the organization and thus constitutes a compulsory prerequisite for the successful implementation of process-based IT systems such as ERP, SCM and CRM. However, there is a lack of an enabling science to guide the generation of an individualized process model for a particular enterprise. Conceptually, content based enterprise process modeling – itemizing the processes carried out within the enterprise – is based on the assumption of similarity between enterprises that operate within a given industrial sector, so that a generic model should be applicable, with some customization, to all enterprises within that sector. Our approach is based upon the premise that enterprises are characterized by their functionalities, rather than by their end products or technologies. We thus propose a method which enables the functionality of a specific enterprise to be defined; and from this definition a unique enterprise process model can be generated to constitute a statement of the business processes of the enterprise.

1 Process Modeling: Structural and Content Frameworks

Modeling the business processes of an enterprise is an essential part of any IT development or implementation process [14], [16]. Due to the large number and different granularity levels of processes, business process models are most commonly described by an hierarchy. An industry common division is into four or five levels (demonstrated by a variety of enterprise software vendors and process standardization organizations) [8], [9], [12]. As there is no cross industry standard of hierarchal structure and terminology of the different levels, the hierarchal nomenclature varies; and terms categorizing processes are associated with the various levels.

The fourth, less detailed layer of business process models enumerates and itemizes all the business processes carried out, or intended to be carried out, within the enterprise. Modeling in this context channels its focus towards the *content* layer of business process models – which we define as the suite of business processes constituting the framework of activity within the enterprise. Few scientific publications deal with such content frameworks [6], [11]. On the other hand, they

W.M.P. van der Aalst et al. (Eds.): BPM 2005, LNCS 3649, pp. 446–451, 2005.

have been extensively developed and applied by practitioners such as enterprise software vendors, IT integrators, consulting firms, and specialized BPM companies. Content models include, for example, SAP's industry and cross-industry Business Solution Maps as depicted by SAP's proprietary "Solution Composer" and "Solution Manager" tools [12], Intentia's ERM (Enterprise Reference Models) managed by their proprietary DB and client [2], and Oracle's OBM (Oracle Business Models) library managed by Oracle's process flows system [9]. Other, collaborative (yet more restricted) content frameworks include Rosettanet [10], which covers a collection of B2B processes, and OAGIS [8], which incorporates a set of processes for logistics and general business administration demonstrated by predefined XML schemas. Within these frameworks the first, second and third layers deal with the categorization of the processes included in the process model (or enterprise process suite) at a major and main level of aggregation. In the SAP business solution maps, for example [12], the top level "solution map" for an industrial sector presents the major processes for that industry (about 8), and the corresponding main functions (about 7) for each major function. We refer to a top-level categorization map (encompassing the first three levels) as a "capstone model". Presumably, practitioners have developed their process repositories on the basis of experience accumulated from implementing IT infrastructures in existing industries. This has led to a paradigm whereby these content frameworks are accepted as being generic – i.e. typical for each industrial sector. However, the existence of many reference models, even for a given sector, indicates a lack of scientific systematization in developing such models and raises the question as to whether these models have become too restricted or vendor oriented in an attempt to create a generic prototype, or reference model.

Despite these concerns, there is little doubt regarding the practical necessity of such business process content frameworks, especially as creating an individualized (enterprise specific) process model ab initio is usually complex and daunting. The amount of detail (processes, entities and interrelations) is large, and this makes the formulation of an enterprise-specific model a challenging, if not cumbersome, task.

2   Content-Based Process Modeling (CBPM): State of the Art

Due to the relatively small amount of academic work carried out in the field, we mainly rely on the world of practice when describing the state of the art. We define content based business process modeling (CBPM) as the itemization of the suite of business processes constituting the framework of business activity within a particular industrial sector, or, alternatively, within a particular enterprise. CBPM is considered a compulsory pre-condition for integration and implementation software projects [14], [15] and is usually carried out by ERP/SCM/CRM vendors such as SAP [12], [17]and Oracle; system integrators such as EDS, IBM BCS (Business Consulting Services), and Accenture; and BPM specific companies such as Staffware, Pegasystems, FileNet and others. CBPM is based on the assumption of significant similarity between enterprises that operate within a certain industry. Oracle corporation for example, offers process flows that cover 19 industrial branches [9]; SAP offers Business

Solutions for 24 industrial branches [12]; and other ERP/SCM/CRM vendors similarly base their business models on a finite set of predefined business processes, that comprise fixed [4] "industry-specific" reference models. The industry specific templates are introduced to organizations, in order to facilitate a final product in the form of a customized, "enterprise specific" organizational business process model [7]. This approach assumes a (finite) "universe" of n possible business processes. These are allocated to the industrial branches (industries) in a manner that is intended to best represent the activity within each sector and constitute the reference models which encompass the IT support provided by ERP/CRM/SCM software vendors or BPM consultants. When a vendor, integrator or BPM specialist approaches, say, two enterprises x, y within a certain industrial branch (such as manufacturing, utilities, chemicals, healthcare, consumer goods products) both enterprises are first presented with the same reference content model. The next stage is a top-down (or middle-out) customization of the reference model, by eliminating unwanted functionalities or processes, so as to best cater to the needs of the implementing enterprise. The conventional procedure is based on a generic sectorial model: enterprises are presented with an integrator or vendor reference model as basis for developing an enterprise-specific content model that is aimed to have a high fit with the subsequent ERP system. This approach ignores the fact that sectorial classifications reflect the *end-product* of the enterprise, rather than on its *modus operandi*. For example, both Toyota and Aston-Martin would be presented with the automobile industry model, although their operations differ significantly. Production, marketing, sales and post-sales service at Toyota are based on mass production (MtS – make to stock) and anonymous clients. Aston-Martin, on the other hand, operates on the basis of MtO (make to order) and full recognition of its clients and their requirements. It is clear that not only the core production functions differ between these two "automobile" organizations. It is safe to assume that also processes such as procurement, sales, marketing, service provision and CRM are carried out differently when comparing these two cases. This illustrates that focusing on *what* the enterprise produces (or supplies), instead of *how* this production is carried out, can be misleading and may result in inappropriate business process models. Therefore, as opposed to the *experiential* background underlying the scope of a sector-specific suite there seem to be little *scientific* evidence with regard to two basic assumptions: genericity (functions and processes are common to all enterprises in that sector) and additivity (a complete model can be constructed by combining different functionalities and their corresponding processes) of both reference and enterprise models. Moreover, current business process modeling projects require the execution of a cumbersome customization process that is usually carried out by consultants or implementers through a manual process that involves a significant amount of interviews with Subject Matter Experts (SMEs) within the organization. This procedure has made modeling projects notorious for long duration, high costs (25%-50% of ERP/SCM project total cost) and manual intensity [15]. Another concern is that ERP vendors and integrators offer a relatively small number of industry specific business process models, which provide only partial coverage of industries. SAP, for example, offers

24 such industry maps [12] and Oracle offers 19 [9]. In contrast, the Israel Central Bureau of Statistics (CBS) publishes information regarding 79 industry sectors.

3   A Functional Approach to CBPM

Our paper suggests a solution to the above mentioned concerns. Instead of determining *what* organizations are producing and then tagging them according to their industrial classification, we determine in a general and then detailed way *how* organizations are operating, so that they are characterized by the processes which they implement – i.e. by their functionality. In order to establish the enterprise-specific business process suite, we analyze the existing or planned functionalities in the enterprise and create an enterprise-specific model. Our approach begins with a repository of business processes. It is constructed such that each process (level four in the hierarchy) is subsumed under a main process (level three in the hierarchy); and each main process is subsumed under a major process (level two) under a process category (level one). The current repository is gleaned from a survey of the content models published by vendors [12] and process modeling standardization bodies e.g. [10] & [8]. The top-down approach to CBPM is then based on two main principles: (1) separability: (a particular business process is classified under one main process only and a specific main process is classified under one major process only); and (2) addivity (as the functionalities and processes are separable, a model is formed from a conjunction of major processes, thus main processes, and thus business processes and corresponding activity flows). As the converse is not true – that associating a major process with the enterprise implies implementation of all its constituent main processes, and hence all its constituent business processes and activities – we employ an imbedded expert system methodology, based on the probability, assessed by the system, that an organization will carry out certain processes or groups of processes. This approach enables the quasi-automatic generation of a customized model for the enterprise.

3.1   Generation of Enterprise-Specific Business Process Content Models

The procedure for generating an enterprise specific model is as follows:

a)   Using an enterprise analyzer in the form of a comprehensive questionnaire, to identify the general operational characteristics of the enterprise.

b)   Using a matrix that describes the correlation between operational characteristics and business processes that are part of a process repository.

c)   Determining a threshold probability to automatically retrieve those business processes, having a probability equal to or greater than the threshold value.

d)   Based on the data accumulated in stages (a) to (c) automatically generate the initial content model for the enterprise.

e)   Fine tune the model – usually at the process and activity levels (levels four and five) – to ensure that all relevant processes have been included, and unnecessary processes eliminated, and that the suggested workflow (order of activities within a process) caters the need of the implementing organization.

4 Conclusion

We claim that an enterprise must have a process model which faithfully represents and supports the *"hows"* of its particular modus operandi – how the operational aims and objectives are to be achieved. Our function-oriented CBPM paradigm contributes to this is in the following ways:

- Both industry-specific and enterprise-specific models can be constructed on the basis of a common superset of business processes, and a uniform representation.
- Its major processes are *separable* and *generic*, so that any derived model is composed of an *additive* set of major and main processes and their related business processes and activities.
- Enterprises are differentiated mainly by their industrial functionalities
  Some of the challenges for the next generation of CBPM science include:
- Improving the predictive capabilities of the model first by extending the repository's representation level, and next by increasing the accuracy of the correlation matrix.
- Adding holonic (real-time) trade-off capabilities triggered by the inclusion or exclusion of a certain process or group of processes from a content business process model.

References

[1] W.M.P. van der Aalst. Three Good Reasons for Using a Petri-net-based Workflow Management System. In S. Navathe and T. Wakayama, editors, *Proceedings of the International Working Conference on Information and Process Integration in Enterprises (IPIC'96)*, pages 179–201, Cambridge, Massachusetts, Nov 1996.

[2] G. K. Janssens, J. Verelst, B. Weyn. Techniques for Modeling Workflows and their Support of Reuse. In In W.M.P. van der Aalst, J. Desel, and A. Oberweis, editors, *Business Process Management: Models, Techniques, and Empirical Studies*, volume 1806 of *Lecture Notes in Computer Science*, pages 1–15. Springer-Verlag, Berlin, 2000.

[3] Intentia Reference Models, *http://www.intentia.com/WCW.nsf/pub/tools_index*, 2004.

[4] D. Karagiannis, H. Kühn. MetaModeling Platforms. In: K. Bauknecht; A. Min Tjoa, G. Quirchmayer, Editors, *Proceedings of the Third International Conference EC-Web 2002*, volume 2455 of *Lecture Notes in Computer Science*, page 182. Springer-Verlag, Berlin, 2002. (full version in http://www.dke.univie.ac.at).

[5] B. Kiepuszewski, A.H.M. ter Hofstede, and W.M.P. van der Aalst. Fundamentals of Control Flow in Workflows (Revised version). *QUT Technical report*, FIT-TR-2002-03, Queensland University of Technology, Brisbane, 2002. (Also see *http://www.tm.tue.nl /it/research/patterns*).

[6] T. W. Malone, K. Crowston, J. Lee, B. Pentland, C. Dellarocas, G. Wyner, J. Quimby, C. S. Osborn, A. Bernstein, G. Herman, M. Klein, and E. O'Donnell. Tools for Inventing Organizations: Toward a Handbook of Organizational Processes. *Management Science* 45(3) pages 425-443, March, 1999.

[7] T. W. Malone, *The Future of Work: How the New Order of Business Will Shape Your Organization, Your Management Style, and Your Life*. Boston, MA: Harvard Business School Press, 2004.

[8] OAGIS. Best Practices and XML Content for Everywhere-to-Everywhere Integration, http://www.openapplications.org/, 2004.

[9] Oracle.   Business   Models   (OBM),   *http://www.oracle.com/consulting/offerings/ implementation/methods_tools/*, 2004.

[10] Rosettanet. Lingua Franca for Business, *http://www.rosettanet.org/*, 2004.

[11] A.W. Scheer, M. Nüttgens. ARIS Architecture and Reference Models for Business Process Management. In W.M.P. van der Aalst, J. Desel, and A. Oberweis, editors, *Business Process Management: Models, Techniques, and Empirical Studies*, volume 1806 of *Lecture Notes in Computer Science*, pages 376-390. Springer-Verlag, Berlin, 2000.

[12] SAP.   Business   Maps   and   Solution   Composer,   *http://www.sap.com/solutions/ businessmaps/composer/*, 2004.

[13] G. Yang. Towards a Library for Process Programming. In W.M.P. van der Aalst, A.H.M. ter Hofstede, and M. Weske, editors, *BPM 2003*, volume 2678 of *Lecture Notes in Computer Science*, pages 120-135. Springer-Verlag, Berlin, 2003.

[14] C.P. Holland, B. Light, *A critical success factors model for ERP implementations*, IEEE Software 16 (1999) 30–35.

[15] B. Light, *The maintenance implications of the customization of ERP software*, J. Software Maintenance: Res. Practice 13 (2001) 415–429.

[16] M. Krumbholz, N. Maiden, *The implementation of enterprise resource planning packages in different organizational and national cultures*, Inf. Systems 26 (2001) 185–204.

[17] J. Ghosh, *SAP Project Management*, McGraw-Hill, New York, 2000. Eng. 41 (2000) 180–193.

Introducing Business Process into Legacy Information Systems

Marcos R.S. Borges[1,#], A.F. Vincent[1], Mª Carmen Penadés[2],
and Renata M. Araujo[1,3]

[1] Graduate Program in Informatics, Federal University of Rio de Janeiro, Brazil
mborges@nce.ufrj.br, andre.vincent@mba.berkeley.edu
[2] Department of Computer Science (DSIC), Technical University of Valencia, Spain
mpenades@dsic.upv.es
[3] Department of Applied Informatics, Federal University of the Rio de Janeiro State, Brazil
renata@nce.ufrj.br

Abstract. The majority of legacy information systems running today were built without adopting a business process approach. In these systems, the control over the execution of the process activities is partial, leaving out all those activities that have not been automated. Moreover, the activities that constitute the process are not formally interconnected, causing loss of the overall business process context. This paper presents a method for gradually integrating the underlying business processes supported by these systems, without disrupting the automation they already support. The method is particularly attractive for legacy systems that are expected to last a long time and whose redevelopment costs are high.

1 Introduction

Many organizations have invested in a new approach for the development of information systems that is based on business process analysis [7] [9]. There are, however, many information systems that were developed before this new approach became popular and which are still fully operational [8] [10]. They are called legacy systems. They remain in operation because they provide the functionality needed and their redevelopment requires a significant investment [8]. On the other hand, these systems do not provide an explicit view of the underlying processes. In general, a legacy system supports some of the activities of the associated process, but without explicitly establishing the order of these activities (the process).

Being aware of the best practices in new product development is one issue, but implementing these approaches in a company to reduce time to market is a different one [2]. Unless the entire company accepts the new working methods, improvements may not be achieved [5]. Therefore, regardless of strong recommendations against this practice [7] [9], many information systems have been and continue to be developed focusing only on the activities they automate. This still occurs in spite of the strong methodological support and tools that are now available [6]. The reasons

---

[#] On sabbatical leave at DSIC-Technical University of Valencia, Spain.

W.M.P. van der Aalst et al. (Eds.): BPM 2005, LNCS 3649, pp. 452–457, 2005.

behind this attitude are mainly cultural, but lack of training and scepticism about the return on investment are also strong. How should these barriers be overcome? Can the preservation of legacy systems and the business process approach be combined? A revolutionary approach is an alternative, but past experience indicates a high rate of failure when the organization is not culturally prepared.

This paper presents an alternative for introducing a business process culture while preserving the legacy systems with minimum changes. The redevelopment of new systems is delayed until the process approach has been assimilated by the organization. The method aims at preserving the existing systems while introducing the process culture. Our approach will reduce resistance within the organization and make it better prepared for future developments. It is important to emphasize that we believe that an effective approach to business architecture is very much needed and must include an integrated and seamless method of business engineering and software development [7]. We are simply proposing an alternative for those organizations that are not yet prepared to move towards a revolutionary approach. This is particularly true in small and medium-sized enterprises in which the benefits of a business process approach may differ significantly from their large enterprises [5].

The hypothesis of this work is that it is always possible to identify a subset of processes and integrate it with the existing information systems. This partial integration will generate some limited but important benefits that are associated with the business process approach. The most relevant problems caused by the absence of an explicit process-centric system development are: the difficulty that users have in evaluating the impact of their decisions and activities on the organization as a whole, because they don't know the entire process [4]; the lack of process monitoring, that difficulty to identify opportunities for improvement; the control of manual activities; the absence of documentation for training of a new professional, because most knowledge about the organization's processes is tacit.

The method proposed in this paper integrates activities that are supported by legacy systems and business process activities. It is not a method aimed at supporting the development of new information systems based on a business process definition. Several initiatives on this direction already exist [3] [12]. Our proposal preserves the existing systems while introducing an integrated view of the business process. This approach extends the life of successful systems and, at the same time, prepares the organization for the business process approach.

The rest of this paper is organized as follows. Section 2 describes our evolutionary approach. Section 3 presents the method for transforming legacy systems into a process support system; the first phase of our approach. Finally, Section 4 presents the conclusions of the paper.

2   The Proposed Approach and Method

In order to promote the integration between business process and legacy systems, first it is necessary to identify the processes and their activities. Some of these activities are manual, i.e., they are not automated by the legacy systems. Others are either totally or partially embedded in the legacy system. Our approach proposes the

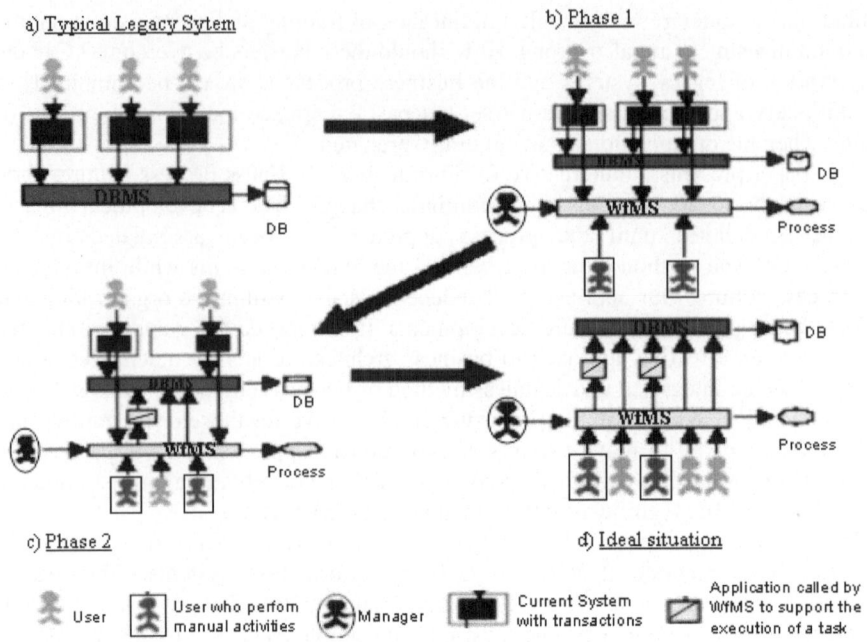

Fig. 1. Summary of the evolutionary phases

modeling of theses processes and the implementation of their control by a Workflow Management System (WfMS) [13]. Although this is similar to a process-centric approach for new development, there are some significant differences. First, the processes are modeled and implemented as they are. Second, the WfMS is not in control of all activities. The workflow will receive notification from the system for automated activities, but it will not control them. Finally, in our approach, we keep the initial system changes to a minimum; they are performed gradually as activities are moved to the control of the WFMS.

The complete evolutionary approach is summarized in Figure 1. Figure 1a represents the typical situation found in legacy systems. There is neither a workflow tool nor an explicit process representation. The process transactions are implemented in one or more systems. Users interact directly with the system interface. Figure 1d, on the other hand, represents what we consider to be an ideal situation. Processes are explicitly defined. The WfMS controls the process execution, and users interact directly with its interface. Whenever needed, the WfMS invokes an application, and the user interacts directly with it.

The situation depicted in Figure 1a is quite different from the one depicted in Figure 1d. Moving from the typical to the ideal situation is not an easy task. In our evolutionary approach, we keep the current applications and introduce the business process approach gradually. We distinguish two phases.

The first phase (Figure 1b) requires the introduction of a WfMS to control selected processes. The manual activities of theses processes are controlled directly by the WfMS, but those activities that are performed through the system are not under the control of the WfMS. In this case, in order to proceed to the next step, the WfMS must be notified by the system upon completion of the activity. This is the change required in the legacy system and, in general, is very simple to perform. The system operations that are not involved in the process are not affected.

The second phase (Figure 1c) is the transfer of control of some automated activities to run under the WfMS. The goal of this transfer is to move all activities that are embedded in a subprocess. The user does not have to interact with both the WfMS and the system for the same subprocess. Parts of the application will need to be redeveloped. With the continuity of the replacement procedure, the workflow will gradually take control of all the process activities. At the same time, the legacy system will reduce its routing function.

3   The Transformation Method

The method to support the evolutionary approach consists of a set of steps which are carried out by development teams. These teams are composed of specialists playing roles that are different from the typical system development process. The teams are: Project Committee, Users Team, Systems Team, WfMS Team and Process Team. Legacy Organization represents that all the teams are working together. The Project Committee should follow the project as a whole (defining the project goals, providing the necessary resources and following its development). The Users Team should provide the information about the current system and use the new function. The Systems Team should provide information about the system architecture and perform the changes required by the new approach. The WfMS Team should select the Workflow technology, translate the process models into workflow models, and configure the WfMS. Finally, the role of the Process Team is to elicit the processes and to represent them in a model.

Figure 2 shows an overview of the first phase of the proposed method. The initial definitions made by the Project Committee refer to the scope and duration of the project, the priority areas and processes involved, the configuration of the teams, and the systems affected. Once these definitions are available, the other teams can start working. The Process Team can elicit the business processes selected. The process elicitation should identify the flow of all activities, both manual and automated and should not be based on the system functions. The Users Team should evaluate the impact the workflow control will generate on the business activities. The procedure changes and the need for training are part of this impact analysis. The System Team should perform an analysis on the complexity of the legacy system functions which serve as the basis for estimating the costs required to perform the system changes.

With all the previous data at hand, all the teams can define an implementation strategy. The strategy should define the process and the activities affected by the first phase of the project, estimate the cost to perform these changes, and, include an impact and a risk analyses. This strategy will probably require approval from the executive board. If it is not approved, the scope of the project should be reduced.

After approval, the next three steps can be performed at the same time. The Process Team can elicit in detail the processes involved; the System Team can define in detail

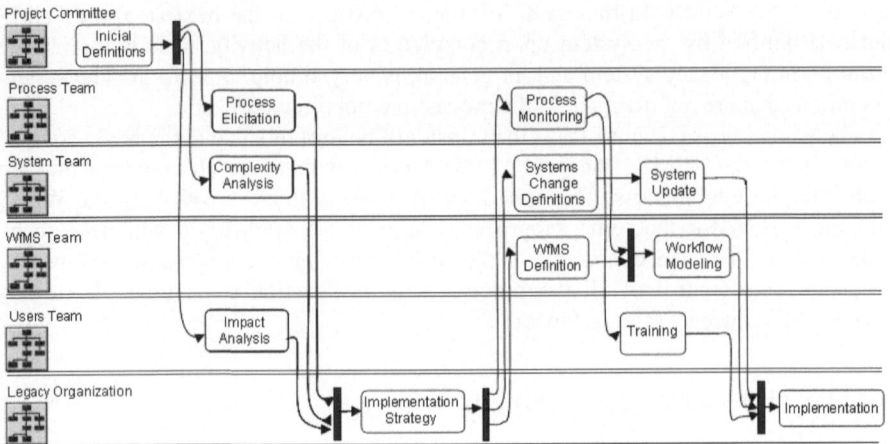

Fig. 2. Overview of the proposed method activities

the changes required in the legacy system; and the Workflow Team will have the elements to choose the workflow technology that will support the process control.

The Workflow Team translates the process model generated by the Process Team to a workflow model. Unfortunately, this translation is not always straightforward as the modeling tools of workflow systems are not capable of representing all business situations. Actually, the richness of the model supported by the Workflow tool should be an important element in selecting workflow technology. The user training and the system changes can also be performed at the same time [1].

Finally, when all pieces have been completed, it is time to put them together and simulate the process as a preparation for the new environment. It is important to check the correct interaction between the legacy system and the WfMS and to check if the business processes are supported as expected.

These steps can be repeated as many times as necessary until all the target processes have been integrated. The reaction of the users should be followed very closely as it will help support the decision to invest in a new process-driven system. We expect, however, that if the legacy system is running well, it will be redeveloped only if new functions are required.

4 Conclusions

This paper has presented an alternative to the revolutionary approach for introducing the BPM approach with new information systems development. For organizations with a high investment in legacy systems and which show signs of resistance to a BP approach, we suggest an evolutionary approach where business processes are introduced gradually with minimum change in existing systems. We consider that this will postpone the investment in system redevelopment and will increase the chances of success when complete process redefinition occurs. The alternative is described in the form of a method with a description of its activities and the roles played by the specialists in charge of them. We believe that this method can be applied with simple adaptations to many organizations where legacy systems play an important role.

A pilot experiment has been performed in a software company [11]. We carried out the first part of the method by adapting the existing systems to report their activities to a WfMS that was based on a process model designed by the process team. The results were in keeping with those predicted by our approach.

Acknowledgements. We would like to thank the CICYT (Project DYNAMICA-PRISMA TIC2003-07776-C02-02) for partial funding of this work. Marcos R.S. Borges was partially supported by a grant from the "Secretaria de Estado de Educación y Universidades" of the Spanish Government.

References

1. Van Der Aalst, W.M.P., Ter Hofstede, A.H.M., Kiepuszewski, B., Barros, A.P.: Workflow Patterns, Distributed and Parallel Databases, 14:1 (2003) 5-51
2. Aversano, L., Canfora, G., De Lucia, A., Gallucci, P.: Business process reengineering and workflow automation: a technology transfer experience, Journal of Systems and Software 63:1 (2002) 29-44
3. Baresi, L. et al.: WIDE Workflow Development Methodology. In: International Joint Conference on Work activities Coordination and Collaboration. San Francisco, USA. (1999)
4. Borges, M.R.S., Pino, J.A., Araujo, R.M.: Bridging the Gap between Decisions and their Implementations, In: Lecture Notes in Computer Science 3198 (2004) 153-165
5. Dwyer, T. Revisiting the Process-centric Company. BPM Institute (2004) http://www.bpminstitute.org/article/article/revisiting-the-process-centric-company.html
6. Fingar, P.: Component-based frameworks for e-commerce. Communications of the ACM. 43:10 (2000) 61-66
7. Jacobson, I., Ericsson, M., Jacobson, A.: The Object Advantage – Business Process Reengineering with Object Technology. USA. Addison Wesley (1994)
8. Rubinstein, D.: Breathing New Life Into Legacy Systems. Software Development Times: News & Top Stories. October (2003) http://www.sdtimes.com/news/088/story9.htm
9. Smith, H., Fingar, P.: Business Process Management: The Third Wave. Meghan-Kiffer Press (2003)
10. Ulrich, W.M.: Legacy Systems: Transformation Strategies. Prentice Hall (2005)
11. Vincent, A.F.: A method to make processes explicit in legacy information systems. M.Sc. Dissertation, Graduate Program in Informatics, Federal University of Rio de Janeiro (2002)
12. Weske, M. et al.: A Reference Model for Workflow Application Development Processes. In: International Joint Conference on Work activities Coordination and Collaboration. San Francisco, USA (1999)
13. Workflow Management Coalition: Workflow Client Application (Interface 2) – Application Programming Interface (WAPI) Specification – Doc # TC-1009 – Issue 1.1. May (1996)

Spheres of Isolation: Adaptation of Isolation Levels to Transactional Workflow

Adnene Guabtni, François Charoy, and Claude Godart

LORIA - INRIA - CNRS, BP 239, 54506 Vandouvre-lès-Nancy Cedex, France
{guabtni, charoy, godart}@loria.fr

Abstract. In Workflow Management Systems (WFMSs), transaction isolation is managed most of the time by the underlying database system using ANSI SQL strategies. These strategies do not take sufficiently into account process aspects. Our work consists in studying with more depth the relation between isolation strategy and process dimension as well as the real isolation needs in workflow environments. To carry out these needs, we define 'spheres of isolation' inspired from 'spheres of control' proposed by C. T. Davies. Spheres of isolation take into account real workflow isolation needs with separation of concerns between workflow design and the specification of its transactional properties.

1 Introduction

The specification of transactional constraints in business processes is always a paramount stake especially in co-operative processes or distributed and composed e-services. In WFMSs, transactions are usually implemented by teh DBMS. Those systems generally use standard ANSI SQL [1] to define the isolation's constraints of a transaction. The problem lies in the fact that these isolation's constraints cannot always satisfy those of a workflow process. The process dimension in atomicity has been already analysed in [4] and give more capabilities to transactional WFMS. Isolation [11] has been already studied in cooperative process environment in a recent past (Contracts [9] and Coo [6]) but has never been generalized to workflow processes.

To carry out that problem, we take as a starting point the approach of 'Spheres of control' proposed by C.T. Davies in [8]. This approach was re-used in [4] to introduce spheres of atomicity allowing customised specification of atomicity in transactional workflow. We follow the same approach to define spheres of isolation in order to allow a customized specification of isolation constraints in transactional workflow. We consider a process as being the concurrent execution of sets of activities which can have various constraints regarding isolation. We want to allow the workflow designer to decide on the degree of isolation necessary for a group of activites. Our approach introduces also a separation of concerns between process and transactional properties definition. The definition of the process should reflect the real organization of work in the company. Transactional properties should reflect technical aspects of the execution and consistency needs and should not influence the process definition.

W.M.P. van der Aalst et al. (Eds.): BPM 2005, LNCS 3649, pp. 458–463, 2005.

In the following sections of the article, we analyze the stakes and needs of isolation in workflow systems compared to database systems. Next we develop our approach based on 'spheres of isolation' to allow customized isolation in transactional workflow.

2   Transactions in Workflows: Current Approaches

Advanced transaction models were introduced to enhance transaction support in WFMSs and provide more flexibility compared to traditional database transactions (ACID). Their implementation in workflows was studied in [2]. These models included process dimension on transaction management but were focused mainly on atomicity property. Current implementations of transaction models in WFMS are so heterogeneous and complex that a real taxonomy of transactional workflow implementations was defined in [5]. This taxonomy is a representation of the real practice of transactional properties in WFMSs.

Our approach is to study the real needs of WFMS for isolation properties. These properties are usually confused with database transactional needs. Atomicity needs in WFMS has been already established in [4]. The crucial difference between these needs and those of database systems is the definition of atomicity constraints to groups of activities called spheres of atomicity. In the next section, we perform a similar approach to study the real isolation needs in WFMSs.

3   Isolation Requirements in WFMS

3.1   Isolation Levels in Traditional Transactions

The isolation problem occurs when several transactions access to the same data. In several information systems, the isolation stake grows when the data used are accessed by more and more concurrent transactions and increasingly independent transactions. The problem that occurs in this case is the lack of flexibility of the isolation strategy. In database systems, isolation is guaranteed via isolation levels [1]. There are four isolation levels: Read Uncommitted, Read Committed, Repeateable Read and Serializable. These levels make it possible to provide more or less undesired phenomena (dirty read, fuzzy read and phantom [1]).

Isolation levels suggested in (ANSI SQL, 1992)[1] were criticized in [3]. However they are largely used in current databases systems. Other approaches based on timestamps were studied and are based on optimistic locking systems. Nevertheless, all existing approaches do not express isolation requirements adapted to transactional workflows. In the next section, we expose what are process dimension based isolation requirements in WFMSs.

3.2   Isolation Requirements for Transactional Workflow

Isolation problems are more and more obvious depending on the data visibility. Indeed, the data used in WFMS were classified in 7 types according to their visibility according to workflow data patterns[10]. The need of process support in isolation strategy depends on the workflow data visibility starting at task data visibility where there is no isolation needs and continue with this order : block, scope, case, workflow, environment, multiple instance data visibility. Multiple instance execution [7] produces the highest need of process dimension support.

The goal of our work is to adapt isolation levels to workflow. That becomes possible if we take into account not only needs of single activities but also needs of a group of activities of the process (collaborative work, distributed or composed e-services). We identified two main needs consisting in the control of cohesion and coherence of a group of activities.

Cohesion means the fact that activities of the same group use the same reference for data access. Activities can then use data with ensurance that all of them are using the same version and are seeing only changes made by them. External activities (not part of the group) that may want to modify the same data during the execution of the group will not influence the referencial used by the group. The referential can be seen as a view of data, readable and writable only by activities of a restricted group.

Coherence of data is another important need. Indeed, a group of activities usually needs to ensure that the impact of its execution do not introduce some mistakes or inconsistencies. These inconsistencies are usually due to the use of temporary or uncommitted data produced by the group.

The coherence concerns the external environment of the group while cohesion concerns the internal one. Based on these two main needs, we will introduce in the next section the notion of 'isolation spheres'.

4   Our Approach: Isolation Spheres

In the last few years, some work has been inspired from the sphere of control proposed by Davies [8] to enhance expressivity of transactional properties, especially in [4] where the notion of atomicity sphere has been developed. In our work, we take the same approach to define 'spheres of isolation' as follows:

Definition : An isolation sphere represents a group of activities in a workflow process working in concurrency on some data. The sphere ensure the cohesion (constraints on reference data) and the coherence of the sphere refering to concurrent activities or other spheres. The cohesion and coherence constraints allow a process support in isolation strategy.

All or a part of the data used by sphere activities represents the data that have to be controlled (data concerned by isolation on which necessary locks need to be applied). To ensure cohesion and coherence on these data, we introduce some cohesion levels and some coherence levels. Before introducing these levels we need to define some notations:

A process (or workflow process) represents tasks called activities and these tasks are executed following an execution order established throw a control flow between activities. A sphere is defined as part of a process. In WFMSs, a sphere is composed of activities of the workflow.

Let S the set of spheres and $s \in S$.

Δ_s is the set of data concerned by isolation sphere $s \in S$. This set of data is defined by the workflow designer and is a subset of the data used by the group.

$A(s)$ is the set of activities of $s \in S$.

The state of a data δ changes over time due to activities execution and takes several values $\{\delta_0, \delta_1, ..., \delta_n\}$ corresponding respectively to several instants $\{t_0 < t_1 < ... < t_n\}$. If the value δ_i was written by an activity α, we note it δ_i^α. We note δ^α the value validated (committed) of δ written by α.

4.1   Properties of Isolation Spheres

Isolation spheres properties are cohesion and coherence. Cohesion means the fact that all activities of the sphere have the same view on data they access. The view represent a reference data that all activities of the sphere will read or update. External activities updates will not be visible from the sphere view. This common view represents the basis of cohesion in a group of activities but there are different possible cohesion levels based on the initial view isolation constraints. These levels of cohesion are as follows: Let $s \in S$ and $\delta \in \Delta_s$,

Level 0 : Read Uncommitted: if an activity of the sphere s reads δ it can read only $max(\delta^s, \ \delta_i^\alpha)$ such as $\alpha \in A(s)$ and δ^s corresponds to the value of δ read the first time by an activity belonging to the sphere.

Level 1 : Read Committed: if an activity of the sphere reads δ then it can read only the $max(\delta^s, \ \delta_i^\alpha)$ such as $\alpha \in A(s)$ and δ^s corresponds to the validated (committed) value of δ read the first time by an activity belonging to the sphere.

Level 2 : Repeatable Read: same case of Read Committed except that it is also concluded that the value of δ is not modified by an activity external to the sphere as long as the sphere did not finish its execution yet. The end of the execution of a sphere occurs when all its activities finished their execution.

Level 3 : Serializable: emulates an execution in series of the sphere and its external environment (activites, spheres or processes). This level ensure a serialisability between the sphere and the external environment of the sphere but does not ensure a serialisability between the activities of the sphere.

Coherence of sphere represents how activities share their data with their external environment. Different levels of coherence can be defined as follows:

Level 0 : Atomic coherence: all the values of a data written by the activities of the sphere are visible outside of the sphere. If an activity α of the sphere writes δ then all δ_i^α are visible outside the sphere.

Level 1 : Selective coherence: only the **validated** values written by the activities of the sphere are visible outside of the sphere. If an activity α of the sphere writes δ then only δ^α is visible outside the sphere.

Level 2 : Global coherence: only the **last validated** value written by an activity of the sphere is visible outside. If activities of the sphere s write δ then only δ^α is visible outside the sphere, α being the last activity of s to write δ.

4.2   Phenomena Significance in Isolation Spheres Context

The undesired phenomena noted in database systems don't have the same significance when we use isolation spheres. Both cohesion and coherence release isolation constraints and the significance of each phenomenon differs from a classic transaction to an isolation sphere as follows:

For a classic transaction χ:
Dirty Read : Read of δ_i^α and α rollbacks
Fuzzy Read : Read of δ^α such as $\delta^\alpha < \delta_i^\beta < \delta^\chi$
Phantom : Ask for a request and the result is modified during execution by insertion of new data by another transaction
For isolation Sphere s:
Dirty Read : Read of δ_i^α such as $\alpha \in A(s)$ and α rollbacks
Fuzzy Read : Read of δ^α such as $\delta^\alpha < \delta_i^\beta < \delta^s$ and $\beta \notin A(s)$
Phantom : Ask for a request and the result is modified during execution by insertion of new data by activity external to the sphere

The control of the two dimensions (cohesion + coherence) makes it possible to define in a finer way isolation requirements for groups of activities. The choice of cohesion and coherence levels influences the degree of divergence and the degree of data exchange flexibility between activities of the sphere and its environment. Divergence increases from (cohesion3/coherence0) to (cohesion0/coherence2). Flexibility increases from (cohesion3/coherence2) to (cohesion0/coherence0).

4.3   Advanced Organization of Isolation Spheres: Nested Isolation Spheres

Activities of a sphere are able to execute without worrying if somebody of the outside environement will obstruct their work. However it is inevitable to have requirements on isolation inside the sphere itself. A sphere can then contain others sub soheres that have different isolation needs. Thus we introduced nested isolation spheres. A sub sphere ensure its own cohesion and define its coherence with the immediate top sphere. We think that this kind of organization increases considerably the expressivity in term of isolation in a transactional workflow.

5   Conclusion and Perspective

In this article, we have focused on isolation in transactional workflow. Existing approaches use techniques of isolation adapted to databases and not really to workflow context. We have made a specific adaptation of isolation levels to transactional workflow increasing expressivity in term of isolation and allowing process to get rid of long blocking due to database isolation methods. Our study of the problem revealed that the basic isolation entity in current transactional workflow systems is the single activity. We have established the importance of isolation properties for groups of activities. Two main isolation properties have been established for groups of activities in transactional workflow : Cohesion and Coherence. Our approach to make these two propeties realizable is based on 'Isolation Spheres' inspired from 'Spheres of control'.

This work requires to be continued in order to consider several aspects as the relation between the declaration of isolation spheres and the control flow governing the workflow, a simple way to easy choose coherence and cohesion levels, and finally an implementation of 'isolation spheres' functionalities must be carried out in a WFMS in order to validate the feasibility of this approach.

References

1. Ansi x3.135-1992, american national standard for information systems - database language - sql. November 1992.
2. Gustavo Alonso, Divyakant Agrawal, Amr El Abbadi, Mohan Kamath, Roger Günthör, and C. Mohan. Advanced transaction models in workflow contexts. In Stanley Y. W. Su, editor, *Proceedings of the Twelfth International Conference on Data Engineering, February 26 - March 1, 1996, New Orleans, Louisiana*, pages 574–581. IEEE Computer Society, 1996.
3. Hal Berenson, Phil Bernstein, Jim Gray, Jim Melton, Elizabeth O'Neil, and Patrick O'Neil. A critique of ansi sql isolation levels. In *Proceedings of the 1995 ACM SIGMOD international conference on Management of data*, pages 1–10. ACM Press, 1995.
4. Wijnand Derks, Juliane Dehnert, Paul Grefen, and Willem Jonker. Customized atomicity specification for transactional workflow. In *Proceedings of the Third International Symposium on Cooperative Database Systems for Advanced Applications (CODAS'01)*, pages 140–147. IEEE Computer Society, 2001.
5. Paul W. P. J. Grefen. Transactional workflows or workflow transactions? *DEXA*, pages 60–69, 2002.
6. Daniela Grigori, François Charoy, and Claude Godart. Coo-flow: a process technology to support cooperative processes. *International Journal of Software Engineering and Knowledge Engineering - IJSEKE Journal*, 14(1), 2004.
7. Adnene Guabtni and François Charoy. Multiple instantiation in a dynamic workflow environment. In Anne Persson and Janis Stirna, editors, *Advanced Information Systems Engineering, 16th International Conference, CAiSE 2004, Riga, Lavtia*, volume 3084 of *Lectures Notes in Computer Science*, pages 175–188. Springer, Jun 2004.
8. Charles T. Davies Jr. Data processing spheres of control. *IBM Systems Journal 17(2): 179-198*, 1978.
9. Andreas Reuter and Friedemann Schwenkreis. Contracts - a low-level mechanism for building general-purpose workflow management-systems. *IEEE Data Eng. Bull.*, 18(1):4–10, 1995.
10. Nick Russell, Arthur H. M. ter Hofstede, David Edmond, and W.M.P. van der Aalst. Workflow data patterns. Technical Report FIT-TR-2004-01, Queensland University of Technology, Brisbane, Australia, April 2004.
11. Heiko Schuldt, Gustavo Alonso, Catriel Beeri, and Hans-Jörg Schek. Atomicity and isolation for transactional processes. *ACM Trans. Database Syst.*, 27(1):63–116, 2002.

Verification of SAP Reference Models

B.F. van Dongen and M.H. Jansen-Vullers

Department of Technology Management, Eindhoven University of Technology,
P.O. Box 513, NL-5600 MB, Eindhoven, The Netherlands
{b.f.v.dongen, m.h.jansen-vullers}@tm.tue.nl

Abstract. To configure a process-aware information system (e.g., a workflow system, an ERP system), a business model needs to be transformed into an executable process model. Due to similarities in these transformations for different companies, databases with reference models, such as ARIS for MySAP, have been developed. The models stored in such a database can be customized to generate an executable model. Since these customized models are typically used on an execution level, it is of the utmost importance that both the reference models and their customizations are free of erroneous constructs.

In this paper, we analyze a reference model for SAP R/3 that is stored in the ARIS for MySAP database, and we verify whether it is correct. Since the model is stored as an Event-driven Process Chains (EPC), we use a verification approach tailored towards the verification of this language to check for errors in the model. We show that using this approach adds value to a set of reference models, such as ARIS for MySAP, since modelling errors are discovered at an early stage and can be avoided on an execution level.

Keywords: Event-driven Process Chains, Verification, SAP, Reference Models.

1 Introduction

Nowadays, *process-aware information systems* such as Enterprise Resource Planning (ERP) [9] systems and Workflow Management (WFM) [1,10] systems are used to support a wide range of operational business processes. On an operational level, these systems are often configured on the basis of a process model. The design of such a process model is a complicated and error prone task. Furthermore, the process models that are designed in difference companies are often very similar. For this reason, databases with process models for many different applications have been developed. These databases can be used as a reference during process design, hence the term *reference models*.

Together with the business model of a company, a reference model is selected that best fits the process under consideration. During the process model design phase, a designer customizes that reference model to fit the business model of the company. The result of this customization phase is an informal specification of a process in terms of a customized process model. In the implementation

W.M.P. van der Aalst et al. (Eds.): BPM 2005, LNCS 3649, pp. 464–469, 2005.

phase, this model is used to implement an executable specification for a specific information system, such as SAP R/3. Since all the steps between selecting a reference model and producing an executable specification are performed by humans, errors are likely to be introduced.

The use of reference models does not eliminate the possibility of introducing errors into the process model. It should, however, assist the designer in such a way that errors are less likely to be introduced. Therefore, it is of the utmost importance that the selected reference model is correct. Especially since, usually, many processes are modelled independently of each other. When considering the real life processes however, process models are highly dependent. Furthermore, when errors in process models are implemented in an executable specification in the implementation phase, they can have severe operational consequences.

To find errors in process models, many authors have developed verification methods. Basically, all of these verification methods can be used to check whether a process model is *correct*, in other words, they can be used to check for *correctness* of a process model. In this paper, we focus on the correctness of reference models for a specific information system, SAP R/3. The reference models are available in the ARIS for MySAP database in the ARIS Toolset, a commercial product of IDS-Scheer. As a modelling language, the ARIS Toolset uses *Event-driven Process Chains* (EPCs) [8,9,14]. We selected SAP R/3, since EPCs are used in a large variety of systems, including SAP R/3. Moreover, SAP R/3 is market leader in the field of Enterprise Resource Planning systems. Furthermore, many verification approaches exist for EPCs. The verification method we chose looks at verification from a designers point of view and assumes the process designer to know what he intends to model.

In the remainder of this paper, we take the SAP reference models as a starting point, and use the verification approach presented in [5] as our verification method. Using this combination, we analyze the "internal procurement" model and show that this model contains an error. In Section 2 however, we first discuss related work with respect to the verification of process models. In Section 3 we describe our domain of analysis: SAP R/3 reference models. Next, in Section 4, we describe the approach for the verification of these models as implemented in the ProM framework[1], and described in [4,5]. Following this approach we are able to evaluate the SAP reference models in Section 5. Finally, in Section 6, we draw some conclusions.

A longer version of the work presented here, including more extensive analysis results, is available as an internal report [6].

2   Related Work

Since the mid-nineties, a lot of work has been done on the verification of process models, and in particular workflow models. In 1996, Sadiq and Orlowska [13] were among the first to point out that modeling a business process (or workflow)

---

[1] See www.processmining.org for details.

can lead to problems like livelock and deadlock. In their paper, they present a way to overcome syntactical errors, but they ignore the semantical errors. Nowadays, most work that is conducted focusses on semantical issues, i.e. "will the process specified always terminate" and similar questions. The work that has been conducted on verification in the last decade can roughly be put into three main categories, namely "verification of models with formal semantics", "verification of informal models" and "verification by design". In [6] we present these categories and give relevant literature for each of them.

In this paper, we use the technique presented in [5] that can be seen as a combination of the first two categories. It assumes *the designer* to be able to decide whether or not a specification is semantically correct. This technique has been implemented in the Process Mining (ProM) Framework[2], that is able to import EPCs defined in the ARIS Toolset[3] and provides the designer with feedback about possible problems. Since SAP reference models are available in the ARIS Toolset format, and the users of these reference models are typically consultants that have a deep knowledge about the process under consideration, we found this to be the best approach for the verification of the SAP R/3 reference models.

3   SAP R/3 Reference Models

Several authors researched the area of reference models before, see e.g. [2,3,7,11,12,15,16,17,18]. In this paper we use the definition of reference models based on [12]. Furthermore, we use the ARIS for MySAP reference databases, that contains hundreds of reference models, all modelled as Event-driven Process Chains (EPCs). These EPCs that can be used in many different situations, from "asset accounting" to "procurement" and "treasury". Since we cannot discuss all these models here, we focus on one of the models, consisting of 12 events, 5 functions and 8 connectors. We analyze this model using the approach described in [5]. Before we show the result in Section 5, we first briefly introduce this verification approach in Section 4.

4   Verification Approach

For the verification of the EPCs in our reference model database, we use the approach described in [5]. This verification approach is tailored towards the verification of Event-driven Process Chains and it assumes *the designer* of an EPC to be able to decide whether or not the EPC is correct. The approach is implemented in the ProM framework ([4]) and it is freely available for download.

The verification process described in [5] consists of several steps. In the first step, the designer of the EPC has to provide the tool with all combinations of initial events that could initiate the modelled process. Using this, the tool calculates all the possible outcomes of the process (in terms of events that occurred

---

[2] See www.processmining.org for details.
[3] See www.ids-scheer.com for information about the ARIS toolset.

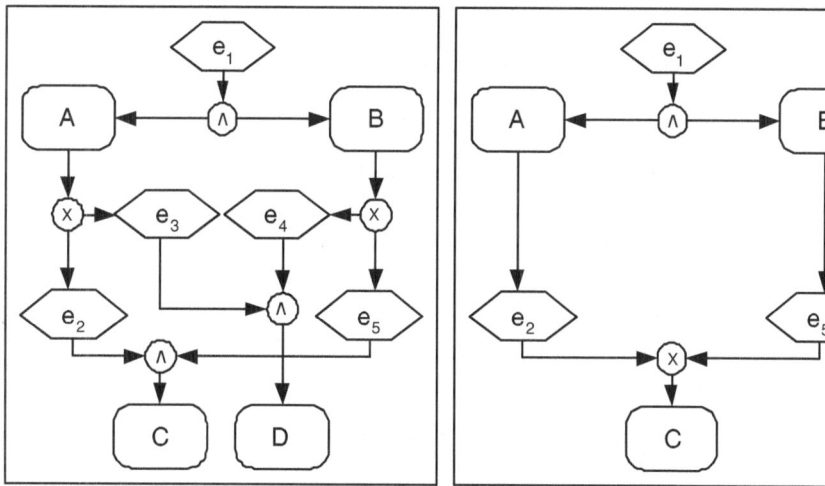

Fig. 1. EPC with choice synchronization   **Fig. 2.** EPC with erroneous routing

and have not been dealt with). Then, the tool requires the designer to divide those outcomes in two groups, the first of which contains all the outcomes that represent the desired behavior of the process. The second group contains the undesired behavior.

The tool finally presents the designer with three possible answers. First, an EPC can be *semantically correct*. Models that are semantically correct are models of processes that, when started in any allowed state, will *always* terminate in one of the allowed termination states, i.e. all choices can be made locally. Second, an EPC can be *syntactically correct*. Syntactically correct EPCs are models of processes that, when started in any allowed state, will always *have the possibility to* terminate in one of the allowed termination states. In other words, not all choices can be made locally, instead, the execution history limits the available options. An example of such a construct can be found in Figure 1, where the choices after functions A and B have to be synchronized in order to allow function C or D to execute. Finally, an EPC can be *incorrect*. These models contain syntactical errors, such as an AND-split followed by an XOR-join or the other way around. An example of such an incorrect model is shown in Figure 2, where functions A and B originate from an AND-split, and are later joined by an XOR-join. As a result, function C will be carried out twice.

5   Verification of the "Procurement" Model

The application of the verification approach presented in Section 4 is based on a basic assumption: It assumes that the designer of a model has a good understanding of the actual business process that was modelled, and he knows which combinations of events may actually initiate the process in real life. Typically,

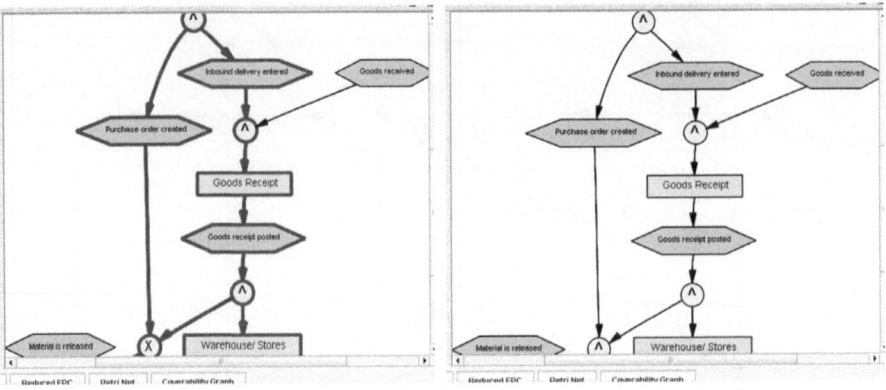

Fig. 3. Erroneous "Internal Procurement"   **Fig. 4.** Repaired "Internal Procurement"

reference models are used by consultants that do indeed have a good understanding of the process under consideration. Besides, they know under what
circumstances processes can start, and which outcomes of the execution are desired and which aren't. Therefore, the approach seems to be well suited for the
verification of the SAP reference models.

As stated in Section 3 we focus on the "Internal Procurement" model of
the ARIS for MySAP reference model database. We analyzed the model using
the approach presented in Section 4. Surprisingly, the "Internal Procurement"
model contains structural errors. In Figure 3, we show a part of a screenshot of
the verification tool used. It shows part of the EPC in which an AND-split is
later joined by an XOR join. Recall Figure 2, where we have shown that this
is clearly incorrectly modelled. As a result, if this model would *not* be repaired,
payments could be made for goods that were never received. Obviously, this is
not desirable. In Figure 4 we show the repaired model, i.e. the XOR-join has
been changed into an AND-join. Now, the model is semantically correct, which
means that it can be used in a business environment without problems.

6   Conclusion

In this paper, we only looked at one EPC from the reference model database.
However, using the more extensive results from [6], we can draw two important
conclusions. First of all, it seems that problems are more easily introduced into
larger models than into smaller ones. The reason that we did not find many
problems in low level models can probably be explained by the fact that these
models are typically very small. However, when these models are combined by
higher level models, errors are easily introduced. As we saw in Section 5, these
errors can lead to severe complications, such as invoices being paid for goods
that were never received.

Second, the errors we found with our verification approach were all trivial to
repair. Therefore, we feel that the use of such a verification tool in the early stages

of process modelling, or reference model development would greatly improve the effectiveness and applicability of these models in later stages.

References

1. W.M.P. van der Aalst and K.M. van Hee. *Workflow Management: Models, Methods, and Systems*. MIT press, Cambridge, MA, 2002.
2. P. Bernus. *GERAM: Generalised Enterprise Reference Architecture and Methodology*.
3. T. Curran and G. Keller. *SAP R/3 Business Blueprint: Understanding the Business Process Reference Model*. Upper Saddle River, 1997.
4. B.F. van Dongen, A.K.A. de Medeiros, H.M.W. Verbeek, A.J.M.M. Weijters, and W.M.P. van der Aalst. The ProM framework: A new era in process mining tool support. In *accepted tool presentation at ATPN 2005*, Lecture Notes in Computer Science. Springer-Verlag, Berlin, 2005.
5. B.F. van Dongen, H.M.W. Verbeek, and W.M.P. van der Aalst. Verification of EPCs: Using reduction rules and Petri nets. In *Conference on Advanced Information Systems Engineering (CAiSE 2005)*, Lecture Notes in Computer Science, pages 372–386. Springer-Verlag, Berlin, 2005.
6. B.F. van Dongen and M.H. Vullers-Jansen. EPC Verification in the ARIS for MySAP reference model database. BETA Working Paper Series, WP 142, Eindhoven University of Technology, Eindhoven, 2005.
7. P. Fettke and P. Loos. Classification of Reference Models - a methodology and its application. *Information Systems and e-Business Management*, 1(1):35–53, 2003.
8. G. Keller, M. Nüttgens, and A.W. Scheer. Semantische Processmodellierung auf der Grundlage Ereignisgesteuerter Processketten (EPK). Veröffentlichungen des Instituts für Wirtschaftsinformatik, Heft 89 (in German), University of Saarland, Saarbrücken, 1992.
9. G. Keller and T. Teufel. *SAP R/3 Process Oriented Implementation*. Addison-Wesley, Reading MA, 1998.
10. F. Leymann and D. Roller. *Production Workflow: Concepts and Techniques*. Prentice-Hall PTR, Upper Saddle River, New Jersey, USA, 1999.
11. M. Rosemann. *Application Reference Models and Building Blocks for Management and Control (ERP systems)*, pages 595–616. Springer-Verlag, Berlin, 2003.
12. M. Rosemann and W.M.P. van der Aalst. A Configurable Reference Modelling Language. QUT Technical report, FIT-TR-2003-05, Queensland University of Technology, Brisbane, 2003.
13. W. Sadiq and M.E. Orlowska. Modeling and verification of workflow graphs. Technical Report No. 386, Department of Computer Science, The University of Queensland, Australia, 1996.
14. A.W. Scheer. *Business Process Engineering, Reference Models for Industrial Enterprises*. Springer-Verlag, Berlin, 1994.
15. A.W. Scheer. *Business Process Modelling*. 3rd edition, 2000.
16. L. Silverston. *The Data Model Resource Book, Volume 1, A Library of Universal Data Models for all Enterprises*. revised edition, 2001.
17. L. Silverston. *The Data Model Resource Book, Volume 2, A Library of Data Models for Specific Industries*. revised edition, 2001.
18. U. Frank. Conceptual Modelling as the Core of Information Systems Discipline - Perspectives and Epistemological Challanges. In *Proceedings of the America Conference on Information Systems - AMCIS '99*, pages 695–698, Milwaukee, 1999.

Author Index

Lecture Notes in Computer Science

For information about Vols. 1–3526

please contact your bookseller or Springer

Vol. 3576: K. Etessami, S.K. Rajamani (Eds.), Computer Aided Verification. XV, 564 pages. 2005.

Vol. 3575: S. Wermter, G. Palm, M. Elshaw (Eds.), Biomimetic Neural Learning for Intelligent Robots. IX, 383 pages. 2005. (Subseries LNAI).

Vol. 3574: C. Boyd, J.M. González Nieto (Eds.), Information Security and Privacy. XIII, 586 pages. 2005.

Vol. 3573: S. Etalle (Ed.), Logic Based Program Synthesis and Transformation. VIII, 279 pages. 2005.

Vol. 3572: C. De Felice, A. Restivo (Eds.), Developments in Language Theory. XI, 409 pages. 2005.

Vol. 3571: L. Godo (Ed.), Symbolic and Quantitative Approaches to Reasoning with Uncertainty. XVI, 1028 pages. 2005. (Subseries LNAI).

Vol. 3570: A. S. Patrick, M. Yung (Eds.), Financial Cryptography and Data Security. XII, 376 pages. 2005.

Vol. 3569: F. Bacchus, T. Walsh (Eds.), Theory and Applications of Satisfiability Testing. XII, 492 pages. 2005.

Vol. 3568: W.-K. Leow, M.S. Lew, T.-S. Chua, W.-Y. Ma, L. Chaisorn, E.M. Bakker (Eds.), Image and Video Retrieval. XVII, 672 pages. 2005.

Vol. 3567: M. Jackson, D. Nelson, S. Stirk (Eds.), Database: Enterprise, Skills and Innovation. XII, 185 pages. 2005.

Vol. 3566: J.-P. Banâtre, P. Fradet, J.-L. Giavitto, O. Michel (Eds.), Unconventional Programming Paradigms. XI, 367 pages. 2005.

Vol. 3565: G.E. Christensen, M. Sonka (Eds.), Information Processing in Medical Imaging. XXI, 777 pages. 2005.

Vol. 3564: N. Eisinger, J. Małuszyński (Eds.), Reasoning Web. IX, 319 pages. 2005.

Vol. 3562: J. Mira, J.R. Álvarez (Eds.), Artificial Intelligence and Knowledge Engineering Applications: A Bioinspired Approach, Part II. XXIV, 636 pages. 2005.

Vol. 3561: J. Mira, J.R. Álvarez (Eds.), Mechanisms, Symbols, and Models Underlying Cognition, Part I. XXIV, 532 pages. 2005.

Vol. 3560: V.K. Prasanna, S. Iyengar, P.G. Spirakis, M. Welsh (Eds.), Distributed Computing in Sensor Systems. XV, 423 pages. 2005.

Vol. 3559: P. Auer, R. Meir (Eds.), Learning Theory. XI, 692 pages. 2005. (Subseries LNAI).

Vol. 3558: V. Torra, Y. Narukawa, S. Miyamoto (Eds.), Modeling Decisions for Artificial Intelligence. XII, 470 pages. 2005. (Subseries LNAI).

Vol. 3557: H. Gilbert, H. Handschuh (Eds.), Fast Software Encryption. XI, 443 pages. 2005.

Vol. 3556: H. Baumeister, M. Marchesi, M. Holcombe (Eds.), Extreme Programming and Agile Processes in Software Engineering. XIV, 332 pages. 2005.

Vol. 3555: T. Vardanega, A.J. Wellings (Eds.), Reliable Software Technology – Ada-Europe 2005. XV, 273 pages. 2005.

Vol. 3554: A. Dey, B. Kokinov, D. Leake, R. Turner (Eds.), Modeling and Using Context. XIV, 572 pages. 2005. (Subseries LNAI).

Vol. 3553: T.D. Hämäläinen, A.D. Pimentel, J. Takala, S. Vassiliadis (Eds.), Embedded Computer Systems: Architectures, Modeling, and Simulation. XV, 476 pages. 2005.

Vol. 3552: H. de Meer, N. Bhatti (Eds.), Quality of Service – IWQoS 2005. XVIII, 400 pages. 2005.

Vol. 3551: T. Härder, W. Lehner (Eds.), Data Management in a Connected World. XIX, 371 pages. 2005.

Vol. 3548: K. Julisch, C. Kruegel (Eds.), Intrusion and Malware Detection and Vulnerability Assessment. X, 241 pages. 2005.

Vol. 3547: F. Bomarius, S. Komi-Sirviö (Eds.), Product Focused Software Process Improvement. XIII, 588 pages. 2005.

Vol. 3546: T. Kanade, A. Jain, N.K. Ratha (Eds.), Audio-and Video-Based Biometric Person Authentication. XX, 1134 pages. 2005.

Vol. 3544: T. Higashino (Ed.), Principles of Distributed Systems. XII, 460 pages. 2005.

Vol. 3543: L. Kutvonen, N. Alonistioti (Eds.), Distributed Applications and Interoperable Systems. XI, 235 pages. 2005.

Vol. 3542: H.H. Hoos, D.G. Mitchell (Eds.), Theory and Applications of Satisfiability Testing. XIII, 393 pages. 2005.

Vol. 3541: N.C. Oza, R. Polikar, J. Kittler, F. Roli (Eds.), Multiple Classifier Systems. XII, 430 pages. 2005.

Vol. 3540: H. Kalviainen, J. Parkkinen, A. Kaarna (Eds.), Image Analysis. XXII, 1270 pages. 2005.

Vol. 3539: K. Morik, J.-F. Boulicaut, A. Siebes (Eds.), Local Pattern Detection. XI, 233 pages. 2005. (Subseries LNAI).

Vol. 3538: L. Ardissono, P. Brna, A. Mitrovic (Eds.), User Modeling 2005. XVI, 533 pages. 2005. (Subseries LNAI).

Vol. 3537: A. Apostolico, M. Crochemore, K. Park (Eds.), Combinatorial Pattern Matching. XI, 444 pages. 2005.

Vol. 3536: G. Ciardo, P. Darondeau (Eds.), Applications and Theory of Petri Nets 2005. XI, 470 pages. 2005.

Vol. 3535: M. Steffen, G. Zavattaro (Eds.), Formal Methods for Open Object-Based Distributed Systems. X, 323 pages. 2005.

Vol. 3534: S. Spaccapietra, E. Zimányi (Eds.), Journal on Data Semantics III. XI, 213 pages. 2005.

Vol. 3533: M. Ali, F. Esposito (Eds.), Innovations in Applied Artificial Intelligence. XX, 858 pages. 2005. (Subseries LNAI).

Vol. 3532: A. Gómez-Pérez, J. Euzenat (Eds.), The Semantic Web: Research and Applications. XV, 728 pages. 2005.

Vol. 3531: J. Ioannidis, A. Keromytis, M. Yung (Eds.), Applied Cryptography and Network Security. XI, 530 pages. 2005.

Vol. 3530: A. Prinz, R. Reed, J. Reed (Eds.), SDL 2005: Model Driven. XI, 361 pages. 2005.

Vol. 3528: P.S. Szczepaniak, J. Kacprzyk, A. Niewiadomski (Eds.), Advances in Web Intelligence. XVII, 513 pages. 2005. (Subseries LNAI).

Vol. 3527: R. Morrison, F. Oquendo (Eds.), Software Architecture. XII, 263 pages. 2005.